MASERS & LASERS

Second Edition

T0133824

MASERS
& LASERS

AN HISTORICAL APPROACH

Second Edition

Mario Bertolotti

CRC Press
Taylor & Francis Group
Boca Raton   London   New York

CRC Press is an imprint of the
Taylor & Francis Group, an **informa** business

CRC Press
Taylor & Francis Group
6000 Broken Sound Parkway NW, Suite 300
Boca Raton, FL 33487-2742

First issued in paperback 2021

© 2015 by Taylor & Francis Group, LLC
CRC Press is an imprint of Taylor & Francis Group, an Informa business

No claim to original U.S. Government works

ISBN 13: 978-0-367-78357-0 (pbk)
ISBN 13: 978-1-4822-1777-3 (hbk)

Visit the Taylor & Francis Web site at
http://www.taylorandfrancis.com

and the CRC Press Web site at
http://www.crcpress.com

Contents

Contents

Contents

Foreword to the First Edition

Masers, and especially lasers, are by now familiar devices that are very widely used in many fields of science and technology. To a few, lasers seem to offer fulfilment of one of mankind's oldest dreams of technological power, an all-destroying energy ray. That may be the basis for the ancient, probably apocryphal, legend that Archimedes was able to set enemy ships on fire by using large mirrors to reflect and focus sunlight. The English novelist H.G. Wells, in his 1898 story *War of the Worlds*, had Martians nearly conquering the Earth with a heat ray. In the 1920s, the Soviet novelist Alexei Tolstoi wrote *The Hyperboloid of Engineer Gorin*, which device also was described as producing an intense light beam. Newspaper comic strips such as *Buck Rogers* in the 1930s made "disintegrator guns" familiar ideas.

Yet as scientists learned more about how light is produced and absorbed, these novelists' dreams seemed all the more unlikely. Thermal emitters of light, which were all we had, seemed to absorb the light that they emitted, if we tried to make them thicker to get more intensity. Yet it turned out that this and many other apparent difficulties could be overcome. Indeed, when the first lasers were operated, I and other scientists close to the research were surprised at how easy it turned out to be. We had assumed that, since lasers had never been made, it must be very difficult. But once you knew how, it was not at all difficult. Mostly what had been lacking were ideas and concepts.

This book blazes a new trail, in retracing the history and expounding the theory and experiments as they were discovered. This is a complex task, as there is no earlier book of comparable cope to use as starting point. Inevitably, there are many points in the discussion which I would state differently, and some which I would have to dispute. But Professor Bertolotti's long experience in this field, to which his research contributed much, has enabled him to produce a sound outline of the way things developed.

The treatment could serve very well as an introduction to the theory of masers and lasers, since these matters are thoroughly discussed in a sequence appropriate to the historical presentation.

Arthur L. Schawlow
Department of Physics
Stanford University
October 1982

Foreword to the Second Edition

After the explosion in the last 20 years in the number of books with the word "laser" in their title, I found myself muttering "what—yet another book?..." when I learned about Mario Bertolotti's second edition of *Masers and Lasers*. Upon reading it, however, I was left with an altogether different reaction to the book, which I can only describe as—a great story, well told.

There is little doubt that the laser is shaping up as one of the great technological and scientific inventions of the twentieth century, and its practical and theoretical ramifications are reshaping our lives. A book that manages to clearly grasp and convey the human historical background of this invention (with names like Townes, Bloenbergen, Basov, Glauber, plus many others, and their individual stories), as well as the chronology and significance of the key development in the field, is a major social document.

This need is excellently served by *Masers and Lasers*, and I would like to congratulate Mario Bertolotti on this major endeavor, and on the elegant and entertaining telling of the story. I found myself learning some new facts, for example, the early pre-laser period where concepts such as "negative absorption" by inverted population of atomic levels and contributions by scientists such as Einstein and Ladenburg were but a step away from proposing optical amplification and then oscillation.

I also found that some parts of the book could serve as a good introduction to the topic even in graduate level courses. One such example is Chapter 10, "On the statistical properties of light," which is obviously a topic close to the author's area of research.

In conclusion, I found the book highly instructive and enjoyed as well the personal/historical background material. I will make it a "Highly Recommended" reference for my future classes in quantum electronics and nonlinear optics.

December 9, 2014

Amnon Yariv, PhD
Martin and Eileen Summerfield
Professor of Applied Physics and Professor of Electrical Engineering
California Institute of Technology

Preface

Preface to the Second Edition

This second edition takes advantage of the many recollections that the first protagonists of the initial development of lasers have written. It was also necessary to update some developments. In particular, a general survey of the development of nonlinear optics and how lasers have evolved in the last 40 years has been included. Several arguments that were omitted in the first edition have been added, together with some new developments, such as micromasers, lasers without inversion, nano-lasers, spasers, and so on. Lasers and their applications have experienced an exponential grow both in interest and development. I have, therefore, been obliged to compress the materials into a synthetic presentation considering the most important advances, adding references to more in-depth discussions that are now available, focused on single cases. I believe this revised edition is a reasonable good start for more extended study. In any case, I apologize for any omission and will be grateful to readers who provide me with additions and suggestions.

A special acknowledgment goes to my wife who allowed me to spend all my time on this work. I would also like to thank the Taylor & Francis editorial staff, Luna Han, Robert Sims, and Karthick Parthasarathy from Techset Composition, for help in publication and editing the manuscript.

Mario Bertolotti
Roma
June 2014

Preface to the First Edition

Nowadays masers and especially lasers are very popular devices. When masers were invented in 1951, they offered a completely new and revolutionary method of producing microwaves. The theoretical foundations necessary to understand the way they work and actually to build them, however, had already been well established in the 1930s. It took people 20 years to get rid of the old traditional schemes of producing electromagnetic waves and to find a completely revolutionary path of achieving them and, as often occurs in these cases, at the moment the time was ripe and the same idea occurred to many scientists almost simultaneously.

Lasers were the natural extension of this idea to light. Mostly popularized, invented even before they actually worked by science fiction writers as the "death ray" and other similar names, lasers were for some time ironically defined as a "solution in search of a problem."

Nowadays, there is scarcely any physical laboratory which does not own at least one. More applications are discovered every day. In fact, lasers have not been fully exploited yet, and probably their best applications are to come. I think it both interesting and instructive to trace out the history of how these devices developed since the first basic principles at their origin were established. And now let me say, as Agamemnon does in Troilus and Cressida by William Shakespeare (Act I, Scene II):

> Speak, Prince of Ithaca; and be't of less expect,
> That matter needless, of importless burden
> Divide thy lips, than we are confident,
> When rank Thersites opens his mastiff jaws,
> We shall hear music, wit, and oracle.

Before starting, let me first acknowledge help from many people who provided me with documentation and discussion. Notably Professors N Bloembergen, B Crosignani, P Di Porto, H Gamo, R J Glauber, S F Jacobs, B A Lengyel, V S Letokhov, S I Nishira, A M Prokhorov, M Sargent, A L Schawlow, C P Slichter, C H Townes, and V Vavilov.

Special thanks are due to Dr. Roy Pike who kindly turned my Anglo-Italian into English, to Miss A De Cresce who, with great patience and ability, typed the manuscript and to Mrs F Medici and C Sanipoli for preparing drawings and diagrams.

Mario Bertolotti
Roma
October 1982

Author

Mario Bertolotti received his degree in physics from the University of Roma, Italy, and has been full professor of physics and optics at the Engineering Faculty of the University of Roma La Sapienza from 1970 until 2008. He is the author of more than 500 peer-reviewed publications. A major part of his research has been in lasers and their applications, as well as non-linear optics and nano-optics. He is an elected Fellow of the Institute of Physics, Optical Society of America (OSA), and European Optical Society (EOS), of the Italian Physical Society (SIF) and the Italian Optics and Photomic Society (SIOF) of which he was the president.

He is currently enjoying his retirement with his wife, Romana de Angelis (an archeologist and a writer). They have two children, Alessandro, a television director and writer, and Elena, who died too soon. His inspiring cat Einstein passed on at the ripe age of 21, although the spirit of his illustrious homonymous Albert Einstein pervades this entire book.

1

Introduction

THIS BOOK AIMS AT tracing out the history of the development of the fundamental ideas on which masers and lasers are based.

Although this development has been deeply impressed in the minds of all of the early researchers in this field, it may well escape the knowledge of young researchers approaching this area for the first time. Graduate students in physics and electronics can profit from such knowledge, and the level of exposition has been chosen accordingly. Whenever possible, original authors have been allowed to speak for themselves through their papers, and short biographies of the leaders in the field have been included. In this second edition, the development of lasers has been retrieved up to the present, including nonlinear optics, ultrashort pulses, and some special issues.

1.1 Principle of Operation

Masers and lasers have in common their principle of operation, which is based on the use of *stimulated emission* of electromagnetic radiation in a medium of molecules or atoms, with more particles in the upper (*excited*) state than in the lower state (i.e., with an *inverted population*).

An electron bound to a molecule or an atom may change its energy state by jumping from one energy level to another with the emission or absorption of a photon, in which process it will have, respectively, lost or gained energy.

If the particle is initially in an upper state with respect to the fundamental ground state, it decays spontaneously to the ground state by the emission of a photon of energy

$$h\nu = \Delta E, \tag{1.1}$$

where ΔE is the energy difference between the two levels, ν the frequency of emission, and h Planck's constant.

This is the normal emission process (*spontaneous emission*) which takes place every time a material de-excites itself having been suitably excited. Due to the random nature of spontaneous decay, photons are emitted by various particles in an independent way and the resultant emission is incoherent.

The probability of spontaneous emission increases with the cube of frequency. It is, therefore, negligible in microwave transitions wherein thermal relaxation processes are predominant.

If electromagnetic radiation is present with a frequency so as to fulfil Equation 1.1, two processes can be distinguished:

1. The photon interacts with a particle which is in its lower energy level: in this case, radiation is absorbed and the particle is forced to go to the upper level.

2. The photon interacts with a particle which is already in an upper state. In this case, the particle is forced to go to the lower energy state by emitting another photon of the same frequency as the incident photon. This process is called stimulated emission.

In spontaneous emission, both the direction and polarization of photons are randomly distributed, whereas in the case of stimulated emission they coincide with those of the incident photon.

In general, the particles of an ensemble in equilibrium are more in the lower energy level, according to the Maxwell–Boltzmann distribution law. By making the system interact with radiation of frequency equal to the difference between two levels (the ground and the excited levels), processes of type (1) and (2) take place simultaneously, with a prevalence of the former because more particles are in the lower state. However, if, in some way, the distribution between levels is altered, so that more particles are present in the excited state than in the ground state (one usually says that an *inversion of population* has been obtained), then in the interaction process between the radiation and the particles, a net excess of emitted photons will take place over the absorbed photons, that is, an amplification process occurs.

The radiation emitted this way will be monochromatic—because it is emitted in correspondence with a well-defined transition—and coherent—because it is a forced emission which produce at the end an amplification of the wave.[1]

This is the fundamental working principle common to masers and lasers.

1.2 The Devices

Masers are devices emitting in the microwave region and, therefore, molecular roto-vibrational states or atomic Zeeman levels are usually used. Lasers are devices emitting in the infrared to ultraviolet region and they can, therefore, use either molecular or atomic levels.

Once suitable energy levels have been chosen, the next problem is how to obtain an appreciable amount of radiation and how then to couple that radiation efficiently to the particles. This is achieved by the use of a suitable resonant cavity, that is, a microwave cavity for masers and an optical cavity for lasers. The effect of such a cavity is not only to increase the residence time of photons in contact with the ensemble of excited atoms so as to increase the probability of de-excitation, but also to provide the feedback necessary to make the emitted wave grow coherently.

The logical steps in building up such devices are consequently:

1. The understanding of the role played by stimulated emission and thermodynamic equilibrium conditions in an ensemble of particles interacting with an electromagnetic field;

2. The creation of a suitably inverted population; and

3. The appreciation of the role of a resonant cavity.

Once these three basic features were well integrated into one single concept, first masers and later lasers could be constructed.

1.2.1 Masers

In masers, one may consider two fundamental methods for inverting population:

1. *A preselection of excited molecules in a gas.* This was the first method used by Townes, in which an excess of excited molecules from ammonia gas was introduced into the resonant cavity. A very small signal was therefore greatly amplified at the molecular transition. This kind of maser is very suitable as a low-noise, single-frequency amplifier or as a frequency standard.

2. *The use of paramagnetic levels in a solid material.* This was a later proposal made by Bloembergen. The three-level maser is the best example of this type: its working principle is shown with reference to **Figure 1.1**. A suitable paramagnetic ion is put into a magnetic field so that a set of Zeeman levels is created. Let us consider three of these levels. With a suitable *pump* frequency, particles are excited from the lower, ground level 1 to the upper level 3. If the energy between levels is large compared with kT, level 2 is approximately empty and therefore an inversion of population is created between levels 3 and 2. If the ions are put into a resonant cavity, amplification of radiation at frequency ν_{32} occurs. Relaxation from 2 to 1 is finally necessary to maintain the needed population inversion between 3 and 2.

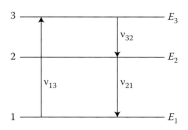

FIGURE 1.1 The three-level diagram proposed by Bloembergen for a solid-state paramagnetic maser.

The distance between levels is a function of the strength of the external magnetic field. Therefore, frequency ν_{32} can, to some extent, be tuned. This maser is therefore much more flexible than masers of type 1.

1.2.2 Lasers

Lasers were developed later. Initial proposals were for gas media, but the first realization was in solids. This was followed by devices exhibiting laser operation in a myriad of different systems.

The most important obstacle to the progression from masers to lasers was the choice of a suitable resonant cavity at optical frequencies. The solution proved to be an open cavity in the form of a Fabry–Perot device, two parallel plane mirrors at a distance apart from each other. Light traveling along the axis of the system is forced to reflect back and forth, the principal condition being that in the distance of separation between the two mirrors there is an integer number of wavelengths.

The principle of operation of a laser is the same as for masers. For example, in the ruby laser, which was the first to be successfully operated, the excited population is produced by illumination of Cr ions in corundum with a broad band of visible light from a flash lamp. The light in the green part is absorbed by the ions which are then excited to a band of higher levels. From this band, they decay, without emission of light, simply by giving their excess energy to the lattice as heat, to an intermediate level (actually a doublet) which can therefore become more populated than the fundamental level.

From this intermediate level (or rather, from one of the two doublet levels), Cr ions finally decay via strong light emission at 6943 Å at room temperature to the fundamental level, and this is the radiation which is amplified in a suitable Fabry–Perot cavity.

1.3 Applications

In **Figure 1.2**, the chronology of coherent wave generation is shown—an impressive visual display. The frequencies obtained are seen to follow an approximate exponential law up to the invention of lasers. From that point on, the trend has a marked change in slope, which shows the attainment of very short wavelengths to be very difficult.

In the case of microwaves, the maser followed many other devices for their production and although it has unique properties, mainly high monochromaticity and an extremely low noise temperature, it has received not so many applications, being superseded by the progress in electronic devices.

In the case of the infrared–visible–ultraviolet—x-ray region, the laser constituted a big revolution. With it, for the first time, man was able to have in the optical region a fully coherent source with enormous brightness and monochromaticity.

By just using these two properties, a number of fields in optics received a large boost. Holography, which had already been invented by Gabor in 1948, and nonlinear optics, which was just the optical counterpart of well-known

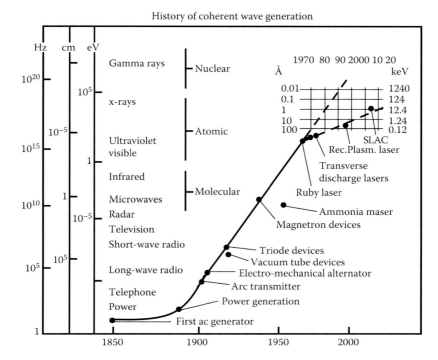

FIGURE 1.2 History of coherent wave generation. (Courtesy of Dr. D.J. Nagel, Naval Research Laboratories, Washington, DC, modified.)

phenomena in radio waves, were two domains, for example, in which rapid growth was made practicable by the availability of the laser. Two other fields which, although well known and in operation well before the advent of laser, received by its use a new and extraordinary boost were interferometry and all light-scattering techniques. At the end, but not the least, we must mention the application to the transmission of information which together with the use of optical fibers has made impressive changes in the field of communications.

However, the revolution was not complete until it was appreciated that the properties of light emitted by lasers were completely different from those of light emitted by an ordinary source, such as a lamp or an ordinary gas discharge.

The use of stimulated emission in an inverted population is a way of producing radiation in conditions far from equilibrium and therefore the statistical properties of the emitted light are completely different from those of any other light source known before. This point, which initiated a new activity in coherence theory, has importance, for example, in scattering and detecting light. The recast of the theory of coherence in the frame of quantum mechanics has originated the new field of quantum optics.

1.4 Development

In this book, an effort is made to retrace the conceptual path which has brought the invention of first masers and then lasers.

The history of the development of these devices can be divided roughly into four periods. The first period starts with the introduction of the concept of stimulated emission and ends with its experimental demonstration. It extends from 1916 to 1953 and is covered by Chapters 2 and 3.

The second period begins with the construction of the first maser in 1954 and ends in 1960 with intensive speculation concerning the extension of the art of the maser from the microwave to the optical region. This maser period is covered by Chapters 4 and 5.

The laser or third period opens with the achievement of the first operating laser. It is characterized by an explosive growth of research into and development of many different kinds of laser. This period extends from 1961 to roughly 1970.

The last or second-generation laser period is characterized by the development of many different kinds of laser which have the characteristics necessary to make them of use for scientific, industrial, medical, and general technical research. It opens about 1970 and is still open today. These last two periods are covered in Chapters 6, 7, and 9.

The subject of applications is omitted, since this would have more than doubled the space required. However, in this new edition, two exceptions have been made: nonlinear optics and ultrashort pulses down to attosecond (Chapter 8). Eventually, Chapter 10 is devoted to the statistical properties of laser light and to the latest developments in coherence theory, in the light of the previous discussion.

Note

1. Strictly speaking, in a classical treatment, conservation of energy requires that radiation be emitted in advance of a quarter of period with respect to the stimulating field. The quantum problem is considered, for example, in M Sargent III, M O Scully and W E Lamb Jr., *Laser Physics* (Addison-Wesley, 1974), Chapter 3.

2

Stimulated Emission
Could the Laser Have Been Built More than 80 Years Ago?

2.1 Stimulated Emission

After struggling in various ways[1] with the blackbody problem for more than 10 years, in November 1916, Einstein (1879–1955) (**Figure 2.1**) wrote to his friend Michele Besso (1873–1955):[2] "A splendid light has dawned on me about the absorption and emission of radiation." In that year and in the following one, Einstein[3] published a new, extremely simple, and elegant proof of Planck's law of radiation and, at the same time, obtained important new results concerning the emission and absorption of light from atoms or molecules. In his paper, for the first time, the concept of *stimulated emission*, which is basic to the laser effect, is introduced. Our story can therefore begin by analyzing the methods used and conclusions reached by Einstein. He skillfully combined "classical laws" with the new concepts of quantum mechanics, which were at that time growing up. The line of reasoning he followed was more or less similar to the one adopted by W. Wien (1864–1928) in his derivation of the radiation law;[4] Einstein, however, adapted it to the new situation created by Bohr's spectral theory. At the beginning of his paper, he writes

> Not long ago I discovered a derivation of Planck's formula which was closely related to Wien's original argument and which was based on the fundamental assumption of quantum theory … [Later on] I was led to these hypotheses by my endeavour to postulate for the molecules, in the simplest possible manner, a quantum-theoretical behaviour that would be the analogue of the behaviour of a Planck resonator in the classical theory.[4]

His reasoning is the following.[5] Let us consider some well-defined molecule, without taking into account its orientation and translational movements. According to the postulates of the quantum theory already developed at that time, it could have only a discrete set of states $Z_1, Z_2, \ldots, Z_n$, whose internal energies can be labeled with $\varepsilon_1, \varepsilon_2, \ldots, \varepsilon_n$. If a large number of such molecules belongs to a gas at temperature T, the relative frequency W_n of the state Z_n is given by the formula of Gibbs' canonical distribution modified for discrete states, that is,[6]

FIGURE 2.1 Albert Einstein in 1921.

$$W_n = g_n \exp\left(-\frac{\varepsilon_n}{kT}\right),$$

where g_n is a number, independent of T and characteristic for the molecule and its nth quantum state, which can be called the *statistical weight* of this state.

Let us now take into account the radiative exchanges of energy (i.e., processes with emission or absorption of electromagnetic waves). They are treated by considering a Planck resonator in the classical scheme. Einstein identifies three different interaction mechanisms. The first one is a radiative emission process which we now call *spontaneous emission*. To describe this process, he assumes that the probability of a single molecule in state Z_m going in a time dt, without being excited by external agents, to the state of lower energy Z_n, with the emission of radiant energy $\varepsilon_m - \varepsilon_n$, to be

$$A_m^n \, dt, \tag{2.1}$$

where A_m^n is a given constant.

The jumps from one energy level to another that in the Bohr theory give rise to the spontaneous emission of radiation are taken as being analogous to spontaneous radioactive disintegrations, and Einstein in writing Equation 2.1 assumed that radiative transitions of free atoms are governed by a probability law similar to the one postulated in the elementary theory of radioactivity. τ_m^n, the reciprocal of A_m^n, which is the spontaneous lifetime of the upper level with respect to the lower one.

He then considers the absorption of radiation and writes:[7]

> If a Planck resonator is located in a radiative field, the energy of the resonator is changed through the work done on the resonator by the electromagnetic field of the radiation; this work can be positive or negative, depending on the phases of the resonator and of the oscillating field. We correspondingly introduce the following quantum-theoretical hypothesis. Under the influence of a radiation density ρ of frequency ν a molecule can make a transition from state Z_n to state Z_m by absorbing radiation energy $\varepsilon_m - \varepsilon_n$ according to the probability law

$$dW = B_n^m \rho \, dt. \tag{B}$$

We similarly assume that a transition $Z_m \to Z_n$, associated with a liberation of radiation energy $\varepsilon_m - \varepsilon_n$ is possible under the influence of the radiation field, and that it satisfies the probability law

$$dW = B_m^n \rho \, dt, \tag{B'}$$

where B_n^m and B_m^n are constants. We shall give both processes the name *changes of state due to irradiation*.

Process (B') is the one we now call *stimulated emission* (induced emission) and is introduced here for the first time.

The term *stimulated emission* does not appear in the above quotation and was introduced at a later time by John van Vleck (1899–1980) in 1924.[8]

Once these fundamental hypotheses have been established, Planck's law is derived at once by assuming that the energy exchange between radiation and molecules does not perturb the canonical distribution of states given before.

Therefore, averaged over unit time, as many elementary processes of type (B) as of emission (2.1) and (B') must take place:

$$g_n \left[\exp\left(-\frac{\varepsilon_n}{kT} \right) \right] B_n^m \rho = g_m \left[\exp\left(-\frac{\varepsilon_m}{kT} \right) \right] \left(B_m^n \rho + A_m^n \right). \tag{2.2}$$

Let us now assume that, by increasing T, ρ also increases tending towards infinity. From Equation 2.2, we obtain

$$g_n B_n^m = g_m B_m^n \tag{2.3}$$

and

$$\rho = \frac{A_m^n / B_m^n}{\exp\left[(\varepsilon_m - \varepsilon_n)/kT \right] - 1}. \tag{2.4}$$

This is Planck radiation law. It gives asymptotically Rayleigh's law for large wavelengths and Wien's law for small wavelengths if we take

$$\varepsilon_m - \varepsilon_n = h\nu \tag{2.5}$$

and

$$A_m^n = \left(\frac{8\pi h \nu^3}{c^3} \right) B_m^n. \tag{2.6}$$

In this and in all other derivations up to the time of Bose's paper on statistics in 1924, the probability factor between A and B was obtained by appealing at one point or another to classical electromagnetic theory. The quantity $8\pi\nu^2/c^3$

represents the number of normal modes of the radiation per unit volume and per unit frequency interval.[9]

Equations 2.3 and 2.6 by Einstein appeared for the first time in this work and are fundamental to the theory of energy exchange between matter and radiation.[10] The probabilities that Einstein assumed for each of the elementary processes suffered by a molecule are today indicated as *transition probabilities* between states. Bohr's quantum theory did not give any indication of the laws governing such transitions and the concept of transition probability originated in Einstein's paper. Einstein was not able to express them in terms of the characteristic parameters of the atom. Such an expression would be given more than 10 years later by P A M Dirac (1902–1984),[11] utilizing the quantum mechanics which at that time was fully developed.

One of the principal problems of quantum mechanics at that time was to calculate these coefficients from data pertaining to atoms and molecules.

Equation 2.6 was experimentally verified through a comparison of the intensities of absorption and emission lines. The constants B were obtained from measurements of the intensity of multiplet components in spectra by L S Ornstein and H C Burger.[12]

Another important result established in Einstein's work is connected with the exchange of momentum between atoms or molecules and the radiation. Einstein showed that when a molecule (atom), making a transition from Z_n to Z_m, receives energy $\varepsilon_m - \varepsilon_n$, it also receives momentum $(\varepsilon_m - \varepsilon_n)/c$ in a defined direction. However, when a molecule (atom) in the transition from Z_m to the lower state Z_n emits radiant energy $\varepsilon_m - \varepsilon_n$, it gains momentum $(\varepsilon_m - \varepsilon_n)/c$ in the opposite direction. Therefore, the emission and absorption processes are *direct processes*; emission or absorption of spherical waves is not likely to occur. He writes, "Outgoing radiation in the form of spherical waves does not exist."[13]

The field required for thermal equilibrium turns out to obey Plank law. Atoms interacting with radiation at temperature T assume Maxwell–Boltzmann velocity distribution due to the interaction even in the absence of collisions.

The subsequent A Compton (1892–1962) experiment on the scattering of x-rays provided the first experimental confirmation of these predictions.[14]

Einstein's theory of emission and absorption allowed M Wolfke[15] to argue that the cavity radiation, looked at from the point of view of Einstein's light quantum hypothesis, consists of light molecules $h\upsilon$, $2h\upsilon$, $3h\upsilon$, …, which are mutually independent in space. W Bothe (1891–1957)[16] gave an instructive calculation of the number of radiation quanta $h\upsilon$ of a black body that are associated as "photo-molecules" in pairs $2h\upsilon$, triplets $3h\upsilon$, and so on. He considered a cavity filled with blackbody radiation at a temperature T in which a large number of gas molecules were enclosed, each one being in one of the two states Z_1 and Z_2, where $\varepsilon_2 - \varepsilon_1 = h\upsilon$, their relative mean numbers being given by the canonical distribution law at temperature T. He assumed that when a single quantum $h\upsilon$ of the radiation causes a stimulated emission, the emitted quantum moves with the same velocity and in the same direction as the stimulating quantum, so that they become a pair of quanta $2h\upsilon$. If the exciting quantum itself already belongs to a pair, then a triplet $3h\nu$ is produced, and so on.

The absorption of one quantum by a pair of quanta leaves a single quantum, and spontaneous emission produces a single quantum. By writing the conditions for the mean number of single quanta, of quanta pairs, and so on to be constant in time and by using Einstein's relations between the stimulated, spontaneous, and absorption coefficients, he obtained a system of equations from which the mean number of single quanta united in s-quanta molecules $sh\nu$ in the cavity volume and in the frequency range $d\nu$ is derived as

$$\left(\frac{8\pi\nu^2}{c^3}\right)\left\{\exp\left[-\left(\frac{sh\nu}{kT}\right)\right]\right\}d\nu. \tag{2.7}$$

The mean total energy per unit volume in the range $d\nu$ is therefore

$$\left(\frac{8\pi h\nu^3}{c^3}\right)\left\{\sum_{s=1}^{\infty}\exp\left[-\left(\frac{sh\nu}{kT}\right)\right]\right\}d\nu \tag{2.8}$$

or

$$\left(\frac{8\pi h\nu^3}{c^3}\right)\left\{\frac{h\nu}{\exp(h\nu/kT)-1}\right\} \tag{2.9}$$

in agreement with Planck's radiation law.

These considerations, which will also be useful later when examining the statistical properties of the radiation field, are useful here to establish another important point. Already, in 1923, when Bothe presented these results, one of the most fundamental properties of stimulated emission was clear, namely that the quantum emitted in the stimulated emission process, besides having exactly the same energy as the stimulating quantum also has the same momentum, that is, travels in the same direction as the incident quantum. This type of behavior is exactly what is needed in order to have an amplification process.

For about 30 years, however, the concept of stimulated emission was used only in theoretical works, as will be shown in the following section, and received only marginal attention from the experimental point of view. Even in 1954, the classical monograph by W H Heitler (1904–1981) on the quantum theory of radiation[17] gives very little space to this argument while dedicating considerable attention to phenomena such as resonance fluorescence and Raman scattering.

2.2 Role of Stimulated Emission in the Theory of Light Dispersion

Einstein's theory of emission and absorption coefficients allowed theoretical physicists to build up a satisfactory quantum theory of scattering, refraction, and light dispersion in a few years.

Masers and Lasers

Let us first consider a few elementary definitions relating to the concept of absorption lines. If light from a source emitting a continuum spectrum over some range of frequencies is made parallel and sent through an absorption cell filled with a monatomic gas, the intensity of the transmitted light I_v may have a frequency distribution as shown in **Figure 2.2**. When this happens, the gas is said to have an absorption line at frequency v_o, v_o being the line center frequency.

The absorption coefficient k_v of the gas is defined by the relation

$$I_v = I_o \exp(-k_v x), \tag{2.10}$$

where x is the thickness of the absorbing layer. From Figure 2.2 and Equation 2.10, k_v can be obtained as a function of frequency as shown in **Figure 2.3**. The total width of the curve where k_v falls to half its maximum value k_{max} is called the *half linewidth* or *halfwidth*, Δv. In general, the absorption coefficient of a gas is given by an expression involving a function of v and a definite value of k_{max} and Δv, which depend on the nature of the gas molecules, their motion, and the interactions between themselves or with other molecules.

The classical theory of dispersion and absorption of radiation is due principally to P Drude (1863–1906)[18] and W Voigt (1850–1919)[19] and was elucidated fully by HA Lorentz (1853–1928).[20]

In the classical theory, the atom is considered to be formed by some oscillators whose frequencies are equal to the absorption frequencies, v_i. If such oscillators are treated as particles with charge e and mass m, under the action of an oscillating electric field E, we may write for the position vector of the generic oscillator with proper frequency v_o:

$$\ddot{\mathbf{r}} + \gamma\dot{\mathbf{r}} + (2\pi v_o)^2\,\mathbf{r} = \left(\frac{e}{m}\right)\mathbf{E}, \tag{2.11}$$

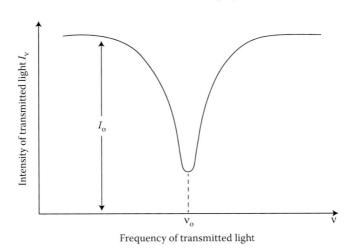

FIGURE 2.2 Absorption line centered at a frequency v_o.

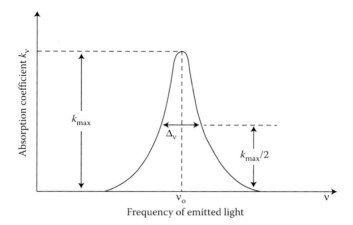

FIGURE 2.3 Absorption coefficient k_ν as a function of frequency.

where γ is a damping factor. For $E = E_o[\exp(2\pi i \nu t)]$, a solution of Equation 2.11 is

$$\mathbf{r} = \frac{(e/m)\mathbf{E}}{[4\pi^2(\nu_o^2 - \nu^2) + 2\pi\nu\gamma i]}, \tag{2.12}$$

and the oscillator has an electric dipole moment

$$\mathbf{p} = e\mathbf{r}. \tag{2.13}$$

The polarization $\mathbf{P}$ produced by the vibrating electric field is therefore written as

$$\mathbf{P} = eN\mathbf{r} = \alpha\mathbf{E},$$

where N is the number of oscillators (atoms) per unit volume and the polarizability α is a complex number (Drude's formula)

$$\alpha = \frac{(e^2/m)N}{4\pi^2(\nu_o^2 - \nu^2) + 2\pi\nu\gamma i}. \tag{2.14}$$

If ν is not very near to ν_o, the imaginary term in the denominator can be neglected.

In a gas, by introducing a complex refractive index $\tilde{n}$ defined by

$$\varepsilon = 1 + \left(\frac{4\pi P}{E}\right) = (\tilde{n})^2 = (n - ik)^2, \tag{2.15}$$

where ε is the complex dielectric constant and n and k are the real and imaginary part of $\tilde{n}$, respectively, we therefore have

$$(\tilde{n})^2 = (n^2 - k^2) - i(2nk) = 1 + 4\pi\alpha. \tag{2.16}$$

By using Equation 2.14, we have finally

$$(\tilde{n}) = (n^2 - k^2) - i(2nk) = 1 + 4\pi[\text{Re}(\alpha) + \text{Im}(\alpha)]$$

$$= 1 + 4\pi \left\| \left(\frac{e^2 N}{m}\right) \left\{ \frac{(v_o^2 - v^2)}{4\pi^2(v_o^2 - v^2)^2 + v^2\gamma^2} \right\} \right. \tag{2.17}$$

$$\left. -i\left(\frac{e^2 N}{m}\right) \left\{ \frac{v\gamma}{2\pi[4\pi^2(v_o^2 - v^2)^2 + v^2\gamma^2]} \right\} \right\|.$$

In the case of a gas of not too high density, we can put

$$k \ll 1, \quad |n - 1| \ll 1, \quad n^2 - 1 = 2(n - 1),$$

so that we obtain

$$n = 1 + \left(\frac{2\pi e^2 N}{m}\right) \left\{ \frac{v_o^2 - v^2}{4\pi^2(v_o^2 - v^2)^2 + v^2\gamma^2} \right\} \tag{2.18}$$

and

$$k = \left(\frac{2\pi e^2 N}{m}\right) \left\{ \frac{\gamma v}{2\pi[4\pi^2(v_o^2 - v^2)^2 + v^2\gamma^2]} \right\}. \tag{2.19}$$

The behavior of k and n near the resonant frequency is shown in **Figure 2.4a** and **b**.

Far from resonance, the refractive index n increases with increasing frequency. The corresponding negative wavelength coefficient $dn/d\lambda$ is called *normal dispersion*.

The first observations of the dispersion phenomena were by Marcus in 1648 and Grimaldi (1613–1703) in 1665.[21]

Near resonance, however, the dispersion changes its sign; it is now called *anomalous dispersion* and in this region absorption is also appreciable. In the ideal case of oscillation without damping ($\gamma = 0$) the dispersion, instead of having a maximum and a minimum, tends toward $+\infty$ or $-\infty$ according to whether v_o is approached from the lower or higher frequency side, respectively. **Figure 2.4c** shows the curve $\varepsilon = n^2$ for $\gamma = 0$. P Le Roux (1832–1907) was, in 1862, the first to observe anomalous dispersion.[22]

The range where $\varepsilon < 0$ is characterized by total reflection. Qualitatively, Equations 2.18 and 2.19 are in agreement with the measured profiles of absorption lines in gases. To have quantitative agreement, however, the concept of effective number of oscillators has to be introduced. Equation 2.18 is accordingly written as

$$n - 1 = \left(\frac{2\pi e^2}{m}\right) \sum_i \frac{N_i f_i(\omega_i^2 - \omega^2)}{[(\omega_i^2 - \omega^2)^2 + \omega^2\gamma^2]}, \tag{2.20}$$

N_i being the number among the N atoms which are in state i.

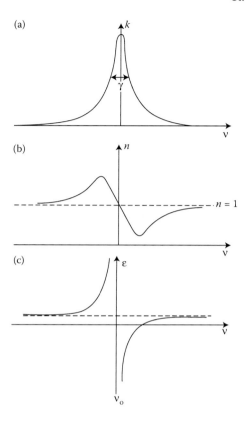

(a)

(b)

(c)

FIGURE 2.4 Behavior of (a) k, (b) n, and (c) ε as a function of frequency.

The factor f_i was interpreted by Drude as the *number of dispersion electrons per atom*. W Pauli (1900–1958)[23] called it the *strength* of the oscillator.

The classical equations considered above were in fairly good agreement with experiment and gave a satisfactory interpretation of dispersion and also of absorption when the imaginary part of the refractive index was considered. However, when Bohr's theory of stationary states superseded the classical theory of elastically bound electrons, these formulae, notwithstanding their *de facto* validity, lost their theoretical justification completely. The first attempts to formulate a dispersion theory in terms of quantum mechanical concepts, made by P Debye[24] (1884–1966), A Sommerfeld[25] (1868–1951), and C J Davisson[26] (1881–1958), were unsatisfactory, mainly due to the application of classical perturbation theory to Bohr's atomic model. This application leads to the inevitable conclusion that the Fourier frequencies are also the frequencies of the optical resonance lines—contrary to experience.

The first correct step towards the formulation of the quantum mechanical interpretation of dispersion was taken by Ladenburg (**Figure 2.5**). Rudolf Walther Ladenburg plays an important role in our history: as we shall see,

FIGURE 2.5 Rudolf Ladenburg (1882–1952).

he came very close to discovering amplification by stimulated emission. He was born in Kiel, Germany, on June 6, 1882 and died in Princeton, New Jersey, on April 3, 1952. He was the third of three sons of the eminent chemistry professor Albert Ladenburg. After school in Breslau, where his father was chemistry professor at the University, Ladenburg went to Heidelberg University in 1900. In 1901, he came back to Breslau and then in 1902 went to Munich, where he took his degree in 1906 with a thesis on viscosity, under Roentgen. From 1906 to 1924, he was in Breslau University first as a *Privat Docent* and from 1909 as *Extraordinary Professor*. In that period, he wrote a paper on the photoelectric effect[27] to which his older brother Eric Ladenburg had given important experimental contribution—accepting Einstein's view of the existence of the photon, at least as a means of explaining the photoelectric effect results.

He married in 1911 and 3 years later served in the army as a cavalry officer, but later on, during the 1914–1918 war, he did research on sound ranging. In 1924, at the invitation of F Haber (1868–1934), he went to the Kaiser Wilhelm Institute at Berlin, the prestigious Institute where Einstein also served, and where he stayed as Head of the Physics Division until 1931 when he went to Princeton.

After the World War I, Ladenburg was looking for the theoretical relations connecting the number of oscillators by means of the new method based on Bohr's theory of describing emission and absorption. In 1921, he first introduced the concept of *dispersion electrons* and gave an expression in terms of the Einstein A coefficient.[28] He obtained this number by calculating the energy emitted and absorbed by an ensemble of molecules in thermal equilibrium with radiation, on the basis of classical theory, on the one hand, and of quantum theory, on the other hand, so making an important application of the correspondence principle formulated by Bohr.

To do this, he considered $\mathcal{N}$ dispersion electrons per cm^3, able to oscillate freely with frequency ν_1. For a harmonic oscillator of frequency ν_1, if the displacement at the generic time t is $x_0 \cos 2\pi\nu_1 t$, the mean value of the total energy

$$\mathcal{U} = \frac{1}{2}m\left[\left(\frac{dx}{dt}\right)^2 + 4\pi^2\nu_1^2 x^2\right] \tag{2.21}$$

is

$$\mathcal{U} = 2\pi^2 m\nu_1^2 x_0^2. \tag{2.22}$$

The mean energy radiated per second by each such electronic oscillator is[29]

$$\left(\frac{16\pi^4 e^2}{3c^3}\right)\nu_1^4 x_o^2 \quad \text{or, therefore,} \quad \left(\frac{8\pi^2 e^2 \nu_1^2}{3mc^3}\right)\mathcal{U}. \tag{2.23}$$

The energy radiated per second by the $\mathcal{N}$ dispersion electrons is then

$$J_{\text{el}} = \left(\frac{8\pi^2 e^2 \nu_1^2}{3mc^3}\right)\mathcal{N}\mathcal{U}. \tag{2.24}$$

If the molecules are in equilibrium with radiation at temperature T, and if we consider the electrons as spatial oscillators with three degrees of freedom, then between $\mathcal{U}$ and the radiation density ρ there exists the relation (Planck)

$$\mathcal{U} = \left(\frac{3c^3}{8\pi\nu_1^2}\right)\rho. \tag{2.25}$$

Therefore,

$$J_{\text{el}} = \left(\frac{\pi e^2}{m}\right)\mathcal{N}\rho. \tag{2.26}$$

The energy absorbed at equilibrium is of course equal to the radiated energy.

In the quantum theory of Bohr and Einstein, the radiation emission from a molecule, as we have seen, is produced in two ways: the spontaneous emission from state k to state i and the stimulated emission. For every transition, a quantum of energy $h\nu$ is emitted and therefore the total energy emitted per second is given by

$$J_Q = h\nu_{ik}N_k(A_k^i + B_k^i\rho_{ik}), \tag{2.27}$$

where N_k is the number of molecules in state k. The absorbed energy is

$$A_Q = h\nu_{ik}N_i B_i^k\rho_{ik}, \tag{2.28}$$

where N_i is the number of molecules in state i. At equilibrium, these two quantities must be equal and by using Einstein's relation, Equation 2.6, between the emission and absorption coefficients and Equation 2.3, one readily finds

$$A_Q = J_Q = N_i\left(\frac{c^3}{8\pi\nu_{ik}^2}\right)A_k^i\rho_{ik}\left(\frac{g_k}{g_i}\right), \tag{2.29}$$

and by equating J_Q to J_{el}, we finally have[30]

$$\mathcal{N} = N_i\left(\frac{mc^3}{8\pi^2 e^2 \nu_{ik}^2}\right)A_k^i\left(\frac{g_k}{g_i}\right). \tag{2.30}$$

Equation 2.30 expresses the constant $\mathcal{N}$ (which can be experimentally deduced from emission, absorption, anomalous dispersion, and magnetic rotation measurements, and in the classical theory is interpreted as the dispersion electron number) in terms of quantum quantities N_i and A_k^i. Therefore, from measurements, for example, of anomalous dispersion at different lines of a spectral series, information on the probability of different transitions can be deduced.

Ladenburg applied Equation 2.30 to explain his experiments on hydrogen and sodium.[28] In the latter element, the oscillator density was about equal to the number of atoms per cm³. In hydrogen, an approximate value of about 4 was found for the ratio of $\mathcal{N}$ relative to the lines H_α and H_β of the Balmer series.

Although Ladenburg does not make explicit mention of it, his work relates to and generalizes the classical Drude formula of dispersion in which the atomic frequencies were the absorption frequencies of the atom.[31]

Now let us again take Equation 2.20, neglecting the damping term and substitute $N_i f_i$ by $\mathcal{N}$ as given by Equation 2.30. We finally find a formula developed by Ladenburg and Reiche (1883–1969):[32]

$$n - 1 = \left(\frac{2\pi e^2}{m} \right) \sum_i \left\{ \frac{N_i \left(mc^3 / 8\pi^2 e^2 \nu_{ik}^2 \right) A_i^i (g_k / g_i)}{\omega_i^2 - \omega^2} \right\}. \tag{2.31}$$

This formula is not yet complete. The existence of negative terms in the dispersion formula was not seen before the work by Kramers and Heisenberg.

Hendrik Anthony Kramers (**Figure 2.6**) was born on December 17, 1894, in Rotterdam, where his father was a physician. He studied at Leyden University, principally with P Ehrenfest (1880–1933), who in 1912 had succeeded

FIGURE 2.6 G. Uhlenbeck, H.A. Kramers (center), and S. Goudsmit around 1928.

H A Lorentz. In 1916, Kramers went to Copenhagen to work with Niels Bohr (1885–1962). When, in 1920, the Bohr Institute of Theoretical Physics opened, Kramers was at first assistant, and then in 1924, Lecturer. In 1926, he accepted the theoretical physics chair at Utrecht and in 1934 he returned to Leyden as the successor of Ehrenfest, who committed suicide in September 1933.

From 1936 until his death on April 24, 1952, Kramers taught at Leyden and paid a number of visits to other countries, including the United States. During his years at Copenhagen, he worked on dispersion problems. In an early publication,[33] he wrote the following expression for the polarization, P:

$$P = E \sum_i f_i \left(\frac{e^2}{m} \right) [4\pi^2(\nu_i^2 - \nu^2)]^{-1} \qquad (2.32)$$

and observed that a formula of this kind, where ν_i are equal to the atomic absorption frequencies, represents the experimental results fairly well.

However, the formula does not satisfy the condition, required by the correspondence principle, that, in the region of high quantum numbers, the interaction between the atom and the radiation field tends to coincide with what is expected from classical theory.

To satisfy this condition, Kramers proposed another expression which also contains negative terms corresponding to emission frequencies. To obtain this new expression, Kramers considered the case of an excited atom and proposed to treat it taking into account not only the stationary state i with energy levels higher than the state 1, but also the states j which have lower energy levels than state 1, so that the formula becomes[34]

$$P = \left(\frac{c^3 E}{32\pi^4} \right) \left\{ \sum_i \frac{A_i^1}{\nu_i^2(\nu_i^2 - \nu^2)} - \sum_j \frac{A_1^j}{\nu_j^2(\nu_j^2 - \nu^2)} \right\}, \qquad (2.33)$$

where

$$\nu_j = \frac{E_1 - E_j}{h}. \qquad (2.34)$$

Equation 2.33 of course relates to a single atom, and a factor has to be adjoined to represent the number of atoms in this state.

In order to derive Equation 2.33, Kramers made use of the concept of the "virtual oscillator," a concept suggested first by J C Slater (1900–1976)[35] and elaborated by Bohr, Kramers, and Slater[36] in a celebrated work which suggested that the energy conservation principle could not be valid in elementary processes. Although this idea did not find any immediate confirmation and was later disproved, in the following years the work had a strong influence. It emphasized the notion of virtual oscillator associated with quantum transitions.[37]

According to this point of view, the dispersion is not to be calculated by considering the real orbit (the stationary state) reacting classically to the exciting wave. Instead, the stationary states appear to be unaffected, except for occasional quantum leaps, so the dispersion is rather to be computed as being due to a set of hypothetical linear oscillators whose frequencies are the spectroscopic ones rather than those of the orbits. Now, as we have seen in classical theory, the atoms behave as electric dipoles of amplitude

$$\frac{e^2 E}{4\pi^2 m(v_1^2 - v^2)}.$$

By considering Equation 2.32, we see that according to quantum mechanics, the atom behaves with respect to the incident radiation as if it contains a number of linked electrical charges constituting the harmonic oscillators as in classical theory, with each one of these oscillators corresponding to each possible transition between the atomic state and another stationary state.

We may therefore describe the behavior of a dispersion atom by means of a doubly infinite set (i.e., dependent on two quantum numbers m and n) of virtual harmonic oscillators, with the displacement of the oscillator (m,n) represented by

$$q(m,n) = Q(m,n)\exp[2\pi i v(m,n)t], \tag{2.35}$$

where $v(m,n)$ indicates the frequency of this oscillator. The set of these virtual harmonic oscillators was called the *virtual orchestra* by A Landé (1888–1975).[38] The virtual orchestra is then a classical formalism substitution for the radiation and so indirectly it becomes the representation of the quantum radiator itself.

In place of the classical e^2/m, we have $c^3 A_i^1/8\pi^2 v_i^2$ for one of the *absorption oscillators*, that is, the ones corresponding to transitions between state 1 and the higher states, but we have the value $-c^3 A_i^j/8\pi^2 v_j^2$ for one of the *emission oscillators*, that is, the ones corresponding to transitions between state 1 and the lower states. There is, therefore, a kind of negative dispersion arising from emission oscillators that can be considered analogous to the *negative absorption* represented by the Einstein B_2^1 coefficient.

In his work,[33] Kramers does not say how he derives the dispersion formula. In a second work[39] (which was a reply to a note by G Breit published in the same volume of *Nature*), Kramers gives an account of the derivation. A complete demonstration of the formula is contained in a work[40] written with Heisenberg (1901–1976) who spent the winter 1924–1925 in Copenhagen working with Bohr and Kramers. The final form of the main formula of this work was suggested by Heisenberg, but the origin of the work is entirely due to Kramers, as Heisenberg himself stated.[41]

In another work, Kramers[42] writes:

> If the atom is in one of its higher states, also terms belonging to the second sum inside the brackets of [our Equation 2.33] appear. In the neighborhood of the frequency v_{em} of an emission line, the atom will then give rise to an anomalous dispersion of similar kind as in the case of an absorption line, with the difference that the sign of P is reversed. This so-called *negative dispersion* is closely connected with the prediction made by Einstein that the atom for such

a frequency will exhibit a *negative absorption*, that is, light waves of this frequency, passing through a great number of atoms in the state under consideration, will increase in intensity.

Induced emission and negative absorption started to be mentioned in several papers.

A clarification and an extension of the correspondence principle was first given by J H van Vleck, in the paper *A correspondence principle for absorption*,[43] and then in more detail in another work.[8] John Hasbrouch van Vleck (**Figure 2.7**) was one of the most prominent American theoretical physicists among the founders of the modern theory of solids and regarded as the founder of modern quantum mechanical theory of magnetism.[44] He obtained his PhD at Harvard with the first American thesis on quantum mechanics in 1922 and was awarded the Physics Nobel Prize in 1977 with N F Mott and P W Anderson for his quantum mechanical description of the magnetic properties of matter. J H van Vleck's idea is the following: if we wish to calculate the absorption by means of the correspondence principle, we must compare absorption, calculated classically, with the difference between absorption and induced emission, calculated from Einstein's formula. In the limit of large numbers, this difference must be equal to the classical absorption.

In van Vleck's paper in *Physical Review*, the term *induced emission* appears for the first time (§3, Note 8). Moreover, he uses freely the expression *negative absorption* and speaking of the emission Einstein's terms writes, "… we shall call [it] the spontaneous emission while the second or remaining term proportional to the energy density we shall call the induced emission, although it is sometimes called the 'negative absorption' in distinction from the true or positive absorption."

FIGURE 2.7 Van Vleck receives the Lorentz Medal from H B G Casimir at the Royal Netherlands Academy of Arts and Sciences, Amsterdam.

He continues

> The existence of the induced emission term in the quantum theory may at first
> sight seem strange, but it is well known that it is qualitatively explained in that
> with the proper phase relations a classical electric wave may receive energy
> from an atomic system, although on the average (i.e., integrating over all pos-
> sible phase relations) it contributes more than it receives in exchange. It is there-
> fore the excess of positive absorption over the induced emission which one must
> expect to find asymptotically (for large quantum numbers) connected to the net
> absorption in the classical theory.

In the same period, the American physicist Richard C Tolman (1881–1948)
(**Figure 2.8**)—a relativity and statistical mechanics scholar, the discoverer of
an effect that demonstrates the existence of free electrons in metals—in the
paper *Duration of molecule in upper quantum states*[45] writes that

> The possibility arises, however, that molecules in the upper quantum state may
> return to the lower quantum state in such a way as to reinforce the primary
> beam by "negative absorption."

Tolman deduced "from analogy with classical mechanics" that the negative
absorption process "would presumably be of such a nature as to reinforce the
primary beam."

After having so clearly prepared the basis for the invention of the laser,
Tolman said that "for absorption experiments as usually performed the
amount of 'negative absorption' can be neglected."

In the same paper, he derived a relation between the integral of the absorption coefficient and the Einstein coefficient $A = 1/\tau$ in an explicit form,
although neglecting the effect of negative absorption. A similar relation was
already found implicitly in 1920 by Fuchtbauer[30] (1877–1959). It was also

FIGURE 2.8 Tolman with Einstein.

derived independently by E A Milne[46] (1896–1950) by means of the following argument and this time taking negative absorption into account.

Let us consider a parallel light beam of frequency in the range between v and $v + dv$ and intensity I_v traveling in the positive x-direction through a sheet of atoms limited by two planes at x and $x + dx$. Let us assume that there are N normal atoms per cm^3, δN_v of which are able to absorb in the frequency range between v and $v + dv$, and N' excited atoms of which $\delta N'_v$ are able to emit in this frequency range. By neglecting the effect of spontaneous re-emission that takes place in all directions, the decrease in energy of the beam is given by

$$-d\left[I_v\delta v\right] = \delta N_v\, dx\left(\frac{hv}{c}\right)B_1^2\left(\frac{I_v}{4\pi}\right) - \delta N'_v\, dx\left(\frac{hv}{c}\right)B_2^1\left(\frac{I_v}{4\pi}\right), \quad (2.36)$$

where $I_v/4\pi$ is the intensity of equivalent isotropic radiation. By resolving Equation 2.36, one has

$$-I_v^{-1}\left(\frac{dI_v}{dx}\right)\delta v = \left(\frac{hv}{4\pi c}\right)(B_1^2\delta N_v - B_2^1\delta N'_v). \quad (2.37)$$

By integrating over the whole absorption line and neglecting the small variation in v over the line, we have (remembering Equation 2.19)

$$\int k_v\, dv = \left(\frac{hv_o}{4\pi c}\right)\left(B_1^2 N - B_2^1 N'\right), \quad (2.38)$$

where v_o is the line center frequency. By using the relation between Einstein coefficients, we have finally

$$\int k_v\, dv = \left(\frac{\lambda_o^2 g_2 N}{8\pi g_1 \tau}\right)\left[1 - \left(\frac{g_1}{g_2}\right)\left(\frac{N'}{N}\right)\right]. \quad (2.39)$$

This integral was often used to measure τ. The integral $\int k_v\, dv$ over the 2537 Å resonance line of Hg was measured by Fűchtbauer et al.[47] in the presence of extraneous gases at pressures between 10 and 50 atm and they found it decreased. An explanation of this in terms of an elevated population in the higher state was given by Mitchell and Zemansky.[48]

The astrophysicists A S Eddington[49] (1882–1944), R H Fowler[50] (1889–1944), E A Milne,[51] and others used in the 1920s the Einstein coefficients to calculate the radiative transfer through stellar atmospheres. In 1937, the American astrophysicist D H Menzel (1901–1976) studied the total emission and absorption of radiation by atomic hydrogen in gaseous nebulae[52] pumped by the light of some star existing in the nebula taking into account the effect of "stimulated emission" which he counted as "negative absorption" and wrote:

> Outside of thermodynamic equilibrium, the condition may conceivably arise when the value of the integral [the total energy absorbed] turns out to be negative. The physical significance of such a result is that energy is emitted rather than absorbed. This energy must be distinguished, however, from that arising

in random emission. The process merely puts energy back into the original beam, as if the atmosphere had a negative opacity.

He then added:

This extreme will probably never occur in practice.

To conclude, we may add that Vitalij Lazarevic Ginzburg (1916–2009), a Russian astrophysicist who had a complex and adventurous life in the revolutionary Russia, exposed by him in an autobiography,[53] and a winner of the Physics Nobel Prize 2003 together with A A Abrikosov (1928–) and A J Leggett (1938–) for his researches in superconductivity, in his autobiography remembers that Saul Maksimovich Levy—a Lithuanian who had worked with Ladenburg[54] collaborating with him in negative dispersion experiments, then emigrated from Hitlerian Germany to the USSR, and later to the United States, after being accused in 1937 to have been a German worker—who was Ginzburg's thesis advisor in the 1930s—often was saying him that stimulated emission could be used for amplification:

"Create an overpopulation at higher atomic levels and you will obtain an amplifier; the whole trouble is that it is difficult to create a substantial overpopulation of levels," Levy was saying.

2.3 Experimental Proofs of Negative Dispersion

It was necessary to wait some years before the first experimental proofs of negative dispersion were found. An expression given by Ladenburg[55] for the dispersion is particularly useful:

$$n - 1 = \left(\frac{e^2}{4\pi mc^2} \right) \left[\frac{\lambda_{kj}^3}{\lambda - \lambda_{kj}} \right] N_j f_{kj} \left[1 - \left(\frac{N_k}{N_j} \right) \left(\frac{g_j}{g_k} \right) \right], \qquad (2.40)$$

where k and j refer to any two stationary states (k being the highest one) with statistical weights g_k and g_j, respectively, N_k and N_j are the numbers of atoms in the two states, λ_{kj} is the wavelength of radiation emitted in the $k \rightarrow j$ transition and

$$f_{kj} = \left(\frac{mc\lambda_{kj}^2}{8\pi^2 e^2} \right) \left(\frac{g_k}{g_j} \right) A_k^j. \qquad (2.41)$$

The term $(1 - N_k g_j / N_j g_k)$ was indicated as a *negative dispersion term*. Near resonance, the effect of all the other absorption lines becomes negligible and, if the gas excitation is low, so that the negative dispersion term can be taken equal to 1, Equation 2.40 can be simplified as

$$n - 1 = \left(\frac{e^2 Nf}{2\pi mc^2} \right) \left[\frac{\lambda_o^3}{\lambda - \lambda_o} \right], \qquad (2.42)$$

$$\lambda_{21} = \lambda_o, \quad f_{21} = f.$$

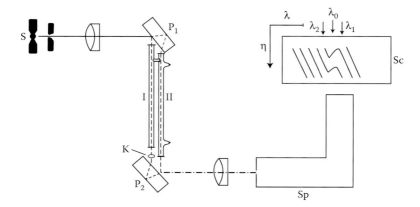

FIGURE 2.9 The hook method for anomalous dispersion. S, continuum source; I, path through the excited medium; II, path through the normal medium; K, compensator; P_1, P_2, Jamin interferometer plates; Sp, spectrograph; Sc, anomalous dispersion pattern.

This is the well-known formula of anomalous dispersion used in experiments to obtain precise measurements of f.

The experimental method used was the so-called *hook method* developed by Rozhdestvenskii (1876–1940).[56] It was based on the use of a Jamin interferometer as shown in **Figure 2.9**. Tube I can be filled with a gas at known pressure and tube II is kept evacuated. A continuous radiation source is used and the resulting beam is focused on the slit of a spectrograph.

With both tubes evacuated and with the compensating plate K removed, the continuous spectrum is crossed by horizontal interference fringes. With the compensating plate in position, the interference fringes are oblique (see **Figure 2.10**).[57] If the separation in wavelengths of a convenient number of fringes in the immediate neighborhood of λ_o is measured, an important constant, D, of the apparatus can be calculated

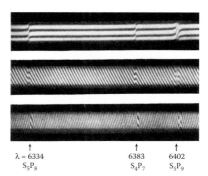

FIGURE 2.10 Pictures of anomalous dispersion near some neon lines.[57]

$$D = -\lambda_o \left(\frac{\text{fringe number}}{\text{separation in wavelengths of these fringes}} \right). \qquad (2.43)$$

If a gas with an absorption line at λ_o is now introduced into tube I, the oblique interference fringes take the form of a hook symmetrically shaped on both sides of the absorption line.

If A represents the separation in wavelengths of two hooks located symmetrically with respect to the absorption line, then the theory of this method, in connection with Equation 2.42, gives

$$f = \left(\frac{\pi m c^2}{e^2 \lambda_o^3 N l} \right) D A^2 \tag{2.44}$$

from which f can be calculated once N and l (thickness of the column of the gas) have been determined.

A simple demonstration of this formula is given in a paper by Ladenburg and Wolfsohn.[58]

When the excitation is strong enough, the number of atoms in the higher state N_k can become an appreciable fraction of the number of atoms in the lower state N_j and the expression $(1-N_k g_j/N_j g_k)$ in Equation 2.40 becomes appreciably different from unity. With the hook method, Ladenburg and his collaborators tested the validity of the negative dispersion term during research performed between 1926 and 1930 and published in volume 48 (1928) and 65 (1930) of *Zeitschrift für Physik*.[55,59,60] The most interesting of these studies is the subject of three papers published by Kopfermann and Ladenburg on the study of dispersion of gaseous neon near the red emission lines.[59] Neon was excited in a tube by means of an electric discharge, and dispersion was studied as a function of the discharge current intensity.

The authors excited tubes of 50 and 80 cm in length and of 8–10 mm in diameter by means of a 20-kV DC generator with currents between 0.1 and 700 mA. The quantity measured was

$$F_{kj} = N_j A_k^j \left(\frac{g_k}{g_j} \right) \left(\frac{mc^3}{8\pi^2 e^2 v_{kj}^2} \right) \left[1 - \left(\frac{N_k g_j}{N_j g_k} \right) \right], \tag{2.45}$$

where j and k are the lower and higher levels, respectively, producing the spectral line under investigation, N_j and N_k are the numbers of atoms/cm^3 in these levels, which are of statistical weights g_j and g_k, respectively.

At low values of the excitation current up to 60 mA, they found that the quantity $Q = N_k g_j/N_j g_k$ was negligible with respect to unity and the value of F_{kj} for different lines increased with current (**Figure 2.11**). In **Figure 2.12**, the values of F_{kj} are shown for different lines of Figure 2.11 which belong to the same lower level S$_5$, normalized to their maximum value.

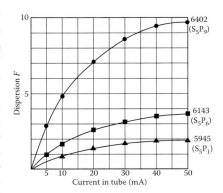

FIGURE 2.11 Behavior of F with current in the tube for different S$_5$ lines of neon. (From R Ladenburg, *Rev. Mod. Phys.* **5**, 243, 1933.)

The values so reduced coincide and therefore Ladenburg deduced that the population N_{S5} of levels S_5 (common to the different lines) changes with current.

The increase in population of level S with increasing current was justified through considerations of the atomic excitation and de-excitation mechanisms by invoking the existence of a statistical equilibrium between excited atoms and electrons.

By increasing the current beyond 100 mA, Ladenburg and Kopfermann (1895–1963)[59,61] found a decrease in F. The results of these experiments are shown in **Figure 2.13**, where the values of F for different neon lines having the lower level S_5 in common are shown for currents up to 700 mA.

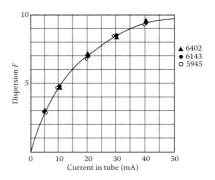

FIGURE 2.12 F values of different S_5 lines of neon as a function of current, reduced to the same scale. (From R Ladenburg, *Rev. Mod. Phys.* **5**, 243, 1933.)

If the values of F for different lines are again reduced to the same scale, these reduced values of F coincide up to 60 mA (see **Figure 2.14**). With higher current values, the reduced F values no longer coincide but separate considerably from one another. Those of the longest wavelength decrease most and those in the shortest wavelength decrease least, that is, the smaller the difference in energy between the common level S_5 and the different upper states P_k, the larger is the decrease in F. This result was correctly interpreted by observing that the greater the current is, the larger is the number of atoms in the upper level P_k. At the same time, the number of atoms in the lowers state S_5

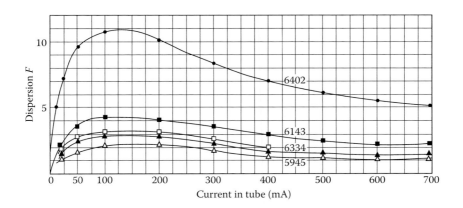

FIGURE 2.13 Behavior of F values of S_5 lines with higher currents. (From R Ladenburg, *Rev. Mod. Phys.* **5**, 243, 1933.)

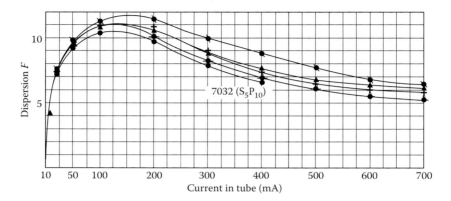

FIGURE 2.14 *F* values of Figure 2.13 reduced to the same scale. (From R Ladenburg, *Rev. Mod. Phys.* **5**, 243, 1933.)

does not increase, but rather decrease a little above 100 mA. Therefore, with increasing current, the ratio N_k/N_j increases.

These experiments gave the first experimental proof of the existence of negative terms in the dispersion equation.

Ladenburg ended his paper[57] in 1933 with these words:

> All these different experiments prove without doubt that the population of the P levels does increase with the current, and so much so that above 100 mA, the ratio $Q (= N_k g_j / N_j g_k)$ has appreciable value. Therefore the experiments shown by the curves of [Figure 2.14] prove the influence of the negative term in the dispersion formula. This "negative dispersion" corresponds to the negative absorption of the theory of radiation and to the term -1 in the denominator of Planck's formula for the radiation of a black body as is easily shown by Einstein's derivation of this formula.

If Ladenburg and his coworkers in the experiments with neon had continued by using more intense discharge currents, they would have found that the curve persisted in decreasing and becomes negative.

The curve of anomalous dispersion (Figure 2.4) is reversed if the absorption is negative.[62] Schawlow observed in a Conference in Megeve[63] that people did not continue Ladenburg and Kopfermann's studies on anomalous dispersion because they believed so firmly in equilibrium that they thought it was impossible to go so far away from it as to have negative absorption.

2.4 More on Negative Absorption

In 1940, Fabrikant[64] observed:

> For molecular (atomic) amplification, it is necessary that N_2/N_1 be greater than g_2/g_1. Such a situation has not yet been observed in a discharge even though

such a ratio of population is in principle attainable. ... Under such conditions we would obtain a radiation output greater than the incident radiation and we could speak of a direct experimental demonstration of the existence of negative absorption.

It is interesting to observe that the laser could have been invented, even through "casually" from 1947 onwards. W E Lamb Jr. (1913–2008) and R C Retherford (1912–1981) were able, by using the methods of radiofrequency spectroscopy, to measure reliably the difference in energy between the $2S_{1/2}$ and $2P_{1/2}$ states of hydrogen. The two states should have been exactly degenerate according to the Dirac equation but a splitting was expected, caused by the coupling of the radiation field with the atom.

The measured splitting (*Lamb shift*) was first published in 1947[65] and Lamb was awarded the Nobel Prize in 1955 for these researches. In 1950, a subsequent paper on the same subject was published in *Physical Review*,[66] where the microwave investigation of the fine structure of hydrogen was fully treated.

In this paper, there is an appendix where the authors analyzed the conditions observed in a Wood discharge. They wrote:

> The absorption of radio waves by excited hydrogen atoms in a Wood's discharge tube depends on the population of the various states. These in turn depend on the rates of production and decay of the excited atoms. It would involve a lengthy program of research to make quantitative calculations of these, and we shall be content here with the roughest sort of estimate.

Lamb and Rutherford applied their reasoning to the $n = 1$ and 2 levels of atomic hydrogen, concluding that the effective decay of some of these levels, under particular conditions, was much less than that due to natural lifetime. Consequently, the population of these states increased correspondingly, by favoring the occurrence of induced emission when these levels were bathed in radiation.

In particular, they considered the transitions induced by radio waves between the $2^2S_{1/2}$ and $2^2P_{3/2}$ levels of hydrogen, and wrote:

> Let us consider the transitions between $2^2S_{1/2}$ and $2^2P_{3/2}$ induced by radiowaves. If these states are populated in accordance with their statistical weights (equipartition), there will be no appreciable net absorption of RF since the induced emission exactly cancels the induced absorption. (Spontaneous transitions between $2^2P_{3/2}$ and $2^2S_{1/2}$ occur at negligible rate.)
>
> If the population of $2^2S_{1/2}$ is increased relative to $2^2P_{3/2}$, there will be a net absorption of RF. If, on the other hand, $2^2P_{3/2}$ is more highly populated, there will be a net induced emission (negative absorption).
>
> On the basis of the preceding discussion alone, one would expect that the 2p levels would be about five to ten times more populated than 2s levels. In that case, one would expect to find negative absorption and as estimated below, a large one.

Some calculation on the attenuation coefficient followed and they concluded,

> In view, however, of the extreme crudeness of the numerical estimates, it is possible that some appreciable departure from equipartition may exist, and that an absorption or induced emission could be detected. It is therefore highly desirable that a search for such effects should be made, especially under discharge conditions which do not favour equipartition.

Lamb wrote later:[67]

> The concept of negative absorption was new to us at the time, and we were unaware of the earlier references. I think that we understood that the radiation would be coherent, as was the input signal. However, we did not associate negative absorption with self-sustained oscillation.
>
> Even if we had so, at least three factors would have kept us from inventing the maser: (1) our interest was centred on the fine structure of hydrogen, (2) the smallness of the expected absorption (gain), and doubt as to its sign, and (3) the ready availability of oscillators at the frequency used.

Notes

1. A Einstein and L Hopf, *Ann. Phys.* **33**, 1026 (1910); *ibidem* **33**, 1105 (1910); A Einstein and O Stern, *Ann. Phys.* **40**, 551 (1913). See also P W Milonni, *The Quantum Vacuum. An Introduction to Quantum Electrodynamics* (Academic Press, Boston, 1994), Chapter 1.
2. Letter to M Besso, November 18, 1916, *Albert Einstein–Michele Besso Correspondence 1903–1955* edited by P Speziali (Hermann, Paris, 1972), p. 78.
3. A Einstein, *Mit. Phys. Ges., Zurich* **16** no. 18, 47 (1916). This work was subsequently published in *Z. Phys*, **18**, 121 (1917) and is translated into English in the book *Sources of Quantum Mechanics* edited by B L van der Waerden (North-Holland: Amsterdam, 1967) and in *The Old Quantum Theory* (Pergamon: Oxford and New York, 1967) edited by D. ter Haar. Other translations exist as for ex. in *The World of the Atom*, edited by H A Boorse and L Motz, Vol. II (Basic Books, 1966), p. 884. See also A S Eddington, *Phil. Mag.* **1**, 803 (1925), L Ornstein and F Zernike, *Versl. Akad. Amsterdam* **28**, 280 (1919).
4. W Wien, *Verh. Dtsch. Phys. Ges.* **18**, no. 13/14, 318 (1916).
5. A good critical reading of Einstein's paper can be found, for example, in D Kleppner, *Phys. Today*, February 2005, p. 30.
6. This expression had already been introduced in a previous work by Einstein, *Verh. Dtsch. Phys. Ges.* **16**, 820 (1914). See also Einstein's work on specific heat in *Ann. Phys. Lpz.* (4) **22**, 180 (1907).
7. Actually, if one averages over all the phases, the classical theory would not have predicted stimulated emission. Einstein either paid no attention to this or understood it was not important. See also W Heitler, *The Quantum Theory of Radiation, 2nd edn* (Oxford University Press; Oxford, 1944), chapter 1, section 5.
8. J H van Vleck, *Phys. Rev.* **24**, 330 (1924).

9. According to relation 2.6, spontaneous emission increases with ν^3 and this behavior reflects the circumstance that excited atoms can decay spontaneously in any of the $8\pi\nu^2\,d\nu/c^3$ modes per unit volume, while stimulated emission must occur in the single mode of the incident photon. Relation 2.6 can therefore be written as

$$A_m^n\,d\nu = \left(\frac{8\pi\nu^2\,d\nu}{c^3}\right)h\nu B_m^n,$$

which tells us that the spontaneous transition probability is equal to the product of the number of modes and the absorption rate for a quantum per mode. In other words, the spontaneous emission probability per mode is equal to that of (negative or positive) absorption in the presence of a quantum in that mode.

10. Einstein's formulae were extended to the case of non-sharp energy levels by R Becker, *Z. Phys.* **27**, 173 (1924) and to the interaction laws between radiation and free electrons by A Einstein and P Ehrenfest, *Z. Phys.* **19**, 301 (1923).

11. P A M Dirac, *Proc. Roy. Soc. Lond.* **A114**, 243 (1927).

12. L S Ornstein and H C Burger, *Z. Phys.* **24**, 41 (1924).

13. However, it is well known that the light emitted by an atom resembles a spherical wave in many of its properties. This, for example, is the case in the discussion of many interference and scattering phenomena. Einstein's argument about unidirectional emission at first sight seems in conflict with this view. The argument was discussed in the following years and comes under close examination in a paper by G Breit (*Rev. Mod. Phys.* **5**, 91 (1933), §3). The reconciliation of the two demands is found in Heisenberg's uncertainty principle. In Einstein's discussion, the momentum of the light quantum and the momentum of the atom are definitely known. According to the uncertainty principle, this automatically excludes knowledge of the position of the atoms and thus makes a discussion of interference impossible. On the other hand, in discussing interference, we suppose the position of the atoms to be known so that the momentum cannot be ascertained and under these circumstances the atom may be said to emit spherical waves. If the atom were held fixed, it would emit a spherical wave. Breit shows that, due to recoil, the spherical waves emitted at each point of its recoil trajectory by the atom under integration over the recoiling time interfere and give rise to unidirectional quanta. For the unidirectional emissions, also see F W Milonni, *The Quantum Vacuum. An Introduction to Quantum Electrodynamics* (Academic Press, Inc.: Boston, 1994), pp. 20–25.

14. M Jammer, *The Conceptual Development of Quantum Mechanics* (McGraw-Hill: New York, 1966), chapters I and IV.

15. M Wolfke, *Phys. Z.* **22**, 375 (1921).

16. W Bothe, *Z. Phys.* **20**, 145 (1923).

17. W H Heitler, *The Quantum Theory of Radiation*, 3rd edn (Oxford University Press: Oxford, 1954).

18. P Drude, *Ann. Phys. Lpz.* (4) **1**, 437 (1900).

19. W Voigt, *Magneto-Elektrooptik* (G B Teubner: Leipzig, 1916).

20. H A Lorentz, *Theory of Electrons* (G B Teubner: Leipzig, 1916).

21. See S A Korff and G Breit, *Rev. Mod. Phys.* **4**, 471 (1932).

22. F P Le Roux, *C.R. Acad. Sci. Paris* **40**, 126 (1862).

23. W Pauli, *Quantentheorie, Handbuch der Physik*, vol. 23 (Springer: Berlin, 1926), p. 87.

24. P Debye, *Munchener Berichte* (1915), pp. 1–26.

25. A Sommerfeld, *Ann. Phys. Lpz.* **53**, 497 (1917).

26. C Davisson, *Phys. Rev.* **8**, 20 (1916).

27. R Ladenburg, *Jahrbuch der Radioaktivitat und Elektronik* **6**, 425 (1909).

28. R Ladenburg, *Z. Phys.* **4**, 451 (1921). This work is translated into English in *Sources of Quantum Mechanics*, edited by B L van der Waerden (North-Holland: Amsterdam, 1967).

29. Expressions 2.23 were simple applications of the well-known Larmor formula; J Larmor, *Phil. Mag.* **44**, 503 (1897).

30. See also Ch Főchtbauer, *Phys. Z.* **21**, 322 (1920).

31. Compare on this point van der Waerden's comment "… it follows that Drude's formula [Equation 2.14 here] is valid for a set of classical harmonic oscillators. Ladenburg does not write out his derivation, but it was essentially known at his time, and we may safely assume that he had such a derivation in mind. Hence we may say that Ladenburg replaced the atom, as far as its interaction with the radiation field is concerned, by a set of harmonic oscillators with frequencies equal to the absorption frequencies v_i of the atom. This idea is not explicitly formulated in Ladenburg's paper, but it is implicitly contained in it, and Ladenburg's contemporaries realised this …" [in van der Waerden, *Sources of Quantum Mechanics* (North-Holland: Amsterdam, 1967), p. 11].

32. R Ladenburg and F Reiche, *Naturwiss.* **11**, 584 (1923).

33. A H Kramers, *Nature* **113**, 673 (1924).

34. The formula given by Kramers contained an additional factor 3; this was a consequence of his hypothesis that free oscillations were parallel to the incident field, while Equation 2.33 in the text here assumes that all atomic orientations are equiprobable. See also the reference in Note 8.

35. J C Slater, *Nature* **113**, 307 (1924).

36. N Bohr, H A Kramers and J C Slater, *Phil. Mag.* **47**, 785 (1924).

37. A discussion of the importance of this work can be found, for example, in M Jammer, *The Conceptual Development of Quantum Mechanics* (McGraw-Hill: New York, 1966), section 4.3.

38. A Landé, *Naturwiss.* **14**, 455 (1926).

39. H A Kramers, *Nature* **114**, 310 (1924).

40. H A Kramers and W Heisenberg, *Z. Phys.* **31**, 681 (1925). An English translation is available in *Sources of Quantum Mechanics*, edited by B L van der Waerden (North-Holland: Amsterdam, 1967).

41. In *Sources of Quantum Mechanics*, edited by B L van der Waerden (North-Holland: Amsterdam, 1967), p. 16.

42. H A Kramers, *Skand. Mat. Kongr.* (1925) 145, reproduced in H A Kramers *Collected Scientific Papers* (North-Holland: Amsterdam, 1956), p. 321.

43. J H van Vleck, *J. Opt. Soc. Am.* **9**, 27 (1924).

44. He wrote an important textbook on the argument: J H van Vleck, *Electric and Magnetic Susceptibilities* (Oxford University Press: New York, 1932).

45. R C Tolman, *Phys. Rev.* **23**, 693 (1924).

46. E A Milne, *Mon. Not. R, Astron. Soc.* **85**, 117 (1924).

47. C Főchtbauer, G Joos and O Dinkelacker, *Ann. Phys., Lpz.* **71**, 204 (1923).

48. See A C G Mitchell and M W Zemansky, *Resonance Radiation and Excited Atoms* (Cambridge University Press: Cambridge, 1934), pp. 113–114.

49. A S Eddington, *Internal Constitution of the Stars* (Cambridge University Press: Cambridge, 1926).

50. R H Fowler, *Phil. Mag.* **47**, 257 (1924).

51. E A Milne, *Mont. Not. Royal Astron. Soc.* **88**, 493 (1928).

52. D H Menzel, *Asp. J.* **85**, 330 (1937).

53. V I Ginzbutg, *About Science, Myself and Others* (IoP: Bristol, 2004); see also V I Ginzburg, *The Physics of a Lifetime* (Springer: Heidelberg, 2001).

54. R Ladenburg and S Levy, *Z. Phys.* **65**, 189 (1930).

55. R Ladenburg, *Phys. Z.* **48**, 15 (1928).

56. D Rozhdestvenskii, *Ann. Phys. Lpz.* **39**, 307 (1928) and *Trans. Opt. Inst- Leningrad* 2, no. 13 (1921); see also J Jamin, *Ann. Chem. Phys.* **52**, 163 (1858); L Puccianti, *Nuov. Cim.* **2**, 257 (1901); R Ladenburg and St Loria, *Z. Phys.* **9**, 875 (1908).

57. R Ladenburg, *Rev. Mod. Phys.* **5**, 243 (1933).

58. R Ladenburg and G Wolfsohn, *Z. Phys.* **63**, 616 (1930).

59. R Ladenburg and H Kopfermann, *Z. Phys.* **48**, 26 and 51 (1928) and *Z. Phys.* **65**, 167 (1930). See also H Kopfermann and R Ladenburg, *Nature* **122**, 438 (1928).

60. R Ladenburg and S Levy, *Z. Phys.* **65**, 189 (1930); A Carst and R Ladenburg *Z. Phys.* **48**, 192 (1928). All these works were synthesized in the beautiful paper by R Ladenburg, *Rev. Mod. Phys.* **5**, 243 (1933).

61. H Kopfermann and R Ladenburg, *Z. Phys. Chem.* **139**, 378 (1928).

62. Compare A Kastler, *Ann. Phys., Paris* **7**, 57 (1962).

63. A Schawlow in a discussion after a paper at the Second International Conference in Laser Spectroscopy, Megeve, June 23–27, 1975.

64. V A Fabrikant, *Thesis* (1940) quoted in F A Butayeva and V A Fabrikant, *Investigations in Experimental and Theoretical Physics, A Memorial to S G Landsberg* (USSR Academy of Science Publications: Moscow, 1959), pp. 62–70.

65. W E Lamb Jr. and R C Rutherford, *Phys. Rev.* **72**, 241 (1947).

66. W E Lamb Jr. and R C Rutherford, *Phys. Rev.* **79**, 546 (1950).

67. W E Lamb Jr. Physical concepts of the developments of the maser and lasers, in: *Impact of Basic Research on Technology*, edited by B Kursunoglu and A Perlmutter (Plenum Press: New York, 1973).

3

Intermezzo
Magnetic Resonance and Optical Pumping

3.1 Introduction

There is no immediate explanation for why more than 20 years had to elapse before the invention of masers and lasers, notwithstanding the fact that the concepts of stimulated emission and negative absorption had been well established since the 1930s. One may argue that one of the reasons for this may be that until the 1950s efforts to produce coherent radiation were directed essentially towards radio waves, which are, in practice, always emitted coherently.

The concept of coherence, moreover, was not yet fully understood: nor was the connection with stimulated emission completely appreciated.

We may also observe that the 20 years between the 1930s and the 1950s were not lost insofar that an acquisition of knowledge took place which was later used in masers and lasers. Moreover, these efforts were directed towards problems which, although they intrinsically contained ideas basic to the making of stimulated emission devices, had, however, a completely different end in mind. Among the arguments then under consideration were studies on magnetic resonance and optical pumping. During the World War II, the main efforts were directed toward the production and detection of microwaves for radars. The technical problems which were solved, among others, were the development of high-power generators, called *magnetrons*, to produce the radar signal; the construction of sensitive crystal detectors to detect the echo; the development of electronic methods of distinguishing the echo over the background noise; and the perfection of narrow-band amplifiers, lock-in detectors, and other noise-reducing circuits to increase the sensitivity of the radar system.

New fields of science—such as microwave and semiconductor device engineering—grew, and the resultant new techniques were invaluable in the development of electronic and nuclear resonance.

Magnetic resonance opened the way both to the understanding of many concepts and to the use of many techniques which were later to be used in masers and lasers. The parallel development of spectroscopy finally led to the optical detection of magnetic resonance and to the techniques of optical pumping.

However, in the subsequent development of masers and lasers, magnetic resonance and optical pumping had noticeably different weights. The studies of magnetic resonance led to a consideration of the possibility of changing the population of the various energy levels and of introducing population inversion through the concept of negative temperature, so leading the way to new approaches and methodologies which had, as a natural output, the maser principles. Optical pumping, on the other hand, although clearly showing the way to extend the same concepts to the optical domain, had little influence on the later developments in the direction, and lasers came out more as a natural extension of masers to shorter wavelengths rather than as a possibility offered by optical pumping.

Much later, A Kastler, the inventor of optical pumping, who was awarded the Nobel Prize for the work in 1966, said himself that he and his collaborators had "never worked on induced emission problems which are at the base of laser operation."[1]

3.2 The Resonance Method with Molecular Beams

Magnetic resonance involves the reorientation of a magnetic dipole as a whole in an external field, or the reorientation of one of the magnetic dipoles existing in an atom or a nucleus with respect to the others. In the first case, transitions are produced between Zeeman components of an energy level, and, in the second, the transitions are among fine-structure or hyperfine components. To observe the effect, two magnetic fields are needed: one static, which removes degeneracy by splitting the energy levels, and the other oscillating, to induce transitions between two states. In this way, absorption (or emission) of radiation takes place which produces changes in the equilibrium distribution of energy levels.

The phenomenon is, in some way, analogous to the electric dipole transition case, but it is much more involved. In the electric case, the levels between which transitions take place (i.e., the oscillating electrical dipole in the classical representation) always exist, being the energy levels of electrons in the atom. In the magnetic case, magnetic energy levels must first be created by some suitable external field and, in general, their spacing is proportional to this field. The study of these magnetic phenomena, in the matter, is also complicated by the circumstance that paramagnetic susceptibilities are lower than the electrical ones by several orders of magnitude. Moreover, the theory of paramagnetism was developed rather late.[2]

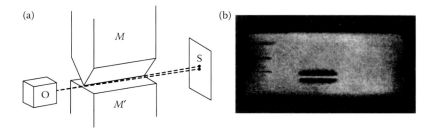

FIGURE 3.1 The Stern and Gerlach experiment. (a) A beam of silver atoms emitted by the oven O passed through the highly inhomogeneous magnetic field *MM'* and was split into two beams according to the two different orientations of the magnetic moments of the atoms. (b) The trace of the split beam. At the time the electron spin was unknown and the two beams were explained assuming the atom having an angular momentum that according to the Bohr–Sommerfeld theory was $\frac{1}{2}\hbar$.

In the case of an atom or molecule, if they have a magnetic moment, according to quantum mechanics, under a static external magnetic field, the magnetic moment vector can have only specific orientations with respect to the external field, corresponding to definite energy values (energy levels). By using this property, Otto Stern (1888–1969) and Walther Gerlach (1889–1979) gave the first experimental proof of space quantization (**Figure 3.1**), being able to measure the angular moment of certain atoms using the deflection effect of an inhomogeneous magnetic field on the magnetic dipole of the atom.[3] By improving the technique, in a series of experiments performed between 1933 and 1937, Stern succeeded in measuring the magnetic moment of the proton and of the deuteron.[4]

E Majorana (1906–1938)[5] and I I Rabi (1898–1988)[6] discussed theoretically the magnetic resonance absorption and Rabi and his collaborators[7] detected magnetic resonance in atomic beams[8] in which transitions between energy levels corresponding to different nuclear spin orientations in a strong constant magnetic field were induced by a radio-frequency magnetic field.[9] Studying the process of orientation of a magnetic dipole in the presence of a magnetic field varying with time, Rabi introduced the concept of "Rabi frequency," that is, the frequency of the oscillatory motion of the dipole under the changing magnetic field.[10]

Isidor Isaac Rabi was born in Poland, but his parents emigrated to the United States when he was a boy and he grew up in the Jewish community of New York where his father owned a drugstore. In 1927, he earned a PhD from Columbia University to which, after 2 years spent in Europe, he returned for the rest of his life until he retired in 1967. In 1927, in Germany, Rabi worked with Otto Stern and his attention was attracted by the experiment Stern did with Walther Gerlach; therefore, when he came back to Columbia, he continued to work on atomic and molecular beams, inventing a method of magnetic resonance described below. By using this method, after the World War II, he was able to measure the electron's magnetic moment with an exceptional accuracy, offering a very good test to check the validity of quantum electrodynamics. The method had huge applications in atomic clocks, in nuclear magnetic resonance and then in masers and lasers. During the World War II, he worked on the development of microwave radar.

In a famous paper written in 1937, Rabi[11] described the fundamental theory for magnetic resonance experiments. At this time in his laboratory, the magnetic moments of many nuclei were under measurement with the method of the inhomogeneous magnetic field used by Stern. The measurements started in 1934 and continued until 1938. Rabi, however, wanted[12] to improve them and therefore studied the effect of the precession motion of the spin around a magnetic field, but did not pay attention to the resonance phenomenon that can be produced if radiation that has exactly the frequency corresponding to the energy difference between one level and the other is used. In September 1937, the Dutch physicist C J Gorter (1907–1980), then at Groeningen University in Holland, paid a visit to Rabi and described to him his unsuccessful attempts to observe nuclear magnetic resonance effects in solids.[13] In the course of his discussion with Gorter, Rabi began to appreciate the resonant nature of the phenomenon and immediately with his collaborators modified his instrumentation. So, in 1939, the method underwent a notable improvement that allowed the reorientation of the atomic, molecular, or nuclear moments with respect to a constant magnetic field, superposing an oscillating magnetic field. When the frequency of the oscillatory field is equal to the energy difference between two levels in the magnetic field divided by Planck's constant, a reorientation may occur that in this case is resonant and may involve an absorption from a lower to a higher level or, in competition a stimulated emission process, from a higher to a lower level. To detect these reorientations, Rabi and his collaborators, J M B Kellog, N F Ramsey, and J R Zacharias (1905–1986), used an ingenious system consisting of two regions in which an inhomogeneous magnetic field acted on the beam (**Figure 3.2**).

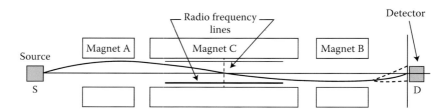

FIGURE 3.2 Molecular beam magnetic resonance experiment. The beam from the source crosses two regions, A and B, with inhomogeneous magnetic field, which deflect the beam in opposite directions. If the molecule does not change its spin state as it comes through region C, it has no deflection. An oscillating field is introduced into region C. When the frequency of the oscillator is the same of the Larmor precession frequency, it changes the spin orientation of the molecule and the beam intensity at the detector drops sharply. (From Ramsey, *Phys. Today* p. 41, October 1993.)

In the first region (A), the field deflected the molecular beam in one direction, while in the second region (B) the inhomogeneous field was applied in the reverse direction, deflecting the beam in the opposite direction so as to refocus it on a detector. Because the focusing and defocusing effects depend in the same way on the velocity of the molecules, at the end all molecules, regardless of their velocity, were refocused on the detector.

A strong magnetic field was applied at the center of the structure (C) that produced a Larmor precession of the magnetic moments of the molecules and in the same region a weaker alternating magnetic field was superposed. If the frequency of this alternating field was equal to the Larmor frequency, it was able to reorient the magnetic moment of the molecule and therefore the particle in the second region was deflected in a different way and was no longer focused. So if the intensity of the beam is observed, by making the intensity of the strong magnetic field constant and slowly changing the frequency of the weak alternating field, a curve such as the one shown in **Figure 3.3** is obtained, which shows a minimum in the signal from the detector when the frequency of the oscillation is equal to the frequency corresponding to the jump between the two levels.[14]

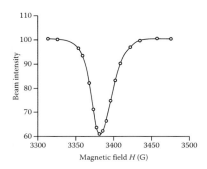

FIGURE 3.3 Resonance curve of Li[7] observed in LiCl. (From Rabi et al., *Phys. Rev.* **55**, 526 (1939).)

The experiment was followed by an accurate measurement of the proton and deuteron magnetic moment and the discovery of the electric quadrupole moment of the deuteron which in turn shows the existence of a tensor force between the proton and the neutron.[15]

For these experiments, Rabi was awarded the Nobel Prize in 1944.

He was highly reputed in the scientific world and many people asked his advice on how to direct their own research. In 1937, he discouraged Ramsey, at the time his PhD student at Columbia University, to continue research on molecular beams because there was little future in the field. Ramsey boldly ignored the advice of his master and provided important contributions in that area, some of which found application in the building of atomic clocks. Yet Rabi himself, 7 years later, earned the Nobel Prize just for the development of the beam-resonance method.

Norman Ramsey was born in Washington, DC, in 1915. His father was an officer in the Army Ordnance Corps and his mother, the daughter of German immigrants, had been a mathematical instructor at the University of Kansas. He entered Columbia College, New York, in 1931 and graduated in 1935, then went to Cambridge University, UK. In the summer of 1937, after 2 years at Cambridge, he returned to Columbia to work with Rabi. Rabi had just invented his method, and Ramsey was the only student to work with him and his colleagues on one of the first experiments. Norman Ramsey died on November 4, 2011.

In 1949, then at Harvard, while studying how to improve Rabi's method to perform measurements with greater accuracy, Ramsey invented the method of separate oscillatory fields in which the single oscillating magnetic field in the center of the Rabi device is replaced by two oscillating fields at the entrance and exit, respectively, of the space in which the nuclear magnetic moments are to be investigated.[16] The method offered a number of advantages and found many applications.[17] For example, with it, more accurate measurements of quadrupole moment of deuteron[18] were performed. The argument was treated in detail by Ramsey.[19]

A very similar method to the one introduced by Rabi was used by L W Alvarez (1911–1988) and F Bloch in 1940 for measuring the magnetic moment of the neutron,[20] which they determined with an accuracy of 1%. Because their papers describing the experiment appeared 3 years after the first magnetic resonance papers from Rabi, their technique is usually presumed to be an adaptation of Rabi's method. However, Bloch had the idea independently to use magnetic resonance by employing an oscillating magnetic field. Alvarez was awarded the Nobel Prize in 1968 for the development and use of a device for the detection of elementary particles in nuclear physics (the *bubble chamber*).

3.3 Magnetic Relaxation Phenomena in Solids

If we now consider a solid or liquid material, the different magnetic moments of nuclei or of electrons, in the presence of an external field, may sum up to give a total magnetic moment that produces a magnetization of the material (paramagnetism).

The problem of how fast the average magnetic moment in a paramagnetic substance responds to a sudden change in the magnetic field in which the substance is placed[21] had already been faced in the 1920s by W Lenz (1888–1957),[22] P E Ehrenfest (1880–1933),[23] and G Breit (1899–1981) and H Kamerling Onnes (1853–1906)[24] before the electronic and magnetic resonance techniques were developed. In the 1930s, the problem of how a magnetic system reaches thermal equilibrium received great attention and this interest was stimulated by the first experiments on adiabatic demagnetization and magnetic relaxation.

The Swedish physicist Ivar Waller (1898–1991)[25] in a paper, now famous, which appeared in 1932, had already distinguished in solids the two main mechanisms through which the system reaches equilibrium after being disturbed (*relaxation phenomena*): the interaction of the spins of electrons or nuclei with each other (*spin–spin interaction*), and the mechanism of interaction among the spins and the lattice (*spin–lattice interaction*). If the interaction of spins with each other and with the lattice did not exist, the application of a static magnetic field would produce only the precession motion of individual spins, independent from one another, without producing a collective motion. Conversely, the effect of the interactions of spins with each other and with the lattice and the presence of the external magnetic field produce energy exchanges between spins and among spins and the lattice and, because atoms with different orientation of their spins may assume different energy levels, a distribution of levels results. After a relaxation time (which is longer, the smaller the interactions are), a thermodynamic equilibrium is reached and we may therefore apply Boltzmann's distribution law which shows that the probability of finding a nucleus or an atom with its spin parallel to the field (lower energy state) is greater than the probability of finding it with anti-parallel spin.

The important concept of spin–lattice relaxation was later taken up again in 1937 by H B G Casimir (1909–2000) and F K Dupré.[26] They observed that concerning the interaction among spins and the lattice, one may consider that electrons, being light and fast, interact strongly and rapidly with each other, reaching a spin equilibrium corresponding to some temperature, in a very short time of about a tenth of millisecond (*spin–spin relaxation time*). The magnetic crystal could therefore be considered as being divided into two systems, each possessing its own temperature. One system contains the magnetic degrees of freedom and goes into thermal equilibrium at a temperature T_M in a very short time (the spin–spin relaxation time $t_2 \sim 10^{-10}$ s). The other system, the lattice or phonon system, contains all the other degrees of freedom and is at a temperature T_L which may be different from T_M. The time needed to establish thermal equilibrium between the two systems is the *spin–lattice relaxation time* τ_{SL} or t_1 (typically in the order of milliseconds). The nuclear spin systems have spin–lattice relaxation times enormously greater than the electronic spin–lattice relaxation times.[27]

Excitation of magnetic levels was at that time done in order to study atomic levels and nuclear spins. C J Gorter had considered in the late 1930s and early 1940s[28] the possibility that nuclear magnetic moment precession in an external field could give rise to macroscopic effects. In 1936, he attempted to detect nuclear resonance in solids by observing an increase in temperature and he showed remarkable insights by attributing the negative results of his experiment to a long spin–lattice relaxation time ($t_1 > 10^{-2}$ s). Later on, he tried again, this time by measuring magnetic dispersion; but again he had no result. The early history and development of the field of paramagnetic relaxation studies until the end of the Second World War has been reviewed in a monograph by Gorter,[29] in which he gives a detailed account of theoretical and experimental results known to him at that time. He discussed these experiments with Rabi in 1937.

3.4 Magnetic Resonance: Bloch, Purcell, and Zavoisky

The first successful experiments[30] to detect magnetic resonance in matter by electromagnetic effects were carried out independently by F Bloch at Stanford,[31] E M Purcell at Harvard,[32] and E Zavoisky in the USSR.[33]

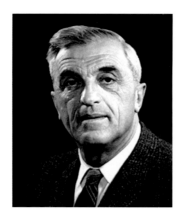

FIGURE 3.4 Felix Bloch (1905–1983).

Bloch and Purcell were awarded the Nobel Prize in 1952 for their researches. They introduced magnetic resonance through two different paths which were, however, substantially similar. Zavoisky was the first to observe transitions between fine-structure levels of the fundamental state in paramagnetic salts (*paramagnetic electronic resonance*).

Felix Bloch (**Figure 3.4**) was born in Zurich, Switzerland, on October 23, 1905. He entered the Federal Institute of Technology (Eidgenössische Technische Hochschule) in Zurich in 1924. After one year's study of engineering, he decided instead to study physics and changed to the Division of Mathematics and Physics at the same institution. During the following years, he studied under Professor P J Debye (1884–1966), P H Scherrer (1890–1969), H Weyl (1885–1955), and E Schrödinger (1887–1961). He was interested initially in theoretical physics. After Schrödinger left Zurich in the autumn of 1927 under Peter Debye's suggestion he worked on his thesis in Leipzig where W Heisenberg (1901–1976) would soon join the faculty. So Bloch became Heisenberg's first graduate student. Under Heisenberg's tutelage, he attached the problem of conductivity in metals and succeeded the summer of 1928 in his PhD thesis in finding the solution of the quantum theory of metals. This led in turn to the modern quantum theory of solids. His results on the motion of electrons in the periodic field of a crystal[34] are today recognized as *Bloch functions* and *Bloch theorem* and have been extended to the case of photons in periodic structures.[35] In the years that followed, he worked with W Pauli (1900–1958), Kramers, Heisenberg, Bohr, and Fermi. In 1930, in Leipzig with Heisenberg he worked on ferromagnetism and established the nature of the boundaries between domains, which have subsequently been known as *Bloch walls*.

After Hitler's ascent to power, Bloch left Germany in 1933 for his native Zurich. A year later, he accepted a position at Stanford University, California. There he started experimental research. In 1936, he published a paper[36] in which he showed that the magnetic moment of free neutrons could be measured through the observation of scattering in iron and showed that in this way polarized neutrons could be obtained.

In 1939, he measured, with Luis Alvarez, the magnetic moment of the neutron.[20] During the war, he was also engaged in the early stages of the work on atomic energy at Stanford University and Los Alamos and later in countermeasures against radar at Harvard University. Through this latter work, he became acquainted with the modern developments in electronics which, toward the end of the war, suggested to him, in conjunction with his earlier work on the magnetic moment of the neutron, a new approach to the investigation of nuclear moments in solids. In 1945, immediately after his return to Stanford, he began the study of *nuclear induction*, as he was later to call it.

His contribution to the theory of magnetism in matter was broad and important[37]; he also held important scientific positions. In 1954, he was the first Director General of CERN in Geneva, the large European organization for high-energy research. He died in Zurich on September 10, 1983.

Edward Mills Purcell (**Figure 3.5**) was born in Taylorville, Illinois, on August 30, 1912. In 1929, he entered Purdue University in Indiana where he graduated in electrical engineering in 1933. His interest had already turned to physics and K Lark-Horovitz (1892–1958)—the great professor to whom solid-state physics in the United States is so indebted—allowed him to take part in experimental research in electron diffrac-

FIGURE 3.5 Edward Mills Purcell (1912–1997).

tion. After spent 1 year in Germany at the Technische Hochschule, Karlsruhe, where he studied under Professor W Weizel, he entered Harvard University, and received a PhD in 1938. After serving 2 years as instructor in physics at Harvard, he joined the Radiation Laboratory at MIT, which was established in 1940 for military research and the development of microwave radar. He became Head of the Fundamental Development Group in the Radiation Laboratory which was concerned with the exploration of new frequency bands and the development of new microwave techniques. The discovery of *nuclear resonance absorption*, as he called it, was made just after the end of the war and at about that time Purcell returned to Harvard as Associate Professor of Physics, becoming Professor of Physics in 1949. In the same year in which he performed the experiments on nuclear resonance absorption, he realized that because the rate of emission or absorption of an electric dipole depends on the density of electromagnetic modes in its environment, enclosing the dipole (it could be an atom or a molecule) into an electromagnetic cavity could change the rate. This is today a well-known effect called *Purcell effect* (see Note 94 in Chapter 4). He died in 1997.

Eugenii Konstantinovich Zavoisky was born in Kazan in 1907 in a doctor's family. He studied and then worked at Kazan University. He was interested almost from his student days in the use of radio-frequency electromagnetic fields for the study of the structure and properties of matter. Commencing in 1933, he performed exploratory experiments on the resonant absorption of radio-frequency fields by liquids and gases. In 1941, he became the first to use the modulation of a constant magnetic field by an audio-frequency field in such experiments. In 1944, he discovered electron paramagnetic resonance, which became the subject of his doctoral dissertation. In 1945–1947, he performed a series of important experiments, recording paramagnetic dispersion curves in the resonance range and obtaining electron paramagnetic resonance in manganese. Later on, he became associated with the Kurchatov Institute of Atomic Energy in Moscow where he worked for more than 20 years.

He made contributions in various fields of nuclear physics—developed, among other things, the scintillation chamber in 1952—plasma physics—discovered magneto-acoustic resonance in 1958. He was awarded the Lenin and State Prizes. He died in 1976. His studies became known in the West only after the Second World War.

The announcement of the first experiments on magnetic resonance was given independently by Bloch and Purcell within 1 month of each other. In the January 1946 issue of the *Physical Review*, E M Purcell, H C Torrey (1911–1998), and R V Pound (1919–2000)[32] in a short letter to the editor, received on December 24, 1945, announced that they had observed absorption of radio-frequency energy, due to transitions induced between energy levels which corresponded to different orientations of the proton spin in a constant applied magnetic field in a solid material (paraffin). In this case, there are two levels the separation of which corresponds, in a field of 7100 Oe, to a frequency, ν, of 29.8 MHz.

They observed:

> Although the difference in population of the two levels is very slight at room temperature ($h\nu/kT \sim 10^{-5}$), the number of nuclei taking part is so large that a measurable effect is to be expected providing thermal equilibrium can be established. ... A crucial question concerns the time required for the establishment of thermal equilibrium between spins and lattice. A difference in the population of the two levels is a prerequisite for the observed absorption, because of the relation between absorption and stimulated emission. Moreover, unless the relaxation time is very short the absorption of energy from the radio-frequency field will equalise the population of the levels more or less rapidly, depending on the strength of this RF field.

The experimental arrangement consisted of a resonant cavity adjusted to resonate at about 30 MHz. The inductive part of the cavity was filled with par-

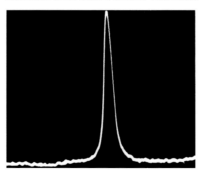

affin and the resonator was placed in the gap of the large cosmic-ray magnet in the Research Laboratory of Physics at Harvard. Radio-frequency power was introduced into the cavity at an extremely low level (10^{-11} W), with its magnetic field everywhere perpendicular to the steady field. When the strong magnetic field was varied slowly, an extremely sharp resonance absorption was observed (**Figure 3.6**, originally Figure 1 from Note 38) at a frequency of 29.8 MHz which corresponded to the energy difference between the two energy levels of the two possible orientations of the proton (spin 1/2) under the static magnetic field.

FIGURE 3.6 Proton resonance (absorption) curve in ferrite nitrate solution as derived from Bloembergen et al. (From N Bloembergen, E M Purcell and R V Pound, Phys. Rev. 73, 679 (1948).)

Purcell's discovery may be considered as the natural follow-on to the effort made during the war at the MIT Radiation Laboratory to decrease radar wavelengths to 1.25 cm. That wavelength happened to fall upon a strong absorption band of atmospheric water vapor and was therefore absorbed during its propagation in air, thus precluding practical radar operation. Purcell was interested in developing precise methods to measure absorption bands and in coherence with this he called his technique *nuclear magnetic resonance absorption.*

In the next issue of the *Physical Review*, again in letters to the editor, there appeared a short note by F Bloch, W W Hansen (1909–1949), and M Packard (1921)[39] which had been received on January 29, 1946. In it was written:

> The nuclear magnetic moments of a substance in a constant magnetic field would be expected to give rise to a small paramagnetic polarisation, provided thermal equilibrium be established, or at least approached. By superposing on the constant field (z direction) an oscillating magnetic field in the x direction, the polarisation, originally parallel to the constant field, will be forced to precess about that field with a latitude which decreases as the frequency of the oscillating field approaches the Larmor frequency. For frequencies near this magnetic resonance frequency one can, therefore, expect an oscillating induced voltage in a pick-up coil with axis parallel to the y direction. Simple calculation shows that with reasonable apparatus dimensions the signal power from the pick-up coil will be substantially larger than the thermal noise power in a practicable frequency band.
>
> We have established this new effect using water at room temperature and observing the signal induced in a coil by the rotation of the proton moments.

Therefore, at resonance the macroscopic magnetization was detected by an induction coil placed normal to both fixed and alternating fields. With Bloch's method, the sign of nuclear magnetic moments can be determined, it being possible to observe the sense of rotation of the Larmor precession of the nuclear magnets. The experiment was later fully described.[40]

In this research, Bloch was motivated by the desire to find techniques to make more accurate magnetic field measurements.

Purcell and Bloch's experiments used nuclear spin levels, whereas Zavoisky's experiment was carried out using electron magnetic levels. The essential difference is that, in the case of electron spin resonance, the applied external magnetic fields are acting on the magnetic moment associated with the electron, which in general comes from the contribution of its spin and orbital motion, and the incoming electromagnetic radiation induces transitions between energy states of the electron.

3.5 Bloch Equations

The theoretical explanation of the Bloch experiment was given shortly afterwards in a famous work entitled *Nuclear Induction*, published in the *Physical Review*,[41] in which Bloch gave a phenomenological explanation of magnetic resonance by taking into account relaxation effects and using an entirely classical treatment.

He considered a paramagnetic sample in a magnetic field H. The equation of motion for the macroscopic magnetization vector $\mathbf{M}$ describes the precession of $\mathbf{M}$ around $\mathbf{H}$:

$$\frac{d\mathbf{M}}{dt} = \gamma(\mathbf{M} \times \mathbf{H}), \tag{3.1}$$

where γ is the gyromagnetic ratio. In this equation, the interactions among spins and between spins and the lattice are not considered. The precession frequency of $\mathbf{M}$ around $\mathbf{H}$, that is, the Larmor frequency, is given by[42]

$$\omega_L = \gamma H. \tag{3.2}$$

Without interaction, there is no change in the component of $\mathbf{M}$ along the direction of $\mathbf{H}$, which we shall take parallel to the z-axis.

Bloch then considered what happens if the magnetization is not in equilibrium and described phenomenologically the relaxation of $\mathbf{M}$ toward equilibrium through simple exponential laws with two characteristic times.[43] One time, t_2, describes how fast the transverse components M_x or M_y die out, and the other one, t_1, describes how fast the component along z attains the equilibrium value

$$M_{zo} = \chi_o H. \tag{3.3}$$

The resulting phenomenological equations were written as

$$\frac{dM_{x,y}}{dt} = \gamma(\mathbf{M} \times \mathbf{H})_{x,y} - \frac{M_{x,y}}{t_2},$$

$$\frac{dM_z}{dt} = \gamma(\mathbf{M} \times \mathbf{H})_z - \frac{M_z - M_{zo}}{t_1}, \tag{3.4}$$

which are usually referred to as *Bloch equations*. Then Bloch considered the existence of a weak oscillating radio-frequency field along the x-direction of the kind

$$H_x = 2H_1 \cos \omega t. \tag{3.5}$$

The solution of Equation 3.4 was found by replacing this field with a rotating field around the z-direction[44]

$$H_x = H_1 \cos \omega t, \quad -H_y = \pm H_1 \sin \omega t, \quad H_z = H_o, \tag{3.6}$$

with the sign of H_y, and therefore the sense of rotation, being negative or positive, depending upon whether the sign of γ is positive or negative. By calling the x and y components of the magnetization vector in the rotating system u and v, respectively, we have

$$M_x = u\cos\omega t - v\sin\omega t, \tag{3.7a}$$

$$M_y = v\cos\omega t + u\sin\omega t. \tag{3.7b}$$

The introduction of rotating coordinates[45] is equivalent to replacing the magnetic field H with an effective field of constant direction equal to

$$H_{\text{eff}} = \left[H_o + \left(\frac{\omega}{\gamma} \right) \right] \mathbf{k} + H_1 \mathbf{i}, \tag{3.8}$$

where $\mathbf{i}$ and $\mathbf{k}$ are unit vectors along the x- and y-axes, respectively, of the rotating system. The angle θ between H_{eff} and H_o which goes from 0 to π is given by

$$\tan\theta = \frac{H_1}{H_o + (\omega/\gamma)} = \frac{\omega_1}{\omega_o - \omega},$$

with $\omega_1 = -\gamma H_1$. In this rotating system, the motion of the magnetic moment M is a Larmor precession around the effective field H_{eff}, with angular velocity

$$\gamma H_{\text{eff}} = \gamma \left\{ \left[H_o \pm \left(\frac{\omega}{\gamma} \right) \right]^2 + H_1^2 \right\}^{1/2}. \tag{3.9}$$

Bloch introduced the set of abbreviations

$$\omega_o = |\gamma| H_o, \quad \omega_1 = |\gamma| H_1, \quad \beta = \frac{1}{\omega_1 t_2}, \quad \alpha = \frac{1}{\omega_1 t_1}, \quad \delta = \frac{\omega_o - \omega}{\omega_1}, \quad \tau = \omega_1 t.$$

In terms of these quantities, Equation 3.4 in the rotating system become

$$\frac{du}{d\tau} + \beta u + \delta v = 0,$$

$$\frac{dv}{d\tau} + \beta v - \delta u + M_z = 0, \tag{3.10}$$

$$\frac{dM_z}{d\tau} + \alpha M_z - v = \alpha M_o.$$

In terms of these equations, Bloch was able to deal with both his nuclear induction and the absorption experiments of Purcell. These experiments were done by making the field H_o along z constant and changing the frequency of the field H_1, passing through the resonance, or else by making the frequency constant and causing H_o to vary around the resonance value. These variations were considered by Bloch in the two limiting cases in which the passage through resonance takes place in a time much shorter than the relaxation times involved (*adiabatic fast passage*) or in the opposite case in which the passage time is much longer than these times (*slow passage*).

In the first case, the variation of δ is slow, and both quantities α and β are assumed to be small compared to unity. The first condition implies

$$\left|\frac{d\delta}{d\tau}\right| \ll 1.$$

In order to have $\alpha \ll 1$, $\beta \ll 1$, it is necessary that either the relaxation times t_1 and t_2 are sufficiently large or that the amplitude $2H_1$ of the oscillating field is sufficiently large.

With the three quantities $|d\delta/d\tau|$, α, and β small compared to unity, a particular solution can be written in the convenient form

$$M_x = \left[\frac{M}{(1+\delta^2)^{1/2}}\right]\cos\omega t, \tag{3.11a}$$

$$-M_y = \pm\left[\frac{M}{(1+\delta^2)^{1/2}}\right]\sin\omega t, \tag{3.11b}$$

$$M_z = \frac{M\delta}{(1+\delta^2)^{1/2}}. \tag{3.11c}$$

The quantity M depends, in a rather involved manner, on the nuclear relaxation times. Under favorable conditions, it may be expected to be of the order of the equilibrium polarization M_o. While its absolute value $|M|$ still represents the instantaneous magnitude of the polarization, the quantity M itself is not necessarily positive but may have both signs, depending on the positive or negative values which δ had assumed in the past. The amplitude of M_y is therefore proportional to $1/(1+\delta^2)^{1/2}$.

The opposite limiting case is that of "slow passage" through resonance or short relaxation times. In this case, Bloch found for arbitrary values of α and β:

$$u = \left\{\frac{|\gamma|H_1 t_2^2 \Delta\omega}{1+(t_2\Delta\omega)^2+(\gamma H_1)^2 t_1 t_2}\right\}M_o,$$

$$v = -\left\{\frac{|\gamma|H_1 t_2}{1+(t_2\Delta\omega)^2+(\gamma H_1)^2 t_1 t_2}\right\}M_o,$$

$$M_z = \left\{\frac{1+(t_2\Delta\omega)^2}{1+(t_2\Delta\omega)^2+(\gamma H_1)^2 t_1 t_2}\right\}M_o, \tag{3.12}$$

where

$$|\gamma|H_o - \omega = \Delta\omega.$$

For nuclear induction experiments, it is evidently favorable to have u as large as possible.

In this case, all three components of polarization vanish at resonance. The amplitudes of M_x and M_y are proportional to the function [neglecting $(1/\gamma H_2 t_2)^2$]

$$f(\delta) = \frac{\delta}{\delta^2 + t_1/t_2}. \tag{3.13}$$

Both functions $1/(1 + \delta^2)^{1/2}$ and $f(\delta)$ are shown in **Figure 3.7**.[46]

From Equation 3.11b and from the definition of M, it is evident that the magnitude of the signal induced by the M_y component of nuclear polarization depends not only on M, but also, in a rather involved way, on the relaxation times and the magnitude and variation in the velocity of δ. In the special case of rapid passage, an estimate of the induced RF voltage is obtained by considering that a receiver coil with N turns around a cross-sectional area A of the sample has a magnetic flux through it given by

$$-\Phi(B) = -NAB_y = -NA4\pi M_y = \pm\frac{4\pi NAM \sin\omega t}{(1 + \delta^2)^{1/2}}. \tag{3.14}$$

The induced voltage V across the terminals of the coil is

$$V = -\left(\frac{1}{c}\right)\left(\frac{d\Phi}{dt}\right) = \pm\left(\frac{4\pi}{c}\right)\frac{NAM\omega\cos\omega t}{(1 + \delta^2)^{1/2}}, \tag{3.15}$$

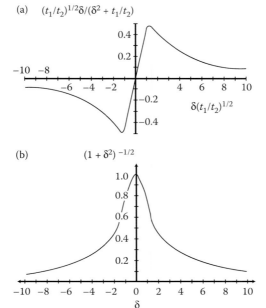

FIGURE 3.7 Plot (a) of the function $f(\delta)$ in normalized form and (b) of the function $1/(1 + \delta^2)^{1/2}$ as a function of δ.

where the variation of δ has been considered sufficiently slow that its time derivative can be neglected when compared to that of cos ωt.

Finally, it can be shown immediately that an expression for magnetic susceptibility can be derived from Equation 3.12. In fact, from the relation

$$M_x = \chi H_x = (\chi' - i\chi'')H_x, \qquad (3.16)$$

where χ' and χ'' are the real (connected to dispersion) and imaginary (connected to absorption) parts of χ, respectively, we have

$$\chi' = \frac{u}{H_1}, \quad \chi'' = -\frac{v}{H_1}. \qquad (3.17)$$

The real and imaginary parts of χ have, therefore, the same behavior as the functions u and v and are qualitatively represented by the curves of Figure 3.5.

The Bloch and Purcell experiments allow the observation of either χ' and χ'' separately or various combinations of these quantities.

These different possibilities had been discussed earlier by Bloch.[41] He had noted in this paper that in order to observe nuclear induction, it was evidently advisable to have u as large as possible.

Its maximum is obtained for

$$\Delta\omega = \left(\frac{1}{t_2}\right)[1 + (\gamma H_1)^2 t_1 t_2]^{1/2}, \qquad (3.18)$$

and has the value

$$u_{max} = \left\{\frac{|\gamma|H_1 t_2}{[1 + (\gamma H_1)^2 t_1 t_2]^{1/2}}\right\} M_o. \qquad (3.19)$$

This value again increases monotonically with H_1 and, for

$$H_1 \gg \frac{1}{|\gamma|(t_1 t_2)^{1/2}}, \qquad (3.20)$$

becomes

$$(u_{max})_{max} = \left(\frac{t_2}{t_1}\right)^{1/2} M_o. \qquad (3.21)$$

To obtain maximum absorption, it is necessary, on the other hand, to make v as large as possible, since it is this quantity which, through Equation 3.7a determines the out-of-phase part of M_x. v has its maximum

$$v_{max} = -\left\{\frac{|\gamma|H_1 t_2}{[1 + (\gamma H_1)^2 t_1 t_2]}\right\} M_o, \qquad (3.22)$$

for $\Delta\omega = 0$. Unlike u_{max}, this quantity does not increase monotonically for increasing H_1, but decreases for large values of H_1. This phenomenon is today known as *saturation*. The best possible choice is

$$H_1 = \frac{1}{|\gamma|(t_1 t_2)^{1/2}},$$ (3.23)

which yields

$$(v_{max})_{max} = \left(\frac{t_2}{t_1}\right)^{1/2} M_o.$$ (3.24)

Magnetic resonance soon became a field of its own and as such is treated in several excellent textbooks.

3.6 Experimental Proof of Population Inversion

Bloch undertook his experiments on nuclear induction with the collaboration of W W Hansen and M Packard.[40] This followed the theoretical treatment[41] in the same issue of *Physical Review*. One of the experiments described is of particular interest to us. After having determined that the relaxation time of the substance with which they were working (water) was between 1/2 s and 1 min, they did the following experiment to determine its value more precisely:[40]

Starting at a time t_1, with H_{dc} held for a considerable previous time above the resonance field H^* a positive signal was observed on the right-hand side of the oscillogram ... as presented on trace a of Figure 8 [our **Figure 3.8**]. Thereupon the field was quickly (i.e., during about one second) lowered to a value sufficiently below resonance to make the signal appear on the left-hand side of the oscillogram and then was held fixed at this new value.

As was expected ... the signal was originally still positive. However, during the following few seconds it was observed to decrease in magnitude, then to disappear and therefore to grow again with negative values until after several seconds it had reached its full negative value ... as presented on trace c of Figure 8 [our Figure 3.8]. This extraordinary reversal of the signal under fixed external conditions represents actually a direct visual observation of the gradual adjustment of the proton spin orientation to the changed situation caused by the

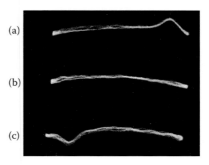

FIGURE 3.8 Photographic record of the proton signal from water. The three traces from top to bottom correspond to an AC modulation of the magnetic field superimposed upon a DC value which in (a) is above, (b) at and (c) below the resonant field H^*. (From Bloch et al., *Phys. Rev.* **70**, 474 (1946).)

previous change of the magnet current. The fact that it takes place during a time interval of several seconds evidently indicates the relaxation time, likewise, to be of the order of a few seconds.

During the inversion time of the signal, the spin population was inverted, but apparently Bloch paid no attention to this, concentrating instead on the problem of the determination of the relaxation time, its exact meaning, and its value. The inversion of population obtained this way (*adiabatic fast passage*) was later used (1958) to create population inversions in two-level solid-state masers (see Chapter 4).

The next year, Bloembergen, a young Dutch physicist, whom we shall discuss in detail later, together with Purcell and Pound, published a paper[38] wherein important considerations on relaxation times were developed and calculations presented on the absorption line shape in an experiment of nuclear magnetic resonance absorption. This paper also included, for the first time, a discussion in terms of populations of the various magnetic levels, which were to prove fundamental to the description of the behavior of masers and lasers. The authors considered a substance containing N_o cm^{-3} nuclei, of spin I and magnetic moment μ, placed in a strong uniform magnetic field H_o along the z-axis, and subjected to a weak oscillating field $H_x = 2H_1 \exp(2\pi i \upsilon t)$, $H_y = 0$. The probability of a single transition in which the magnetic quantum number m_i changes to m_i' can be found with the aid of the standard formula for magnetic dipole transitions:

$$W_{m_i \to m_i'} = \left(\frac{8\pi^3}{3h^2}\right)\left|\langle m_i | M | m_i'\rangle\right|^2 \rho_\upsilon, \tag{3.25}$$

where M is the magnetic moment operator. Ordinarily, ρ_υ represents the energy density, in unit frequency range, in the isotropic unpolarized radiation field. We have to deal here with radiation of a single frequency from levels of a finite width, which we describe by the observed shape of the absorption line, $g(\upsilon)$. The shape function $g(\upsilon)$ is to be normalized so that

$$\int_0^\infty g(\upsilon)d\upsilon = 1. \tag{3.26}$$

The radiation field in this case consists simply of an oscillating magnetic field of single polarization. The equivalent isotropic unpolarized radiation density is calculated as

$$\rho_\upsilon = \frac{3H_1^2 g(\upsilon)}{4\pi}. \tag{3.27}$$

One has finally

$$W_{m \to m-1} = \left(\frac{\pi}{3}\right)\gamma^2(I+m)(I-m+1)\rho_\upsilon$$
$$= \left(\frac{1}{4}\right)\gamma^2 H_1^2 g(\upsilon)(I+m)(I-m+1), \tag{3.28}$$

where m is the magnetic quantum number.

Equation 3.28 gives the probability for a transition $m \rightarrow m - 1$, involving the absorption from the radiation field of the energy $h\upsilon = h\gamma H_o/2\pi$. If the spin system is initially in equilibrium at the temperature T, the population of each level m exceeds that of the next higher level, $m - 1$, by

$$N_m - N_{m-1} \sim \left[\frac{N_o}{2I + 1}\right]\left(\frac{h\upsilon}{kT}\right). \tag{3.29}$$

The approximation 3.29 is an extremely good one, for, in the cases considered, $h\upsilon/kT \sim 10^{-6}$. The net rate at which energy is absorbed is now

$$P_a = \left[\frac{N_o}{(2I + 1)}\right]\left[\frac{(h\upsilon)^2}{kT}\right]\sum_{m=1}^{-I+1} W_{m \rightarrow m-1} = \frac{\gamma^2 H_1^2 N_o (h\upsilon)^2 I(I + 1)g(\upsilon)}{6kT}. \tag{3.30}$$

This is also a calculation of the imaginary part of the magnetic susceptibility χ'', being

$$P_a = 4\pi\chi''\upsilon H_1^2. \tag{3.31}$$

The expression so derived is exact, provided that the original distribution of population among the levels remains substantially unaltered. If the strength of the field H_1 increases, we may expect to have a redistribution of populations in the various levels.

Bloembergen, Purcell, and Pound applied this reasoning to the case $I = 1/2$ so that they had to deal with two levels.

Let n denote the surplus population of the lower level: $n = N_{+1/2} - N_{-1/2}$, and let n_o be the value of n corresponding to thermal equilibrium at the lattice temperature. In the absence of the radio frequency field, the tendency of the spin system to come in thermal equilibrium with its surrounding was described by an equation of the form

$$\frac{dn}{dt} = \left(\frac{1}{t_1}\right)(n_o - n), \tag{3.32}$$

where the characteristic time t_1 is the spin–lattice relaxation time.

The presence of the radiation field requires the addition to Equation 3.32 of another term

$$\frac{dn}{dt} = \left(\frac{1}{t_1}\right)(n_o - n) - 2nW_{1/2 \rightarrow -1/2}. \tag{3.33}$$

A steady state is reached when $dn/dt = 0$, or, using Equation 3.28, when

$$\frac{n}{n_o} = \left[1 + \left(\frac{1}{2}\right)\gamma^2 H_1^2 t_1 g(\upsilon)\right]^{-1}. \tag{3.34}$$

If the maximum value of $g(\upsilon)$ is expressed in terms of a quantity t_2^* defined by

$$t_2^* = \left(\frac{1}{2}\right) g(\upsilon)_{max}, \tag{3.35}$$

the maximum steady-state susceptibility in the presence of the RF field is thus reduced, relative to its normal value, by the "saturation factor,"

$$\left[1 + \gamma^2 H_1^2 t_1 t_2^*\right]^{-1}.$$

The quantity t_2^* defined by Equation 3.35 is a measure of the inverse linewidth.

Effects due to having more particles in excited states were later discussed by W E Lamb,[47] as we saw in Chapter 2. In the same year, Pound[48] showed the possibility of having a variation in population of levels saturated in an RF field.

The next year (1951), Purcell and Pound, in a very short note in the *Physical Review*[49] entitled *A Nuclear Spin System at Negative Temperature*, introduced the concept of *negative temperature* and showed the existence of a negative absorption.

They considered a nuclear absorption experiment and reasoned in the following manner. At field strengths which allow the system to be described by its net magnetic moment and angular momentum, a sufficiently rapid reversal of the direction of the magnetic field should result in a magnetization opposed to the new sense of the field. The reversal must occur in such a way that the time spent below a minimum effective field is so small compared with the period of the Larmor precession that the system cannot follow the change adiabatically.

They found that in a LiF crystal, a zero field resonance occurred at about 50 kHz and the relaxation time was rather long. Therefore, they put the sample in a magnetic field and, after equilibrium was reached, suddenly inverted the direction of the magnetic field. The inversion time was made shorter than the spin–lattice relaxation time and so the configuration of nuclear spins had no time to change during the field inversion.

During the short time in which spins stayed inverted, a negative absorption (i.e., an emission) occurred.

The effect is shown in **Figure 3.9**[49] which is one of the records obtained by sweeping the impressed frequency periodically back and forth through the resonance frequency. The peak at the extreme left is the normal resonance curve, before the field is reversed. Just to the right of this sweep, the field has been reversed and the next resonance

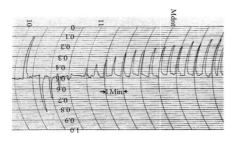

FIGURE 3.9 A typical record of the reversed nuclear magnetization. (From Purcell and Pound, *Phys. Rev.* **81**, 279 (1951).)

peak is seen to point downwards, corresponding to negative absorption. The negative peaks get weaker until finally the state is reached where the positive and negative absorption cancel each other out because there is then equal population of the upper and lower states. The gradually increasing positive peaks show the re-establishment of the thermodynamic equilibrium population.

3.7 The Concept of Negative Temperature

We have already observed how Casimir and Dupré[26] considered that, in the phenomenon of spin–lattice relaxation, it might be convenient to describe the spin system as being a separate system from the lattice, each of the two systems being itself in thermodynamic equilibrium, but not reaching equilibrium with the other system until later. In this case, the spin system can be described by its separate temperature (spin temperature). The probability that a given spin has some energy E is therefore given as

$$p(E) = \exp\left(-\frac{E}{kT}\right),$$

where T can be taken as the temperature of the system. The concept of *spin temperature* was discussed by several authors.[30,50] The distribution function for a system with two temperatures is, for example, described by

$$\exp\left[\left(-\frac{h_1}{kT_1}\right) - \left(\frac{h_2}{kT_2}\right)\right],$$

where h_i is the Hamiltonian function of the ith system described by its temperature T_i; this holds as long as the two systems remain separate from each other.

Purcell and Pound pointed out that the population in the Zeeman levels immediately after the field reversal in their experiments could still be described by the usual Boltzmann distribution if a negative temperature $-T_o$ was considered. In their very short note, they did not give any detailed account of this new concept, so a few remarks are given here. Negative temperature is only possible for a system whose energy levels have an upper bound. It allows population inversions to be described. If we consider the Boltzmann distribution law between two levels 1 and 2, we obtain

$$N_2 = N_1 \exp\left[-\frac{E_2 - E_1}{kT}\right]. \tag{3.36}$$

This law can be used at any instant where T is the so-called *spin temperature*. Equation 3.36 can be considered as defining an instantaneous temperature, T_s, of the spin system in terms of the instantaneous populations N_1 and N_2.

At thermal equilibrium, this spin temperature equals the ambient temperature T. Using it, a population inversion $N_2 > N_1$ is then described by a negative temperature.

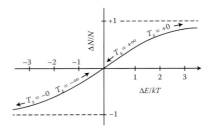

FIGURE 3.10 Relative population difference $\Delta N/N = (N_1 - N_2)/N$ as a function of $\Delta E/kT$.

The ratio

$$\frac{\Delta N}{N} = \frac{N_1 - N_2}{N} = \exp\left[-\frac{E_2 - E_1}{kT}\right] - 1,$$

is shown in **Figure 3.10** as a function of $\Delta E/kT = (E_2 - E_1)/kT$. At very low temperatures, $T_s \rightarrow +0$, all the spins are frozen in the lower level and $\Delta N \sim N_1$ because there is insufficient thermal energy present to lift any spin into level 2. As the spin temperature increases, more spins are thermally excited into the upper level until, as $T_s \rightarrow +\infty$, the populations of the two levels approach equality, $N_1 \sim N_2$ and $\Delta N \rightarrow +0$.

When the population difference becomes negative (inversion), $N_2 > N_1$ and this is possible only if T becomes a negative number. From Figure 3.10, it is apparent that the condition $T = +\infty$ passes over continuously into the condition $T = -\infty$, leading to the observation that negative temperatures are essentially "hotter," not "colder," than positive temperatures. As one increases the spin temperature, passing through $T_s = +\infty$ to $T_s = -\infty$ and so on toward $T_s = -0$, the population of the upper level continuously increases, until at $T_s = -0$ all the spins are "heated" into the upper level and $\Delta N = -N_2$. It is clear that only systems with an upper limit to their energy spectrum can have negative temperatures.

A complete explanation of the meaning of negative temperature was given 7 years after its first introduction in an experiment by Abragam and Proctor[51] where they demonstrated the identities of both spin temperature and thermodynamic temperature.

A comment by Purcell, reported by N Bloembergen[52] on these experiments, was "it is like receiving a marriage licence seven years after the child is born."

Coming back to Purcell and Pound's experiment, a particularly striking feature is that, regardless of how the LiF sample is oriented when it is placed back in the strong field, the spins always remember to point in the "wrong" direction, or in other words, the negative temperature is always achieved, showing clearly that the negative temperatures have a physical reality. This is due to the circumstances that immediately after the reversal of the field the spin–spin and Zeeman temperatures are not equal, but one is in fact the negative of the other. If they were to have remained equal during the reversal of the field, the experiment would have been equivalent to adiabatic demagnetization down to $H = 0$ and then re-magnetization with the field applied in the opposite direction. The resulting temperature would then always be positive, and the spins would always manage to point in the "right" direction.

Purcell and Pound's paper gave only the barest outline of the experiments. The concept of negative temperature was subsequently treated and completed by a number of researchers.[53] The purely philosophical or conceptual aspects were discussed in a later paper by Ramsey.[54]

In a paper which appeared in 1952, Ramsey[55] had already written:

> Pound, Purcell and Ramsey[56] performed a series of experiments with LiF crystals which have a very long relaxation time. They found, among other things, that the spin system is essentially isolated for times which vary from 15 to 5 min and that, for times short compared to these, the spin system can be placed in a state of negative temperature. In a negative temperature state the high-energy levels are occupied more fully than the low, and the system has the characteristic that, when radiation is applied to it, stimulated emission exceeds absorption.

In the Purcell and Pound experiment, the signal they observed was produced by the decay of the inverted population in Zeeman levels. Nobody paid any attention to this method which allowed inversion to be obtained nor to the fact that systems at negative temperature, when in connection with a microwave cavity or a waveguide, could give coherent amplification through the stimulated emission processes. Probably this was due to the circumstance that the inversion method used gave only transient population inversions.

The method of adiabatic magnetization eventually used to obtain population inversions useful for building up paramagnetic masers was not proposed until much later, by Strandberg[57] in 1956 and by Townes and his collaborators,[58] who did not then succeed in making it work. The method was, however, made to work 2 years later[59] (see Chapter 4).

3.8 The Overhauser Effect

In the autumn of 1951, Albert Warner Overhauser (1925–2011) came to the University of Illinois, Urbana, as a postdoctoral student from Berkeley, where he had just completed his PhD thesis with Charles Kittel (1916–). For his thesis, Overhauser had calculated the spin–lattice relaxation time of conduction electrons in metals.[60] No one had actually observed the electron spin resonance of conduction electrons at that time.

Subsequent to completing his thesis, Overhauser had noted a striking result contained in his calculations of the contribution of nuclear spins to relaxing the conduction electron spins: if the conduction electron spin populations were equalized (e.g., by saturating their electron spin resonance), the population difference of the nuclei would be greatly enhanced.

He already had this idea by the time he arrived at Illinois and discussed a possible experiment with Slichter (1924–), then a young Assistant Professor there. The first step in the experiment was to detect the conduction electron spin resonance, and Slichter proposed this argument as a thesis to one of his students, Don Holcomb. They searched unsuccessfully for several months in all of the good metals such as copper, silver, gold, and also in sodium. Eventually, Don Holcomb switched to studying the nuclear resonance in lithium metal and they gave up.

Then in the 15 November issue of the *Physical Review*, Griswold, Kip, and Kittel[61] announced their discovery of the conduction electron spin resonance at microwave frequencies in lithium and sodium.

Slichter immediately repeated the experiment, and with Tom Carver, who had joined Slichter's group in the meantime, immediately set to work to try to verify Overhauser's nuclear polarization prediction.

At this time, Overhauser made his first public announcement of his polarization scheme in a 10-min contribution talk at the Washington meeting of the American Physical Society in April 1953.[62] Present at the talk were, among others, Purcell, Rabi, Ramsey, and Bloch all of whom later entered into a vigorous discussion with Slichter who had been mentioned by Overhauser as attempting with Carver to verify his scheme. A watcher said later to him "You got a Nobel grilling"![63]

Overhauser predicted that if the spin resonance of conduction electrons in a metal is saturated, this should increase by a factor of several thousand the nuclear polarization for metals in which the nuclei reach thermal equilibrium with the lattice by means of the magnetic hyperfine interaction with the conduction electrons. The explanation of this rather surprising result involved solving the problem of the interaction between the electron spin magnetic moment and the nuclear spin magnetic moment, as we shall see later. A full discussion appeared in the 15 October issue of the *Physical Review*.[64] Overhauser's proposal was agreed with scepticism and there was a belief that his scheme violated the second law of thermodynamics.[65]

In the Overhauser effect, one deals with a paramagnetic metal with nuclear spin, usually a feebly paramagnetic alkali, in which the paramagnetism arises from the conduction electrons. A strong constant magnetic field H is applied, whose direction we take as the z-axis. The electrons are exposed to a microwave field of a frequency which is resonant to the value of H employed, and which is powerful enough to give a considerable degree of saturation. Overhauser proposed a method which involved observing the shift of the ESR frequency brought about by the polarization of the nuclei.[66]

In a few months, Thomas R. Carver and Charles P. Slichter eventually got the experimental proof. They looked at the strength of the nuclear resonance absorption, which is proportional to the population difference between adjacent nuclear Zeeman levels, by observing the enhancement of the nuclear resonance in metallic lithium produced by electron saturation.[67] The experiment was performed in a static magnetic field of 30.3 G. The sample containing small pieces of lithium dispersed in oil was placed in the tank coil of a 50-W oscillator operating at 84 MHz, the Larmor frequency for the electrons in the magnetic field. The nuclear resonance was observed using a 50-kHz crystal-controlled oscillator on an oscilloscope. **Figure 3.11** summarizes the results. The top line shows the appearance of the ordinary lithium nuclear resonance, which is so weak at the considered frequencies to be completely lost in noise.

The second line was photographed after the electron saturating oscillator was turned on and the Li resonance now appears strongly. For comparison, the proton line in glycerin (also at 50-kHz) is shown in the bottom line.

The result of this experiment which strikingly confirmed Overhauser's theory gave the opposite behavior to which one would naively expect. Saturation tends to equalize the population of the upper and lower Zeeman states and so produces a higher effective temperature. One might anticipate that the nuclear resonance is correspondingly weakened, but in fact the reverse is found.

The explanation of this paradox is that the electron and nuclear Zeeman temperatures are not the same and that when one is raised, the other is lowered. This is a case in which systems at different temperatures coexist when the thermal contacts between them are weak. The explanation of the effect can be found by considering that, in metals, the direct spin–lattice relaxation time for the nuclear spins is extremely long. By far the most dominant relaxation mechanism for the nuclei is a kind of cross-relaxation or double-spin-flip process in which the spin vectors of a nucleus and an electron simultaneously flip in opposite directions. The interaction between the electron spin magnetic moment and the nuclear spin magnetic moment is that of hyperfine coupling with an interaction Hamiltonian

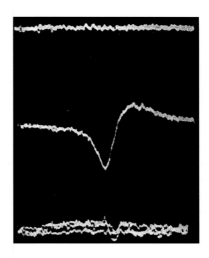

FIGURE 3.11 Demonstration of the Overhauser effect in ^{7}Li. The oscilloscope picture shows nuclear absorption plotted vertically as a function of the magnetic field strength on the abscissa. The magnetic field swing is about 0.2 G. The top line shows the normal ^{7}Li nuclear resonance (lost in noise at the 50 kHz frequency of the nuclear magnetic resonance apparatus). The middle trace shows the ^{7}Li nuclear resonance enhanced by saturating the electron spin resonance. The bottom line shows the proton resonance from a glycerin sample containing eight times as many protons under the same experimental conditions used as reference, from which one concludes that the ^{7}Li nuclear polarization was increased by a factor of 100. (From Carver and Slichter, *Phys. Rev.* **92**, 213 (1953); © The American Physical Society.)

$$\mathcal{H}_{\text{int}} = \left(\frac{8\pi}{3}\right)\beta\beta_{\text{N}}gg_I(I_j \cdot S_k)\delta(r_{jk}), \qquad (3.37)$$

where β and β_{N} are, respectively, the electron and nuclear Bohr magnetons, g and g_I the gyromagnetic ratio of the electron and the nucleus, respectively, and r_{jk} is the distance from nucleus j to electron k. The expression 3.37 is of the "Fermi" or "contact" type and represents an interaction in which $I_z + S_z$ is conserved.

A transition in which $\Delta S_z = +1$ implies $\Delta I_z = -1$, and vice versa; and this fact is essential to the existence of the Overhauser effect.

In treating the interaction of a conduction electron with a nuclear spin, the relevant distribution function can be written as

$$\exp\left[-\frac{k^2 h^2}{8\pi^2 m^* kT_k}\right]\exp\left[-\frac{g\beta H S_z}{kT_z}\right]\exp\left[\frac{g_I \beta_N H I_z}{kT_I}\right],$$

where T_z is the electron Zeeman temperature, T_k is the temperature of the electronic translational motion, which may be identified with the room temperature, and T_I is the Zeeman temperature for the nuclei which, taking into account the interaction 3.37, turns out to be

$$\frac{1}{T_I} = \frac{g\beta + g_I \beta_N}{g_I \beta_N T_k} - \left(\frac{g\beta}{g_I \beta_N T_z}\right). \tag{3.38}$$

If the electronic spin is completely saturated (equal population in its upper and lower states), the temperature T_z can be regarded as infinite. Furthermore, β is enormously greater than β_N, so Equation 3.38 reduces to

$$T_I = T_k \left(\frac{g_I \beta_N}{g\beta}\right).$$

In the case of ^{7}Li, $g\beta$ is 1690 $g_I\beta_N$ and therefore one should have

$$T_I = \frac{T_k}{1690}.$$

The Overhauser effect thus offers the possibility of an enormous reduction in T_I, and hence an enormous enhancement of the nuclear resonance absorption, which is proportional to

$$N_I - N_{I-1} \approx \frac{g_I \beta_N H}{kT_I(2I+1)}.$$

The enhancement factor observed by Carter and Slichter was about 140. The reason for this is that the electron spin resonance was not being completely saturated and because nuclei exchange energy with the lattice.

The Overhauser effect is therefore based on the creation, by pumping, of a high temperature in part of the magnetic system, while a good spin–lattice relaxation keeps another part of the system at low temperature.

The thermodynamic interpretation of the Overhauser effect was first given by Brovetto and Cini[68] and further developed by Barker and Mencher.[69] The effect influenced Bloembergen in his later proposal of the three-level maser (see Section 4.5 in Chapter 4).

In some cases, in fact, the Overhauser effect is very much like a maser action, with the electronic transitions as the pump transition, and the nuclear population strongly inverted. This depends on the sign of g_I. In lithium, g_I is positive and the experimental effect was a greatly enhanced but not inverted nuclear population difference. Inversion of a nuclear magnetic resonance signal via Overhauser effect was obtained much later in 1960 in silicon.[70]

3.9 Spin Echo

In the course of nuclear resonance experiments, many of the concepts which later became essential for the understanding of masers and lasers were introduced and discussed.

Bloembergen, Purcell, and Pound[38] considered in their paper the different kinds of homogeneous and inhomogeneous broadening of the absorption line with the subsequent creation of holes in inhomogeneous line. The concept of *homogeneous* and *inhomogeneous* lines became very important for the understanding of magnetic resonance, masers, and lasers. It was clearly stated later by Portis.[71] As applied to a spin system, *homogeneous* broadening mechanisms broaden the response of each individual spin over the whole linewidth, while an *inhomogeneous* effect spreads the resonance frequencies of different individual spins over some range, widening the overall response of the spin system. The main point is that any excitation applied to one spin in a homogeneous broadened system is immediately transmitted to and shared with all the other spins; in the inhomogeneous case, those spins having one particular resonance frequency can be excited without transferring this excitation to other spins having slightly different resonance frequencies under the overall linewidth. Spin–lattice relaxation, dipolar broadening between like spins, and exchange interaction are all homogeneous broadening effects. Hyperfine interaction, crystalline defects, and inhomogeneous magnetic fields are all inhomogeneous broadening mechanisms.

In atomic or molecular lines, the Doppler effect, due to thermal motion of the particles, produces a homogeneous broadening, while collision effects in a gas give rise to an inhomogeneous broadening. In a real situation, factors producing both homogeneous and inhomogeneous broadening may be present at the same time.

Spin echoes were discussed by E L Hahn, when he was still a graduate student, at a meeting of the American Physical Society in Chicago in November 1949.[72]

In nuclear magnetic resonance phenomena, any continuous Larmor precession of the spin ensemble which takes place in a static magnetic field is finally interrupted by field perturbations due to neighbors in the lattice. The time for which this precession maintains phase memory has been called the spin–spin, or total relaxation, time t_2. If, at the resonance condition, the ensemble at thermal equilibrium is subjected to an intense RF pulse which is short compared to t_2, the macroscopic magnetic moment due to the ensemble acquires a non-equilibrium orientation after the driving pulse is removed. Bloch,[41] on this basis, pointed out that a transient nuclear induction signal should be observed immediately following the pulse as the macroscopic magnetic moment precesses freely in the applied static magnetic field. E L Hahn[73] first verified this behavior: he was then led to consider a closely related effect which he named *spin echo*[72] which may be described as follows. A resonant oscillatory pulse is applied for a short time and then removed. Signals from the previous free Larmor precession are observed for a short time, but these rapidly disappear in a time in the order of t_2. At a time τ later where $t_2 \ll \tau \ll t_1$, a second resonant radio-frequency pulse is applied and then removed. It is then found that at a time τ after the application of the second pulse, a radio-frequency pulse is induced in the receiving circuit. The signal, which apparently arises spontaneously at a time 2τ after the initial pulse is applied, he called *spin echo*.

The origin of this spin echo can most easily be understood by considering the special case in which the initially applied pulse is of just sufficient magnitude and duration to redirect the resultant magnetization through 90° from being parallel to the external magnetic to being perpendicular to that field. After a further time t_2, however, because the transverse relaxation phenomena or perhaps field inhomogeneities, those nuclear moments which had just been aligned in one direction in a plane perpendicular to the external field will be pointing in all directions in that plane. For intervals of time of short duration compared with t_1, however, the moments will still remain aligned in that plane. If, at time τ after the first pulse, a second pulse is applied whose magnitude (to simplify the discussion) is just double that of the original pulse, then this entire plane will be rotated through 180° (twice the original 90°).[74] The nuclear moments will therefore tend to unwind their loss of phase exactly (resulting from the transverse relaxation phenomenon) after a time τ, that is, after the time at which they originally got out of phase.

The method was convincingly compared by N Ramsey[54] with having a number of runners all of whom run at different but constant speeds. If they were started in one direction, they would soon be spread out because of their different speeds. However, if, at a time τ after the start, each runner simultaneously reversed his direction, one would find that, at a time τ after the start, all were neatly drawn up abreast at the starting line.

In **Figure 3.12**, the formation of a spin echo by means of a $\pi/2$–π pulse sequence is viewed in the rotating reference frame. In **Figure 3.12a**, before the application of the pulse, the magnetization M_o is in thermal equilibrium lying along the z-direction (i.e., the direction of the static magnetic field). Immediately after the $\pi/2$ pulse, the magnetization M_o has rotated to lie along the positive y-axis (**Figure 3.12b**). During the time τ, the various spins dephase. A group of them, with magnetization δM is shown in **Figure 3.12c**, precessed by an extra angle θ. **Figure 3.12d** shows the effect of the π pulse on δM.

Looking at the orientation of δM, we see immediately that during a second time interval τ, δM will again advance through the same angle θ, which will bring it exactly along the positive y-axis at $t = 2\tau$ (**Figure 3.12e**). The argument applies to all spins, because the result does not depend on the angle of advance.

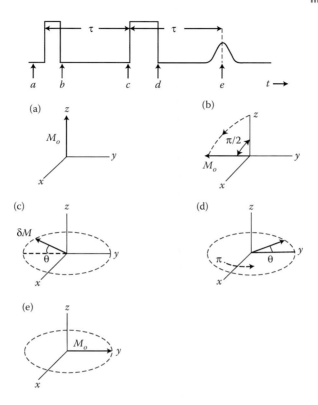

FIGURE 3.12 The formation of a spin echo by means of a $\pi/2-\pi$ pulse sequence viewed in the rotating reference frame. On top, the two pulses and the echo as a function of time. (a–e) The remaining frames show the position of magnetization at the different times marked on the upper time scale.

If one takes into account the relaxation times t_1 and t_2, one can see that during the time interval 2τ, the magnetization decays exponentially, so that the size of the magnetization producing the echo signal is

$$M(t) = M_o \exp\left(-\frac{2\tau}{t_2}\right).$$

Due to the relaxation t_1 after the first pulse, a z-component of the magnetization develops that is subsequently inverted after the second π pulse and does not contribute to the component of M in the x–y-plane existing at time τ. If the second pulse is a $\pi/2$ pulse, as in Hahn's original experiment, it can be seen that several echoes are produced.

In liquids, the diffusion motion allows a nucleus to move between different parts of the sample, where the precession rates may differ. As a result, during a spin echo, the dephasing during the first interval τ may differ from the rephasing during the second interval τ, and so the echo is diminished. This effect, which is of great practical use as a means of measuring diffusion rates in liquids, was discussed by Hahn in his publication, where he showed that diffusion led to a decay of the echo peak magnetization M, given by

$$M(t) = M_o \exp\left[-\gamma^2 \left(\frac{\partial H}{\partial z} \right)^2 \left(\frac{2D\tau^3}{3} \right) \right],$$

where D is the self-diffusion coefficient of the spin-containing molecule. Hahn discovered spin echoes experimentally, but was soon able to derive their existence from the Bloch equations. This solution, which showed the exponential decay of the echo amplitude with t_2, provided a way of measuring linewidths much narrower than the magnetic inhomogeneity. Understanding the physical basis of echo formation has led to a much deeper insight into resonance phenomena, in general, and pulse work, in particular.

Later, in 1957, Feynman (1918–1988), Vernon, and Hellwarth (1930–)[75] introduced a formalism which establishes a formal similarity between the response of a two-level atomic system subjected to a harmonic electric field and that of a magnetic spin in a combined DC and RF magnetic field. In consequence, Bloch equations assumed a wider importance being able to describe also the behavior of two-level atoms.[76]

With their *geometrical representation*, as they called it, it is possible to visualize the atomic dipolar behavior in terms of the conceptually more simple spin precession.

The Schroedinger equation was written, after a substantial transformation, in the form of the real three-dimensional vector equation

$$\frac{d\mathbf{r}}{dt} = \omega \times \mathbf{r},$$

where the components of the vector $\mathbf{r}$ uniquely determine the wavefunction ψ of a given system and the component of ω represent the perturbation. In this paper, the method was shown to enable the analysis of masers and radiation damping, but it can be applied to the understanding of Dicke superradiance and photon echoes, as well as being a quite general one. Photon echoes in optical transitions were also detected much later.[77]

3.10 Medical Application of Nuclear Magnetic Resonance (NMR)

Nuclear magnetic resonance (NMR), initially conceived as a means of studying the magnetic behavior of matter, has become an indispensable medical technique. Because it can measure the shift in the resonance frequency produced in the local nuclear environment, it has become a powerful method of chemical analysis that allows for the identification of chemical compounds and the study of their structure. An important application is in medical diagnostics. The nuclear magnetic resonance allows, in fact, for the position of nuclei-bearing magnetic moments to be identified as a result of the presence of their characteristic absorption spectrum. Nuclei that yield strong signals are, for example, hydrogen, deuterium, carbon, and phosphorous. These nuclei can be identified by their nuclear resonance spectra and with special techniques their position in space can be found, so enabling three-dimensional images to be obtained.

The first spectra from living tissue were obtained only about 30 years ago. The reason why so much time was needed to develop this technique may be found by observing that in nuclear magnetic resonance, very small energies are associated with the transitions, and therefore to obtain sufficiently high signals very strong constant magnetic fields are required. These field must, moreover, be extremely homogeneous over the region of interest which may be rather extended, like, for example, a human body. The use of superconducting magnets has overcome the difficulty. The medical applications of NMR today allows images of anatomic parts of the human body to be obtained and to identify chemical compounds in the organism. For these applications, commercial devices exist whose use has entered the current practice of many large hospitals. The technique may replace the traditional use of X-rays, with improved sensitivity, and without producing the damaging collateral effects of the exposure to X-radiation.

All initial experiments were unidimensional and lacked spatial information. Nobody could determine exactly where the NMR signal originated within the sample. In 1974, Paul C Lauterbur[78] (1929–), working at SUNY Stony Brook, USA, and Peter Mansfield[79] (1933–), working at the University of Nottingham in England, quite independently, described the use of magnetic field gradient for spatial localization of NMR signals. Their discovery laid foundation for magnetic resonance imaging (MRI) and the two authors were awarded the Nobel Prize in Physiology and Medicine in 2003 "for their contribution concerning magnetic resonance imaging."

Their studies set in motion the transformation of NMR technology from a spectroscopic laboratory discipline to a clinical imaging technology. The idea was that by changing the spatial value of the uniform magnetic field in an NMR spectrometer allows to localize from where the signal come. In fact, because the resonance frequency of a spin depends on the uniform magnetic strength, if this strength is different in different points, the resonant frequency is different and therefore the knowledge of the value of the magnetic field in each point allows for the localization of the nuclei whose transitions produce the NMR signal. In 1973, Lauterbur published the first spatially resolved image. In 1974–1975, Mansfield and co-workers developed techniques for scanning samples rapidly and in 1976 obtained the first image of a living human body. The Nobel Prize emphasized their contributions to speeding up the acquisition and display of localized images which was essential to the development of NMR as a useful clinical technique.

The following significant advance was made in 1975, by Richard R. Ernst (1933–)[80] who described the use of Fourier transform of phase and frequency encoding to reconstruct 2D images. This technique is the basis of today's MRI. Ernst was awarded the 1991 Nobel Prize in chemistry.

Whole body magnetic resonance imaging appeared on the scene about 1980. Compared with those vague images, today's pictures are spectacular. Details of organs with a resolution of a few cubic millimeters that vividly display them can be obtained.[81]

Magnetic field gradients had been used previously in NMR to provide some spatial information in one dimension. Erwin Hahn at the University of Illinois, and Purcell and Carr at Harvard had exploited field gradients in the 1950s to study the flow and diffusion of fluids. At Cornell in 1972, D Lee, D Osheroff, and R Richardson used NMR with a gradient to localize different phases of helium 3 in their discovery of superfluidity. But Lauterbur was the first to undertake actual imaging.

3.11 Electronic Paramagnetic Resonance

Electronic paramagnetic resonance does not differ substantially from nuclear resonance except for the fact that the energy levels created by the external magnetic field are not produced by nuclear spins, but are the Zeeman levels produced by the effects of the magnetic field on the motion of electrons in the atom. The application of an external magnetic field to an atom removes the energy degeneration of the atomic orbits, and each electronic energy level is split into many sub-levels separated by a small amount of energy which typically corresponds to microwave frequencies, while the separation among the levels depends on the external magnetic field strength.

The principle of the method is very simple. A constant magnetic field is applied to the substance so that the electronic levels experience Zeeman splitting. Simultaneously, a small radio-frequency field is incident and its frequency is varied. An absorption peak results when the frequency of the variable field corresponds exactly to the energy separation between two Zeeman levels. An example can be discussed by making reference to chromium ions present in ruby. These ions have unpaired electrons that contribute to a total spin equal to 3/2. The electric field, created by all the atoms present in the crystal, blocks all the other angular moments and therefore in the presence of an external magnetic field the behavior arises only from the unpaired electrons. The diagram of the created levels depends on the orientation of the external magnetic field with respect to the principal symmetry axis of the crystal (**Figure 3.13**). This effect is a notable one, as can be seen from **Figure 3.13b,c** that shows the energy levels as a function of the strength of the magnetic field for two different orientations. It is worth observing that by choosing a suitable value of the external magnetic field, the desired separation between levels is obtained. This is the fundamental operating principle of the three-level maser.

3.12 Atomic Clocks

In 1949, Ramsey invented the resonance technique with separated oscillatory fields that in 1955 was applied by Zacharias, J V Parry, Louis Essen, and others

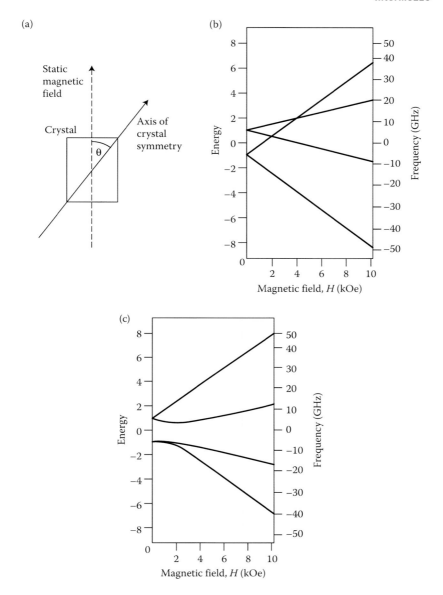

FIGURE 3.13 Energy levels in ruby under the application of an external magnetic field. In (a), the orientation of the ruby crystal symmetry axis with respect to the direction of the magnetic field. The energy levels of ruby are shown for (b) $\theta = 0°$ and (c) $\theta = 90°$. (From J S Thorp, *Masers and Lasers: Physics and Design* (Macmillan, London, 1967).)

to build atomic clocks and frequency standards. For this method, Ramsay was awarded the Physics Nobel Prize in 1989 together with H Dehmelt (1922–) and W Paul (1913–1993) who developed sophisticated techniques to study isolated atoms or molecules by means of the ion trap technique.

The development of atomic theory and, in particular, the conclusion that atoms of a given chemical element were identical to one another, suggested the frequency of an atomic transition as a unit of time. The problem was to convert the very high frequency of the line to more suitable frequencies. C H Townes[82] at Bell Laboratories and R V Pound[83] at the MIT Radiation Laboratory suggested using microwave absorption to stabilize an oscillator.

The solution proposed by Townes and Pound was to use an ammonia molecule which has an absorption that is a function of frequency with a maximum at exactly 23.8 GHz which does not change in time for any reason. The idea was to direct microwave radiation at a frequency that can be varied around 24 GHz into a microwave cavity filled with ammonia and measure the output power, that is, the absorption by ammonia. By varying the frequency, the absorption is maximum just at the central frequency of the ammonia line (23.8 GHz). When this absorption is maximum, we know that the frequency of the microwave we are sending into the cavity is just 23.8 GHz, and therefore we have achieved a very precise frequency. If for any reason, the oscillator suffers a small change in its frequency, the absorption decreases and with a suitable feedback system we may re-adjust the frequency to its proper value.

In the years 1947–1948, Townes at Bell Labs and three other researchers succeeded in assembling four similar devices that stabilized the frequency emitted by a klystron using ammonia.[84] To have a clock, however, the obtained high frequency needs to be reduced into the megahertz region in order to contrast it with suitable oscillators.

Harold Lyons (1913–1991), responsible for the division for microwave standards at the American Bureau of Standards, pushed by Townes, in August 1948 built a frequency standard which was stabilized using the ammonia transition. In 1952, his group successfully obtained a stability of about 10^{-8}, which was however, not enough to compete with other time standards.[85] Strenuous efforts by K Shimoda[86] at the University of Tokyo succeeded in improving the stability to 10^{-9}.

Already in 1948, when he was still assembling his first frequency standard using ammonia absorption, Lyons had also started a program for assembling a true clock that utilized atomic transitions, that is an "atomic clock." Because he was not an expert in spectroscopy, he asked Isidor Rabi for assistance. Rabi's right hand man, Polykarp Kush (1911–1993), prepared a conceptual design in which the use of a cesium atomic beam, extensively studied by Rabi's group, was envisaged.

In the summer of 1951, the apparatus started to operate and in the spring of 1952, the device worked in a modified version using Ramsey's two-field technique to accurately tune the frequency of the microwave field. With it the frequency of the hyperfine transition in the isotope of mass 133 of cesium, the only one that is found to be stable in nature, was measured and found to be 9.192631800 GHz.

These positive results also pushed Britain's National Physical Laboratory to build its own version of the device, still based on the same technique of Ramsey's double resonance. By contrasting their measurements with accurate astronomical measurements, performed at Greenwich, Luis Essen and J V L Parry established[87] the radiation frequency more exactly as 9.192631770 GHz which was adopted in 1967 by the General Conference of Weights and Measures and since then has been accepted as the official definition of the second (in one second there are 9,192,631,770 oscillations of the cesium atom).[88] This definition made atomic time agree with the second based on Ephemeris Time.[89]

Although there are many atoms that could be used for atomic frequency standards, cesium, rubidium, and hydrogen were the most used for many technical reasons.

The hydrogen maser, which Ramsey developed together with Daniel Kleppner in 1961 and which was the first atomic maser, oscillating at 1420 MHz, has been valuable for short-term or comparative measurements such as very-long-baseline radio interferometry. In 1980, by comparing the frequency of an earthbound hydrogen maser clock with one that had been sent aloft on a rocket, R Vessot et al.[90] tested the predictions of general relativity to a precision of 10^{-4}.

All the devices described until now were not proper clocks, as they did not allow a direct comparison with the much lower frequencies that could be used to count seconds. The final step was performed at the MIT Radiation Laboratory by Jerrold Zacharias who, in 1955, succeeded in building a version of an atomic clock, still using cesium, that was to be commercialized under the name "atomichron."[91]

Jerrold Zacharias (1905–1986) entered Columbia as an engineering student and completed his thesis research in 1931, arranging to work with I I Rabi as an unpaid research associate, joining Kellog on measuring the nuclear magnetic moments of hydrogen and deuterium.[92] These observations provided the first indications that the proton has a complex internal structure. In 1937, he participated to the first successful experiment with the Rabi's magnetic resonance method. During the World War II, he developed radar systems at MIT and nuclear weapons at Los Alamos. At the war's end, Zacharias returned to MIT as professor of physics and director of the newly established Laboratory for Nuclear Science and Engineering. At this time, he became interested in developing atomic clocks. After 1956, he devoted himself to education reform. He established the Physical Sciences Study Committee (PSSC) that became a model for educational reform in mathematics and other sciences.

In a typical cesium frequency standard, a beam of cesium atoms emerges from an oven, is collimated and directed through a Stern–Gerlach magnet which deflect and focuses, through a hole, those Cs atoms that are in the correct state in the microwave cavity where the microwave interrogation fields are spatially separated. Atoms leaving the region pass through another Stern–Gerlach magnet. Atoms that have changed state as a result of the microwave interaction are directed to a hot-wire ionizer and detected. Maximizing the current induced in the hot-wire ionizer maximizes the number of atoms making the transition and thus assure that the frequency of the microwave matches the atomic resonance frequency.

Replacing the state selection magnets with lasers that optically pump the cesium atoms into specific energy states, thereby making state selection and detection more efficient, the frequency inaccuracy is in the order of 3×10^{-15}.

Detailed accounts of atomic clocks up to the mid-1980 were given by Norman Ramsey[93] and Paul Forman.[91]

Hyperfine transitions of Rb^{87} were also of use. The use of the rubidium clock and its importance in the GBS system has been described by J Camparo.[94]

The stability and accuracy of atomic clocks have experienced a decisive advance in the 1990s through the development of trapping and cooling atoms by means of lasers. Complex geometries have permitted to build what are called *fountain clocks*. The duration of the atom–wave interaction in an atomic clock is finite and broadens the resonance peak as a result of the Heisenberg time–energy uncertainty principle. Other effects affect the frequency, limiting the accuracy of the clock. With laser techniques, atoms can be cooled relatively easily down to 1 μK. At such temperatures, they have a thermal velocity of only a few millimeters per second, instead of the 100 m/s they have at room temperature. In the so-called *atomic fountain*, the atoms are launched vertically upwards using lasers and eventually fall back down again due to gravity, like a water fountain. The system is so designed that atoms interact with the electromagnetic field both on the way up and the way down, so increasing the interaction time. With a high of 1 m, the interaction time in current atomic fountains approaches 1 s.

The idea of a fountain clock was already considered in 1953 by Jerrold Zacharias, but collisions between atoms in the beam did not allow success. Ramsey provides a brief description of the experiment in his book on molecular beams.[95]

Eventually, in 1989, Steven Chu and his group at Stanford University, California, demonstrated the first fountain using cooled sodium atoms,[96] followed in 1991 by André Clairon at the Laboratoire Primaire du Temps et des Frequences and Christophe Salomon at the Ecole Normale Supérieure in Paris who used caesium.[97] In 1993, Clairon built the first cold-atom fountain which in 1995 achieved an accuracy of less than 1 s in 30 million years.[98]

The need to have such accurate clocks occurs, for example, in radio astronomy or to verify Einstein's relativity theory. The most prominent use of atomic clocks is perhaps in the global positioning system (GPS) of satellites for navigation and monitoring.

The basic operation of a cesium fountain clock starts with a sample of around 10^8 cesium atoms laser-cooled to below 1 μK that are then launched upward at about 4 m/s. The beam travels upward and enters the microwave cavity. The passage through the cavity on the way up provides the first pulse of the two-pulse (Ramsey) microwave interrogation sequence. The atoms reach apogee above the microwave cavity and eventually fall through the microcavity a second time. Atoms that have made a transition in state due to the interaction with the microwave field are detected optically with a laser.

A further step was done using optical transitions. Optical transitions with their higher frequency promise better stability and accuracy. It was recognized for more than 40 years that a frequency standards based on an optical transition in a trapped single ion has the potentiality to reach a precision of 10^{-18}. An optical clock based on a single trapped $^{199}Hg^+$ ion was described by Diddams et al.[99] at the Time and Frequency Division National Institute of Standards and Technology, United States. The clock utilized a mode-locked femtosecond laser. In the authors' words

> Fundamental ideas and technical developments principally in three areas have brought us to the point where we are now able to demonstrate an optical frequency standards representative of clocks of the future: (i) the idea[100] and demonstration of laser cooling of atoms,[101] (ii) the frequency stabilization of lasers,[102] and (iii) the concept[103] and demonstration[104] that femtosecond mode-locked lasers combined with nonlinear fibers can provide a simple, direct, and phase-coherent connection between radio frequencies and optical frequencies. Although most of these concepts have existed for some time, and preliminary demonstrations of optical clocks have even been made,[105] only now have the techniques and tools advanced to the level required for optical frequency standards to move beyond the benchmark results of the microwave standards.

Later, the mercury ion optical clock contrasted versus the NIST-F1 cesium fountain standard[106] reached a fractional uncertainty of 9.1×10^{-16}. The situation at the outset of the twenty-first century was described by Diddams et al.[107]

3.13 Optical Pumping

Optical pumping is directly connected with the problems we have just been considering. It is a method for producing important changes in the population distribution of atoms and ions between their energy states by optical irradiation. Optical pumping techniques were developed by Kastler, Brossel, and others, but it is usually agreed that Kastler was the leading scientist in the field.

Alfred Kastler was born in Guebwiller, Alsace, on May 3, 1902. He studied at the Ecole Normale Supérieure, 1921–1926, then taught in secondary schools. In 1931, Daure of Bordeaux University offered him an assistant professorship in his laboratory. There Kastler took his *doctorat des sciences* in 1936. From 1938 to 1941, he was Professor of Physics at Bordeaux University, and then returned to the Ecole Normale Supérieure in Paris. He was awarded the Nobel Prize for Physics in 1966. He died at Bandol, France, on January 7, 1984.

In 1949, in collaboration with Jean Brossel (1918–2003), he described[108] an optical method which enabled the redistribution of the populations in the sub-levels of an excited atomic level.[109] The following year, he named this method *optical pumping* in a publication in the *Journal de Physique et le Radium*.[110] He used this term to describe a process which produced in a stationary form, a situation in which the population of a set of atomic sub-levels (Zeeman levels or hyperfine levels) of the fundamental state was different from the normal Boltzmann distribution. The technique uses a cycle which entails the absorption of optical resonance light followed by its spontaneous re-emission: the basic principle involves the conservation of angular momentum in both the matter and radiation interactions.

The changes in population can be monitored by noting either the change in intensity of the light transmitted by the sample in which optical pumping is produced or the change in either intensity or polarization of the scattered resonance light. The methods of optical pumping and of optical detection can also be used either together or separately to investigate excited states of atoms.

This was accomplished by Brossel and Bitter (1902–1967) in 1952[111] who obtained selective excitation of Zeeman sub-levels of the excited state 6^3P_1 of a mercury atom and detected the change in polarization of the re-emitted resonance radiation. This method of studying excited states was called *double resonance* by its authors.[112]

In order to understand the principle of the method proposed by Kastler, let us consider the case of a sodium atom in its ground state, which is $^2S_{1/2}$, with its electron spin 1/2 split by a magnetic field into two Zeeman sub-levels: $m = -1/2$ and $m = +1/2$. For simplicity, let us disregard nuclear spin. By absorption of optical resonance radiation (the D_1 and D_2 lines of sodium), the atom is raised to the $^2P_{1/2}$ and $^2P_{3/2}$ states which are the excited states nearest to the ground state.

Figure 3.14a and **b**[113] shows the Zeeman structure of the levels involved and the spectral transitions between them. **Figure 3.14a** is an energy scheme: the energy of the state is given by the height of the horizontal line representing the state. Spectral transitions are indicated by vertical arrows. **Figure 3.14b** is a polarization scheme: magnetic sub-levels of the same state are represented by equidistant points on a horizontal line. The arrows indicate the Zeeman transitions.

In this scheme, vertical arrows correspond to $\Delta m = 0$ or π transitions, arrows with a positive slope to $\Delta m = +1$ or σ^+ transitions, and arrows with a negative slope to $\Delta m = -1$ or σ^- transitions. The numbers indicated by TP are the relative transition probabilities on an arbitrary scale. Only $\Delta m = 0$ and $\Delta m = \pm 1$ transitions are allowed.

Suppose we illuminate the atoms of an atomic beam of sodium with the circularly polarized yellow light of a sodium lamp. Suppose also that the incident light contains only the D_1 line. Only the Zeeman component σ^+ is exciting the atoms, and in the excited states only $m > 0$ states will be reached. From there, the atoms may fall back to the state from which they came. Others will instead make the transition to the lower $m = +1/2$ state.

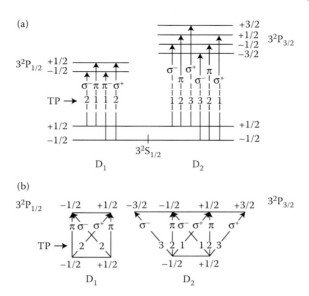

(a)

(b)

FIGURE 3.14 Pertinent levels of Na for the D_1 and D_2 lines: (a) the energy levels; (b) a polarization diagram. (From C Cohen-Tannoudji and A Kastler, *Progress in Optics*, edited by E Wolf (North-Holland, Amsterdam, 1966), Vol. 5, p. 3; reprinted by permission.)

If this process is repeated several times, atoms of the ground state will leave the $m = -1/2$ level and will accumulate in the $m = +1/2$ level. This change in population can be detected optically.

Take a Na beam (**Figure 3.15**) and orient the atoms in the A region by illuminating with σ^+ light. In the B region, the atoms are illuminated again with π light, and we measure the ratio of the intensities, $I\sigma^+$ and $I\sigma^-$, emitted. This ratio gives the degree of orientation.

The number of atoms excited during time Δt can then be calculated: it is

$$\Delta n_1' = B_{mm'}\rho_\nu n_m \Delta t,$$

where n_m is the number of atoms in the initial level m, ρ_ν is the spectral density of the incident radiation, and $B_{mm'}$ the Einstein absorption probability of the $m \to m'$ Zeeman transition.

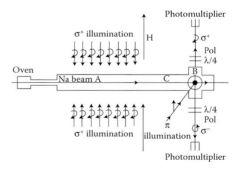

FIGURE 3.15 Production and detection of atomic orientation in a beam of Na atoms. (From A Kastler, *J. Opt. Soc. Am.* **47**, 460 (1957); reprinted by permission of the Optical Society of America.)

Masers and Lasers

The number of atoms which fall back to the fundamental state is

$$-\Delta n_2' = n_m' \Delta t \sum_m A_{m'm} = \left(\frac{n_{m'}'}{\tau}\right) \Delta t,$$

$A_{m'm}$ being the Einstein emission probability of a transition $m' \to m$ and τ the lifetime of the excited level.

At equilibrium, under the influence of steady radiation, we have

$$\Delta n_1' + \Delta n_2' = 0,$$

from which

$$\frac{n_{m'}'}{n_m} = \rho_v \left(\frac{B_{mm'}}{\sum_m A_{m'm}}\right) = \rho_v \left(\frac{c^3}{8\pi h\nu^3}\right)\left(\frac{A_{m'm}}{\sum_m A_{m'm}}\right).$$

For the excitation to the sub-level $m' = -1/2$ of the excited state of atoms coming from the sub-level $m = -1/2$ of the fundamental state, the expression $A_{m'm}/\Sigma A_{m'm}$ is equal to 2/3 for D_1 and to 1/3 for D_2. Therefore, the ratio $n_{m'}'/n_m$ of excited atoms depends essentially on ρ_v, that is, on the intensity of the effective incident irradiation. By taking the factor $\rho_v c^3/8\pi h\nu^3 = 10^{-4}$, the number $\Delta n''$ of atoms which make the transition during the time Δt towards the lower sub-level m'' (different from the initial level m) was calculated by Kastler[110] as

$$\Delta n'' = n_{m'}' \Delta t A_{m'm''} = n_{m'}' \left(\frac{\Delta t}{\tau}\right)\left(\frac{A_{m'm''}}{\sum_m A_{m'm}}\right),$$

where $A_{m'm''}/\Sigma A_{m'm}$ for the $m' = -1/2 \to m'' = +1/2$ transition is equal to 1/3 for D_1 and to 2/3 for D_2. The increase in the number of atoms in level $m = +1/2$ of the fundamental state turns out to be $\Delta n'' = (2/9) \times 10^{-4} n_m \Delta t/\tau$, and it is equal to $-\Delta n$, the simultaneous decrease in the number of atoms in the level $m = -1/2$.

By admitting that the two exciting lines D_1 and D_2 have the same intensity, the number of atoms of this level decreases under the influence of irradiation, according to the differential relation

$$\frac{\Delta n}{n} = -\left(\frac{4}{9}\right) \times 10^{-4}\left(\frac{\Delta t}{\tau}\right).$$

Kastler defined

$$\tau' = \left(\frac{9}{4}\right) \times 10^4\,\tau,$$

$$n_{-1/2} = n_o \exp\left(-\frac{t}{\tau'}\right), \quad n_{+1/2} = n_o\left[2 - \exp\left(-\frac{t}{\tau'}\right)\right].$$

While the number of atoms in level $m = -1/2$ decreases exponentially, the number of atoms in the other level, $m = +1/2$, tends exponentially to twice the initial number. After a time in the order of τ', the ratio $n_{+1/2}/n_{-1/2}$ will surpass 1. The population asymmetry will be considerable.

The lifetime of level 3^2P of Na is in the order of $\tau = 10^{-8}$ s. Therefore, $\tau' = 2 \times 10^{-4}$ s. Kastler considered that, near 100°C, the thermal velocity of sodium atoms is in the order of 500 m s^{-1}. At time τ', these atoms travel a distance of some 12 cm. It is therefore possible, by irradiating an atomic beam from the side, to obtain, after a passage of several centimeters, an important population asymmetry of level m of spatial quantification.

It may be worthwhile at this point to recall briefly the state at this time of research into the detection of radio-frequency resonances by using optical resonance transitions of atoms.

The first researchers to observe the effect of a radio-frequency field on the optical resonance radiation of atoms were Fermi (1901–1954), Rasetti (1901–2001),[114] Breit (1899–1981), and Ellett,[115] who in 1925 observed the change in population of resonance fluorescence of mercury vapor by changing the frequency of an applied alternating magnetic field.

If we assume the mercury atom to be a damped classical oscillator, it will have a precession motion around the magnetic field with a Larmor frequency

$$\omega_L = \frac{eH}{2mc}.$$

If the reciprocal of the lifetime of the oscillator (for mercury, it is $1/\tau = 10^7$ s^{-1}) is of the same order of magnitude as the Larmor frequency, this precession motion in a static magnetic field makes it possible to obtain emission with the electric field of the wave vibrating in different directions, if excitation of fluorescence is achieved with linearly polarized light. In other words, one has depolarization.

If an alternating field is used at a frequency much larger than the Larmor frequency, ω_L, no large effect on polarization will occur, the precession being first in one direction and then in the opposite direction, depending on the direction of the alternating field, and also being very small in both directions due to the rapid changing of field. If, on the other hand, the frequency is lower than ω_L, the oscillator will have time to precess in the field before its direction changes and consequently a depolarization will appear.

In the Fermi and Rasetti experiments,[114] the magnetic field could be changed from 1.13 to 2.13 G (1 G gives a Larmor precession frequency of about 1.4×10^6 s^{-1} for a classical oscillator). The frequency of the field could be changed from 1.2 to 5.10^6 s^{-1}. At 1.13 G and at a frequency of 5×10^6 s^{-1}, they found that practically no depolarization was present, whilst at 1.87 G and at the same frequency a depolarization was observed. At 2.13 G, the depolarization was as large as in a static field. If the field strength had been larger by a factor 3/2, the results obtained could have been explained satisfactorily on the basis of the classical oscillator model.[116]

They therefore measured the *g* factor, which is just the 3/2 factor, and the lifetime of the excited states. These results were described in a short letter of 17 lines to *Nature*.[114] A more detailed description of the experiment was published later.[117]

Rasetti remembers the experiment as follows:[118]

Wood and Ellett[119] and Hanle[120] had announced their remarkable discovery of the effect of weak magnetic fields on the polarisation of resonance radiation of mercury. Rasetti had observed these effects in Florence. When Fermi came to that university a few months later, he was greatly interested in the phenomenon, whose only theory at the time was a classical one based on the concept of Larmor precession. Fermi pointed out that, since the mercury resonance line showed an anomalous Zeeman effect with a Landé factor of 3/2, the mercury atom should more likely precess with a frequency 3/2 times higher than the Larmor frequency. The choice between the two alternatives might be decided by investigating the behaviour of the population under magnetic fields, of the intensity of about one gauss and a frequency of a few megacycles per second in approximate resonance with the precession frequency of the atom.

Rasetti had experience with the spectroscopic technique, but neither of the experimenters had any with radio-frequency circuits. However, Fermi calculated the characteristics of a simple oscillator circuit which should produce fields of the proper strengths and frequencies. Fortunately some triodes were discovered in an instrument cabinet and pronounced by Fermi apt to operate the projected circuit. The laboratory also possessed several hot-wire ammeters to measure the current in the coils, in order that the magnetic field strength could be determined. Had these instruments not been available, the experiment could not have been performed, since the research budget of the laboratory was exceedingly meagre and did not allow the purchase of costly equipment. Another consequence of this financial situation was the fact that the building was never heated, since it was easy to calculate that one month's heating would have absorbed the entire annual budget of the Physics Department. The temperature in the building from December to March ranged from 3 to 6°C. Unfortunately, in the spring, when the experiments were performed, the room temperature had risen to 12°C, more comfortable for the inhabitants, but somewhat too high to ensure a sufficiently low density of the saturated mercury vapor.

Inductance coils and other simple parts were built by the experimenters and when the circuit was assembled, it instantly worked as Fermi had predicted. The experiments were readily performed; unfortunately the accuracy was poor, due to the high temperature and the photographic method employed for measuring the polarisation. Still the results clearly showed that the precession frequency of the atom agreed with the prediction based on the Landé factor.

In the following years, much experimental work was done on atomic fluorescence.[121,122] Then microwave and radio-frequency spectroscopy started to catch the attention of physicists and chemists. The magnetic resonance which had been detected in atomic and molecular beams was now detected in bulk materials; the molecular rotational transitions and the nuclear electric quadrupole interactions began to be observed in the microwave spectra of the vapors of various compounds. These new experiments produced many advances in the elucidation of atomic and molecular structure. These works permitted the determination of the nuclear spins and moments which, at that time, were two of the few quantities which could be obtained experimentally and so could be used in any discussion on nuclear structure.

In 1949, Bitter (1902–1967) called attention to the possibility of optical detection of a resonance at radio-frequency.[123] However, Pryce[124] showed that the detection method proposed by Bitter was not possible. It was at this time that Brossel and Kastler[109,110] proposed a method for the optical re-orientation of atoms and discussed the different ways of detecting orientation. Jean Brossel recalled those days:[125]

It was the end of 1948, I had already been in England for three years, at the Tolansky Laboratory in Manchester, where I had become acquainted with interferometry problems and with the optical methods of studying hyperfine structures. One of our American colleagues told Kastler of the existence of a bursary for a young physicist who might be interested in any aspect of spectroscopy … I had to make up my mind very quickly and we agreed I would go to work at Francis Bitter's laboratory.

The work Brossel was called upon to do was on an experiment which could allow the extension of the magnetic resonance methods to the study of excited atomic states. Brossel tells us that after hesitations and changes.

I prepared the experiment. A few days before it was ready, M H L Pryce (to avoid a misadventure to some experimentalist) published a paper in *Physical Review* where he gave the right answer to the problem posed by Bitter, and where he found that the experiment proposed by Bitter was not valid.

It was no longer worthwhile to proceed in this direction; everything had to begin again. It was too late to go back on my decision to stay for one more year at MIT. I had to make the best use of my time.

I began to read texts again, particularly the books by Pringsheim[122] and Mitchell and Zemansky[121] on fluorescence and optical resonance. And when I began to read afresh the chapter on the polarisation of optical resonance light, I understood that there was a very simple and powerful method for the observation of the magnetic resonance of excited levels; it was only necessary to prepare the system by exciting it with polarised light. A situation is then obtained in which large differences exist in the populations of the Zeeman sub-levels of the excited state.

At resonance $\omega = \omega_o$ the radio-frequency field reorient the kinetic moment and equalises the populations. This results in a depolarisation of fluorescence light; no optical resolution is needed. I immediately told Francis of this. We decided to perform the experiment immediately.

During this period I was in constant contact with Kastler in Paris, keeping him in touch, through regular correspondence, with our progress and work. As often is the case where the situation is ripe for discovery, our ideas on the solution of the problem had evolved in parallel. By coincidence, we had arrived independently and at the same time at an identical conclusion. A week after I had discussed it with Francis, I received a letter from Paris in which Kastler was proposing exactly the same experiment.

He was developing his idea in connection with the sodium atom, employing an excitation with circularly polarised light (which led him a few months later to propose the idea of optical pumping) whilst I had "seen" the effect on the 2537 line of the even isotope of Hg with π excitation.

As we were sure of the success of the experiment we published the results immediately in *Comptes Rendus de l'Accademie des Sciences*, Paris, not omitting to point out that excitation by a direct beam of slow electrons would allow the extension of the method to all excited atomic states.

In a few months the experiment performed at MIT with Paul Sagalyn completely confirmed the preceding conclusions. The lack of space, which is a factor in every laboratory, had forced us to operate in highly uncomfortable conditions, being only too happy to be able to make use of a kind of loft without windows that was used as a box-room and that had long been abandoned, under the exit stairway of the Eastman Building, towards Walker's Hall ... During the time all this work was being undertaken at MIT, Kastler was trying to extend the optical detection method of magnetic resonance to ground states and invented optical pumping ... Kastler quickly started to assemble an experimental set-up to put into practice, on an atomic jet, the production of an atomic orientation of the ground state of Na by optical pumping. It was a failure. The same happened to me when, a few months before I left MIT, I tried to produce nuclear orientation of ^{199}Hg by pumping the vapour using the 2537 line. The intensity of my sources was too low. A few years later in Paris, when I built up some very bright sources, I did this experiment again with Cagnac, and we succeeded without difficulty.

The experiments described by Brossel were performed with the 3P$_1$ state of the isotopes of mercury by Brossel and Bitter[111] and on the fundamental state of ^{23}Na by Brossel, Kastler, and Winter[126] and Hawkins and Dicke.[127] The experiment with Sagalyn was published in 1950[128] and the one with Cagnac in 1959.[129]

It is also worth noting that by 1936 in his doctoral thesis on grating (echelon) excitation of mercury vapor, Kastler[130] showed that selective excitation of Zeeman sub-levels of excited states could be obtained by a suitable polarization of monochromatic exciting radiations, and was already proposing and realizing experiments whose basic ideas were not very different from the ones that conducted him, 14 years later, to the discovery of optical pumping.

However, Kastler had never had it in mind to use optical pumping for population inversion. In the summary of his paper, he writes:

By illuminating the atoms of a gas or of an atomic beam with oriented resonance radiations (light beams having a determined direction) which are suitably polarised, it is possible, when these atoms in the ground state are paramagnetic (quantum numbers $J \neq 0$ or $F \neq 0$) to obtain an unequal population of the different m sub-levels that characterise the spatial or magnetic quantification of the ground level. A rough evaluation shows that, with present irradiation facilities, this population asymmetry can become very important. From the examination of the transition probabilities of Zeeman transitions π and σ it can be seen that illumination with natural or linearly polarised light enables the concentration of atoms either in the m sub-levels of the middle ($m = 0$) or, on the other hand, on the external sub-levels (m maximum).

The use of circularly polarised light enables the creation of a population asymmetry between negative and positive m levels, with the sign of this asymmetry able to be reversed by reversing the direction of the circular polarisation of the incident light.

This creation of asymmetry can be obtained either in the absence of any external field or in the presence of a magnetic or electric field. In the presence of an external field the different sub-levels m (in the case of a magnetic field) or $|m|$ (in the case of an electric field) are energetically different, and the creation of a population asymmetry by the optical process corresponds to an increase or a decrease of the "spin temperature."

A population asymmetry of the sub-levels m of the ground state can be detected optically by examination of the intensity of polarization of optical resonance radiations. The use of photoelectric detectors and a modulation technique enables easy and sensitive detection.

The optical examination of the different branches into which an atomic beam divides itself in the Stern and Gerlach experiment permits the control of quantum level m of the atoms of each of these branches. This optical method allows extension of the magnetic analysis of atoms in the Stern and Gerlach experiment to the study of metastable excited levels.

In the magnetic resonance experiments the transitions induced by the radio-frequency oscillating magnetic field tend to destroy the population inequality of levels m. The study of the magnetic resonance of the atoms of an atomic bream can therefore be undertaken by replacing the nonuniform magnetic field of the Rabi set-up by an optical producer of an asymmetry which precedes the magnetic resonance set-up, and by an optical detector with an asymmetry at the output of the resonator. The optical method permits extension of the study of magnetic resonance to metastable levels.

This method enables one to study the transitions between hyperfine levels in zero field, the hyperfine Zeeman effects in the case of low fields and the hyperfine Paschen-Back effects in strong fields. So, thanks to the connection between the hyperfine Zeeman effect and the hyperfine Paschen-Back effect, one can optically analyse pure nuclear resonance in fields that decouple **J** and **I** vectors. Finally, the study of the Stark effect of an atomic level by the resonance method can equally well be undertaken optically. The process for the optical study of an atomic beam enables large beams with ill defined boundaries to be used. The set-up to be necessary for this study is therefore simple and cheap.

At this point, the various consequences of the method proposed by Kastler with respect to the problems involved in laser developments become apparent. Kastler proposed a method of a magnetic type to change the population of some levels, with the intention of using the changes in the characteristics of the emitted light in the transitions among such levels to study magnetic resonance phenomena. He never did mention the possibility to obtain population inversion in this way nor the possibility of using this inversion to create amplifiers of light generation. Neither did he give any priority to the invention of the laser, although today it is in fact by the use of optical pumping method that some lasers work.

Notes

1. A Kastler, *Cette étrange matière* (Stock: Paris, 1976).
2. The explanation of magnetism in classical terms may be attributed to W Weber (1804–1890) *Leipzig Berichte* **I**, 346 (1847): *Ann. Phys. Lpz.* **LXXIII**, 241 (1848) [translated in Taylor's *Scientific Memoirs* **V**, p. 477]; *Abh. K. Sächs. Ges.* **I**, 483 (1852); *Ann. Phys. Lpz.* **LXXXVII**, 145 (1852) [translated in Tyndall and Francis' *Scientific Memoirs*, p. 163]; *Leipzig Abh. Math. Phys.* **X**, 1 (1873); *Phil. Mag.* **XLIII**, 1 and 119 (1872). The classification of different substances into diamagnetic, paramagnetic, and ferromagnetic and the temperature behavior of paramagnetic substances was discussed by P Curie (1859–1906) in his thesis published in *Ann. Chem. Phys.* **VII**, 5, 289 (1895). The re-statement of these ideas in terms of the theory of electrons was undertaken in 1901–1903 by W Voigt (1850–1919) *Gött. Nach.* 169 (1901); *Ann. Phys. Lpz.* **9**, 115 (1902) and J J Thomson (1856–1940) *Phil. Mag.* **6**, 673 (1903). The general theory of magnetism received a complete formulation by P Langevin (1872–1946) (*Ann. Chem. Phys.* **V**, 70 (1905) and *C.R. Acad. Sci., Paris* **139**, 1204 (1904)). In 1907, Langevin's theory was extended by P Weiss (1865–1940) so to give an account of ferromagnetism (*Bull. Séances Soc. Fr. Phys.* 95 (1907); *J. Physique* **6**, 661 (1907)). A quantum theory of paramagnetism was started by Pauli (1900–1958) (W Pauli, *Z. Phys.* **21**, 615 (1920)) and became feasible after the introduction of the concept of spin (G E Uhlenbeck and S Goudsmit, *Naturwiss.* **13**, 953 (1925)) by the work of P Debye (in *Handb. D. Radiologie* **VI**, 713 (1925)) and L. Brillouin (*J. de Phys.* **8**, 74 (1927)). The need for a quantum theory had already been demonstrated by N Bohr (N Bohr, *Studieren over Metallernes Elektronteri*, Copenhagen, 1911) and H J van Leuven (*Inaugural Dissertation*, Leiden, 1919 and *J. Phys. Radium* **2**, 361 (1921)) (see also J J H van Vleck in *The Theory of Electric and Magnetic Susceptibilities*, Oxford University Press: Oxford, 1932). The correct expression for the diamagnetism was first obtained by Landau (*Z. Phys.* **64**, 629 (1930)). The concepts of ferromagnetism and antiferromagnetism were developed thanks to the exchange field ideas of P A M Dirac, *Proc. R. Soc.* **A112**, 661 (1926); L Néel (*Ann. Phys. Lpz.* **17**, 5 (1932); *J. Physique* **3**, 160 (1932); *Ann. Phys. Lpz.* **5**, 232 (1936); *C.R. Acad. Sci. Paris* **203**, 304 (1936)), W L Heisenberg (*Z. Phys.* **49**, 619 (1928)) and J Frenkel (*Z. Phys.* **49**, 31 (1928)).
3. O Stern, *Z. Phys.* 7, 249 (1921); W Gerlach and O Stern, *Z. Phys.* **9**, 353 (1922); see also B Friedrich and D Herschbach, *Physics Today*, December 2003, p. 53.
4. R Frisch and O Stern, *Z. Phys.* **85**, 4 (1933); I Estermann and O Stern, *Z. Physyk* **85**, 17 (1933); I Estermann, Simpson and O Stern, *Phys. Rev.* **52**, 535 (1937).
5. E Majorana, *Nuovo Cim.* **9**, 43 (1932). The paper is translated into English with the correction of some misprints in *Ettore Majorana: Scientific Papers*, edited by G F Bassani (Springer: Berlin, 2006).
6. I I Rabi, *Phys. Rev.* **51**, 652 (1937); see also I I Rabi, *Phys. Rev.* **49**, 324 (1936).
7. I I Rabi, S Millman, P Kusch and J R Zacharias, *Phys. Rev.* **53**, 318 (1938); *Phys. Rev.* **55**, 526 (1939).
8. Atomic beams were invented by L Dunoyer in 1910–1912 in Paris (L. Dunoyer, *C.R. Acad. Sci. Paris* **152**, 592 (1911); *Le Radium* **8**, 142 (1911)).
9. A historical survey together with an exhaustive discussion of these experiments may be found in H Kopferman, *Nuclear Moments* (Academic Press: New York, 1958).
10. I I Rabi, *Phys. Rev.* **49**, 324 (1936); **51**, 652 (1937); see also F Bloch and I I Rabi, *Rev. Mod. Phys.* 17, 137 (1945).
11. I I Rabi, *Phys. Rev.* **51**, 652 (1937).

12. See N F Ramsey, *Physics Today*, October 1993, p. 40.
13. C J Gorter, *Physica* **3**, 503, 995 (1936).
14. I I Rabi, J R Zacharias, S Millman and P Kusch, *Phys. Rev.* **53**, 318 (1938); **55**, 526 (1939).
15. J M B Kellog, I I Rabi, N F Ramsey and J R Zacharias, *Phys. Rev.* **55**, 318 (1939); **56**, 728 (1939); **57**, 677 (1940); N Ramsey, *Phys. Rev.* **58**, 226 (1940).
16. N F Ramsey, *Phys. Rev.* **76**, 996 (1949); **78**, 695 (1950).
17. N F Ramsey, *Physics Today*, July 1980, p. 25.
18. H G Kolsky, T E Phipps, N F Ramsey and H B Silsbee, *Phys. Rev.* **79**, 883 (1950); **80**, 483 (1950); **81**, 1061 (1951).
19. N F Ramsey, *Nuclear Moments and Statistics in Experimental Nuclear Physics*, ed. E Segrè, vol. I (Wiley: New York, 1953), p. 468.
20. L W Alvarez and F Bloch, *Phys. Rev.* **57**, 111 (1940).
21. For more information on this point, see J C Verstelle and D A Curtis, *Paramagnetic Relaxation* in *Handbuch der Physik*, edited by S Flugge (Springer: Berlin, 1968), Vol. 18/1, p. 1.
22. W Lenz, *Z. Phys.* **21**, 613 (1920).
23. P E Ehrenfest, *Comm. Leiden Suppl.* no. 44b (1920).
24. G Breit and H Kammerling Onnes, *Comm. Leiden* no. 168c (1924).
25. I Waller, *Z. Phys.* **79**, 370 (1932).
26. H B G Casimir and F K Dupré, *Physica* **5**, 507 (1937). The theory of paramagnetic relaxation times was mainly developed by I Waller, *Z. Phys.* **79**, 370 (1932); R Kronig, *Physica* **6**, 33 (1939); J H van Vleck, *Phys. Rev.* **57**, 426 (1940). For a discussion of nuclear paramagnetism, see E Teller and W Heitler, *Proc. R. Soc.* **155**, 629 (1936).
27. The concept of magnetic temperature has been discussed in a number of papers. See, for example, the excellent paper by J H van Vleck, *Nuovo Cim. Suppl.* **6**, 1081 (1957) and Bloembergen's review in *Am. J. Phys.* **41**, 325 (1973) where other useful references may be found. Spin temperature was extensively discussed by N Blombergen, *Physica* **15**, 386 (1949).
28. C J Gorter and L J F Broer, *Physica* **9**, 591 (1942); C J Gorter, *Physica* **3**, 995 (1936).
29. C J Gorter, *Paramagnetic Relaxation* (Elsevier: Amsterdam, 1947): C J Gorter, *Physica* **3**, 995 (1936).
30. More information can be found in J Rigden, *Rev. Mod. Phys.* **58**, 433 (1986).
31. F Bloch, W W Hansen and M Packard, *Phys. Rev.* **69**, 127 (1946); F Bloch, *Phys. Rev.* **70**, 460 (1946); F Bloch, W W Hansen and M Packard, *Phys. Rev.* **70**, 474 (1946).
32. E M Purcell, H C Torrey and R V Pound, *Phys. Rev.* **69**, 37 (1946).
33. E Zavoisky, *J. Phys. USSR* **9**, 211 (1945); **10**, 197 (1946).
34. F Bloch, *Z. Phys.* **52**, 555 (1928).
35. See, for example, K Sakoda, *Optical Properties of Photonic Crystals* (Springer, Berlin, 2005).
36. F Bloch, *Phys. Rev.* **50**, 259 (1936).
37. Bloch's work on magnetism is well known. Fundamental works are the classic paper on the spin-wave approach and the celebrated $T^{3/2}$ relation for the decrease of saturation magnetisation (*Z. Phys.* **61**, 206 (1930)) and the wall concept in a ferromagnet (*Z. Phys.* **74**, 295 (1932)).
38. N Bloembergen, E M Purcell and R V Pound, *Phys. Rev.* **73**, 679 (1948).
39. F Bloch, W W Hansen and M Packard, *Phys. Rev.* **69**, 127 (1946).
40. F Bloch, W W Hansen and M Packard, *Phys. Rev.* **70**, 474 (1946).
41. F Bloch, *Phys. Rev.* **70**, 460 (1946).

42. Equation 3.2 is exactly the Planck condition for the transition between two adjacent Zeeman levels. For a single nucleus of magnetic moment μ and spin I, there are $2I + 1$ levels separated by an energy

$$\Delta E = \frac{\mu H}{I},$$

so that

$$\frac{\Delta E}{\hbar} = \omega = \frac{\mu H}{I\hbar} = \gamma H,$$

where $\gamma = \mu/I\hbar$.

43. A justification of this exponential law was given shortly afterwards by N Bloembergen, E M Purcell and R V Pound, *Phys. Rev.* **73**, 679 (1948).

44. The possibility of representing the oscillating field as the superposition of two fields rotating in opposite directions was pointed out by F Bloch and A Siegert, *Phys. Rev.* **57**, 522 (1940).

45. The use of rotating coordinates for the treatment of magnetic resonance was dealt with by I Rabi, N F Ramsey and J Schwinger, *Rev. Mod. Phys.* **26**, 167 (1954). Here the complete quantum mechanical treatment of the theorem can also be found.

46. From Figures 1 and 2 of F Bloch, *Phys. Rev.* **70**, 460 (1946).

47. W E Lamb and R C Retherford, *Phys. Rev.* **79**, 549 (1950).

48. R V Pound, *Phys. Rev.* **79**, 685 (1950).

49. E M Purcell and R V Pound, *Phys. Rev.* **81**, 279 (1951).

50. L J F Broer, *Physica* **10**, 801 (1943); N F Ramsey, *Nuclear Moments* (Wiley: New York, 1953); N Bloembergen, *Physica* **15**, 386 (1949).

51. A Abragam and W G Proctor, *Phys. Rev.* **109**, 1441 (1958).

52. N Bloembergen, *Am. J. Phys.* **41**, 325 (1973).

53. J H van Vleck, *Nuovo Cim. Suppl.* **6**, 1081 (1957); B D Coleman and W Noll, *Phys. Rev.* **115**, 262 (1959); P T Landsberg, *Phys. Rev.* **115**, 518 (1959); L C Hebel Jr. *Solid State Physics* (Academic Press: New York, 1963), vol. 15.

54. N F Ramsey, *Phys. Rev.* **103**, 20 (1956). See also J H van Vleck, *Nuovo Cim* (Suppl.) **6**, 1081 (1957). Negative temperature has been discussed successively in various systems; see, for example, P Hakonen and O V Lounasmaa, *Science* **265**, 1821 (1994).

55. N F Ramsey, *Ann. Rev. Nucl. Sci.* **1**, 99 (1952); reproduced with permission © by Annual Review Inc.

56. R V Pound, *Phys. Rev.* **81**, 156 (1951); E M Purcell and R V Pound, *Phys. Rev.* **81**, 279 (1951); N F Ramsey and R V Pound, *Phys. Rev.* **81**, 278 (1951).

57. In a footnote on p. 690 of the paper by J Weber, *Rev. Mod. Phys.* **31**, 681 (1959).

58. J Combrisson, A Honig and C H Townes, *C.R. Acad. Sci., Paris* **242**, 2451 (1956).

59. G Feher, J P Gordon, E Buehler, E A Gere and C D Thurmond, *Phys. Rev.* **109**, 221 (1958).

60. A Overhauser, *Phys. Rev.* **89**, 689 (1953).

61. T W Griswold, A F Kip and C Kittel, *Phys. Rev.* **88**, 951 (1952).

62. A W Overhauser, *Phys. Rev.* **91**, 476 (1953).

63. I am indepted for this information on the Urbana work to Professor C P Slichter who kindly provided me with this recollection.

64. A W Overhauser, *Phys. Rev.* **92**, 411 (1953).

65. See F Bloch, *Phys. Rev.* **93**, 944 (A) (1954); C Kittel, *Phys. Rev.* **95**, 589 (1954); J Korrings, *Phys. Rev.* **94**, 1388 (1954): A Abragam, *Phys. Rev.* **98**, 1729 (1955).

66. See J I Kaplan, *Phys. Rev.* **99**, 1322 (1955).
67. T R Carver and C P Slichter, *Phys. Rev.* **92**, 212 (1953). Figure 1 from this reference is our Figure 3.11. See also T R Carver and C P Slichter, *Phys. Rev.* **102**, 975 (1956).
68. P Brovetto and M Cini, *Nuovo Cim.* **11**, 618 (1954).
69. W W Barker and A Mencher, *Phys. Rev.* **102**, 1023 (1956).
70. J Combrisson in *Quantum Electronics*, edited by C H Townes (Columbia University Press: New York, 1960), p. 167.
71. A M Portis, *Phys. Rev.* **91**, 1071 (1953).
72. E L Hahn, *Phys. Rev.* **77**, 746 (1950). A full account of the theory and experiments was given later in *Phys. Rev.* **80**, 580 (1950).
73. E L Hahn, *Phys. Rev.* **77**, 297 (1950).
74. In Hahn's original experiment, two equal pulses were applied (or two pulses at 90°). The application of 90° and 180° pulses was first considered by H Y Carr and E M Purcell, *Phys. Rev.* **94**, 630 (1954).
75. R P Feynman, F L Vernon Jr. and R W Hellwarth, *J. Appl. Phys.* **28,** 49 (1957).
76. See, for example, L Allen and J H Eberly, *Optical Resonance and Two-level Atoms* (Dover Publications, Inc.: New York, 1975).
77. N A Kurnit, I D Abella and S R Hartmann, *Phys. Rev. Lett.* **13**, 567 (1964).
78. P C Lauterbur, *Nature* **242**, 190 (1973).
79. P Mansfield, P K Grannell, *J Phys. C: Solid State Phys.* **6**, L422 (1973); A N Garroway, P K Grannell, P Mansfield, *J. Phys. C: Solid State Phys.* **7**, L457 (1974).
80. R R Ernst, *Rev. Sci. Intsrum.* **36**, 1689 (1965); A Kumar, D Welti and R R Ernst, *J. Mag. Res.* **18,** 69 (1975); Nobel lecture in *Angew. Chem. Int. Ed. Engl.* **31**, 805 (1992).
81. More information on the historical development of magnetic resonance imaging is for ex. T Geva, *J. Cardiovasc. Magn. Resonance* **8**, 573 (2006).
82. C H Townes, *Applications of Microwave Spectroscopy*, memorandum to James Fisk, Bell Labs, circa May 15, 1945.
83. R V Pound, *Rev. Sci. Instrum.* **17**, 490 (1946).
84. W V Smith, J L Garcià de Quevedo, R L Carter and W S Bennett, *J. Appl. Phys.* **18**, 1112 (1947); E W Fletcher and S P Cooke, *Cruft Lab. Tech. Rep.* **64**, 1950; W D Hershberger and L E Norton, *RCA Rev.* **9**, 38 (1948); C H Townes, A N Holten and F R Merritt, *Phys. Rev.* **74**, 1113 (1948).
85. H Lyons, *Annals New York Acad. Sci.* **55**, 831 (1952). See also P Forman, *Proc. IEEE*, **73**, 1181 (1985).
86. K Shimoda, *J. Phys. Soc. Japan* **9**, 378 (1954).
87. L Essen and J V L Parry, *Nature* **176**, 280 (1955).
88. Resolution 1, 13ᵉ Conference Generale des Pois et Mesures, *Metrologia* **4**, 41 (1968).
89. W Markowitz, R G Hall, L Essen and J V L Parry, *Phys. Rev. Lett.* **1**, 105 (1958).
90. R Vessot et al. *Phys. Rev. Lett.* **45**, 2081 (1980).
91. P Forman, *Proc. IEEE* **73**, 1181 (1985).
92. I I Rabi, J M B Kellog, J R Zacharias, *Phys. Rev.* **46**, 157 (1934); **50**, 472 (1936); **56**, 728 (1939); J M B Kellog, I I Rabi, N F Ramsey and J R Zacharias, *Phys. Rev.* **55**, 318 (1939); **57**, 677 (1940).
93. N F Ramsey, *J. Res. Natl. Bur. Stand.* **88**, 301 (1983).
94. J Camparo, *Phys. Today*, Nov 2007, p. 33.
95. N F Ramsey, *Molecular Beams* (Clarendon Press: Oxford, 1956), p. 138. See also R A Naumann and H H Stroke, *Phys. Today*, May 1996, p. 89.

96. M A Kasevich, E Riis, S Chu and R S DeVoe, *Phys. Rev. Lett.* **63**, 612 (1989).
97. A Clairon et al., *Europhys. Lett.* **16**, 165 (1991).
98. A Clairon et al., *IEEE Trans. Instrum. Meas.* **44**, 128 (1995).
99. S A Diddams et al., *Science,* **293**, 825 (2001).
100. T W Haensch and A L Schawlow, *Opt. Commun.* **13**, 60 (1975).
101. D J Wineland and W M Itano, *Phys. Today* **40**, 34 (1967); C Cohen-Tannoudji and W D Philips, *Phys. Today* 43, 33 (1990).
102. J L Hall, *Science* 202, 147 (1978); Ch Salomon, D Hils and J L Hall, *J. Opt. Soc. Am. B* **8**, 1576 (1988); S N Bagaev et al., *IEEE J. Quantum Electron.* **4**, 868 (1968); B C Young et al., *Phys. Rev. Lett.* **82**, 3799 (1999).
103. J N Eckstein, A I Ferguson and T W Haensch, *Phys. Rev. Lett.* **40**, 847 (1978).
104. Th Udern, *Phys. Rev. Lett.* **82**, 3568 (1999); J Reichert et al., *Phys. Rev. Lett.* **84**, 3232 (2000); D J Jones et al., *Science* **288**, 635 (2000); S A Diddams et al., *Phys. Rev. Lett.* **84**, 5102 (2000); R Holtzwarth et al., *Phys. Rev. Lett.* **85**, 2264 (2000); J Stenger et al., *Phys. Rev. A* **63**, 021802R (2001).
105. V P Chebotayev et al., *Appl. Phys. B* **29**, 63 (1982); C O Weiss et al., *IEEE J. Quantum Electron.* **24**, 1970 (1988); J Ye et al., *Opt. Lett.* **25**, 1675 (2000).
106. W H Oskay et al., *Phys. Rev. Lett.* **97**, 020801 (2006).
107. S A Diddams et al., *Science* **306**, 1318 (2004).
108. A recollection of the first experiments was given by A Kastler in his Nobel lecture that has been published in *Phys. Today*, September 1967, p. 34.
109. J Brossel and A Kastler, *C.R. Acad. Sci. Paris* **229**, 1213 (1949).
110. A Kastler, *J. Phys. Radium* **11**, 255 (1950).
111. J Brossel and F Bitter, *Phys. Rev.* **86**, 308 (1952). An excellent review of these experiments was made later by F Bitter, *Appl. Opt.* **1**, 1 (1962).
112. See also A Kastler lecture at the Sommerfeld Tagung, Munich, Sept. 1968, published in *Physics of the One and Two-Electron Atoms* (North-Holland: Amsterdam, 1969); A Kastler, Nobel lecture 1966, published in *Phys. Today*, 20 (September 1967), p. 34.
113. We closely follow here the description given by A Kastler in *J. Opt. Soc. Am.* **47**, 460 (1957).
114. E Fermi and F Rasetti, *Nature* **115**, 764 (1925); *Z. Phys.* **33**, 246 (1925).
115. G Breit and A Ellett, *Phys. Rev.* **25**, 888 (1925).
116. I have followed the description of the experiment given by A C G Mitchell and M W Zemansky, *Resonance Radiation and Excited Atoms* (Cambridge University Press: Cambridge, 1961), p. 270.
117. E Fermi and F Rasetti, *Rend. Lincei* **1**, 716 (1925); **2**, 117 (1925). The work described in the second paper was performed at the Istituto Fisico, Florence.
118. F Rasetti, in the comment on Fermi's works in *Note e Memorie di E Fermi*, vol. 1 (Accademia dei Lincei: Roma, 1962), p. 159.
119. A Ellett, *Nature* **114**, 931 (1924).
120. W Hanle, *Z. Phys.* **30**, 93 (1924).
121. A C G Mitchell and M W Zemansky, *Resonance Radiation and Excited Atoms* (Cambridge University Press: Cambridge, 1934).
122. P Pringsheim, *Fluorescence and Phosphorescence* (Interscience: New York, 1949).
123. F Bitter, *Phys. Rev.* **76**, 833 (1949).
124. M H L Pryce, *Phys. Rev.* **77**, 136 (1950).
125. J Brossel, *Quelques souvenirs … in Polarisation, Matière et Rayonnement, Volume Jubilaire en l'Honneur d'A Kastler* (Presses Universitaires de France: Paris, 1969), p. 143.
126. J Brossel, A Kastler and J Winter, *J. Phys. Radium* **13**, 668 (1952).

127. W B Hawkins and R H Dicke, *Phys. Rev.* **91**, 1008 (1953).
128. P Sagalyn and F Bitter, *Phys. Rev.* **79**, 196, 225 (1950).
129. B Cagnac, J Brossel and A Kastler, *C.R. Acad. Sci. Paris* **246**, 1027 (1958); B Cagnac, *J. Phys. Radium* **19**, 863 (1958); B Cagnac and J Brossel, *C.R. Acad. Sci. Paris* **249**, 77, 253 (1959).
130. A Kastler, *Thèse*, Paris (1936) and *Ann. Phys., Lpz.* **6**, 663 (1936).

4

The Maser

4.1 Introduction

The development of radar during the World War II and nuclear research, culminated with the construction of the atomic bomb, just to mention few examples, showed the enormous potential existed in physical research and the extraordinary applications could be obtained. In 1948, the invention of the transistor and the consequent revolution in the field of electronics provided a decisive step change. Physics was discovered not to be an abstract science for a few of the initiated, but a discipline able to provide basic elements for the development of society or, depending on the approach that one was willing to adopt, for its destruction. Nuclear research and the development of nuclear reactors for energy production, today regarded with suspicion, at the end of the World War II were looked upon with great favor as a means of solving energy problems and thus benefit mankind.

At this time, in the United States began a link with industry, and the creation of industrial research laboratories and industries based on physics research ensued. A huge amount of money was given to research. In this atmosphere, anybody who had a good idea that potentially could lead to applications had the quasi-certainty of finding adequate support to develop it. This ideal environment lasted for the entire wartime period and continued during the Cold War between the United States and Russia until the 1970s. In this period, the development of masers and lasers was helped in America through the interest of both military agencies and industry, which were receiving financial support at a level never seen before. An excellent discussion of this situation has been made by Lisa Bromberg.[1]

In the ex-Soviet Union, the Academy of Sciences could be considered a state within the State, very rich, and powerful.[2] In its institutes distributed though the country, thousands of researchers were working for much higher than average earning compared with other workers. There also existed secret laboratories for military research which, together with the excellent laboratories of the Academy, were the only ones financially capable of employing the necessary instrumentation. The research was conducted over a broad spectrum.

In Europe, large international research enterprises, such as CERN in Geneva, were founded. This large laboratory was established with the financial support of Italy, France, Germany, Great Britain, The Netherlands, and Belgium to build powerful particle accelerators for research into high-energy physics using facilities whose funding was impossible for a single country.

The fact that the development of the new devices, the maser and the laser, took place in the Unites States and in the ex-Soviet Union, the sole countries in which research encompassed a broad range of fields, can be attributed to the fundamental strategy of not focusing research into specific directions.

The maser implies an amplification technique so radically different from the usual techniques that it could not originate as a simple improvement of the electronic techniques already known, but required the development of new fields such as magnetic resonance and microwave spectroscopy.

Moreover, until then electromagnetic radiation was produced and detected using tubes (diodes, triodes, etc.) which today are completely obsolete, that operated using electrons emitted by a filament heated by an electric current. The action of these devices, as well as of magnetrons, klystrons, and so on, was perfectly understandable by applying Maxwell's laws of classical electromagnetism. Engineers had no need to study quantum mechanics, which is indispensable in understanding the operation of the maser, until 1948 when the transistor was invented. To understand how this device works, it was necessary to consider the electronic states in solids which are described by quantum mechanical laws. At this point, engineers discovered quantum mechanics and started to study it.

Population inversions for use in working devices were first obtained in the microwave region, where the spontaneous emission probability, which is proportional to the cube of the frequency, is so small as to be negligible.

Townes has always connected these results with the development, during the World War II, of microwave technology. After the war, a good deal of attention was given to the interaction between microwaves and matter, especially gases. This led to the growth of a new field of research known as *microwave spectroscopy*, which was developed initially in several places (firstly in the industrial laboratories possessing radar apparatus) and then spread quickly to universities.

The maser idea was therefore born as a logical consequence of the resultant detailed knowledge of the interaction between microwaves and matter. In Townes' words[3]

> It was the mixture of electronics and molecular spectroscopy which set appropriate conditions for the invention of the maser.

The physical principles and experimental techniques for its development were thus well established in the period 1945–1950.

The idea originated independently in the United States in the Universities of Maryland and of Columbia and in the Soviet Union at Moscow's Lebedev Institute, during the early 1950s. The importance of the work carried out at

Columbia and at the Lebedev Institute, in the field of both masers and lasers, was recognized by the international scientific community by the award of the Nobel Prize for Physics to C H Townes, N G Basov, and A M Prokhorov in 1964.

As so often happens, after the report of the first maser operation, other concomitant or prior works were found, which we shall consider later.

4.2 Weber's Maser

The first public description of the maser principle (without a working device) was at an Electron Tube Research Conference in Ottawa, Canada, in 1952, a prestigious conference to which participation was only by invitation and at which new ideas for advanced devices were often presented, by Joseph Weber (1919–2000).[4] Weber was then a young electrical engineering professor at the University of Maryland and a consultant at the United States Naval Ordinance Laboratory.

He was born in 1919 in Paterson, New Jersey, and was awarded an appointment to the US Naval Academy where he received a BS degree in 1940. The same year, he was commissioned as an ensign and was posted to the aircraft carrier *Lexington*, which narrowly escaped disaster by steaming out of Pearl Harbor on December 5, 1941. Weber survived the sinking of the ship during the Battle of the Coral Sea on May 8, 1942. He then commanded a submarine chaser and participated in the landing on Sicily in July 1943.

After the World War II ended, he graduated in Annapolis and was first a naval officer from 1945 to 1948, having responsibility, as a specialized microwave engineer, for the section on electronic countermeasures of the Navy. Here he had the opportunity to become acquainted with the technological importance of amplifiers with high sensitivity at microwave and millimeter wavelengths, since the receivers that are employed for countermeasures against enemy radar waves employ such amplifiers to detect very faint radar waves. Information about the employed wavelength and its origin is then utilized to send signals that blind the enemy receivers, hindering their identification of targets.

He resigned from the Navy and entered graduate school at the Catholic University in Washington, DC, where he obtained his PhD in 1951. The idea of the maser came to him after he had attended a seminar on stimulated emission by Karl Herzfeld (1892–1978) while he was studying for his doctorate at Washington.

In his work, Weber considered a system with two energy levels, E_1 and E_2 ($E_2 > E_1$) with populations n_1 and n_2, respectively.[5] By irradiating with radiation of frequency

$$\nu = \frac{E_2 - E_1}{h},$$

(4.1)

the absorbed power in the transition $1 \to 2$ is

$$P_{\text{abs}} = W_{12} h \nu n_1.$$

(4.2)

The power emitted by the particles which decay from the upper state down to the lower state is

$$P_{em} = W_{21} h\nu n_2, \qquad (4.3)$$

where $W_{ij} = B^i{}_j \rho$, with $B^i{}_j$ being the usual Einstein's coefficients. Because $W_{12} = W_{21}$, the net absorbed power is

$$P_{net} = W_{12} h\nu (n_1 - n_2). \qquad (4.4)$$

At equilibrium, the number of particles in state 2 is governed by the Maxwell–Boltzmann law

$$n_2 = n_1 \exp\left(-\frac{h\nu}{kT}\right) \sim n_1\left(1 - \frac{h\nu}{kT}\right). \qquad (4.5)$$

By substituting Equation 4.5 into Equation 4.4, we obtain

$$P_{net} = \frac{W_{12}(h\nu)^2 n_1}{kT}. \qquad (4.6)$$

Weber writes[4]

... and this [P_{net}] is a positive quantity. Thus under ordinary circumstances we get absorption of radiation (ordinary microwave spectroscopy) because the transition probability up is the same as the transition probability down, but since there are more oscillators in the lower state, we get a net absorption .... We could get amplification if somehow the number of oscillators in the upper state could be made greater than the number in the lower states. A method of doing this is suggested by Purcell's negative temperature experiment.

He then considered two ways of obtaining the necessary reversal, a sudden reversal of magnetic field or a pulsed system with polar molecules. By applying an electric field, one obtains a separation of levels, by the Stark effect. If equilibrium is reached and the field is quickly inverted, one obtains $n_2 > n_1$ for a time corresponding to the thermal relaxation time. Weber also suggested making a gas flow through a region of electric field reversal, to have continuous operation.

The work described above was carried out by Weber in 1951 and presented at the conference in 1952. He published it in summary form in 1953.[4] As explained by Weber himself,[6] it was his intention to publish his results in a

... widely read journal. Early in 1953 Professor H J Reich of Yale University wrote to say that he had been chairman of the 1952 Electron Tube Conference program committee, and was also editor of a (not so widely read) journal. As a result the conference summary report was published in the June 1953 issue of Transaction of the Institute of Radio Engineers Professional Group on Electron Devices.

In this work, Weber underlines the fact that the amplification is coherent. The method he proposed for obtaining population inversion has never, in fact, been put into practice and it seems mostly unlikely ever to be so.

Moreover, one may observe that although he uses the word "coherent" in his work, only an amplifier is considered, and nothing is said on self-sustained oscillation. However, the basic maser ideas—using stimulated emission to excite atoms or molecules and invert the population—are clearly stated.

After his presentation of this work at the conference, Weber was asked by RCA to give a seminar on his idea. For this, he received a fee of $50. After the seminar, Townes wrote to him, asking for a preprint of the paper. Weber's work was, however, not quoted in the first of Townes' papers but was referred to later.[7]

Weber's efforts were acknowledged by the IRE when he was awarded a fellowship in 1958 "for his early recognition of concepts leading to the maser." He spent the 1955–1956 year as a fellow of the Institute for Advanced Study in Princeton and immersed himself in general relativity. During the early 1960s, he was interested in gravitational waves, building mechanical detectors which, however, failed to detect them definitively. Weber continued to be interested in the detection of gravity waves,[8] also proposing to do an experiment using laser interferometric techniques. A first experiment was made by Robert Forward. Today interferometric techniques are used in trying to detect gravitational waves all over the world in several projects as the US project for the Laser Interferometer Gravitational wave Observatory (LIGO) or the French-Italian VIRGO.

He died on September 30, 2000.

4.3 Townes and the First Ammonia Maser

The first, experimental, operating maser was built by a group of researchers at Columbia University, headed by C H Townes[9] (**Figure 4.1**).

Charles H Townes[9] was born in 1915 in Greenville, South Carolina. He passed away on January 27, 2015. When he was 16, he entered Furman University. Although he soon discovered his vocation for physics, he also studied Greek, Latin, Anglo-Saxon, French, and German, and received a BA degree in modern languages after 3 years at Furman. At the end of his fourth year, he received a BSc in physics. He next went to Duke University on a scholarship. When he was 21, he finished work on his master's degree, continuing to study French, Russian, and Italian. He then went to the California Institute of Technology in Pasadena, where, in 1939, he received his PhD, after which he accepted an appointment at the Bell Telephone Laboratories. During the war, Townes was assigned to work with Dean Wooldridge, who was then

FIGURE 4.1 C H Townes (1915–2015).

designing radar bombing systems. Although Townes preferred theoretical physics, he nevertheless worked on this practical project.

At that time, people were trying to push the operational frequency of radar higher. The Air Force asked Bell to work on radar at 24,000 MHz. Such a radar would exploit an almost unexplored frequency range and would result in more precise bombing equipment.

Townes, however, observed that radiation of that frequency is strongly absorbed by water vapor. The Air Force nevertheless insisted on trying it. So the radar was built by Townes, who then was able to verify that it did not work. As a result of this work, Townes became interested in microwave spectroscopy.

In 1947, Townes accepted an invitation from Isidor I Rabi to leave Bell Laboratories and join the faculty at Columbia University in which Rabi was working. There was a Radiation Laboratory group in the Physics Department which had continued war-time program on magnetrons for the generation of millimeter waves. This laboratory was supported by a Joint Service contract from the US Army, Navy, and Air Force, with the general purpose of exploring the microwave region and extending it to shorter wavelengths. Among the people active in the sponsorship of this program were Dr Harold Zahl of the Army Signal Corps and Paul S Johnson of the Naval Office of Research. Townes quickly became an authority on microwave spectroscopy and on the use of microwaves for the study of matter.[10]

In 1950, he became a full professor of physics. In the same year, Johnson organized a study commission on millimeter waves and asked Townes to take the chair. Townes worked on the committee for nearly 2 years and became rather dissatisfied with its progress. Then, one day in the spring of 1951, when he was in Washington DC to attend a meeting of the committee, he tells us[11]

> By coincidence, I was in a hotel room with my friend and colleague Arthur L Schawlow, later to be involved with the laser. I awoke early in the morning and, in order not to disturb him, went out and sat on a nearby park bench to puzzle over what was the essential reason we had failed [in producing a millimetre wave generator]. It was clear that what was needed was a way of making very small, precise resonator and having in it some form of energy which could be coupled to an electromagnetic field. But that was a description of a molecule, and the technical difficulty for man to make such small resonators and provide energy meant that any real hope had to be based on finding a way of using molecules! Perhaps it was the fresh morning air that made me suddenly see that this was possible: in a few minutes I sketched out and calculated requirements for a molecular-beam system to separate high-energy molecules from lower ones and send them through a cavity which would contain the electromagnetic radiation to stimulate further emission from the molecules, thus providing feedback and continuous oscillation.

He did not say anything at the meeting, and once back to Columbia in the fall of 1951, when Jim Gordon came to him seeing a thesis project, he started work. In addition to Gordon (1928–2013), Herb Zeiger was asked to join the project because Townes reasoned that someone expert in molecular beam work, on which Herb had just completed a thesis, would be helpful.

Zeiger was supported by a scholarship from the Union Carbide Corporation. The chemical engineer H W Schultz, a couple of years before the work on the maser started, had persuaded his company, the Union Carbide, to give $10,000 per year to someone to study how to create intense infrared radiation, having understood its importance to induce specific chemical reactions. Although Townes insisted he did not know how to solve the problem, even if he was very interested in it, Schulz gave him the money for a post-doctorate assistant and that money was used to support Schawlow and Herbert Zeiger in the years before the work on the maser started. It was so that Zeiger was able to join the project proposed by Townes.

In Townes' design, the resonator was very important. In fact, the cavity was required to confine the electromagnetic energy for the longest possible time so to interact with the molecules, and therefore should have low losses or high Q. Detailed calculations performed in the autumn of 1951 showed it was very difficult to make a cavity for radiation of half a millimeter as Townes had initially thought would be achievable using molecules of ammonia deuterate. Therefore, he decided to turn his attention to radiation of 1.25 cm, emitted by ordinary ammonia, which was the wavelength at which the components necessary for success (the cavity) already existed. The decision was therefore made to switch from a project that had to advance the frontiers of research in the millimeter region to one that demonstrated a new principle for obtaining generation in an already known spectral region.

The active material envisaged by Townes was ammonia gas. In the classical picture, the ammonia molecule (NH_3) is like a triangular pyramid (**Figure 4.2**) with the three hydrogen atoms at the vertices of the base and the nitrogen atom at the apex. A quantum mechanical treatment shows that there are two possible, equivalent positions of the nitrogen atom, either above or below the plane formed by the three hydrogen atoms, and that the potential energy of the nitrogen atom as a function of its distance from this plane is as shown qualitatively in Figure 4.2. Here the broken lines show what would be valid if only one side (or the other) of the plane were accessible. Two states for each value of the allowed energy, one for each side, would exist. Actually, because only a finite potential "hump" is present between the two wells, the two states interact with each other to give two new states. In these sates, we cannot say that the nitrogen atom is on one side or the other of the plane, but rather that it has equal probability of being on either side.

The wavefunction describing one of the two states remains unaltered by an interchange of the two positions (it is symmetric), while the other wavefunction has its sign changed (antisymmetric). The interaction splits the energies of the two new states with the symmetric state being somewhat lower than the antisymmetric. The energy separation between members of a pair increases with increasing pair energy, but for the ammonia molecule it corresponds to frequencies in the microwave range.

Besides these states, which are vibrational in character, the molecule also has rotational states corresponding to rotations around either of two axes, one perpendicular to the hydrogen plane and the other lying in it. The rotational states are identified by the value of the (quantized) angular momenta about

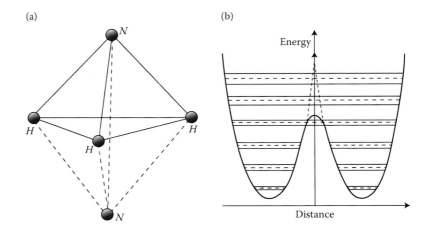

FIGURE 4.2 (a) The pyramidal structure of the ammonia molecule; (b) double minima of the potential curve for ammonia which give rise to the inversion splitting of the vibrational levels.

each of these axes. Rotation alters the vibrational potential energy curves, stretching the molecule with centrifugal force, so that the separation of each vibrational pair depends on the rotational state.[12] Townes was well acquainted of ammonia transitions, having worked with it previously to stabilize microwave oscillators, as we have seen in Chapter 3.

Gordon, Zeiger, and Townes, after some alternation of ideas, decided to look at the transition between the lower vibrational pair in the rotational state with three units of angular momentum about each axis, called the 3–3 state. This transition has a frequency of 23,870 MHz. It must be remembered that another important property of the ammonia molecule is that, although it has no permanent electric dipole moment, an applied electric field will induce such a moment in both members of the rotational pair, but opposite in sign. If the field is inhomogeneous, a force will act on the molecule which has opposite direction for both members of the pair. The project was then to use a strong electrostatic field to obtain a beam of excited ammonia molecules and to focus this beam through a small hole in a box or cavity which had been tuned to resonance at exactly 24,000 MHz.

In the 1950s, Wolfang Paul (1913–1993) together with Helmut Friedburg and Hans Gerd Bennewitz designed special electric and magnetic lenses (quadrupole and exapoles) to focus atomic and molecular beams.[13] Townes thought about using lenses of this kind to separate the molecules.[14] The field created by these lenses were hence used in the ammonia maser and then in the hydrogen maser. Later, Paul developed a three-dimensional version capable of confining ions in a small region (the Paul trap) and for this achievement, which allows the behavior of a single atom to be studied, previously impossible, he was awarded the Nobel Prize for physics in 1989 together with N Ramsey and Hans Dehmelt (who built a similar trap but of different design).

By concentrating the excited ammonia molecules, Townes was hoping to obtain either population inversion in them or else a surplus of excited molecules. After having reached the condition of population inversion, each excited molecule, on decaying into its fundamental state, would trigger other molecules to relax. Therefore, the cavity would emit coherent radiation at 24,000 MHz. It is to the credit of Townes that he clearly understood at the outset the need for a resonant cavity with which to couple the radiation to the excited medium.

Townes said that he felt responsible, particularly for Gordon; "I'm not sure it will work, but there are other things we can do with it if it doesn't." In fact, he told Gordon that if the method did not work, he could use the set-up to investigate the microwave absorption spectrum of ammonia. Gordon thus worked simultaneously on both experiments. In this manner, he was able to study the hyperfine structure of ammonia (the interaction between the electron and the nucleus) with an accuracy higher than had been possible before.[15]

The resulting quarterly reports on this laboratory work had a certain amount of circulation among scientists who were interested in microwave physics. The first public mention of this maser project appears, under the names of Zeiger and Gordon, in a report of December 31, 1951, entitled *Molecular Beam Oscillator*. Preliminary calculations on the design of a molecular beam oscillator were reported there. A description was given on the essential elements of the oscillator: a molecular beam source; a deflecting region for separating an excited state of the beam from a ground state; a resonant cavity, tuned to the frequency of transition from that excited state to the ground state by induced emission; and a detector for observing the radiation emitted from such transitions.

This projected oscillator was intended for use in the long-wavelength infrared ($\lambda \sim 0.5$ mm), and the transition considered was $J = 2, K = 1, M = 2 \rightarrow J = 1, K = 1, M = 1$ in ND_3 which has a transition energy of 20.55 cm^{-1}. After some considerations concerning the focusing system, the total beam flux in the upper state entering the cavity was calculated to be 6×10^{12} molecules per second which, if all these were to undergo induced transitions to the ground state, would deliver approximately 2.4×10^{-9} W of power to the cavity. This was estimated to be sufficient to be detected by a Golay cell.

A theoretical calculation was also made of the quality factor Q for a tuned cylindrical cavity. For a cavity 1 cm in diameter and 1 cm in length, at liquid–air-temperature, and tuned to 20.55 cm^{-1}, the calculated Q was 1.5×10^5. The Q which had been calculated as necessary to maintain oscillations for 2.4×10^{-9} W input was 1×10^5. The authors concluded: "On the basis of these calculations, it therefore seems only barely possible that oscillations will be sustained in the cavity."

In the following quarterly report, the goal of the project was changed. Operation was now in the K band, using the ammonia inversion transition $J = 3, K = 3$ at 1.25 cm wavelength. Subsequent reports concern the details of the vacuum system, of the focus equipment, of the resonant cavity, and of the microwave resonator.

For 2 years, the Townes group worked on. At about this time, two friends called at the laboratory and tried to insist that Townes stop this nonsense and the wastage of government money, for Townes had by then spent about

$30,000 under a Joint Service grant administered by the Signal Corps, the US Office of Naval Research, and the Air Force.

Finally, one day in 1953, Jim Gordon rushed into a spectroscopy seminar that Townes was attending crying: *it works!* The story goes that Townes, Gordon, and the other students (Zeiger had by this time left Columbia to go to the Lincoln Laboratory and T C Wang had replaced him) went to a restaurant both to celebrate and to find a Latin or Greek name for the new device, the latter without success. Only a few days later, with the help of some of the students, they coined the acronym MASER: microwave amplification by stimulated emission of radiation. This name appears in the title of a paper in the *Physical Review*[16] and its meaning was fully spelt out in a subsequent paper by Shimoda et al.[17] (detractors reread it as Means of Acquiring Support for Expensive Research!).

The first mention of the operation of the oscillator was in a report of January 30, 1954, in nearly the same form in which it was published in a letter to *Physical Review*:[18]

A block diagram of the apparatus is shown in Figure 1.[19] A beam of ammonia molecules emerges from the source and enters a system of focusing electrodes. These electrodes establish a quadrupolar cylindrical electrostatic field whose axis is in the direction of the beam. Of the inversion levels, the upper states experience a radial inward (focusing) force, while the lower states see a radial outward force. The molecules arriving at the cavity are then virtually all in the upper states. Transitions are induced in the cavity, resulting in a change in the cavity power level when the beam of molecules is present. Power of varying frequency is transmitted through the cavity, and an emission line is seen when the klystron frequency goes through the molecular transition frequency.

If the power emitted from the beam is enough to maintain the field strength in the cavity at a sufficient high level to induce transitions in the following beam, the self-sustained oscillations will result. Such oscillations have been produced. Although the power level has not yet been directly measured, it is estimated at about 10^{-8} W. The frequency stability of the oscillation promises to compare favorably with that of other possible varieties of "atomic clocks."

Under conditions such that oscillations are not maintained, the device acts like an amplifier of microwave power near a molecular resonance. Such an amplifier may have a noise figure very near to unity.

High resolution is obtained with the apparatus by utilising the directivity of the molecules in the beam. A cylindrical copper cavity was used, operating in the TE011 mode. The molecules, which travel parallel to the axis of the cylinder, then see a field which varies in amplitude as $\sin(\pi x/L)$, where x varies from 0 to L. In particular, a molecule travelling with velocity v sees a field varying with time as $\sin(\pi vt/L)\sin \Omega t$, where Ω is the frequency of the RF field in the cavity. A Fourier analysis of this field, which the molecule sees from $t = 0$ to $t = L/v$, gives a frequency distribution whose amplitude drops to 0.707 of its maximum at points separated by a Δv of $1.2v/L$. The cavity used was twelve centimetres long, and the most probable velocity of ammonia molecules in a beam at room temperature is 4×10^4 cm s^{-1}. Since the transition probability is proportional to the square of the field amplitude, the resulting line should have a total width at half-maximum given by the above expression, which in the present case is 4 kc/sec [4 kHz]. The observed linewidth of 6–8 kc/sec [6–8 kHz] is close to this value.

Consideration as to the use of the device as a spectrometer was given later. In the subsequent paper sent to the *Physical Review* exactly 1 year later,[16] more particulars are given.

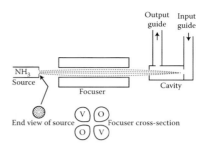

FIGURE 4.3 Simplified diagram of the essential parts of the ammonia maser. (J P Gordon, H J Zeiger and C H Townes, *Phys. Rev.* **99**, 1264 (1955).)

The electrodes of the focuser were arranged as shown in Figure 1 [our **Figure 4.3**]. High voltage is applied to the two electrodes marked V, while the other two are kept at ground. Paul et al.[20] have used similar magnetic pole arrangements for the focusing of atomic beams.

The electric field increases with radial distance from the central axis of the focusing electrodes. Because the force on molecules is $F = -\partial W/\partial r$ and because the molecules in the upper state increase in energy with increasing field, these molecules are forced towards the centre of the focuser. The lower-energy molecules, on the other hand, are turned away from the central axis of the focuser.

The cavity, in practice, worked as the positive reactance in an electronic circuit, amplifying the radiation at its resonant frequency. The basic requirement for microwave generation was therefore to produce a positive feedback through some resonant circuit which ensured that the gain in energy received by the wave through stimulated emission was greater than the circuit losses.[17]

Let us consider a microwave resonant cavity with conducting walls, volume V, and a quality factor Q; this last quantity is defined as the ratio between the incident power and the power loss P_R due to wall resistance

$$Q = \frac{\overline{E^2}V\nu}{4P_R}, \tag{4.7}$$

with $\overline{E^2}$ being the mean square value (for that volume) of the electric field of the mode and ν its frequency. If a molecule in an excited state is put into an electric field E at resonance (when the frequency ν coincides with the energy difference between the levels divided by h), the emitted power is

$$P_e = \left(\frac{\overline{E^2}\mu^2}{\hbar^2}\right)\left(\frac{h\nu}{3\Delta\nu}\right), \tag{4.8}$$

where μ is the matrix element for the transition and $\Delta\nu$ is the half linewidth of the resonance. Therefore for N_b molecules, in the upper state, and N_a in the lower state, the power given by the field to the cavity is

$$P_e = (N_b - N_a) \left(\frac{\overline{E^2}\mu^2}{\hbar^2} \right) \left(\frac{h\nu}{3\Delta\nu} \right). \tag{4.9}$$

If the molecules are distributed uniformly in the volume, it follows that

$$(N_b - N_a) \left(\frac{\overline{E^2}\mu^2}{\hbar^2} \right) \left(\frac{h\nu}{3\Delta\nu} \right) \geq \frac{\overline{E^2}V\nu}{4Q}, \tag{4.10}$$

from which the threshold condition for the onset of oscillations in the cavity is derived as

$$(N_b - N_a) \geq \frac{3hV\Delta\nu}{16\pi^2 Q\mu^2}. \tag{4.11}$$

Townes observed immediately that:[16]

Associated with the power emitted from the beam is an anomalous dispersion .... These two effects can be considered at the same time by thinking of the beam as a polarisable medium introduced into the cavity whose average electric susceptibility is given by $\chi = \chi' + i\chi''$. The power emitted from the beam is

$$P = 8\pi^2 \nu_B W \chi'', \tag{4.12}$$

where W is the energy stored in the cavity.

The connection between the imaginary part of the susceptibility and the absorbed or emitted power had already been considered by both Bloch[21] and Bloembergen, Purcell, and Pound,[22] and various treatments of the maser were achieved by having recourse to circuit analogies.[23]

The principal property of the maser is its extremely low noise, both as an amplifier and as an oscillator. It is therefore able to amplify extremely low-level signals. Only very few photons are emitted randomly by those molecules that de-excite spontaneously and not by stimulated emission and these are the photons which constitute the noise. In many electronic devices, noise arises from fluctuations in the number of electrons that produce the electric current. These fluctuations are proportional to temperature and are independent of the specific device; therefore, engineers have adopted the habit of characterizing the noise of devices by quoting their noise equivalent temperature, that is, the temperature at which an electrical resistance should be brought such that electrons that travel through it produce the observed fluctuations. While for the resistance of an ordinary circuit, the noise temperature is practically room temperature (300 K), for the maser the equivalent noise temperature is very low, in the order of only a few degrees Kelvin.

When it is used as an oscillator, the maser is able to generate monochromatic radiation with very good frequency stability. The maser monochromaticity is described quantitatively by the halfwidth, $\delta\nu$, of its emission spectrum. Townes[16] (using a linear approximation) gave the expression

$$\delta\nu = \frac{4\pi kT(\Delta\nu)^2}{P},\tag{4.13}$$

where $\Delta\nu$ is the halfwidth of the spectral line. This is the theoretical lower limit, although that was not clearly stated. A different expression which did take into account the nonlinear behavior of the maser was given later by Shimoda, Wang, and Townes.[17]

At operational temperatures and at the powers actually used, $\delta\nu$ is several orders of magnitude smaller than $\Delta\nu$. The noise generated in a maser has been the object of a number of studies and measurements.[24] The center frequency, ν_o, of the oscillation was derived in these studies as

$$\nu_o = \nu_B + \left(\frac{\Delta\nu_B}{\Delta\nu_c}\right)(\nu_c - \nu_B),\tag{4.14}$$

where $\Delta\nu_c$ and $\Delta\nu_B$ are the halfwidths of the cavity mode and of the molecular emission line, respectively, and $\nu_c - \nu_B$ is the difference between the cavity resonant frequency ν_c and the line frequency ν_B. (It is worth observing that the coherence of the radiation was never explicitly mentioned!)

The only work published before the successful operation of the maser was by Townes in 1953[25] that resulted from a talk Townes gave at a meeting in Japan where he responding to a question from the audience gave some information on the work on going in his laboratory. Mention of the work on stimulated emission had also been made in May 1951 at a *Symposium on Submillimeter Waves* at the University of Illinois by A H Nethercot of Columbia[7] on behalf of Townes.

It was immediately realized that one important application of molecular beam masers would be molecular spectroscopy. Molecular beams had already been considered by gas spectroscopists in the early 1950s; however, the basic problem had been that, as a consequence of beam formation, the resultant density of molecules in the spectrometer cell was very low. For beams where the molecules may be assumed to be in or very near thermal equilibrium, the absorption and emission processes in the presence of external radiation nearly balance out against a small net absorption which occurs because more molecules are in the lower energy state. If all molecules in the lower state are removed, as in maser operation, then the magnitude of the signal can be enhanced by a factor of $kT/h\nu$ over its thermal equilibrium value which, at microwave frequencies and room temperature, is more than two orders of magnitude. Although the first spectroscopic study had already been undertaken by Gordon[15] in 1951, the full potential of the method was not realized until the early 1970s.[26]

The power of the first maser was only 0.01 μW. It was very meagre, but the device emitted a very narrow line. To find how pure the frequency was,

Townes and his group built a second maser to make a comparison between the frequencies emitted by the two masers and early in 1955 they could say that during a time interval of 1 s the frequencies of the two masers differed between them by only 4 parts in 10^9 and over a longer time interval of 1 h by one part in 10^9.

The results suggested the maser as an optimal candidate to make very precise standards of frequency or build atomic clocks. Research on the ammonia maser became more widespread among other laboratories in universities, government, and industry, both under the push of military requests and through personal contacts between interested scientists; however, even so, there were only around 10 groups with few researchers and very modest support.

4.4 Basov and Prokhorov and the Soviet Approach to the Maser

There is a patent filed on June 18, 1951, by V A Fabrikant of the Moscow Power Institute, together with some of his students, which was only published in 1959 entitled *A method for the amplification of electromagnetic radiation (ultraviolet, visible, infrared, radio wavebands).*[27] The title is so general as to cover nearly anything connected with maser or laser action. However, the work seems to have been principally focused towards lasers, and so will be considered later.

The idea of using a gas as a molecular amplifier—or *molecular generator* as they called it—came to Basov and Prokhorov at the Lebedev Institute, Moscow. They[28] published a theoretical paper[29] a few months before the paper by Townes[25] appeared in the *Physical Review*.

Alexander Mikhailovich Prokhorov was born on July 11, 1916, in Atherton, Australia, to the family of a revolutionary worker who emigrated there from exile in Siberia in 1911. Prokhorov's family returned to the Soviet Union in 1923. In 1939, he graduated from Leningrad University and went to work at the Lebedev Institute of Physics (FIAN) Moscow, one of the most prestigious research institutes of the USSR Academy of Sciences. He started his scientific work in 1939, studying the propagation of radio waves over the Earth's surface.

During the World War II, he was wounded twice and returned to the Institute in 1944. After the war, following a suggestion by V I Veksler, he demonstrated experimentally in his doctoral thesis that the synchrotron can be used as a source of coherent electromagnetic oscillations in the centimeter waveband. He went on to head a group of young researcher workers (among whom was Basov) working on radio wave spectroscopy.

After his work in the field of masers and lasers, which we shall consider in the following, he was in 1960 elected a Corresponding Member of the USSR Academy of Sciences and in 1966 he became Full Member. For his researches, he was awarded the title of Hero of Socialist Labour, the Lenin prize and, in 1964, together with Basov and Townes, the Nobel Prize for Physics. He died in January 8, 2002.

Nikolai Gennadievich Basov was born on December 14, 1922. At the start of the Second Word War, he graduated from a secondary school in Voronezh, and enlisted. He was sent first to Kuibyshev and then to the Kiev school of Military Medicine, from which he graduated in 1943 with the rank of lieutenant in the medical corps. His service began in the chemical warfare defense forces and then continued at the front. A little after the end of the war, following his return from Germany, he realized his dream of studying physics while still in the Soviet Army. He enrolled at the Moscow Institute of Mechanics (now Physics Engineering). Exactly 20 years later, he was elected to the USSR Academy of Sciences.

In 1948, Basov began work as a laboratory assistant in the Oscillation Laboratory of the Lebedev Institute of Physics. The laboratory was headed at that time by M A Leontovich, and Basov later became an engineer there. Later still, in the early 1950s, a group of young physicists under the leadership of A M Prokhorov began work there on the study of molecular spectroscopy. This marked the start of many years of fruitful collaboration in the field of masers and lasers between Basov and Prokhorov, as will be seen later. After the maser work, he also made important contributions to the development of a number of lasers. He became director of the Lebedev Institute in 1973 and was also a member of the Supreme Council of the USSR and the Presidium of the Russian Academy of Sciences. He died on July 1, 2001.

In the 1950s, Prokhorov was heading research on synchrotron light, and Basov started working on this project. Then Sergej Ivanovich Vavilov (1891–1951), the Director of the Institute, asked them to become involved in microwave spectroscopy. So they built a spectroscope and when it worked started a number of experiments.

This group of researchers was interested in both rotational and vibrational molecular spectroscopy. The possibility of using microwave absorption spectra to produce frequency and time standards was also investigated. The operational accuracy of microwave frequency standards is determined by the resolving power of the radio spectroscope. This in turn depends exclusively on the width of the absorption line itself. An effective way of narrowing down the absorption line was found to be to use spectroscopes operating in conjunction with molecular beams. However, the capabilities of molecular spectroscopes were strongly limited by the low intensity of the observed lines, which in turn was determined by the small population difference of the quantum transition investigated at microwave frequency. The idea that it was possible to increase the sensitivity of the spectroscope appreciably by artificially varying the populations in the levels arose at this stage of their work. In a review paper written in 1955,[30] they say that they had already pointed out the theoretical possibility of constructing a device producing microwaves by using stimulated emission at an *All Union Conference on Radio Spectroscopy* in May 1952. However, their first written paper was not published until October 1954.[31] It contained a detailed theoretical study of the use of molecular beams in microwave spectroscopy. In this, the authors showed that molecules of the same kind, present in a beam containing molecules in different energy states, can be separated one from the other by letting the beam pass through a non-uniform electric field. Molecules in a pre-selected energy state can then be sent into a microwave resonator where absorption or amplification takes place. The paper contained detailed calculations on the role of the relevant physical parameters, the effects of linewidth, of cavity dimensions, and so on.

Calculations, applied to the rotational spectrum of CsF—that was chosen because of its large dipole moment—had indicated that the required cavity factor Q for a signal generator could not be obtained in this case. However, the quantitative conditions for the operation of a microwave amplifier and of a generator—called by them a *molecular generator*—were given. In the discussion that followed the presentation at the conference, also ammonia, well known to spectroscopists all around the world, was suggested.

A shorthand record of the proposal which was presented at a workshop on Magnetic Moments of Nuclei held on January 22–23, 1953, exists in the *Archives of the Russian Academy of Sciences* (in Russian) as mentioned by Karlov et al.[28]

The first published paper[29] by Prokhorov and Basov was sent to the *Journal of Experimental and Theoretical Physics* in Russian in December 1953 and was printed in October 1954, therefore after the publication of Townes' paper on the maser. The delay to the publication occurred due to the desire of its authors to correct some numerical errors in the formulae.

In a subsequent paper, submitted for publication on November 1, 1954,[32] Basov and Prokhorov proposed (shortly before Bloembergen[33]) a three-level method, considering in this case a gas to be the active medium.

They examined two possible three-level schemes (**Figure 4.4**). With reference to the first method, **Figure 4.4a**, at equilibrium

$$n_1 < n_2 < n_3. \qquad (4.15)$$

The authors considered the use of a strong radiation source $\nu_{31} = \nu_{ex}$ such that

$$\nu_{31} = \frac{E_1 - E_3}{h}, \qquad (4.16)$$

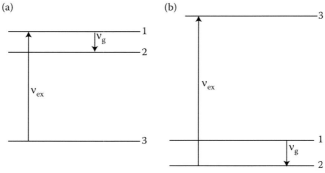

FIGURE 4.4 The three-level configuration of Basov and Prokhorov in gas molecules. The two possible schemes of inversion between the two upper levels (a) or lower levels (b) are considered.

in order to saturate the transition and to have $n_1 \sim n_3$. The reduction of population on level 3 and the increase on level 1 was indicated as producing a situation of inverted population in which

$$n_1 > n_2. \tag{4.17}$$

In **Figure 4.4b**, on the other hand, the case is shown in which a decrease of molecules on level 2 is obtained, which thus makes possible inversion between level 1 (near enough to level 2 so as to be populated thermally) and level 2.

Although these systems could be used as amplifiers, they would lack the tunability of the paramagnetic system proposed soon afterwards by Bloembergen. Moreover, all discussion of relaxation mechanisms was omitted. The methods considered were, in many cases, based on the various rotational and vibrational levels of the molecules, and none of these has ever worked.

Basov and Prokhorov also developed a theory of the molecular oscillator using a semi-classical approach. They began by considering a cavity filled with a medium, with negative losses in the neighborhood of a frequency ω_g. The active medium was characterized by a complex dielectric constant[30,34]

$$\varepsilon = \varepsilon' - i\varepsilon'' = 1 + 4\pi N\chi, \tag{4.18}$$

where χ is the molecular polarizability and N is the density of molecules in the cavity. The value of χ was written as follows:

$$\chi = \left(\frac{|m^m_{zn}|^2 \omega}{\hbar \omega_g} \right) \left\{ \frac{\left[\omega_g - \omega - (i/\tau) \right](\rho° - \rho°_n)}{\left[(\omega - \omega_g)^2 + (1/\tau^2) + |m^m_{zn}|^2 (E/\hbar)^2 \right]} \right\}, \tag{4.19}$$

where ω_g is the central frequency of the atomic line; $\tau = l/\langle v \rangle$, with $\langle v \rangle$ mean velocity of molecules and l cavity length, is the mean time for molecules to travel through the field; $\rho°_k$ is the probability of molecules being on level k at the moment of being subjected to the field inside the cavity; and m^m_{zn} is a dipole matrix element.

If N_0 is the number of active molecules (the difference between the numbers of molecules in the upper and lower levels) traveling through the cavity cross-section S every second, then

$$N = \frac{N_0}{S\langle v \rangle}. \tag{4.20}$$

Substituting Equations 4.19 and 4.20 into Equation 4.18, one obtains the expression for ε' and ε'':

$$\varepsilon' = \frac{1 - A\gamma(\omega/\omega_g)(\omega - \omega_g)\tau}{(\omega - \omega_g)^2 + (1/\tau^2) + \gamma \mid E \mid^2},$$

$$\varepsilon'' = \frac{-A\gamma(\omega/\omega_g)}{(\omega - \omega_g)^2 + (1/\tau^2) + \gamma \mid E \mid^2},$$

(4.21)

where

$$A = \frac{4\pi\hbar N_o}{Sl}, \quad \gamma = \frac{\left| m^m_{zn} \right|^2}{\hbar^2}.$$

(4.22)

By assuming that the electric field intensity in the cavity section is uniform, they then wrote

$$\left(\frac{d^2 E}{dt^2} \right) + \left(\frac{\omega_o}{Q} \right) \left(\frac{dE}{dt} \right) + \left(\frac{\omega_o^2}{e} \right) E = 0,$$

(4.23)

where ω_o is the natural frequency of the cavity without molecules.

For a stationary condition

$$E = E_o \exp(i\omega t).$$

(4.24)

Substituting the expression for ε and E into Equation 4.23 and taking the real and imaginary parts equal to zero, they obtained two equations for E_o and ω:

$$-\omega^2 + \frac{\omega_o^2 \varepsilon'}{(\varepsilon')^2 + (\varepsilon'')^2} = 0,$$

$$\frac{\omega}{Q} + \frac{\omega_o \varepsilon''}{(\varepsilon')^2 + (\varepsilon'')^2} = 0.$$

(4.25)

If $\omega_o = \omega_g \sim \omega$ and $E \to 0$, the condition for self-excitation is obtained as

$$z = \left(\frac{4\pi N_o}{Sl\hbar} \right) |m^m_{zn}|^2 Q\tau^2 \to 1.$$

(4.26)

Otherwise, if $z \gg 1$, the approximate expression for a stationary amplitude oscillation is

$$E_o^2 = \left(\frac{4\pi N_o \hbar \omega_g}{Sl} \right) Q,$$

(4.27)

and the maximum power surrendered by molecules to the cavity is equal to $(1/2)N_o \hbar \omega_g$.

The frequency

$$\omega \sim \omega_g \left[1 - \left(\frac{2Q}{\omega_g \omega_o \tau} \right)(\omega_o - \omega_g) \right], \tag{4.28}$$

results if

$$\frac{\omega_o - \omega_g}{\omega_o} \ll 1. \tag{4.29}$$

If the self-oscillation condition in the molecular oscillator is not fulfilled, the device can be used as a power amplifier, in which case the following equation is valid:

$$\left(\frac{d^2 E}{dt^2} \right) + \left(\frac{\omega_o}{Q} \right)\left(\frac{dE}{dt} \right) + \left(\frac{\omega_o^2}{\varepsilon} \right) E = B\omega^2 \exp(i\omega t), \tag{4.30}$$

where B is the amplitude of the external force. The solution is

$$E = A \exp(i\omega t) \tag{4.31}$$

with, at the resonance condition,

$$A \approx \frac{B}{(1/Q) - (4\pi N_o |m^m_{zn}|^2 \tau^2 / Sl\hbar)}. \tag{4.32}$$

Basov worked actively in the new field of *quantum radiophysics*, as it was called in the Soviet Union, and for his doctoral thesis assembled the first Soviet maser (**Figure 4.5**) a few months after Townes.[35]

FIGURE 4.5 The first maser in Russia. (From N V Karlov et al., *Appl. Opt.* **49**, F32 (2010). With permission.)

FIGURE 4.6 Basov (left in a, right in b) and Prokhorov (right in a, left in b) in 1964–1965. (From N V Karlov et al., *Appl. Opt.* **49**, F32 (2010). With permission.)

Prokhorov met Townes for the first time in Great Britain in 1955 at a conference of the Faraday Society where he presented a work on the maser,[36] while Basov met Townes, Schawlow, Bloembergen, and many others at the first *International Conference on Quantum Electronics* at Shawanga Lodge, NY, in September 1959.

At variance with what occurred in America, neither Basov nor Prokhorov (**Figure 4.6**) were familiar with radar, nor they had worked on it. They arrived at the maser concept from the spectroscopy side and the generic wish to create new kinds of sources in the centimeter wavelength range, which was their primary purpose when working on synchrotron light. In this respect, the Russia school tradition which encourages new ideas without worrying about an immediate practical realization was very helpful to them.

When the first maser was assembled in Moscow, a stream of visitors from all over the Soviet Union came to see it, and the group built three masers to study their frequency stability. Also a maser was built with two beams that originated from opposite directions which allowed a stability of one part in 10^9 to be obtained. This stability was employed to build a frequency standard which with some improvements was used for a long time at the All Union Institute of Physicotechnical and Radio Engineering Measurements which provided the time standard in the Soviet Union.

Later, after receiving the Nobel Prize, Basov organized the Division of Quantum Radiophysics and became the Director of the entire P N Lebedev Institute. A M Prokhorov organized a new General Physics Institute.

A list of activity of the two scientists may be found in a paper by Karlov et al.[37]

4.5 The Three-Level Solid-State Maser

Townes took a sabbatical year in 1955–1956 that he spent half in Paris and half in Tokyo. When he arrived at the Ecole Normale Supérieure, in the laboratory of Alfred Kastler, in the autumn of 1955, one of his students, Arnold Honig, who was working there with Jean Combrisson on electronic paramagnetic resonance, told him that arsenic ions in silicon crystals at liquid helium temperature, had a very long relaxation time of 16 s. Townes immediately realized that such circumstance would allow the arsenic ions to remain in the magnetic state of higher energy long enough to permit energy to be extracted from them by stimulated emission, and persuaded Combrisson and Honig to perform an experiment. When Townes left for Japan in the spring of 1956, the device did not work but Townes was convinced that it was a promising direction to explore and the three researchers published a paper in which they discussed the possibilities offered by the system.[38]

At about the same time, but independently from Townes, a physicist at MIT, Malcolm Woodrow P Strandberg (1919–), was considering the possibility of building a maser using solid materials instead of a gas.[39] During the war, he had worked with radar and later he became interested in microwave spectroscopy, starting in the early 1950s to work on paramagnetic resonance. On May 17, 1956, at MIT,[40] he ended a seminar on paramagnetic resonance with some observations on the advantage of a solid-state maser. Among the listeners was a young Dutchman, Nicolas Bloembergen, a professor at the Engineering and Applied Physics Department at Harvard.

Bloembergen was born in Dordrecht, The Netherlands, on March 11, 1920. He studied under L S Ornstein (1880–1941) and L Rosenfeld (1904–1974) and received the Phil. Cand. and Phil. Drs. Degrees from the University of Utrecht in 1941 and 1943, respectively, during the German occupation of the Netherlands. Then he escaped to the United States and went to Harvard where he arrived 6 weeks after Purcell, Torrey, and Pound had detected nuclear magnetic resonance. They were busy writing a volume for the MIT Radiation Laboratory series on microwave techniques, and the young Bloembergen was accepted as a graduate assistant and asked to develop the early NMR apparatus, so he started to study nuclear magnetic resonance and in the meantime attended lectures by J Schwinger (1918–1994), J H van Vleck, E C Kemble (1889–1984) and others.

He returned to the Netherlands for a short period after the war and pursued his research in a postdoctoral position at the Kamerling Onnes Laboratory in 1947–1948 at the invitation of C J Gorter who was a visiting professor at Harvard during the summer of 1947. In 1948, he received a PhD at Leyden University with a thesis on nuclear paramagnetic relaxation which was subsequently published as a short book.[41] He then went back to Harvard and joined Purcell and Pound in work on magnetic resonance to which he made important contributions, some of which were referred in Chapter 3. In 1951, he became an Associate Professor and in 1957 Gordon McKay Professor of Applied Physics at Harvard, where he has been the Gerbard Gade University Professor since 1980. His important research in the fields of nuclear magnetic resonance, masers, and nonlinear optics led to the award of the 1981 Nobel Prize for Physics (an award shared with Schawlow and Siegbahn).

After the Strandberg seminar, Bloembergen asked him why he was thinking of a solid system that could not have the spectral purity of the ammonia maser. Strandberg explained he was considering a completely different application for a microwave amplifier with very low noise. Bloembergen was struck by this idea and discussed it with Benjamin Lax, the Head of the Solid State Physics group who introduced him to the work of Combrisson, Honig, and Townes. Both in this work and in the idea of Strandberg, a two-level maser was considered. These devices had to be pulsed to operate, and solids with abnormally long relaxation times were required. A device which did not have these limitations would clearly be most useful, and Bloembergen spent some weeks thinking about how it could be realized.

Bloembergen understood that the most suitable levels to be used for the device were not two naturally pre-existing levels in a molecule but the Zeeman levels created when the substance is submitted to an external magnetic field. This would allow to tune the energy difference among the produced levels and if three levels were used instead of two[42] he would no longer need to separate physically the molecules in the upper energy state.

Therefore, Bloembergen proposed a tunable maser utilizing Zeeman levels in a paramagnetic material.[43] He was influenced by the work of Overhauser and wrote

> Attention is called to the usefulness of power saturation of one transition in a multiple energy level system to obtain a change of sign of the population difference between another pair of levels. A variation in level population obtained in this manner has been demonstrated by Pound. Such effects have since acquired wide recognition through the work of Overhauser.

To understand Bloembergen's proposal, let us recall that a paramagnetic material with spin 1/2 has two states for each electronic configuration. In a static magnetic field H (in oersted), these levels are separated by an energy difference that corresponds to a frequency (in MHz)

$$\nu = 2.8H. \tag{4.33}$$

Atoms or ions with n unpaired electrons have $n + 1$ such levels, which are degenerate in the absence of an external field, but which are separated in a crystalline field or in some other external field. The solid-state maser proposed by Bloembergen used these magnetically separable levels. Tuning of this maser is obtained by varying the strength of the external magnetic field.

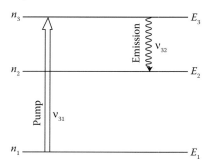

FIGURE 4.7 The three-level configuration of a paramagnetic material.

Let us now consider a material having three relevant, unequally spaced energy levels (**Figure 4.7**). Some paramagnetic ions in single crystals, usually immersed in a magnetic field, have levels separated by transitions at microwave frequencies. At thermal equilibrium, the populations obey the conditions

$$n_{10} > n_{20} > n_{30} \qquad (4.34)$$

but, being in the microwave region, where generally $h\nu < kT$, all three levels are substantially populated.

The system is now subjected to a strong pumping radiation at frequency ν_{31} such as to induce transitions between levels 1 and 3. Because, initially, more atoms are in the fundamental level 1, the system will absorb energy, populating level 3 at the expenses of level 1. The net effect is that the populations n_1 and n_3 tend to become equal. An exact calculation of the populations of each of the three energy levels in a stationary state was made by Bloembergen as follows. Let us assume $E_3 > E_2 > E_1$ and put

$$h\nu_{31} = E_3 - E_1, \quad h\nu_{32} = E_3 - E_2, \quad h\nu_{21} = E_2 - E_1. \qquad (4.35)$$

The transition probability between these spin levels under the influence of the thermal motion of the heat reservoir (lattice) are

$$
\begin{aligned}
w_{12} &= w_{21} \exp\left(-\frac{h\nu_{21}}{kT}\right), \\
w_{13} &= w_{31} \exp\left(-\frac{h\nu_{31}}{kT}\right), \\
w_{23} &= w_{32} \infty \exp\left(-\frac{h\nu_{32}}{kT}\right).
\end{aligned} \qquad (4.36)
$$

The w_{ij} correspond to the inverse of the spin–lattice relaxation times. We denote the transition probability caused by a large saturating field $H(\nu_{31})$ of frequency ν_{31} by W_{31}. Let a relatively small signal of frequency ν_{32} cause transitions between levels 2 and 3 at a rate W_{32}. The number of spins occupying the three levels, n_1, n_2, and n_3, satisfy the conservation law

$$n_1 + n_2 + n_3 = N. \qquad (4.37)$$

For $h\nu_{32}/kT \ll 1$, the populations obey the equations

$$
\begin{aligned}
\frac{dn_3}{dt} &= w_{13}\left[n_1 - n_3 - \left(\frac{N}{3}\right)\left(\frac{h\nu_{31}}{kT}\right)\right] + w_{23}\left[n_2 - n_3 - \left(\frac{N}{3}\right)\left(\frac{h\nu_{32}}{kT}\right)\right] \\
&\quad + W_{31}(n_1 - n_3) + W_{32}(n_2 - n_3), \\
\frac{dn_2}{dt} &= w_{23}\left[n_3 - n_2 + \left(\frac{N}{3}\right)\left(\frac{h\nu_{32}}{kT}\right)\right] + w_{21}\left[n_1 - n_2 - \left(\frac{N}{3}\right)\left(\frac{h\nu_{21}}{kT}\right)\right] \\
&\quad + W_{32}(n_3 - n_2), \\
\frac{dn_1}{dt} &= w_{13}\left[n_3 - n_1 + \left(\frac{N}{3}\right)\left(\frac{h\nu_{31}}{kT}\right)\right] + w_{21}\left[n_2 - n_1 + \left(\frac{N}{3}\right)\left(\frac{h\nu_{21}}{kT}\right)\right] \\
&\quad - W_{31}(n_1 - n_2).
\end{aligned}
$$

$$(4.38)$$

In the steady state, the left-hand sides are zero. If the saturating field at frequency ν_{31} is very large, $W_{31} \gg W_{32}$ and all w_{ij}, the following solution is obtained:

$$n_1 - n_2 = n_3 - n_2 = \frac{hN}{3kT} \frac{-w_{23}\nu_{32} + w_{21}\nu_{21}}{w_{23} + w_{12} + W_{32}}. \tag{4.39}$$

This population difference will be positive, corresponding to negative absorption or stimulated emission at the frequency ν_{32} if

$$w_{12}\nu_{21} > w_{32}\nu_{32}. \tag{4.40}$$

The power emitted by the magnetic specimen is also calculated as

$$P_{\text{magn}} = \left(\frac{Nh^2\nu_{32}}{3kT} \right) \frac{(w_{21}\nu_{21} - w_{32}\nu_{32})W_{32}}{w_{23} + w_{12} + W_{32}}. \tag{4.41}$$

The choice of a paramagnetic substance that Bloembergen made was dependent on the existence both of suitable energy levels and of the matrix elements of the magnetic moment operator between the various spin levels. It is essential that all off-diagonal elements between the three spin levels under consideration be non-vanishing. This can be achieved by putting a paramagnetic salt in a magnetic field in some suitable way. In this way, the state with magnetic quantum numbers m_s are mixed up. The essential role played by relaxation is evident here.

Bloembergen also considered some possible materials, mentioning nickel fluorosilicate and gadolinium ethyl sulfate. The three-level maser was the first maser to offer the practical advantage of continuously tunable amplification with a reasonable bandwidth at microwave frequencies, still maintaining the principal characteristic of a maser (an extremely low noise figure).

Bloembergen remembers[44] he was very much concerned with the possibility of having the off-diagonal matrix elements non-vanishing. He finally realized that a simple way to meet this condition was to use states which are a superposition of several magnetic quantum numbers m_s. This can be obtained by applying the external magnetic field at an arbitrary angle with respect to the axis of the crystal field potential.

At Bell Labs, H E Scovil under suggestion of Rudolf Kompfner (1909–1977), the electronic research director at Bell, started a study on how to make a solid-state maser which operated continuously and on August 7, 1956, presented a memorandum with the proposal to employ gadolinium ethyl sulfate crystal in an approach identical to what Bloembergen was considering and prepared a paper to be sent to the *Physical Review*. Notice of Bloembergen's work arrived in meantime to Bell and so he was invited to present his results at Bell on September 7, 1956. Scovil who did not know about Bloembergen's work attended the seminar and understood that in his own words "Bloembergen had the same idea and effectively had before me." So he did not send his work for publication. Instead, Bell Labs negotiated an agreement to have the use of Bloembergen patent.[45]

Unfortunately, the Harvard group, being interested (for astronomical purposes) in a device working on the interstellar hydrogen line at 1420 MHz failed to obtain the first successful operation of the three-level maser. This took place the following year at Bell Telephone Laboratories, operated by H E D Scovil, G Feher, and H Seidel. They had built the maser using the Gd^{3+} para-magnetic ions in a host lattice of lanthanum ethyl sulfate.[46] A short time later, Alan L McWhorter and James W Meyer at MIT Lincoln Laboratory[47] used Cr^{3+} ions in $K_3Co(CN)_6$ to build the first amplifier. The inversion requirement $w_{32}v_{32} \gg w_{12}v_{12}$ in the Scovil maser was obtained by altering the ratio w_{12}/w_{23} by the introduction of Cr^{3+} into the crystal. This technique of cross-doping was necessary because the energy levels used were such that $v_{12} \sim v_{23}$ for the orientation and field chosen. The preliminary results of electron spin relax-ation times were published by Feher and Scovil in a preceding letter.[48]

While the original ammonia maser was principally useful as a frequency standard or else as a very sensitive detector, the solid-state maser was some-thing which really could be used for communications and radar. It had a larger bandwidth and could be tuned by changing the magnetic field strength.

Not much later, C Kikuchi and his co-workers[49] of Michigan University showed that ruby was a good material for such maser. It was Joseph Geusic who became active in the design and perfection of the ruby maser. He had just gone to Bell Laboratories from Ohio State University where he had written his thesis under J G Daunt dealing, for the first time, with the measurement of microwave resonance in ruby.[50] Paramagnetic resonance in ruby had already been investigated in the Soviet Union in 1955–1956.[51]

During 1957 and 1958, many masers were built in several laboratories, including Harvard,[52] by using Cr^{3+} ions in ruby crystals. Rubies were employed in a great number of types of maser with many different characteristics. For example, at Bell Laboratories,[53] a traveling-wave ruby maser was assembled working below 2 K having a noise temperature that was too low to be mea-sured: it was a most sophisticated laboratory instrument. At Hughes Research Laboratory, on the other hand, a compact lightweight device was built operat-ing at 77 K with a noise temperature of about 93 K,[54] very useful for various applications.

Since 1958, many masers have been built for applications in radioastron-omy or else as components in radar receivers.[55] These masers were almost all of the ruby type. A solid-state maser was used by A Penzias (1933–) and R W Wilson (1936–) in 1965 to discover the 3 K blackbody radiation from the Big Bang.[56] Traveling-wave masers were first suggested by H Motz[57] (1909–1987) in 1957 and discussed by de Grasse et al.[53] in 1959.

The use of two levels of a paramagnetic substance in a magnetic field was tried without success by Townes using Ge.[58] He obtained neither amplification nor oscillation conditions. This proposal was based on the use of adiabatic fast pas-sage, as discussed by Bloch in his fundamental paper.[21] In the solution for the components of magnetization given in Equation 3.11, the sign of M_z depends on the sign of $\delta = [H_o(t) - \omega/|\gamma|]/H_1$. The quantity δ is zero at resonance where the resonant field is $H_o = \omega/|\gamma|$. In the adiabatic fast passage, $\delta(t)$ remains constant and then at time t_o it is quickly increased through resonance (subject to the "adia-batic" condition $|d\delta/dt| \ll |\gamma H_1|$). Thus its sign changes as it goes rapidly but adia-batically through resonance, and according to Equation 3.11c this means that the

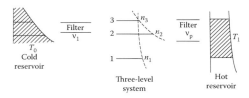

FIGURE 4.8 Three-level system in thermal contact with two heat reservoirs. (From H E D Scovil and E O Schultz-DuBois, *Phys. Rev. Lett.* **2**, 262 (1959).)

sign of M_z has been changed and it is now antiparallel to H_o.

Although the early attempt by Townes[58] was unsuccessful, 2 years later Feher et al.[59] obtained maser action due to adiabatic fast passage using the same material proposed by Townes (paramagnetic electrons associated with the P donors in Si). A volume of 0.3 cm³ of isotopically purified Si²⁸ with a phosphorus concentration of 4×10^{16} atoms/cm³ was used. Radiation at 9000 MHz at 1.2 K was obtained for about 50 μs.

Using neutron-irradiated quartz and magnesium oxide, Chester et al.[60] obtained similar results. Their quartz sample contained 10^{18} spins and the inverted population persisted for 2 μs at 4.2 K. The emission frequency was 9000 MHz and the maser operated as an amplifier of better than 20 dB gain.

A thermodynamic approach to the three-level maser was made in 1959 by Scovil and Schultz-DuBois.[61] Considering **Figure 4.8**, levels 1 and 3 are supposed to be in thermal contact, through a filter passing frequencies in the vicinity of the pumping frequency $\nu_p = \nu_{13}$ and rejecting frequencies in the vicinity of ν_{23} and ν_{21}, with a heat reservoir at temperature T_1. Levels 2 and 3 are in thermal contact with a reservoir at a lower temperature T_o through a filter which passes frequencies in the vicinity of the idler frequency $\nu_i = \nu_{23}$ but rejects those close to ν_p and the signal frequency $\nu_s = \nu_{21}$. During maser operation, for each quantum $h\nu_p$ supplied by the heat source, the energy $h\nu_i$ is passed to the heat sink, and the efficiency of the system is

$$\eta_M = \frac{\nu_s}{\nu_p}.$$

From Boltzmann's distribution law, we find

$$\left(\frac{n_2}{n_1}\right) = \left(\frac{n_2}{n_3}\right)\left(\frac{n_3}{n_1}\right) = \exp\left(\frac{h\nu_i}{kT_o}\right)\exp\left(-\frac{h\nu_p}{kT_1}\right)$$

$$= \exp\left\{\left(\frac{h\nu_s}{kT_o}\right)\left[\left(\frac{\nu_p}{\nu_s}\right)\frac{(T_1 - T_o)}{T_1} - 1\right]\right\}.$$

In this formula, one recognizes the maser efficiency η_M and the efficiency of the Carnot cycle $\eta_C = (T_1 - T_o)/T_1$. Using these, the condition for maser action is

$$\eta_M \ll \eta_C.$$

Maser efficiency equals that of a Carnot engine if the signal transition is at the verge of inversion $n_2 - n_1 \to +0$ or $T_{\text{sig}} \to -\infty$.

With time, microwave technology has advanced so much that the noise levels achieved by cryogenically cooled, conventional semiconductor-based or superconducting amplifiers are comparable to that of masers, but with a much better power and bandwidth performance and far less physical complexity.[62] However, ruby masers are still used in deep space communications. They were used, for example, by NASA to receive images of the Solar System's outer planets and their moons from the Voyager space probes.[63]

4.6 Optically Pumped Masers

Optical pumping techniques for masers have been receiving attention since 1957.[64]

The general idea was to use optical pumping to create a population inversion between some pairs of Zeeman sub-levels of the lower atomic energy level. For various reasons, gaseous systems seemed good candidates for this type of pumping. However, the maser action thereby obtained in gases is inherently very weak because of the low spin density in a gas.

Efforts to observe stimulated emission were unsuccessful[65] until, in 1962, Devor and co-workers[66] operating at 4.2 K made possible maser action in ruby by pumping with a laser. Using a magnetic field at an angle of 67° with respect to the crystalline c-axis, they split the $\bar{E}$ and $^4A_2$ states as shown in **Figure 4.9** (which is Figure 2a of Note 66). At 6700 Oe, the $-1/2(\bar{E}) \rightarrow +1/2(^4A_2)$ transition matched the $\bar{E} \rightarrow \pm1/2(^4A_2)$ component of the pumping ruby laser. Amplification of microwave power was obtained at 22.4 GHz corresponding to the $+3/2(^4A_2) \rightarrow +1/2(^4A_2)$ transition.

The general theory was developed simultaneously by Hsu and Tittel,[67] who considered the three-level configuration shown in **Figure 4.10** (which is Figure 1 of Note 67). The two lower energy levels 1 and 2 both belonging to the ground state are separated by a microwave transition. The third level 3 is separated from the other two by an optical transition.

The condition for stimulated emission between levels 2 and 1 was obtained (by assuming saturation for the pump transition between levels 1 and 3) as

$$\frac{\Delta N}{N_1} = \frac{N_2 - N_1}{N_1} = \left[\frac{w_{32}}{w_{21}} - \frac{h\nu_{12}}{kT} \right] \times \left[1 + \frac{W_{21}}{w_{21}} \right]^{-1} > 0, \quad (4.42)$$

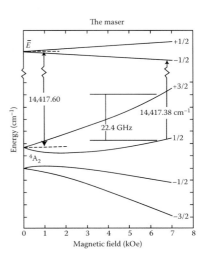

The maser

14,417.60

$\bar{E}$ — +1/2
— -1/2

— +3/2

14,417.38 cm^{-1}

22.4 GHz

— 1/2

$^4A_2$

— -1/2

— -3/2

Energy (cm^{-1})

Magnetic field (kOe)

FIGURE 4.9 Zeeman structure of Cr^{3+} in ruby. The crystalline axis is oriented at 67° with respect to the magnetic field. (From D P Devor, I J D'Haenes and C K Asawa, *Phys. Rev. Lett.* **8**, 432 (1962).)

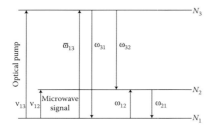

FIGURE 4.10 Three-level configuration used by Hsu and Tittel for discussing optical pumping of a maser. (From H Hsu and F K Tittel, *Proc. IRE* **51**, 185 (1963). Copyright © 1963 IEEE.)

or

$$\left(\frac{w_{32}}{w_{21}}\right) > \left(\frac{h\nu_{12}}{kT}\right). \qquad (4.43)$$

Comparing Equation 4.43 with the equivalent condition derived by Bloembergen in Equation 4.40, one sees that the difference lies in the fact that the former is valid when $h\nu_{32} \gg kT$, and the latter holds for $h\nu_{32} \ll kT$.

From Equation 4.43, a limiting signal frequency can be defined as

$$\nu^\circ_{12} = \left(\frac{w_{32}}{w_{21}}\right)\left(\frac{kT}{h}\right). \qquad (4.44)$$

Above this frequency, the maser ceases to function as a useful device, even with an infinite pumping power.

The excess noise temperature (T_{ex}) was also calculated as a function of operating temperature (T) and the following approximate expression was found:

$$T_{ex} \approx T\left[\left(\frac{\nu^\circ_{12}}{\nu_{12}}\right) - 1\right]^{-1},$$

showing that the excess noise temperature can be much lower than the operating temperature, which is a particular advantage of this kind of maser.

Almost simultaneously and independently, Ready and Chen considered the possibility of optical pumping of masers using the ruby, in a short note in *Proc. IRE*[68] proposing essentially the same scheme as Devor et al.[66]

At last, quite recently, a solid-state maser has been operated at room temperature.[69] The maser's gain medium is pentacene doped with *p*-terphenyl pumped by a pulsed dye laser. Emission is around 1.42 GHz.

4.7 The Hydrogen Maser

One maser that found important applications was the *hydrogen maser*. This maser utilizes free hydrogen atoms operating between the ground-state hyperfine levels. For use as a frequency standard, the maser oscillates on the transition ($F = 1$, $m_F = 0$) → ($F = 0$, $m_F = 0$) at a frequency of approximately 1420 MHz (see **Figure 4.11**).

The hydrogen maser was first demonstrated by Ramsey and co-workers in 1960.[70] It was later developed into a high-stability active oscillator of outstanding robustness and reliability.[71] It uses the hyperfine transition of ground-state atomic hydrogen (the 21 cm line of radio astronomy) and has a very narrow spectral width of 1 Hz.

Because the magnetic dipole transition, the only possible in the ground state of a free atom, has radiation matrix elements which are approximately 100 times smaller than the electric dipole moment characteristic of molecular vibrations, to increase the interaction efficiency, atoms are constrained to move within the volume of a storage box by collisions with the walls.[72] The apparatus is shown in **Figure 4.12**.

Molecular hydrogen is dissociated in the source and is formed into an atomic beam which passes through a state-selecting magnet that is the hexapolar field proposed by Friedburg and Paul.[73] The emergent beam contains only atoms in the state $(F = 1, m = 1)$ and $(F = 1, m = 0)$. The beam passes into a storage bulb which has a specially prepared surface and in which the atoms remain for approximately 0.3 s before escaping. The bulb is located in a cavity tuned to the hyperfine transition frequency. Stimulated emission occurs if the beam flux is sufficiently high and a signal is produced in the cavity. The signal is detected by means of a small coupling loop. The cavity is surrounded by magnetic shields to reduce the ambient field and a small uniform field is produced at the storage bulb by a solenoid.

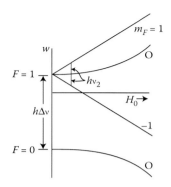

FIGURE 4.11 Energy levels of hydrogen in its ground state. The energy W is depicted as a function of the applied magnetic field H_0. (From D Kleppner et al., *Appl. Opt.* **1**, 55 (1962). With permission.)

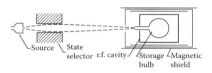

FIGURE 4.12 Schematic diagram of the hydrogen maser. (From D Kleppner et al., *Phys. Rev.* 138, A972 (1965). With permission.)

The hydrogen maser produces a microwave output signal that is highly stable for seconds to hours. Hydrogen masers are large devices but can operate with a minimal maintenance for many years in typical room conditions. They serve as high-stability oscillators for ensembles of atomic clocks in standard laboratories, tests of relativity and fundamental physical laws, very long baseline interferometry for radio astronomy, measurements of continental drift, and—together with the ruby maser amplifier—navigation and tracking of spacecraft in NASA's Deep Space Network of radio antennae in Australia, California, and Spain.

The theory was fully developed in a paper by Kleppner et al.[74]

4.8 The Ancestor of the Free-Electron Laser

In 1958, Javan[75] proposed a method of obtaining amplification using a nonlinear two-photon process with the kind of parametric amplifier which uses Raman scattering and in which it is not necessary to have quantum inversion.

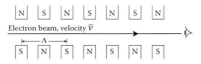

FIGURE 4.13 The undulator structure: the electron beam traverses an arrangement of magnets whose fields alternate in direction with spatial period Λ.

Javan's work was read at a meeting of the American Physical Society and today only a short summary exists. Weber in his review paper[42] makes mention of it on page 692.

Very little attention was paid to important work by Motz who was a pioneer in proposing the production of energy in the millimeter to infra-red portion of the spectrum, through a mechanism which is today employed in the so-called "free-electron beam laser," and which is used for microwaves production in a number of devices (gyrotrons, omichrons, etc.).[76]

As early as 1951, H Motz,[77] then at the Microwave Laboratory, Stanford University, California, described the use of an *undulator* for the production of millimeter and sub-millimeter waves by means of a relativistic beam of electrons passing through the magnetic field produced by an array of magnets[78] with alternating polarity (**Figure 4.13**). With suitable pole pieces, the magnetic field near the axis is approximately sinusoidal, the spatial period of the sine wave being Λ. An observer attached to the electron sees the magnet structure contracted by the Lorentz factor $\gamma = (1 - \beta)^{-1/2}$, where $\beta = v/c$, with v being the velocity of electrons. For such an observer, the wavelength of the electron's oscillations is shortened to $\lambda' = \Lambda/\gamma$, so the frequency at which the electron radiates is increased from $\nu = v/\Lambda$ to $\nu' = \gamma v/\Lambda$.

A different observer in the laboratory reference frame, placed as the eye in Figure 4.13, sees the frequency radiated by the electron which is moving towards him Doppler shifted by a further factor of γ, so that altogether the observed frequency is proportional to γ^2. Actually, the factor turns out to be twice this, as can be seen by more careful analysis.

Undulator radiation may also be regarded as synchrotron radiation emitted from a succession of curved orbits arranged in such a way that the light wave can keep in step with the electron and produce a long, coherent wave train. If electrons can be bunched so as to concentrate them into short packets with a linear size comparable with that of the wavelength, a very large energy can be radiated; and Motz in his 1951 paper calculated that the power which could be emitted coherently in this way is in the order of several tens of kW, in the millimeter band.

Early experiments were done in 1953 by Motz and co-workers[79] at Stanford by using high-energy electrons from a special linear accelerator designed for the injection system of the Mark III Stanford linear accelerator (LINAC). With a 3-MeV electron beam, radiation in the band of wavelength less than 1.9 mm was observed at a power level estimated to be between 10 and 100 W.

Moreover, under the influence of incoming radiation, the beam of electrons can be made to radiate in such a way as to amplify the incoming electromagnetic wave. The incoming wave stimulates the undulating electron in order to amplify it. Classical electronic devices may also often be looked at in this way, and the undulator as an amplifier is closely analogous to the *traveling-wave* tube amplifier invented by Kompfner in 1947.[80] In 1959, Motz realized[81] that coherent amplification may be obtained without a slow-wave structure as in the traveling-wave amplifier, but by means of an undulator.[82]

The further evolution of free-electron lasers is discussed in Chapter 9.

4.9 The Electron Cyclotron Maser

The *electron cyclotron maser*, which can be used to create very high power signals—up to hundreds of thousands of watts—in the millimeter and sub-millimeter wavelength regions was developed from the late 1950s. It belongs to the class of devices that consider the interaction of a beam of charged particles (electrons) with electromagnetic radiation, as the free electron device by Motz or the traveling wave tube. Classically, an electron situated in a homogeneous static magnetic field is assumed to gyrate along a circular trajectory. Quantum mechanics shows that it assumes discrete energy levels with an energy separation $\hbar\omega_H$ where ω_H is the cyclotron resonance frequency

$$\omega_H = eB/mc$$

with B the magnetic field strength and m the electron mass.

Interacting with an alternating electromagnetic field, the electron can transit from one energy level to another. This kind of interaction started to be used in the 1920s in the magnetron as proposed by the Czech physicist Zacek[83] and has evolved in the 1950s to the family of present electron cyclotron masers that are based on a stimulated cyclotron emission process in an electron beam in vacuum involving energetic electrons in gyrational motion.

Generation of electromagnetic radiation by a dc electron beam requires a bunching mechanism and the synchronism with the wave, to obtain amplification.

The *gyrotron* is a type of free electron maser in which the bunching necessary to amplification of radiation by stimulated emission is induced by cyclotron resonance.

A generator as the one shown in **Figure 4.14** considers the motion of electrons submitted to a static magnetic field near a metallic wall. Two different electron trajectories are shown according to the phase of emission with respect to the alternating electric field. There are wrong phase electrons that are accelerated by the RF field at the first cyclotron loop, enlarge their gyration radius, and are immediately captured by the wall. Right phase electrons are decelerated by the RF field at the first cyclotron loop, reduce their gyration radius, and so remain able to give energy to the RF field during a large amount of subsequent RF periods. The cyclotron resonance magnetron whose

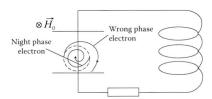

FIGURE 4.14 A scheme of microwave generator. (From M I Petelin, *IEEE Trans. Plasma Sci.* **27**, 294 (1999). © 1999 IEEE. With permission.)

principle of operation is sketched in Figure 4.14 could be called, according to the general definition of maser, the first *cyclotron resonance maser*, though the latter term was introduced only three decades later.[84]

The basic theory for electron cyclotron interaction with an electromagnetic wave, explaining the bunching mechanisms and emission or absorption processes is well resumed in a paper by Chu.[85] A simplified presentation is given by Petelin.[86]

Alternatively, the use of Landau levels generated in the bands of semiconductors submitted to a static magnetic field was considered in 1960 by B Lax[87] and was studied in detail by A A Andronov[88] in the Soviet Union. The paper by Andronov also report experimental results.

A number of theoretical approaches were developed.[89] Application to InSb was considered by A K Ganguly and K R Chu.[90] An experiment was performed by Vertii et al.[91] In the Soviet Union, this kind of generator received considerable attention without, however, to end with a practical device.

4.10 The Rydberg or Single-Photon Maser (Micromaser)

The most basic problem of radiation–matter coupling is the interaction of a single two-level atom with a single-mode electromagnetic field. This problem received great attention shortly after the maser was invented.[92] At that time, the problem was of purely academic interest, since in the experiments it was always necessary to have a large number of atoms and photons. This was due to the fact that it was impossible to detect small amounts of atoms and that the small size of the transition matrix elements resulted in atom-field coupling times much longer than other characteristic times of the system, such as atomic relaxation or the interaction time with the field.

This situation completely changed when the advent of frequency-tunable lasers allowed the study of the highly excited *Rydberg states* of atoms.

Rydberg atoms or *Rydberg states* are obtained when a valence electron of an atom is excited into an orbit with sufficiently high principal quantum number n and therefore far from the ionic core. The energy of these highly excited levels is given by the Rydberg formula: this is the reason that they are called Rydberg states or *Rydberg atoms*.

Rydberg atoms have quite remarkable properties.[93] The energy changes among highly excited states are small compared with the large changes between the lower levels. The transitions between two neighboring Rydberg levels are therefore in the region of microwave or far-infrared radiation (millimeter waves). The radius of the charge distribution of the valence electron scales as n^2, and for $n = 50$, the linear dimension of the atom is already comparable with the wavelength of light in the visible region and compete with the size of large biomolecules. The Rydberg atoms are very sensitive to external electric field which means they already ionize in rather weak fields. The probability of induced transitions between neighboring states of a Rydberg atom scales as n^4. Consequently, a single photon is enough to saturate the transition between adjacent levels. Moreover, the spontaneous lifetime of a highly excited state is very large. The rather large emitted wavelength (millimeters) allows one to build cavities, with low-order modes being sufficiently large too ensure rather long interaction times.

When in a cavity, due to Purcell effect,[94] the spontaneous emission of a single atom is expected to be drastically modified as compared with its behavior in free space. Spontaneous emission is enhanced in a resonant cavity and suppressed if the cavity is off resonance.

The interaction of a two-level atom and the electromagnetic radiation can be studied with a semi-classical model due to Jaynes and Cummings.[92] One fundamental aspect in the strong coupling regime in the Jaynes–Cummings approximation is that at resonance, an atom initially in an excited state $|e\rangle$ and an empty cavity periodically exchange a quantum. The atom field state wavefunction $|\Psi(t)|$ oscillates between an excited atom and no photon state $|e,0\rangle$ and a state in which the atom is on the ground level and a photon is in the cavity $|g,1\rangle$, according to

$$\Psi(t) = \cos(\Omega_o t/2)|e,0\rangle + \sin(\Omega_o t/2)|g,1\rangle, \tag{4.45}$$

where Ω_o is the Rabi[95] frequency and t is the interaction time. So, instead of simply emitting a photon and going on its way, an excited atom in a resonant cavity oscillates back and forth between its excited and unexcited states. The emitted photon remains in the cavity in the vicinity of the atom and is promptly reabsorbed. The atom–cavity system oscillates between the two states, excited atom and no photon, and de-excited atom and a photon trapped in the cavity.

The development of a single-atom maser or a *micromaser* as it was soon dubbed allows a detailed study of the atom–field interaction.[96]

In 1984, at the Max Planck Institute for Quantum Optics in Garching, Germany, the group lead by Herbert Walther (1935–2006) succeeded in operating the first micromaser containing only one atom.[97]

In a micromaser, a population inversion is created by sending a beam of Rydberg atoms in the higher energy metastable level through a resonant cavity that is made of superconducting metal and maintained at very low temperatures. The combination of a strongly interacting electric dipole transition and a cavity that ensure very low losses enable active maser oscillation with very few atoms in the cavity. The injection rate can be such that only one atom is present inside the resonator at any time. Due to the high-quality factor of the cavity, the radiation decay time is much larger than the characteristic time of the atom–field interaction, which is given by the inverse of the single-photon Rabi frequency; in these conditions, a field is build up inside the cavity when the mean time between the atoms injected into the cavity is shorter than the cavity decay time. A micromaser, therefore, allows sustained oscillations with less than one atom on the average in the cavity.

The realization of a single-atom maser or *micromaser* has been made possible due to the enormous progress in the construction of superconducting cavities together with the laser preparation of highly excited Rydberg atoms.

A central feature is the long lifetime of the photons in the cavity, which is demanding for very high Q-factors. The decay time for photons of frequency f is linked to the quality factor Q of the cavity by

$$\tau = Q/f. \tag{4.46}$$

Quality factors Q of the order of 10^{10} at frequencies in the microwave region $f = 10^{10}$ Hz give long lifetimes in the order of a second. A consequence of the high Q value is that the photon lifetime is much longer than the interaction time of an atom with the maser field that means while the atom passes through the cavity.

The Garching group used Rb atoms for their masers.

The set-up for the maser is shown in **Figure 4.15**.[98] A highly collimated beam of rubidium atoms passes through a Fizeau velocity selector. Before entering the superconducting cavity, the atoms are excited into the upper maser level $63p_{1/2}$ by a laser. In this way, a stable beam of excited atoms is obtained. The superconducting cavity is made by niobium cooled by means of a helium 3 cryostat down to 0.3 K. At such a low temperature, the number of thermal photons is reduced to about 0.15 at a frequency of 21.5 GHz. The cryostat is carefully designed to prevent room-temperature microwave photons from leaking into the cavity. The Q of the cavity is 3×10^{10} corresponding to a photon storage time of about 0.2 s.

It is hard to detect low-energy microwave photons, especially when the power radiated by the atoms is only 10^{-18} W. Instead, the micromaser emission is detected indirectly by its effect on the state of the atoms involved. At the exit of the cavity, atoms are subjected to field ionization. The Rydberg atoms in the upper and lower maser levels are detected in two separate field ionization detectors. The field strength is adjusted so as to ensure that in the first detector the atoms in the upper level are ionized, but not those in the lower level. Atoms that have made the transition to the lower micromaser state are therefore counted and their number suddenly increases when maser action start.

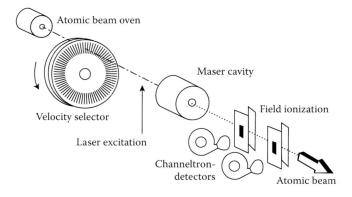

FIGURE 4.15 Scheme of the micromaser. (From Rempe et al., *Phys. Scripta* **T 34**, 5 (1991). With permission.)

For maser operation, the cavity is tuned over the $63p_{1/2} \to 61d_{3/2}$ transition (corresponding to a frequency of 21.50658 GHz) and the flux of atoms in the excited state is recorded simultaneously. The transitions are detected by a reduction of the electron count rate.

Maser operation manifests itself in a decrease in the number of atoms in the excited state. The flux of excited atoms N governs the pump intensity.

The results are shown in **Figure 4.16**.[98] A reduction in the signal is clearly seen as a reduction in the count rate of excited atoms. In the case of measurements at a cavity temperature of 0.5 K, a reduction in the signal is clearly seen for atomic fluxes as small as 1750 atoms s^{-1}. Over the range from 1750 to 28,000 atoms s^{-1}, the field ionization signal at resonance is independent of the particle flux which indicates that the transition is saturated. This and the observed power broadening increasing with the flux show that there is a multiple exchange of photons between Rydberg atoms and the cavity field.

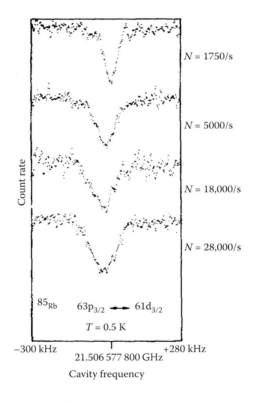

FIGURE 4.16 Maser operation of the one-atom maser manifests itself in a decrease in the number of atoms in the excited state at the resonance frequency of 21.506577800 GHz. The four curves for the flux of excited atoms N in the range from 1750 to 28,000 atoms s^{-1} show that the decrease is independent of the particle flux which indicates that the transition is saturated. (From Rempe et al., *Phys. Scripta* **T** 34, 5 (1991). With permission.)

For an average transit time of the Rydberg atoms through the cavity of 50 µs and a flux of 1750 atoms s^{-1}, approximately 0.09 Rydberg atoms are in the cavity on the average. According to Poisson statistics, this implies that more than 90% of the events are due to single atoms and demonstrates that single atoms are able to maintain a continuous oscillation of the cavity with a mean number of photons between unity and several hundreds.

The one-atom maser is a generator of nonclassical light. The Fizeau velocity selector preselects the velocity of the atoms, hence the interaction time is well defined. This has a very important consequence when the intensity of the maser field grows, as more and more atoms give their excitation energy to the field. Even in the absence of dissipation, this increase in photon number is stopped when the increasing Rabi frequency leads to a situation where the atoms reabsorb the photon and leave the cavity in the upper state. For any photon number, this can be achieved by appropriately adjusting the velocity of the atoms. In this case, the maser field is not changed anymore and the number distribution of the photons in the cavity is sub-Poissonian.[99]

A review of a earlier key work leading up to the micromaser was made by Haroche and Raimond.[100] A description of some experiments was given by H Walther.[101]

Steady-state operation of the one-atom maser has been studied both theoretically[102] and experimentally and has been used to demonstrate many quantum phenomena of the radiation field such as sub-Poissonian statistics,[103] the collapse and revival of Rabi oscillations,[104] and entanglement between the atoms and the cavity field.[105]

Because the average number of atoms in the microcavity can be 1 or less while maintaining active maser oscillation, the one-atom maser is a powerful tool in the field of cavity quantum electrodynamics, for example, in the preparation of pure-photon number states.[106] The one-atom maser is currently being applied to the study of new techniques and protocols for manipulating quantum information, such as controllably creating quantum correlations among chains of atoms. It is an ideal instrument for the study of the resonant interaction of a single atom with a single mode of a superconducting cavity.[107]

In the years 1980s and 1990s, a great number of proposals were done on possible production of states with defined number of photons by using the interaction of atoms with quantum microcavities.[108]

A *number state*, that is, a state with a well-defined photon number, can be generated[109] using a cavity with a high enough quality factor. To experimentally achieve a state of the field where the number of photons is exactly known, two conditions have to be fulfilled. The first condition concerns the temperature. Thermal photons have to be suppressed because they induce decay and influence the statistics so that a superposition of number states is obtained. Thermal photons can be eliminated by cooling the cavity to a low enough temperature. The mean number of thermal photons for a frequency of about 20 GHz is 3×10^{-5} at $T = 0.1$ K. As a second condition, one must not lose photons stored in the cavity for the duration of the experiment, that is, a cavity is needed in which losses can be neglected for this time. Because the quality factor Q can reach 10^{11}, at 20 GHz, this results in photon lifetimes of several seconds.

With the micromaser assembled at Max-Planck Institute in Garching, Walther's group obtained experimentally single quantum states (*Fock states*) in the interaction of atoms and the field confined in the microcavity.

One of the methods they used considered pumping the cavity with a pulsed beam and using the trapping condition to stabilize the photon number, so producing Fock states on demand.[110] *Trapping states* are a feature of low-temperature operation of the micromaser, for which the steady-state photon distribution closely approximates a Fock state under certain conditions, and occur in the micromaser as a direct consequence of the quantization of the cavity field.[111] They are typical of strongly coupled systems and occur when the atom field coupling, Ω, and the interaction time, t_{int}, are chosen such that in a cavity field with n photons each atom undergoes an integer number k of Rabi cycles or

$$\Omega t_{int}\sqrt{(n+1)} = k\pi. \tag{4.47}$$

When Equation 4.47 is fulfilled, the cavity photon number is left unchanged after the interaction of an atom and hence the photon number is "trapped," regardless of the atomic pump rate.

The first demonstration of trapping states in the maser field was described by Weidinger.[112] The relevant quantity was the atomic inversion I defined as

$$I(t_{int}) = P_g(t_{int}) - P_e(t_{int}), \tag{4.48}$$

where $P_e(t_{int})$ and $P_g(t_{int})$ are the probability of finding an excited-state and a ground-state atom, respectively, for a particular interaction time t_{int}. The operation of cw micromaser leads to Fock states, with the micromaser field evolving into a number state, provided the atomic velocity is conveniently tuned.

A Fock state shows ideal sub-Poissonian statistics. Consequently, under the influence of a trapping state, the atomic statistics should also be sub-Possonian. As an atom in the ground state must have emitted a photon into the cavity, the lower-state atom statistics is strongly related to cavity photon statistics. When the maser is under the influence of a trapping state, an atom cannot emit into the cavity until a photon decays from the mode. When this occurs, the next excited-state atom entering the cavity will emit a photon with a high probability, returning the cavity to the trapping state. This results in a regular spacing of ground-state atoms and a reduction of fluctuations in the count rate. The Fano-Mandel Q function

$$Q(t) = [\langle n(t)^2 \rangle - \langle n(t) \rangle^2]/\langle n(t) \rangle - 1, \tag{4.49}$$

where $n(t)$ is the number of counts in an interval of length t, may be used to assess the kind of statistics. If $Q(t) = 0$, the beam is Poissonian; sub-Poissonian statistics produces $Q(t) < 0$, and super-Poissonian statistics has $Q(t) > 0$. Indications of one-photon and two-photon states were clearly obtained by Weidinger.

Using the one-atom maser, the first experimental evidence for the operation of a reliable and robust source of photon Fock states was presented by

Brattke.[113] In the experiment, a beam of rubidium atoms in Rydberg states entering the microcavity cooled at 300 mK was used. The group claimed[114] a success rate of Fock state production of 85%.

4.11 Two-Photon Maser

An intriguing variation of the micromaser is the two-photon maser source. Quantum amplifiers or oscillators on two-photon atomic transition were proposed in 1964[115] and the theory of these systems was studied in many papers.[116] The extension to the visible is discussed in Chapter 9. In the microwave, such a device was operated for the first time in 1988 at ENS by the Haroche group.[117] The chosen transition was one that is not possible by single-photon emission. Haroche and co-workers made atoms to pass through a ultrahigh-Q superconducting cavity tuned to half the frequency of a transition between two Rydberg levels. Under the influence of the cavity radiation, each atom is stimulated to emit a pair of identical photons, each bringing half the energy required for the atomic transition. The maser field builds up as a result of the emission of successive photon pairs.

The presence of an intermediate energy level near the midpoint between the initial and the final levels of the transition helps the two-photon process to occur. To put it simply, an atom goes from its initial level to its final one via a virtual transition during which it jumps down to the middle level while emitting the first photon; it then jumps down again emitting the second photon. The intermediate step is virtual because the energy of the emitted photons, whose frequency is set by the cavity, does not match the energy differences between the intermediate level and either of its neighbors. This may happen due to the Heisenberg principle, provided the time of the emission is very short, that is, inversely proportional to the energy surplus required. For a mismatch of a few BeV, it is a few nanoseconds. The micromaser cavity inhibits single-photon transitions and strongly enhances the emission of the photon pair. The chosen transition was between the $40S_{1/2}$ and $39S_{1/2}$ state of ^{85}Rb at 68.41587 GHz. The rubidium atom beam was first excited into the $40S_{1/2}$ state by using laser beams. The two-photon emission was monitored by measuring the level populations of the atoms leaving the cavity, with the use of the field-ionization method.

The maser was operated with only about 30 photons stored in the cavity. The flux of excited atoms necessary to maintain the radiation against the cavity loss is so low that only five atoms need be in the cavity at any one time. The two-photon maser has remarkable quantum characteristics: it is triggered by the vacuum fluctuations of the quantized cavity field. If the flux of excited atoms is increased until the threshold is reached, the maser turns on but only after a delay, which could be even seconds. This delay occurs because the emission is initiated by quantum fluctuations in the cavity and not by a spontaneously emitted photon.

Also the theory was developed by the group.[118] Using a master equation approach for describing the complete state of the field in the cavity, they found that the photon statistics is sub-Poissonian for a wide range of atomic injection times and hence the device could be used to generate squeezed states of the electromagnetic field.

4.12 Maser Action in Nature

An important advance in the science of radioastronomy was made in 1951 with the observation of radio emission at 1420 MHz originating from the gaseous H interstellar clouds.[119] This particular radio emission represents the spontaneous emission, or what could be called *microwave atomic fluorescence*, of a particular transition in atomic hydrogen, coming from hydrogen atoms at thermal equilibrium at a fairly low temperature (less than 100 K). As such, it has none of the characteristics of maser amplification, but only the usual features of fluorescence emission. A catalog of sources containing ionized hydrogen, comprising mostly massive, star-forming Galactic gas clouds, was constructed by G Westerhout[120] from continuum observations at 1390 MHz, and its sources were classified with the letter W and a number.

To this discovery followed the identification of other molecules by optical methods.

More than 10 years later in 1963, the hydroxyl radical (OH) was the first interstellar molecule detected at radio frequencies.[121] Since then, an incredible number of inorganic and organic molecules have been detected.

Two years later, in 1965, from the Radio Astronomy group led by Professor H Weaver at Berkeley, California, using the Hat Creek radiotelescope built in a mountain meadow near Mount Lassen in far northern California, came news[122] of a surprising observation of radio emission with strange properties at about 1670 MHz coming from OH molecules located near some stars.[123] This emission, in reality, should consists of four transitions at 1612, 1665, 1667, and 1720 MHz. If emission on these lines had originated spontaneously, the four lines would have had intensities in the ratios 1:5:9:1. However, the intensity ratios of some of the lines observed by Weaver were quite different and changed fairly quickly with time (with a time-scale of months). Weaver et al. found sources with 1665 MHz emission, without the 1667 MHz line and referred to the substance producing this line as *misterium*, without giving any explanation. The emission frequency profile of each line was not a smooth one, but sometimes contained very narrow spectral components. The linewidths were such that the temperature of the source would have to be lower than 50 K to have a Doppler broadening as small as that observed. At the same time, the emission intensity was so strong as to correspond to a very high source temperature. This temperature is called *brightness temperature* and is defined as

$$T_b = (S_v/\Omega_s)(\lambda^2/2k),$$

where S_v is the observed flux density, Ω_s is the solid angle of the emission region, λ its wavelength, and k is a constant. T_b can be determined only when the source size Ω_s is known.

Masers and Lasers

A E R Rogers, A Barrett, and J Moran at MIT organized an interferometer to measure the angular dimension of the OH sources and the size was finally resolved after a series of efforts of building interferometers with high resolution, culminated in the very long baseline interferometer (VLBI) that is a dedicated interferometer network operated by the National Radio Astronomy Observatory of the United States, with baselines of thousands of kilometers and milli-arc-second resolution construction.[124] The OH emission was found to have a source size of a few milli-arc-second and the brightness temperature was determined $T_b = 10^{12}$ K.[125]

The results of anomalous emissions by OH sources were confirmed also by other researchers. The emission apparently originated either from extremely narrow point sources or else appeared in the form of highly directional beams. The different spectral lines were also often polarized in a well-defined way, with polarizations changed over time and were generally different from those which would be expected to result from spontaneous emission.

The only reasonable explanation of these results was that such radiations originated by spontaneous emission in some part of an OH cloud and were then strongly and selectively amplified by a maser amplifier while they were going through other regions of the OH cloud.[126]

Such amplification could explain the anomalous intensity ratios and the high brightness and directionality of this emission. Appropriate Zeeman splitting due to intergalactic magnetic fields could also help to explain the properties of polarization.[127]

It also seemed reasonable that the maser gain properties can change quickly with time, although changes both in the total quantity of OH and in the associated spontaneous emission should not be expected on the time scales observed.

The pumping mechanism responsible for the apparent population inversion was not clear: a number of suggestions centered around the optical selective pumping of OH molecules by strong stellar IR radiation—other possibilities were collisions with electrons and chemical reactions. Lethokhov[128] suggested a generation mechanism for the production of intense narrow-band radiation in populated inverted space maser media which applied also to astronomical masers. Feedback in this case can be formed on account of the nonresonant scattering of radio emission by electrons or microscopic dust particles, or as a result of resonance scattering by the active molecules themselves. The model seems to have not been developed further.

Also the regions where the maser action developed were the object of many hypotheses.[129] Turner[130] divided the OH sources into two classes: type I sources, like the original detection by Weaver et al. come from sites of formation of massive stars, as marked by nearby H II regions (regions in which ionized hydrogen exists) and are dominated by the main lines, more commonly by 1665 MHz. Type II sources are brightest at 1720 MHz and are spatially associated with supernova remnants[131] (type IIa) or 1612 MHz (type IIb) and come from singly evolved stars which are undergoing rapid mass loss and are encompassed by dense circumstellar shells.

The 18-cm lines of OH correspond to four possible transitions between hyperfine levels in the Λ-doublet of the ground rotational state $^2\Pi_{3/2}$ ($J = 3/2$)

with frequencies 1612, 1665, 1667, and 1720 MHz and the corresponding line strengths in the ratio of 1:5:9:1.

In principle, all four transitions could show maser action, which is essentially never observed. This indicate the inversion mechanism is not simple.[132] Several other transitions exist in OH and were found to maser.[133]

An amplifier maser model, based on ultraviolet continuum pumping from nearby stars, was proposed by Litvak et al.[134] The model was not able to explain the anomalous greater intensity of the 1665 line but may be considered the first good attempt to explain the mechanism of astronomical maser formation.

Important contribution was later given by Peter Goldreich et al.[135] who wrote a few seminal papers on the theory, studying also polarization effects. A number of contributions followed and a first summary of results was published in the book by Elitzur.[136]

Anyway, it was appreciated that cosmic masers exist because, in contrast to terrestrial conditions, the interstellar gas density is very low so that the level population in molecules is typically not in thermal equilibrium, and sometimes inverted. In the interstellar medium, maser emission occurs naturally, originating typically in high-density (n(H II) $> 10^7$ cm^{-3}, which on the other hand is extremely low compared to terrestrial conditions) gas near an excitation source or a source of energy. This is because in the vast volume outside stars, the conditions are typically out of thermal equilibrium. Under such conditions, where the gas density is below the critical density for collisional de-excitation, excited atoms or molecules decay radiatively and can end up with a population inversion. Moreover, the path length in such clouds is large enough (AU to pc scale[137]) that amplification could occur even if the gain is meagre. For these reasons, cosmic masers are not uncommon natural phenomena in the cosmos.

In fact, a great number of masers were discovered since then. Townes had turned his attention to microwave astrophysics since the beginning of 1960s and looked at microwave emission from some molecule. After having discovered ammonia emission, he turned his attention to water using the Hat Creek facilities. It was so that a few years after the discovery of the OH sources, water vapor maser (H_2O) was discovered in Orion.[138] C Townes in an after-dinner speak at the Washington maser conference of 1992 mentioned that one of the original observers commented: "It's raining in Orion," and this sentence has from then become a classic. Confirmations came immediately.[139]

The 22.235 GHz (1.35 cm) emitted by H_2O masers is due to transition between the 6_{16} and 5_{23} levels that happen to lie very close to each other in energy. Population inversion can result from collisional pumping.[140] Detailed calculations confirmed this mechanism.[141]

Cosmic OH and H_2O masers (emitting at 18 and 1.35 cm, respectively, or 1.6 and 22.2 GHz) are two of the most common found in the Galaxy. They were discovered serendipitously because of their unexpected large flux density.

In 1968, other substances were found to emit through maser processes. The next strong maser effect was discovered in emission from the

SiO molecule. Also this was an accidental discovery in Orion star-forming molecular cloud by Snyder and Buhl in 1974.[142] Most SiO masers reside in the extended atmosphere of highly evolved giants and supergiants.[143] Today, more than 36 molecules and nearly 200 transitions are known (CH_3OH, NH_3, HC_3N, H_2CO, CH, HCN, SiS, Co, H, etc.).[144] The first millimeter and submillimeter masers on hydrogen recombination lines were discovered at the end of 1980s.[145]

Pump mechanisms were divided into collisional or radiative. In collisional pumping, population inversion of the maser levels can be achieved by either of the following two cycles:

i. Collisions bring the molecules to high-lying states from which the spontaneous radiative decay into the upper maser level is faster than the lower maser level;

ii. The spontaneous decay from the upper maser level to low-lying levels is slower than the spontaneous decay from the lower maser level, and collisions bring the molecules back to the upper maser level.[146]

In radiative pumping, there can be more transitions to high-lying states out of the upper maser level than from the lower maser level via absorption of IR photons or, given equal absorption, from the maser levels to the high-lying levels, subsequent de-excitation can overpopulate the upper maser level.

Polarization of maser radiation remains one of the most controversial points. The magnetic interpretation was first addressed by Goldreich, Keeley, and Kwan.[147] The problems connected with the polarization properties are discussed.[148]

In the Galaxy, OH, H_2O, SiO, and CH_3OH masers are widespread and can be classified as either interstellar or circumstellar.[149] The interstellar masers originate in dense molecular gas in star-formation regions[150] closely associated with ultra-compact HII regions, embedded IR sources, hot molecular cores, Herbig-Haro objects, and outflows. Both the OH and H_2O maser emission originate from very compact spots of dimension about 10^{13} cm spread over an extent about 10^{17} cm in the immediate vicinity of IR sources or compact HII regions, both excited by very young stellar objects. OH masers with only 1720 MHz emission appear to be associated with supernova remnants.[151] The supernova remnants that support 1720 MHz masers are old ($>10^4$ years) placing them in the radiative phase of expansion. The masers form in the dense post-shock gas and have brightness temperatures 10^8–10^9 K, suggesting moderate saturation. A pumping model was developed by Elitzur,[152] involving radiative transitions and hard-sphere collisions. This pumping model was revisited by Lockett et al.[153]

Circumstellar masers of SiO, H_2O, and OH originate in the molecular circumstellar envelope of evolved giant and supergiant stars. All maser sources are found in high-density molecular gas, in the vicinity of an energy source that provides the energy for pumping the population inversion, either by radiation or collisions, often mediated by shocks in the medium.

OH emission at 18-cm originates from the ground state so that absorption against a radio continuum source is easy. Thus, extragalactic OH was initially detected in absorption.[154] Improving sensitivity and spectral resolution, OH masers were found also in the cores of extended galaxies.[155]

The first extragalactic H_2O emission was detected by Curchwell et al.[156] with emission properties very similar to the interstellar masers found in the Galaxy. The group used the 100-m telescope at the Max Planck Institute in Bonn, to survey the nearby spiral galaxy M33 in search of water vapor masers. This and subsequent studies detected eight masers similar to the ones in our Galaxy.

Unexpectedly, in 1979, very powerful H_2O and OH maser emission in the order of 10^6 times more luminous than typical Galactic masers were found in the nuclear regions of external galaxies.[157] Marques dos Santos and Lepine[158] and Gardner and Whiteoak[159] found powerful masers in NGC4945 and the Circinus galaxy, respectively, with much greater strengths than the strongest one in our Galaxy. Baan et al.[160] discovered an extremely luminous OH maser in the galaxy Arp 220 that is the prototype far-infrared ultra-luminous galaxy. It was suggested that the powerful masers arose from extended regions of high-density and high-radiation fields, powered by hot young stars, in a large molecular cloud at the galactic nucleus.[161] Initially, because the observations were made with low-angular resolution radio telescopes, the high luminosity of the nuclear maser emission was attributed to a large number of interstellar masers in the nuclear star-burst regions[162] and the results did not seem particularly unusual.

On 1984, Claussen, Heiligman, and Lo[163] with improved instrumentation surveyed the nuclei of 29 nearby spiral galaxies and found four new H_2O masers. Two of them are extremely luminous: the galaxy NGC1068 has a very powerful maser with a luminosity of 350 times that one of our Sun, and NGC4258 harbors a maser with a luminosity of 120 times the Sun. With the very-large-array-telescope, Claussen et al. found that all the emission was confined to a region having a diameter of less than 0.07 arcsec, corresponding to a size of 2 pc. The region in NGC4258 is too small to contain the many massive stars to pump the maser and some other mechanism was to be devised. Claussen et al. found that all the powerful extragalactic masers are associated with galaxies having high infrared luminosity and indications of activity in their nuclei.

These masers with luminosity of 100–10,000 times the sun luminosity or 10^6 times more luminous than typical Galactic maser sources have been called *mega masers*.[164]

The model that the luminous water maser emission is due to the sum of a large number of independent masers, each powered by one of the large number of young stars in a starburst region was ruled out and it was proposed that the luminous water maser could originate in dense gas clouds in a circumnuclear disk excited by mass outflow from an active nucleus.[165] Confirmation was given later[166] based on observations of maser emission from NGC 4258. This is a spiral galaxy at a distance of 6.6 Mpc with unusually 7-kpc-long bipolar jets or bubbles of radio continuum emission. The combination of the high-brightness temperature of circumnuclear water maser emission in NGC 4258 and the high-angular resolution of the very long base interferometer-enabled probing of an AGN.[167] The maser was within 1 parsec (3×10^{18} cm) of its nucleus.[174] Miyoshi[168] suggested that the high maser emission might be from masers orbiting a massive central black hole.

The ultimate energy source in this case is the active galactic nucleus (AGN) instead of stars as in the case of the Galactic masers: these megamasers have been used to probe the mass of the central engine.[169]

Lonsdale et al.[170] in 1994 showed that also the OH maser in Arp 220 originates in a structure <1 pc across.

Active galactic nuclei are thought to be powered by gas falling into a massive black hole; the different types of active galaxies arise because they are viewed through a thick torus of molecular gas at varying angles of inclination. Water megamaser may exist in regions with plentiful molecular gas. Koekemoer et al.[171] reported a water vapor with a maser luminosity of about 6000 times that one of our Sun in the radio galaxy TXFS2226-184 at 100 Mpc distance. The maser luminosity was about seven to eight orders of magnitude higher than that observed in a typical Galactic star-forming region. The strength of the emission suggests a way to determine distances to galaxies outside the usual distance ladder[172] and provide an independent calibration of the Hubble constant.[173]

Water megamasers are rare. The megamasers are found in high-density molecular gas located within parsec of active galactic nuclei in the case of H_2O megamasers, or within the central 100 pc of nuclear star-burst regions in the case of OH megamasers.

Water maser emission in some galaxies (NGC 1068 and NGC 1052, f.e.) is associated with radio jets. The high-pressure region produced by a jet impinging on the surrounding medium drives a radiative shock ahead of it, compressing and heating the post-shock material to produce the requisite conditions for maser emission.[174]

Comets harbor masers. When a comet approaches the Sun, the rising rate of heating causes ice evaporation and the production of OH, whose population distribution is controlled by interaction with solar ultraviolet radiation. Because the heliocentric velocity of the comet varies during its motion around the Sun, the Doppler effect shifts different solar lines into and out of match with the OH transition frequencies. At some location, the net outcome is population inversion, and at other is the opposite. As a result, the detected OH lines can oscillate between emission and absorption as the comet moves around the Sun (this behavior is named *swing effect*). Since 1973, these striking oscillations have been observed in many comets.[175] Mies in 1974[176] predicted the effect to be observed in the comet Kouhoutek in January 1974, unaware that observations were already made by Biraud et al.[177] and Turner[178] who independently gave also the correct explanation. Further discussion of the mechanism of emission was made by Strelnitskii.[179] Observation of the maser emission was found in 16 comets[180] and has been the object of a debate.[181]

An interesting effect consisting in the amplification of a background source at 1667 MHz through the coma of the comet Hale-Bopp was discovered by Galt.[182]

Cosmic masers bear some distinct difference from manmade masers. First of all, they are exclusively gaseous masers; moreover, their dimensions are huge ranging from 10^7 m (in the atmosphere of comets) to 10^{19} m (the core of an entire galaxy) and the radiation they emit can extend in many different directions.

Cosmic masers give us unique information about the chemistry of comets, circumstellar matter, star-forming clouds, active galactic nuclei, and the measurement of the Hubble constant. With the development of radiomapping techniques, masers have become an important astronomical tool, allowing to study these regions on milliarcsecond angular scales.

Maser sources in astrophysics with the history of their discovery have been updated in an exhaustive book by Malcolm Gray.[183]

Notes

1. J L Bromberg, *The laser in America 1950–1970* (The MIT Press: Cambridge, 1991).
2. See, for example, A Vucinich, *Empire of Knowledge: The Academy of Sciences of the USSR (1917–1970)* (University of California Press, 1984).
3. See, for example, C H Townes, *Science* **159**, 699 (1968); Copyright 1968 by the American Association for the Advancement of Science.
4. J Weber, *Amplification of microwave radiation by substances not in thermal equilibrium, Trans. IRE Professional Group on Electron Devices* **PGED**-**3** (June 1953).
5. In his paper, Weber makes no distinction between levels and states (which is correct provided the degeneracy of states is 1).
6. *Masers—Selected Reprints with Editorial Comments*, edited by J Weber (Gordon & Breach: New York), p. 51.
7. C H Townes himself in a work written in collaboration with L E Alsop, J A Giordmaine and T C Wang (*Phys. Rev.* **107**, 1450 (1957)) felt the need to give some historical references. After having recalled the importance of the experiment by E M Purcell and R V Pound (*Phys. Rev.* **81**, 279 (1951)) which we considered in Chapter 3, he remembered the works by Weber and by Basov and Prokhorov and mentioned a statement made by A H Nethercot on his behalf at the *Symposium on Sub-Millimeter Waves* at the University of Illinois in May 1951.
8. A short resumè of these efforts is in J Hough, S Rowan and B S Sathyaprakesh, *J. Phys. B: At. Mol. Opt. Phys.* **38**, S497 (2005).
9. Townes has written two books in which he recollect his life and activity in masers and lasers. C H Townes, *How the Laser Happened: Adventures of a Scientist* (Oxford University Press: New York, 1999); C H Townes, *A Life in Physics*, Interviews conducted by S.B Riess in 1991–1992 (University of California: Berkeley, 1994).
10. Later, Townes together with A L Schawlow wrote a well-known book: C H Townes and A L Schawlow, *Microwave Spectroscopy* (McGraw-Hill: New York, 1955).
11. C H Townes, in *Laser Focus*, August 1978, p. 52.
12. The inversion spectrum of ammonia was the first and most thoroughly studied microwave spectrum. It was originally observed by C E Cleeton and N H Williams (*Phys. Rev.* **46**, 235 (1934)). Its fine structure was observed by B Bleaney and R P Penrose (*Nature* **157**, 339 (1946)) and independently by W E Good (*Phys. Rev.* **70**, 213 (1946)).
13. H Fiedburg and W Paul, *Naturwiss* **38**, 159 (1951); H C Bemmewitz and W Paul, *Z. Phys.* **139**, 480 (1954); H C Bennewitz, W Pau and Ch Schlier, *Z. Phys.* **141**, 6 (1955).

14. More information on this device was given by F O Vonbun, *J. Appl. Phys.* **29**, 632 (1958).

15. J P Gordon, *Phys. Rev.* **99**, 1253 (1955).

16. J P Gordon, H J Zeiger and C H Townes, *Phys. Rev.* **99**, 1264 (1955).

17. The Townes' maser theory, already partly developed in the works mentioned in Notes 16 and 18, was completed in the paper by K Shimoda, T C Wang and C H Townes, *Phys. Rev.* **102**, 1308 (1956).

18. J P Gordon, H J Zeiger and C H Townes, *Phys. Rev.* **95**, 282 (1954).

19. The figure reproduced here (our Figure 4.3) is in fact Figure 1 from the reference in Note 16, but the two figures are substantially identical.

20. H Friedburg and W Paul, *Naturwiss.* **38**, 159 (1951); H G Bennewitz and W Paul, *Z. Phys.* **139**, 489 (1954).

21. F Bloch, *Phys. Rev.* **70**, 460 (1946).

22. N Bloembergen, E M Purcell and R V Pound, *Phys. Rev.* **73**, 679 (1948).

23. N G Basov and A M Prokhorov, as we shall see in Section 4.4, gave a theoretical treatment of the maser in terms of a complex susceptibility. This method was later used by many authors; see, for example, R W De Gasse, E O Schultz-DuBois and H E D Scovil, *Bell Syst. Tech. J.* **38**, 305 (1959); A M Clogston, *J. Phys. Chem. Solids* **4**, 271 (1957); P W Anderson, *J. Appl. Phys.* **28**, 1049 (1957) and A Javan, *Phys. Rev.* **107**, 1579 (1957).

24. Theoretical studies were performed by K Shimoda, H Takahasi and C H Townes, *J. Phys. Soc. Jpn.* **12**, 686 (1957); R V Pound, *Ann. Phys., NY* **1**, 24 (1957); M W P Stranberg, *Phys. Rev.* **106**, 617 (1957). Measurements were done by J P Gordon and L D White, *Phys. Rev.* **107**, 1728 (1957). The role of spontaneous emission was emphasized by R H Dicke at the *Symposium on Amplification by Atomic and Molecular Resonance*, Asbury Park, New Jersey (1 March 1956) and reported by J Weber, *Rev. Mod. Phys.* **31**, 681 (1959).

25. C H Townes, *J. Inst. Electron. Comm. Eng. (Japan)* **36**, 650 (1953) (in Japanese).

26. See, for example, A Dymanus, *Int. Rev. Sci. Phys. Chem. Ser* 2 **3** (1975); D C Laine, *Adv. Electr. Elecron Phys.* **39**, 183 (1975). In this last reference, advances in molecular beam masers in general are reported.

27. V A Fabrikant, M M Vudynskii and F Butayeva, USSR Patent no. 123209 submitted June 18, 1951, and published in 1959. Supplemented by USSR Patent no. 148441.

28. A history of quantum electronics at the Moscow Lebedev and General Physics Institutes has been given by N V Karlov, O N Krokhin and S G Lukishova, *Appl. Opt.* **49**, F32 (2010) in which also the characters of the two scientists Basov and Prokhorov are reported together with their principal activities in maser and laser research. See also A H Guenther interview of N G Basov, *Sov. J. Quantum Electron.* **15** (3), 301 (1985) [*Kvantovaya Elektron.* (Moscow) **12**, 453 (1985)].

29. N G Basov and A M Prokhorov, *Zh. Eksp. Teor. Fiz.* **27,** 431 (1954).

30. N G Basov and M A Prokhorov, *Dokl. Akad. Nauk* **101**, 47 (1955).

31. N G Basov and A M Prokhorov, *Zh. Eksp. Teor. Fiz.* **27**, 431 (1954); interview of N G Basov, *Sov. J. Quantum Electron.* **15** (3), 301 (1985).

32. N G Basov and A M Prokhorov, *Zh. Eksp. Teor. Fiz.* **28**, 249 (1955) (in Russian) and *Sov. Phys. JETP* **1**, 184 (1955) (in English).

33. N Bloembergen, *Phys. Rev.* **104**, 324 (1956).

34. N G Basov and A M Prokhorov, *Trans. Faraday Soc.* **19**, 96 (1955).

35. N G Basov, *Radioteck. Elektron.* **1**, 752 (1956); see also Basov's doctoral dissertation, *Molecular Oscillators* (Lebedev Institute of Physics: Moscow, 1956).

36. N G Basov And A M Prokhorov, *Trans. Faraday Soc.* **19**, 96 (1955).

37. N V Karlov, O N Krokhin and S G Lukishova, *Appl. Opt.* **49**, F32 (2010).
38. J Combrisson, A Honig and C H Townes, *C.R. Acad. Sci., Paris* **242**, 2451 (1956).
39. M W P Strandberg, *Proc. IRE,* **45**, 92 (1956) (received August 24, 1956).
40. M W P Strandberg, as reported by J Weber in *Masers* (Gordon & Breach: New York, 1967), p. 3, noted in 1955 that the use of low temperatures and paramagnetic ions would result in improved intrinsic amplification.
41. N B Bloembergen, *Nuclear Magnetic Relaxation* (Nijhoff: The Hague, 1948) reprinted by Benjamin, New York, 1961.
42. J Weber in his excellent review paper on masers (*Rev. Mod. Phys.* **31**, 681 (1959)) refers on p. 692 to the fact that Ali Javan was also working on a three-level maser idea.
43. N Bloembergen, *Phys. Rev.* **104**, 324 (1956)).
44. N Bloembergen, *Nobel Lecture*, 8 December 1981.
45. See J L Bromberg, *The Laser in America 1950–1970* (The MIT Press: Cambridge, 1991).
46. H E D Scovil, G Feher and H Seidel, *Phys. Rev.* **105**, 762 (1957).
47. A L McWhorter and J W Meyer, *Phys. Rev.* **109**, 312 (1958).
48. G Feher and H E D Scovil, *Phys. Rev.* **105**, 760 (1957).
49. G Makhov, C Kikuchi, J Lambe and R W Terhune, *Phys. Rev.* **109**, 1399 (1958).
50. J E Geusic, *Phys. Rev.* **102,** 1252 (1956).
51. A A Manenkov and A M Prokhorov, *Sov. Phys. JETP* **1**, 611 (1955); M M Zaripov and I I Shammonin, *Sov. Phys. JETP* **3**, 171 (1956).
52. J O Artman, N Bloembergen and S Shapiro, *Phys. Rev.* **109**, 1392 (1958).
53. R W De Grasse, E O Schultz, DuBois and H E D Scovil, *Bell Syst. Tech. J.* **38**, 305 (1959).
54. H R Sent, F E Goodwin, J E Kiefer and K W Cowans, *IRE Trans. Mil. Electr.* **MIL- 5**, 58 (1961).
55. J J Cook, L G Cross, M E Bair and R W Therune, *Proc. IRE* **49**, 768 (1961); J V Jelley, *Proc. IRE* **51**, 30 (1963) gives a list of various maser installations at that time.
56. A A Penzias and R W Wilson, *Astrophys. J.* **142**, 419, 1149 (1965). See also W J Tabor and J T Sibilia, *Bell Syst. Tech. J.* **42**, 1863 (1963).
57. H Motz, *J. Electr. Control* **2**, 571 (1957).
58. J Combrisson, A Honig and C H Townes, *C.R. Acad. Sci. Paris* **242**, 2451 (1956).
59. G Feher, J P Gordon, E Buehler, E A Gere and C D Thurmond, *Phys. Rev.* **109**, 221 (1958).
60. P F Chester, P E Wagner and J G Castle, *Phys. Rev.* **110**, 281 (1958).
61. H E D Scovil and E O Schultz-DuBois, *Phys. Rev. Lett.* **2**, 262 (1959); *Prog. Cryog.* **2**, 173 (1961). An approach using statistical mechanics and irreversible thermodynamics on a three-level maser was made by W A Barker, *Phys. Rev.* **124**, 124 (1961).
62. J Schleeh, *IEEE Electron. Device Lett.* **33**, 664 (2012); B H Eom, P K Day, H G LeDuc, and J Zmuidzinas, *Nature Phys. B,* **623** (2012); N Wadefalk et al., *IEEE Trans. Microw. Theory Tech.* **51,** 1705 (2002).
63. B A Smith et al., *Science* 215, 504 (1982).
64. W H Carter, *Science* **126**, 810 (1957); H H Theissing, P J Caplan, F A Dieter and N Rabbiner, *Phys. Rev. Lett.* **3**, 460 (1959); I Weider, *Phys. Rev. Lett.* **3**, 468 (1959).
65. V E Derr, J J Gallagher, R E Johnson and A P Sheppard, *Phys. Rev. Lett.* **5**, 316 (1960); N Knable, *Bull. Am. Phys. Soc.* **6**, 68 (1961) claimed to have obtained stimulated emission but the effects were extremely weak.
66. D P Devor, I J D'Haenens and C K Asawa, *Phys. Rev. Lett.* **8**, 432 (1962).
67. H Hsu and F K Tittel, *Proc. IRE* **51**, 185 (1963).
68. J F Ready and D Chen, *Proc. IRE* **50**, 329 (1962).

69. M Oxborrow, J D Breeze and N M Alford, *Nature,* **488**, 353 (2012).

70. H M Goldenberg, D Kleppner, N F Ramsey, *Phys. Rev. Lett.* **5**, 361 (1960).

71. H M Goldenberg, D Kleppner and N F Ramsey, *Phys. Rev. Lett.* **5**, 361 (1960); *Appl. Opt.* **1**, 55 (1962); S B Crampton, D Kleppner and N F Ramsey, *Phys. Rev. Lett.* **11**, 338 (1963); H C Berg, D Kleppner and N F Ramsey, *Bull. Am. Phys. Soc.* **8**, 379 (1963); E N Fortson, D Kleppner and N F Ramsey, *Phys. Rev. Lett.* **13**, 22 (1964).

72. D Kleppner, H M Goldenbeg and N F Ramsey, *Appl. Opt.* **1**, 55 (1962); H M Goldenberg et al., *Phys. Rev.* **5**, 361 (1960); *Phys. Rev.* **123**, 530 (1961).

73. H Friedburg and W Paul, *Naturwiss* **38**, 150 (1951).

74. D Kleppner, H C Berg, S B Crampton, N F Ramsey, R FF C Vessot, H E Peters and J Vanier, *Phys. Rev.* **138**, A972 (1963).

75. A Javan, *Bull. Am. Phys. Soc. Soc. Ser.* 2, **3**, 213 (1958).

76. See, for example, J L Hirshfield and V L Granatstein, *IEEE Trans. Microwave Theory Tech.* **MTT-25**, 522 (1977); J L Hirshfield, *Infrared and Millimeter Waves*, edited by K J Button (Academic Press: New York, 1979), Vol. 1, p. 1; V L Granatstein and P Sprangle, *IEEE Trans. Microwave Theroy Tech.* **MTT-25**, 545 (1977); V A Flyagin, A V Gaponov, M I Petelin and V K Yulpatov, *IEEE Trand. Microwave Theory Tech.* **MTT-25**, 514 (1977); K Mizuno and S Ono, *Infrared and Millimeter Waves*, edited by K J Burton (Academic Press: New York, 1979), Vol. 1 p. 213.

77. H Motz, *J. Appl. Phys.* **22**, 527 (1951).

78. In his paper (Note 77) in a note added in proof, Motz also said that according to reports of the Electronic Research Laboratory at MIT, P D Coleman was working on a scheme similar to the one he (Motz) was suggesting. Prior reference to the problem treated in the paper was also found in a paper by V L Ginzburg, Radiation of microwaves and their absorption in air, *Bull. Acad. Sci. USSR Phys. Ser.* 9, no. 2, **165** (1947) (in Russian).

79. H Motz, W Thon and R N Whitehurst, *J. Appl. Phys.* **24**, 827 (1953).

80. R Kompfner, *Proc. IRE* **35**, 124 (1947).

81. H Motz and M Nakamura, *Brooklyn Polytechnic Symposium on Millimeter Waves* (1959), p. 155.

82. Motz's theory in Note 77 was further developed by N M Kroll, *Novel Sources of Coherent Radiation* (Addison-Wesley: New York, 1978), p. 115.

83. A Zacek, *Zc. Hochfr,* **32**, 172 (1928).

84. J Schneider, *Phys. Rev. Lett.* **2**, 504 (1959); J L Hishfield and J M Wachtel, *Phys. Rev. Lett.* **12**, 533 (1964).

85. K R Chu, *Rev. Mod. Phys.* **76**, 489 (2004).

86. M I Petelin, *IEEE Trans. Plasma Sci.* **27**, 294 (1999).

87. B Lax, *Quantum Electronics Symposium* (Columbia University Press: New York, 1960), p. 428.

88. A A Andronov, *IEEE Trans. Plasma Sci.* **27**, 303 (1999); A A Andronov, Semiconductor cyclotron-resonance maser, *Radiofizika* **29**, 1017 (1986) (in Russian). In the paper, he develops the theory of stimulated emission of cyclotron radiation from semiconductors considering an example of germanium.

89. A M Kalmykov, N Ya Kotsarenko and S V Koshevaya, *Izv. Vyssh. Uchebn. Zaved. Raz. Radioelektron* **18**, 93 (1975).

90. A K Ganguly and K R Chu, *Phys. Rev.* **B18**, 6880 (1978).

91. A A Vertii, I V Ivanenchenko, S V KKoshevaya, G D Fustyl'nik and V P Shestopalov, *Radiophys. Quant. Electron.* **25**, 601 (1982).

92. E T Jaynes and F W Cummings, *Proc. IEEE* **51**, 89 (1963); see also M Tavis and F W Cummings, *Phys. Rev.* **188**, 692 (1969); R Bonifacio and G Preparata, *Phys. Rev.* **A2**, 336 (1970); P L Knight and P P W Miloni, *Phys. Rep.* **66C,** 21 (1980).

93. J A C Gallas, G Leuchs, H Walther and H Figger, *Advances in Atomic and Molecular Physics*, edited by D Bates and B Bederson, Vol. 20 (Academic Press: Orlando, 1985), p. 413.

94. The spontaneous emission of an atom is the result of the interaction between the atom dipole and the vacuum electromagnetic fields. Therefore, it is not an intrinsic property of an isolated emitter but rather a property of the coupled system of the emitter and the electromagnetic modes in its environment. In 1946, Purcell (E M Purcell, *Phys. Rev.* **69,** 681 (1946)) first predicted that nontrivial boundary conditions of an electromagnetic field in the vicinity of an excited atom could alter its decay rate. The rate Γ for spontaneous transitions from an initial state $|i\rangle$ with no photons to a final state $|f\rangle$ with one photon is given by the well-known Fermi golden rule (R Loudon, *The quantum theory of light* (Oxford University Press: New York, 2000)

$$\Gamma = h^2 \rho(\nu_c)|\langle f|H|i \rangle|^2,$$

where H is the interaction Hamiltonian and $\rho(\nu_c)$ is the density of states at the transition frequency ν_c, that for radiation in free space is

$$\rho(\nu_c) = 2(4\pi\nu_c^2/c^3).$$

The spontaneous emission rate can therefore be changed if an atomic system is placed close to a metal surface or a dielectric interface. First experimental demonstrations were carried out by K H Drexhage et al., *Ber. Bunsenges: Phys. Chem.* **72,** 329 (1968); H Kuhn, *J. Chem. Phys.* **53,** 101 (1970); R R Chance et al., *J. Chem. Phys.* **60,** 2744 (1974). Approximate solutions of the electromagnetic boundary value problem were reviewed for ex. by K H Drexhage et al., *Progress in Optics*, edited by E. Wolf (North-Holland: Amsterdam, 1974), Vol. 12, p. 163. More detailed calculations are in W Lukosz and R E Kunz, *J. Opt. Soc. Am.* **67,** 1607, 1615 (1977) and references therein. In a more effective way, the radiation emitted by a source can be altered by suitably modifying the surrounding vacuum fields in a cavity.

The application to a small cavity for which the density of modes may be modified was considered by D Kleppner, *Phys. Rev. Lett.* **47,** 233 (1981). In particular, when the transition frequency ν_c is near resonance with a mode eigen frequency, the spontaneous emission rate can be considerably increased. The effect was experimentally observed (P Goy et al., *Phys. Rev. Lett.* **50,** 1903 (1983)) with a sodium Rydberg atom sent through a resonant superconducting cavity. Also inhibited spontaneous emission was observed by studying the cyclotron motion of a single electron (G Gabrielse and H Dehmelt, *Phys. Rev. Lett.* **55,** 67 (1985)). Therefore, a reduction of the two-photon probability relative to a Poisson distribution may be achieved acting on the density of modes. Periodic dielectric structures in the form of photonic crystals can alter the emission properties (E Yablonovitch, *Phys. Rev. Lett.* **58,** 2059 (1987); S John, *Phys. Rev. Lett.* **58,** 2426 (1987)). The existence of forbidden electromagnetic frequencies inside the structure may inhibit the emission of radiation which is, on the contrary, enhanced at the frequencies where the density of modes is maximum.

Enhanced spontaneous emission by quantum semiconductor boxes in a monolithic optical microcavity, that can be considered the ancestor of quantum dots experiments, was studied, for example, by Gerard et al. (J M Gerard et al., *Appl. Phys. Lett.* **69,** 449 (1996); *Phys. Rev. Lett.* **81,** 1110 (1998)). A general review of 1D photonic structures is, for example, Bertolotti (M Bertolotti, *J. Opt. A: Pure Appl. Opt.* **8,** S1 (2006)). In semiconductor systems, enhanced and inhibited spontaneous emission from GaAs quantum well excitons was demonstrated using a planar microcavity (Y Yamamoto et al., *Coherence and Quantum Optics*, edited by E H Eberly et al. (Plenum Press: New York, 1989), Vol. VI, p. 1249; H Yokoyama et al., *Appl. Phys. Lett.* **57**, 2814 (1990)). The properties of bandgap structures have been largely studied (see, for example, Y Yang and S Y Zhu, *J. Mod. Opt.* **47**, 1513 (2000); Y Yang et al., *Phys. Rev.* **A68**, 043805 (2003)).

95. The Rabi frequency $\Omega = |\mu12E|/\hbar$, where $\mu12$ is the transition dipole moment and E is the electric field.

96. J M Raimond, M Brune, S Haroche, *Rev. Mod. Phys.* **73**, 565 (2001).

97. D Meschede, H Walther and G Mueller, *Phys. Rev. Lett.* **54**, 551 (1985).

98. G Rempe, M O Scully and H Walther, *Phys. Scripta* **T34**, 5 (1991).

99. P Filipowicz, J Javanainen and P Meystre, *Opt. Comm.* **58**, 327 (1986); *Phys. Rev.* **A34**, 3077 (1986); *J. Opt. Soc. Am.* **B3**, 906 (1986); L Lugiato, M O Scully and H Walther, *Phys. Rev.* **A36**, 740 (1987).

100. S Haroche and J M Raimond, *Advances in Atomic and Molecular Physics*, edited by D Bates and B Bederson, Vol. 20 (Academic Press: Orlando, 1985).

101. H Walther, *Phys. Scripta* **T23**, 165 (1988); H Walther, *Proc. R. Soc. Lond.* **A454**, 431 (1998); H Walther, B T H Varcoe, B G Englert and T Becker, *Rep. Prog. Phys.* **69**, 1325 (2006).

102. M O Scully and M S Zubaire, *Quantum Optics* (Cambridge University Press: Cambridge, 1997); P Filipowicz, J Javansinen and P Meystre, *Phys. Rev.* **A34**, 3077 (1986); A M Guzman, P Meystre and EE M Wright, *Phys. Rev.* **A40**, 2471 (1989).

103. G Rempe, F SchmidtKaler and H Walther, *Phys. Rev. Lett.* **64**, 2783 (1990).

104. G Rempe, G H Walther and N Klein, *Phys. Rev. Lett.* **58**, 353 (1987).

105. B Englert, M Loeffler, O Benson, M Weidinger, B Varcoe and H Walther, *Fortschr. Phys.* **46**, 897 (1998).

106. B T H Varcoe, S Brattke, M Weidinger and W Walther, *Nature* **403**, 743 (2000).

107. G Rempe, H Walther and N Klein, *Phys. Rev. Lett.* **58**, 353 (1987); G Rempe and H Walther, *Phys. Rev.* **A42**, 1650 (1990); G Rempe, F Schmidt-Kaler and H Walther, *Phys. Rev. Lett.* **64**, 2783 (1990).

108. For a description of the interactions in microcavities, see P Meystre in *Progress in Optics*, edited by E Wolf (North-Holland: Amsterdam, 1992), Vol. XXX, Chapter V.

109. J Krause, M O Scully and H Walther, *Phys. Rev.* **A36,** 4547 (1987); P Meystre, *Opt. Lett.* **12,** 699 (1987).

110. S Brattke et al., *Opt. Express* **8**, 131 (2001).

111. P Meystre et al., *Opt. Lett.* **13**, 1078 (1988); M Weidinger et al., *Phys. Rev. Lett.* **82**, 3795 (1999); S Brattke et al., *Opt. Express* **8**, 131 (2001).

112. M Weidinger et al., *Phys. Rev. Lett.* **82**, 3795 (1999).

113. S Brattke et al., *Phys. Rev. Lett.* **86**, 3534 (2001).

114. See also S Brattke et al., *J. Mod. Opt.* **50**, 1103 (2003).

115. P P Sorokin and N Braslau, *IBM J. Res. Dev.* **8**, 177 (1964); A M Prokhorov, *Science* **149**, 828 (1965).

116. V S Letokhov, *Pis'ma Zh. Exsp. Teor. Fiz.* **7**, 284 (1968); *JETP Lett.* **7,** 221 (1968); R L Carman, *Phys. Rev.* **A12**, 1048 (1975); H P Yuen, *Appl. Phys. Lett.* **26,** 505 (1975); L M Narducci, W W Edison, P Furcinitti and D C Eteson, *Phys. Rev.* **A16,** 1665 (1977); T Hoshimiya, A Yamagishi, N Tanno and N Inaba, *Jpn. J. Appl. Phys.* **17,** 2177 (1978); N Nayak and B K Mohanty, *Phys. Rev.* **A19,** 1204 (1979); Z C Wang and H Haken, *Z. Phys.* **B55**, 361 (1984); ibidem **56**, 77, 83 (1984).

117. M Brune, J M Raimond, P Goy, L Davidovich and S Haroche, *Phys. Rev. Lett.* **59,** 1899 (1987).

118. M Brune, J M Raimond, and S Haroche, *Phys. Rev.* **A35**, 154 (1987); L Davidovich, J M Raimond, M Brune and S Haroche, *Phys. Rev.* **A36**, 3771 (1987).
119. H I Ewen and E M Purcell, *Nature* **168**, 356 (1951); see also C A Muller and J H Oort, *Nature* 168, 357 (1951).
120. G Westerhout, *Bull. Astron. Inst. Netherlands* **14** (Dec), 215, 1958.
121. S Weinreb, A H Barrett, M L Meeks and J C Henry, *Nature* **200**, 829 (1963).
122. H Weaver, D R Williams, H Nannilou Dieter and T W Lum, *Nature* **208**, 29 (1965); H Weaver, H Nannilou Dieter and D R Williams, *Astrophys. J. Suppl.* **16,** 10, 146 and 219 (1968).
123. Previous preliminary observations had given results inconsistent with thermal equilibrium; for example, B J Robinson et al. (*Nature* **202**, 989 (1964)) had found absorption lines at 1665 and 1667 MHz with an intensity ratio 1.2 rather than 1.8 as expected from theory if the OH lines were in thermal equilibrium. However, the measurements were done with poor resolution instrumentation.
124. A E E Rogers et al., *Science* **219**, 51 (1983).
125. J M Moran, B F Burke, A H Barrett, A E E R Rogers, J C Carter et al., *Astrophys. J. Lett.* **152** L97 (1968).
126. S Weinreb, M L Meeks, J C Carter, A H Barrett and A E Rogers, *Nature*, **208,** 440 (1965); R X McGee, B J Robinson, F F Gardner and J G Bolton, *Nature*, **208,** 1193 (1965); B Zuckerman et al., Nature **208**, 441 (1965); A H Barrett and A E E Rogers, *Nature* **210**, 188 (1966); M M Litvak, A L McWhorter, M L Meeks and H J Zeiger, *Phys. Rev. Lett.* **17,** 821 (1966); F. Perkins, T Gold and E E Salpeter, *Astrophys. J.* **145**, 361 (1966); A H Cook, *Nature* **210**, 611 (1966); B E Turer, *J. Roy. Astron. Soc. Canada* **64**, 221, 282 (1970).
127. P Goldreich, D A Keeley and J Y Kwan, *Appl. J.* **179**, 111 (1973).
128. V S Letokhov, *JETP Lett.* **4**, 321 (1966); *Sov. Astronomy-A J.* **16**, 604 (1973).
129. See, for example, V S Strel'nitskii, *Sov. Phys. Uspeki* **17**, 307 (1975) who was one of the first review papers on the subject.
130. B E Turner, *PASP* 82, 996 (1970).
131. W M Goss, *ApJS* **15**, 131 (1968); W M Goss, B J Robinson, *Astrophys. Lett.* **2**, 81 (1968); See M Wardle and F Yusef-Zadeh, *Science* **296,** 2350 (2002) and references therein.
132. M Elitzur, *Astronomical Masers* (Kluwer Academic Publishers: Dordrecht, 1992).
133. See M Gray, *Maser Sources in Astrophysics* (Cambridge University Press: Cambridge, 2012).
134. M M Litvak, A L McWhorter, M L Meeks and H J Zeiger, *Phys. Rev. Lett.* **17,** 821 (1966); M M Litvak, *Science* **165**, 855 (1969); *Astrophys. J.* **156**, 471 (1969); *Phys. Rev.* **A2,** 2107 (1970); *Astrophys. J.* **182,** 711 (1973).
135. P Goldreich, D A Keely and J Y Kwan, *Astrophys. J.* **179,** 111 (1973); **182,** 55 (1973); *ApJ* 179, 111 (1973).
136. M Elitzur, *Astronomical Masers* (Kluwer Academic Publishers: Dordrecht, 1992); *Annu. Rev. Astron. Astrophys.* **30**, 75 (1992).
137. 1 AU = astronomical unit is the distance of earth from Sun = 150,000,000 km. One parsec (pc) = 3×10^{18} cm.
138. A C Cheung, D M Rank, C H Townes, D D Thornton, W J Welch, *Nature* **221**, 626 (1969); D M Rank, C H Townes, and W J Welch, *Science* **174,** 1083 (1971).
139. S H Knowles, C H Mayer, A C Cheung, D M Rank and C H Townes, *Science* **163**, 1055 (1969); S H Knowles, C H Mayer, W T Sullivan and A C Cheung, *Science* **166**, 221 (1969); M L Meeks et al., *Science* **165**, 180 (1969); D Buhl, L E Snyder, P R Schwartz and A H Barrett, *Astrop. J.* **158**, L97 (1969).

140. T de Jong, *Astron. Astrophys.* **26**, 297 (1973).

141. S Deguchi, *Astrophys. J.* **249**, 145 (1981); M Elitzur, D J Hollenbach and C F McKee, *Astrophys. J.* **346**, 983 (1989); D A Neufeld and G J Melnick, *Astrophys. J.* **368**, 215 (1991).

142. L E Snyder and D Buhl, *Astrophys. J.* **189**, L31 (1974).

143. P J Diamond et al., *Astrophys. J.* **430**, L61 (1994).

144. D M Rank, C H Townes and W J Welch, *Science* **174**, 1083 (1971); A H Cook, *Celestial Masers* (Cambridge University Press: London, 1977). A final list is given in the book by M Gray, *Maser Sources in Astrophysics* (Cambridge University Press: Cambridge, 2012).

145. J Martin-Pintado, R Bachiller, C Thum and M Walmsley, *AA* **215**, L13 (1989).

146. P Golreich and J Kwan, *Astrophys. J.* **191**, 93 (1974).

147. P Goldreich, D A Keeley and J Y Kwan, *Ap. J.* **179**, 111 (1973).

148. M Elizur in V Migenes and M Reid (eds), *Cosmic Masers: From Protostars to Black-holes*, Proc. IAU Symp. N.206 (San Francisco: ASP, 2002), p. 452; and in J Chapman and W Baan (eds), *Astrophysical Masers and their Environment*, Proc. IAU Symp. No. 242 (San Francisco: Asp, 2007), p. 7; see also W D Watson, *Rev. Mexicana AyA* (Serie de Conferencias) **36**, 113 (2009).

149. M J Reid, *IAU Symp.* 206, 506 (2002).

150. R Genzel, D Downes, J Moran, K J Johnston, J Spencer et al., *Astron. Astrophys.*, **66**, 13 (1978).

151. F Yusef-Zadeh, K I Uchida and D Roberts, *Science* **270**, 1801 (1995); L J Greenhill, S P Ellingsen, R P Norris, P J McGregor, R G Grough et al., *Astrophys. J.* **565**, 836 (2002).

152. M Elitzur, *Ap. J.* **203**, 124 (1976).

153. P Lockett, E Gauthier and M Elitzur, *Ap. J.* **511**, 235 (1999).

154. L Weliachev, *Astrophys. J. Lett.* **167**, L47 (1971).

155. J B Whiteoak and F F Gardner, *Astrophys. J. Lett.* **15**, 211 (1974); Nguyen-Q-Rien, *A&A* **52**, 467 (1976).

156. E Churchwell, A Witzel, W Huchtmeier, I Pauliny-Toth, J Roland and W Wieber, *Astron. Astrophys.* **54,** 969 (1977).

157. P M Marques dos Santos and J R D Lepine, *Nature* **278**, 34 (1979); W A Baan, P A Wood, A D Haschick, *Astrophys. J. Lett.* **260**, L52 (1982).

158. P M Marques dos Santos and J R D Lepine, *Nature* **278**, 34 (1979).

159. F F Gardner and J B Whiteoak, *Mon. Not. R. Astron. Soc.* **201,**13P (1982).

160. W A Baan, P A Wood, A D Haschick, *Astrophys. J. Lett.* **260**, L52 (1982).

161. F F Gardner and J B Whiteoak, *Mon. Not. R. Astron. Soc.* **201,**13P (1982).

162. M J Claussen, G M Heiligman and K Y Lo, *Nature* **310**, 298 (1984).

163. M J Claussen, G M Heiligman and K Y Lo, *Nature* **310**, 298 (1984).

164. A review paper on megamasers is K Y Lo, *Ann. Rev. Astron. Astrophys.* **43,** 625 (2005).

165. M J Claussen and K Y Lo, *Astrophys. J.* **308**, 592 (1986).

166. W D Watson and B K Wallin, *Astrophys. J. Lett.* **432**, L 35 (1994).

167. M Miyoshi et al., *Nature* **373**, 127 (1995).

168. N Nakai, M Inoue and M Miyoshi, *Nature* **361**, 45 (1993).

169. M Myoshi et al., *Nature* **373**, 127 (1995).

170. C J Lonsdale, P J Diamond, H E Smit and C J Lonsdale, *Nature* **370**, 117 (1994).

171. A M Koekenoer, C Henkel, L J Greenhill, A Dey, W van Breugel, C Codella and R Antonucci, *Nature* **378**, 697 (1995).

172. J Herrnstein et al., *Nature* **400**, 539 (1999); see also C M Violette Impellizzeri et al., *Cosmic Masers—From OH to Ho—Proceedings IAU Symposium* no. 287, 2012;

R S Booth, EM L Humphreys and W H T Vlemmings eds. (Cambridge University Press: Cambridge, 2012), p. 311.

173. See, for example, C Henkel et al. *Cosmic Masers—From OH to Ho—Proceedings IAU Symposium no. 287*, 2012 R S Booth, EM L Humphreys and W H T Vlemmings, eds. (Cambridge University Press: Cambridge, 2012), p. 301.

174. M Elitzur, D J Hollenbach and C F McKee, *Astrophys. J.* **346**, 983 (1989).

175. J Crovisier et al., *Astron. Astrophys.* **393,** 1053 (2002).

176. F H Mies, *Astrophys. J. Lett.* **191**, L145 (1974).

177. F Biraud et al., *Astron. Astrophys.* **34,** 163 (1974).

178. B E Turner, *Astrophys. J. Lett.* **189**, L137 (1974).

179. V S Strel'nitskii, *Sov. Astron. Lett.* **9**, 99 (1983).

180. L E Snyder, *Astron. J.* **91**, 163 (1986).

181. See, for example, A P Graham, B J Butler, L Kogan, P Palmer and V Strel'nitski, *Astron. J.* **119**, 2465 (2000).

182. J Galt, *AJ* **115**, 1200 (1998).

183. M Gray, *Maser Sources in Astrophysics* (Cambridge University Press: Cambridge, 2012).

5

The Laser

5.1 Introduction

Once the maser had been invented, the idea went around that what was then referred to as an *optical maser*—a light generator based on stimulated emission—could be similarly constructed. In both, the United States and the Soviet Union, researchers were working independently on the problem, Among those in the forefront of this work were, in the USSR, V A Fabrikant and, in the United States, Robert H Dicke (1916–1997). Fabrikant, as we saw in Chapter 4, had first made his proposals back in the 1940s.[1] In 1954, Dicke, as we shall see in Section 5.5, published a paper introducing the concept of *superradiance*,[2] which laid the foundation for his subsequent discussion of the "optical bomb."[3] Dicke, in 1956, suggested the use of the Fabry—Perot interferometer as the resonant cavity and, in 1958, patented this proposal.[4]

It was not until later that concentrated efforts toward using stimulated processes to obtain visible radiation were made. In the Soviet Union, this research was mainly performed at the Lebedev Institute under Basov and Prokhorov, while in the United States the main workers were Gordon Gould (1920–2005)—whose story will be told in Section 5.4—and Charles H Townes and Arthur L Schawlow. The latter were the first to publish detailed and exhaustive proposals which subsequently led to the construction of various types of laser, and so, although the first lasers to become operational were not of the kind which had been considered by them, we shall begin by describing their proposal.

5.2 The Townes and Schawlow Proposal

In 1957, Charles H Townes began to consider the problems connected with making maser-type that devices that work at optical wavelengths. Townes carried out his work in close collaboration with Arthur L Schawlow, then a research physicist at Bell Laboratories.

Arthur L Schawlow was born in Mount Vernon, New York, in 1921 and obtained his BA, MA, and PhD in physics in Toronto, Canada. Soon after the war, he wrote to I I Rabi at Columbia University, who suggested that he apply for a postdoctoral fellowship to work under Townes. This fellowship was given by Carbide and Carbon Chemical Corporation, a division of Union Carbide, for research on the application of microwave spectroscopy to organic chemistry. Schawlow and Townes became friends, often dining together at the Columbia Faculty Club where there was a table reserved for a group of physics and mathematics professors. Schawlow later married Townes' sister.

In 1951, Schawlow accepted a post working in solid-state physics at the Bell Laboratories, where he became interested in superconductivity and nuclear resonance. At this time, he was also writing a book with Townes on microwave spectroscopy[5] and spent nearly every Saturday at Columbia.

Many accounts of his research and his warm and inspiring relationship with his students and coworkers were compiled in the book *Lasers, Spectroscopy, and New Ideas*, which was in honor of his 65th birthday.[6]

In 1957, Schawlow began to think about the possibility of building some kind of infrared maser and writes:[7]

A few weeks later, about October 1957, Charles Townes visited Bell Labs, and we had lunch. Townes had been consulting with the Laboratories for about a year, but his contacts were with the maser people and I had not had any serious discussion with him. He told me then that he was interested in trying to see whether an infrared or optical maser could be constructed, and he thought it might be possible to jump over the far infrared region and go to the near infrared or perhaps even the visible portion of the spectrum. He had some notes and said he would give me a copy. We agreed that it might be worthwhile for us to collaborate on this study and so we began.

This story is confirmed also by Townes who writes:[8]

I discovered that my friend Arthur Schawlow, then at the Bell Telephone Laboratories, had also been thinking along somewhat similar lines, and so we immediately pooled out thoughts. It was he who initiated our consideration of a Fabry–Perot resonator for selection of modes of the very short electromagnetic waves in the optical region.

This very likely had something to do with the fact that Schawlow had first been trained as a spectroscopist and had done thesis with a Fabry-Perot . . . .

So, Townes decided to give his notes to Schawlow. In these notes, calculations had been made using thallium atoms, which, it was intended, should be excited from the ground sate (6p) to a higher one (6d or 8s) through the ultraviolet light of a thallium lamp. Schawlow writes:[7]

Such lamps were in use in Kusch's laboratory at Columbia University for experiments on optical excitation of thallium atoms in an atomic beam resonance experiment.

Townes had discussed with Gordon Gould, a student of Kusch's who was working on the atomic beam experiment, the properties of thallium lamps to find out how much power could be expected from them.

On September 14, 1957, Townes asked a Columbia graduate student, Joseph Antony Giordmaine, to sign a notebook wherein a light resonator was described: this consisted of a glass box with four mirror walls and used a thallium lamp to energize thallium inside the cavity. Schawlow quickly demonstrated that this thallium scheme of Townes would not have worked easily and so they started to search for other materials. In the meantime, Townes calculated the number of excited atoms needed (the equations later being published[9]) and performed some experiments. They then turned their attention to the cavity to be used.

As Townes' report[8] mentioned, Schawlow made an important contribution to the final choice (a Fabry–Perot-type cavity). Schawlow himself recalled[7] that when he had been a student in Toronto, he had become familiar with the use of the Fabry–Perot interferometer during some research he carried out on the hyperfine structure in atomic spectra under the direction of Professor Malcolm F Crawford:

> I had in mind from the beginning something like the Fabry–Perot interferometer I had used in my thesis studies. I realized, without even having looked very carefully at the theory of this interferometer, that it was a sort of resonator in that it would transmit some wavelengths and reject others.

Later he wrote some of his deliberations in a notebook. On January 29, 1958, he asked Solomon L Miller, a graduate student of Townes' working at Bell Laboratories (who later went to IBM) to sign this notebook. Somehow Schawlow had understood that a good resonator, which reduced the large number of modes existing in a cavity and therefore prevented the hopping from one mode to another during operation, was simply one made by two parallel plane mirrors some distance apart, without any other reflecting part.

During the spring, Schawlow and Townes decided to write up this work for publication. It is customary at the Bell Laboratories to circulate manuscripts among colleagues prior to publication in order to obtain technical comments and improvement. A copy was also sent to Bell's patent office to see whether it contained any invention worth patenting.

As a result of this process, colleagues asked them to write more on modes but the patent office at first refused to patent either their amplifier or their optical frequency oscillator because[8] "optical waves had never been of any importance to communications and hence the invention had little bearing on Bell System interests." However, upon Townes' insistence, a patent request was filed and it was delivered in March 1960.[10] The paper itself was received on August 26, 1958, and published in December of the same year in the *Physical Review*.[9] Its authors were later awarded the Nobel Prize; C H Townes in 1964, as we have seen, for his invention of the maser and proposal for the laser, and A L Schawlow in 1981 for a related subject: laser spectroscopy.

5.3 Townes' and Schawlow's Idea

In their paper entitled *Infrared and Optical Masers,* Schawlow and Townes observed that, although it was possible in principle to extend the maser techniques into the infrared and optical region to generate very monochromatic and coherent radiation, a number of new aspects and problems arise which require both a quantitative re-orientation of the theoretical discussion and a considerable modification of the experimental techniques used.

The declared purpose of the paper was to discuss the theoretical aspects of maser-like devices for visible or infrared wavelengths and to outline the design considerations, so as to promote the realization of this new kind of maser, called by them optical maser (later to be called laser, where "L" stood for light). The principal points considered were the choice of a cavity and its mode selection properties, the expression of the gain of the device, and some proposals on active materials.

The most immediate problem was the realization of a resonant cavity. In the case of the maser, an ordinary microwave cavity with metallic walls has been used. By suitable design of such a cavity, it was possible to obtain just one resonant mode oscillating near the frequency corresponding to the radiative transitions of the active system.

In order to obtain such a single, isolated mode, the linear dimension of a cavity needs to be in the order of one wavelength which, at infrared frequencies, would be too small to be practical.[11] Hence, it is necessary to consider cavities of dimensions large compared with a wavelength and which can therefore support a large number of modes within the frequency range of interest. Townes and Schawlow realized then that it was necessary to increase selectively the Q of only certain modes. After some general considerations, the authors' choice was a Fabry–Perot interferometer[12] constructed by two perfectly reflecting plane parallel end walls. In it, a plane wave travels back and forth between the two mirrors and if the relation $L = q(\lambda/2n)$ is fulfilled (where λ is the vacuum wavelength, n is the refractive index, q some integer number, and L the distance between the mirrors), stationary waves are created. The paper deals in depth with the choice of the cavity and its mode selection properties—mode hopping, mode instability due to changes in cavity dimensions, and mode competition were clearly identified phenomena.

Having, it was hoped, generated by this process a light signal, it would then have to be collected. It was suggested that one of the mirrors be partially reflecting, so as to allow a beam of light incident on it to be partially transmitted outside the cavity, The idea of adopting a cavity of dimensions much larger than the wavelength had been suggested not only by the practical impossibility of constructing one of the correct dimensions for resonance, but also by the fact that the resonator should contain a reasonable quantity of active material. As discussed later in Section 5.5, other authors had also proposed the use of such a cavity.

Another problem treated in the paper was the determination of the minimum number of molecules or atoms of the active material which should be at the higher energy level to allow the generation of light by stimulated emission. This problem was simplified by considering a material with only two energy levels, E_1 and E_2, with $E_2 > E_1$, and with a population N_1 and N_2, respectively.

The condition for oscillation was obtained by requiring that the power produced by stimulated emission is as great as that lost on the cavity walls or due to other types of absorption. That is,

$$(\mu'E/\hbar)^2(h\nu N/4\pi\Delta\nu) > (\overline{E^2}/8\pi)(V/t), \qquad (5.1)$$

where μ' is the matrix element for the emissive transition, $\overline{E^2}$ is the mean square of the electric field, $N = N_2 - N_1$, V is the volume of the cavity, t is the time constant for the rate of decay of the energy, and $\Delta\nu$ is the half-width of the resonance at half maximum intensity if a Lorentzian shape is assumed.

The decay time was simply calculated considering a plane wave reflected back and forth many times between the two mirrors of the cavity at a distance D from each other. The wave undergoes reflection every time it travels a distance D, and the rate of loss of energy W is given by the equation

$$dW/dt = -c(1 - \alpha)\,W/D, \qquad (5.2)$$

where α is the reflection coefficient of the cavity wall. The solution of Equation 5.2 is

$$W = W_o \exp\left[-c(1 - \alpha)\,t/D\right],$$

from which the decay time t results,

$$t = D/c(1 - \alpha).$$

The condition for oscillation may be conveniently related to the lifetime τ of the state due to spontaneous emission of radiation by a transition between the two levels in question. This lifetime is given by the well-known theory as

$$\tau = hc^3/64\pi^2\nu^3\mu'^2.$$

Now, the rate of stimulated emission due to a single quantum in a single mode is just equal to the rate of spontaneous emission into the same single mode (cf. Equation 2.6).

Hence the rate $1/\tau$ is multiplied by the number of modes p which are effective in producing spontaneous emission. Assuming a single quantum present in a mode at the resonant frequency, the condition for instability can then be written as

$$N h\nu/p\tau > h\nu/t,$$

or

$$N > p\tau/t,$$

which is equivalent to expression 5.1.

The minimum power that must be supplied in order to maintain N systems in excited states is finally derived as

$$P = Nh\nu/\tau = ph\nu/t.$$

The number of effective modes in this expression is

$$p = \int p(\nu)f(\nu)\, d\nu,$$

where $f(\nu)$ is the line profile, and $p(\nu)d\nu$ is the number of modes between ν and $\nu + d\nu$, which is

$$p(\nu)d\nu = 8\pi\nu^2 V\, d\nu/c^3.$$

For a Lorentzian line shape, it is

$$p = 8\pi^2\nu^2 V\Delta\nu/c^3.$$

For a line broadened by Doppler effects, it is

$$p = 8\pi^2\nu^2 V\Delta\nu/(\pi\ln 2)^{1/2}c^3.$$

The minimum power is therefore, for example, for a Lorentzian line,

$$P = 8\pi^2 h\nu^3 V\Delta\nu/c^3 t,$$

which shows an important property—the scaling of pumping power with ν^3. Monochromaticity of a maser oscillator was also considered, and Schawlow and Townes observed that this property is very closely connected with the noise properties of the device as an amplifier.

The noise was conceptually analogous to the one present in the radiowave range, but of a different nature. In an electronic oscillator, noise is essentially thermal and is spread over the whole frequency range of the useful band of the oscillator. In the laser, noise consists of the spontaneous emission of radiation by the active material: Schawlow and Townes, by modifying the calculation previously developed for masers, discussed in their paper the interaction between an atomic line and one mode of the cavity, whose frequency coincided approximately with the center of that atomic line.

They obtained in this way an analytic expression for the linewidth and found it was in the order of one-millionth of the linewidth corresponding to the cavity mode. This gave an estimate of the spectral purity and demonstrated the excellent monochromaticity of the laser radiation. They obtained as a theoretical lower limit for the noise linewidth

$$\delta\nu = 4\pi h\nu(\Delta\nu)^2/P,$$

which is analogous to that of the maser, where the thermal noise, kT, is replaced by the noise resulting from the stimulated emission of radiation. It is interesting to observe that they did not note that in the optical case the quantity $\Delta\nu$ refers to the half-width of the resonant mode and not to the linewidth of the spectral line, which is generally much larger.

A paragraph in the paper was then devoted to discussing a specific example using optical pumping. As an example of a particular system for an infrared maser, they considered atomic potassium vapor pumped at 4047 Å. Schawlow had even performed some preliminary experiments on commercial potassium lamps and had asked Robert J Collins, a spectroscopist at Bell Laboratories, to measure the power output of these lamps. Another possibility was cesium, which they proposed could be excited with a helium line.

They also considered solid-state devices, although they were not very optimistic. In this case, they wrote

> The problem of populating the upper state does not have as obvious a solution in the solid case as in the gas. Lamps do not exist which give just the right radiation for pumping. However, there may be even more elegant solutions. Thus it may be feasible to pump to a state above one which is metastable. Atoms will then decay to the metastable state (possibly by non-radiative processes involving the crystal lattice) and accumulate until there are enough for maser action. This kind of accumulation is most likely to occur when there is a substantial empty gap below the excited level.

The work by Schawlow and Townes created considerable interest and many laboratories started to search for possible materials and methods for optical masers.

Townes, with his group at Columbia, started efforts to build an optical maser with potassium. He worked with two graduate students, Herman Z Cummins and Isaac Abella.

At the same time, Oliver S Heavens, later Professor of Physics at York University, York, UK, who was then already a world expert on highly reflecting mirrors, joined the group. Townes had in fact realized that cavity mirrors were the most delicate point of the system. The system they were studying was a tube filled with the gas which was energized by an electronic discharge. The tube had internal mirrors to make the resonant cavity and we may today ascribe their operational failure as being due to the degradation of the mirror coatings owing to bombardment from the ions of the electrical discharge in the tube. A report on their work was published in 1961.[13] The cesium laser, which was also considered in the paper, was later realized by S Jacobs, G Gould, and P Rabinowitz in 1961.[14] Schawlow, at Bell, had begun to consider ruby as a possible solid-state material. Others, too, were considering ruby: in Japan, Saturo Sugano and Y Tanabe; Irvin Weider at Westinghouse Research Laboratories; and Stanley Geschwind at Bell Laboratories.

In 1959, Schawlow[15] suggested using ruby, but observed that the R lines were not useful for laser action:

> There is a broad absorption band in the green and others in the ultraviolet. When excited through these bands, the crystal emits a number of sharp bands in the deep red (near 7000 Å). The two strongest lines (at 6919 Å and 6934 Å) go to the ground state, and are not suitable for laser action. However, the strongest satellite line (at 7000 Å) … goes to a lower state which is normally empty at liquid helium temperatures, and might be usable … The structure of a solid state maser could be especially simple. In essence, it would be just a rod with one end totally

reflecting and the other end nearly so. The sides would be left clear to admit pumping radiation.

In the same work, Schawlow quoted the studies by Ali Javan at Bell Laboratories on the energy transfer in collisions between two kinds of gases in a mixture which could give rise to the necessary population inversion.

5.4 The Gordon Gould Story

Schawlow and Townes were not the only ones who foresaw the potential of an optical maser and who had faith that methods would soon be found for the creation of a medium with a negative absorption coefficient in the optical range.

Gordon Gould (1920–2005),[16] who after an MSc at Yale in 1943 had been a graduate student at the Radiation Laboratory of Columbia University where Townes taught, filed a patent application in the United States on April 6, 1959, as evidenced by a series of British patents[17] based on his US patent application. These disclose material similar to that in the Schawlow–Townes patent. Although Gould did not publish his findings in the customary manner, his patent disclosures are interesting because they formed the basis of a legal contest concerning the invention of lasers.

At 21, Gordon Gould earned a Bachelor in physics in 1941 from Union College, and in 1943 a Master in optical spectroscopy at Yale University, where he learned how to use the Fabry–Perot interferometer. When he completed his military service, he decided to dedicate himself to the inventing profession and found part-time work. He began by designing a type of contact lens and other items, including attempts to obtain synthetic diamond. However, at the end, he concluded that to continue he needed a more solid scientific base, and in 1949 enrolled at Columbia University, where in 1951 he started his work for a PhD under Professor Polykarp Kusch. The thesis was concerned with the use of thallium molecular beams of which the excited energy levels were to be studied. By illuminating the thallium atoms with light from suitable lamps, he first excited them into the desired level and then examined how they decayed from this state, with what efficiency the state was populated, and so on. However, the work progressed very slowly, so even in November 1957 Gould still had to write his thesis.

In fact, he had been interested in a project to build an optical maser, which he renamed *laser* by substituting the "m" of maser that stands for microwaves, with the "l" that stands for light. When the first lasers were built, Bell Telephone did not like the name, refused to use it, and did what they could to impose the name *optical maser*, without success.

In October 1957, according to the Gordon Gould deposition at a trial (TRG Inc. versus Bell Laboratories, for the priority of laser invention) he was developing the possibility of using the Fabry–Perot arrangement to use as a laser resonator and he had some talk with Townes in October 25 and October 28. Townes, who had his office near Gould on the 10th floor of the Physical

Building at Columbia, wanted some information on the very bright thallium lamps that Gould had used when he was research assistant at Columbia Radiation Laboratory.[18] These conversations were registered in Townes' notebook. They made Gould very excited and led him to rush to finish his studies as quickly as he could. On Friday November 16, 1957, Gould and his wife, Dr Ruth Frances Hill Gould, Assistant Professor of Radiology at Columbia and graduate in Physics at Yale, went to the proprietor of a candy store (who was a public notary and a friend of Gould's wife and of his family) who then put his seal on the first nine pages of Gould's laboratory notebook, which contained the work *Some Rough Calculation on the Feasibility of Laser Light Amplification by Stimulated Emission of Radiation.*

In the notes—more than 100 pages—he suggested to contain the active substance in a tube 1-m long that was terminated with two reflecting mirrors (that is a typical Fabry–Perot structure). He was also considering the possibility of placing the mirrors externally to the tube, by closing its extremities with two very homogeneous glass windows with an accurately worked planar surface at a fraction of wavelength oriented at the Brewster angle to minimize losses. Gould derived the conditions for maser oscillation, finding the right result, mentioned optical pumping as a possible excitation method and as a possible medium he suggested vapor of an alkali metal, quoting as an example potassium vapor, and then ruby or some rare earths. He also quoted pumping by collisions in a gaseous discharge, mentioning the helium–neon mixture as one of the possible gaseous media to excite. He then went on to discuss a large series of applications to spectrometry, interferometry, photochemistry, light amplification, radar, communications, and nuclear fusion. In the list ranged frequency and length standards made with lasers, profile measurement systems, material treatment, techniques to make holes or cut materials and activation of chemical reactions, all made by means of laser light.

By an irony of fate, it was Townes who introduced Gould to the use of a signed notebook as a method of establishing priority of claim to an invention! Gould's obsession for the laser had already cost him a great deal. His thesis advisor Professor Polykarp Kusch, according to Gould, would never let him substitute his thesis work for this subject, so in March 1958 Gould left Columbia without finishing the thesis and joined Technical Research Group (TRG).

TRG was one of those American companies that came about during the Cold War whose focus was defence and military contracts. It was founded as Technical Research Group in 1953 by three men who each had earned a doctorate, one in electronics, another in physics, and the third in applied mathematics, and at first operated as a consulting agency. In 1955, laboratories and workshop were added. The principal work concerned antennas and radar, nuclear reactor physics, and missile guidance, but it had also a small contract on masers and a programme on atomic frequency standards. Gould was assumed to work on this last project. Having not yet finished his thesis, TRG assured him some free time until July 1958 to complete his work. Gould, however, used this time not to work on his thesis but to work on his laser project.

In September 1958, Lawrence Goldmuntz, president of TRG, became aware that Gould was wasting his time working on a private project. Gould and Goldmuntz discussed the research on the laser and TRG took on Gould's project as its own. On December 16, 1958, Goldmuntz asked for $200,000 from the Aerojet-General Corporation of El Monte, California, which at that time owned 18% of TRG. A further $300,000 was also requested from the Advanced Research Projects Agency (ARPA) of the Pentagon to enable Gould's work on the laser to be used for optical radar, range finding, and communication systems.

By coincidence, Gould and Townes, at almost the same time, each had in their hands each other's work. Gould received a preprint of the Schawlow and Townes paper from Dr Maurice Newstein, a researcher at TRG: Townes, as an adviser to the federal government, was reading the 200-page proposal submitted to ARPA. Other people too saw the latter and gave favorable reply to it: TRG received $998,000 for the project. The increase to the required support was (and still is) very rare, not to say unique; however, the agency was heavily engaged in the problem of antiballistic missile defence, and because the laser, even though it did not yet exist, was one of the possible means one thought could be used, the agency asked TRG to study contemporarily the development of all kinds of proposed lasers and increased the support to $998,000, classifying the work. This was a great misfortune for Gould who was unable to work on this project at TRG because he had not got the necessary security clearance. In fact, during the World War II he was a member of a Marxist group and this was sufficient to deny him permission. As a consequence, he could not lead the project, nor read the reports or participate directly in the experimental work. He was only the internal consultant of the research team.

On April 6, 1959, Gould and TRG filed a patent application[19] for the laser in the USA, after which followed a series of requests for British patents which were granted. Schawlow and Townes had already filed their application in July 1958 and had the patent issued in March 1960. Gould and TRG appealed to the US Customs and Patent Appeals Court arguing that though their request had been made after that of Schawlow and Townes, Gould had the idea first. The principal proof was the notebook Gould had sealed on Friday November 16, 1957, after speaking with Townes. The trial was lost on December 8, 1965.

At the trial held in December 1965 in Washington, Dr Alan Berman, a physicist friend and bridge partner of Gould, who was associate director at Hudson Laboratories Dobbs Ferry, Columbia, testified to a conversation he had with Gould in August 1958 at a beach party at Fire Island, New York. Gould at that time was with TRG and Berman said he had noted Gould was working on the laser to the detrimental of his thesis. He was annoyed at him by observing he did not make any attempt to publish it in properly defined channels, as all physicists do, and added he thought this was an unscientific thing to do. It is worth nothing that if Gould had followed Berman's advice he could have sent the *Physical Review* a paper that could have been published simultaneously with the one by Schawlow and Townes!

Gould presented publicly his ideas only later in one of the first meeting in which one spoke of lasers, the Ann Arbor Conference on Optical Pumping, organized by Peter Franken and Richard Sands at the University of Michigan in June 1959. There Gould presented a paper "The LASER: Light Amplification by Stimulated Emission of Radiation" in which for the first time the name laser was publically used.

The Gould story does not end here. On October 11, 1977, after years of effort, he finally received his patent for an optical pumped laser amplifier. The patent arrived after 18 years of waiting: a record perhaps if one considers that three or four years is the usual waiting time! When the Gould patent went into operation those of Schawlow and Townes, issued in 1960, ended (in the USA, patents at the time ran for 17 years). A second patent staking out three broad claims on laser applications was issued on July 17, 1979, to Gould by the United State Patent Office. Gould in the meantime had left TRG and given the management of his long-awaited patent to Refac Technology Development Corporation of New York which started a campaign to demand royalties, but the laser manufacturing companies, after paying royalties to Bell for nearly 20 years, had no intention of paying any money to Gould and another litigation began which eventually favored Gould.

The laser-patent war raged for 30 years. Had Gould been given good legal advice then, he could have applied for a patent before Townes and Schawlow even at the time of his notes in November 1957 and certainly would have been granted it. However, as we have seen, Gould did not surrender and with TRG received British patents for several different phases of laser technology. These patents never made Gould rich, but they did fuel his desire to keep his American claims alive. When Control Data Corp. bought TRG in the early 1960s after liquidating the subsidiary's assets, Gould talked the company into allowing the contest patent rights to revert to him. In the meantime he remained active in the laser world as a professor at the Brooklyn Polytechnic Institute until 1973, when he left to found Optelecom, Inc. of Gaithersburg, Maryland, an early manufacturer of fiber-optic data links.

The same year, the US Court of Customs and Patent Appeals, in a suit over Q-switch (a technique to produce single laser pulses; see Chapter 7) patents, decided that the Schawlow–Townes patent did not adequately describe optical pumping of a laser medium. During the same period, Gould decided to trade a half-interest in his pending laser patent to the Refac Technology Development Corp, a New York patent-licensing firm headed by Eugene Lang, in exchange for absorbing his legal costs.

So it was that in October 11, 1977, the Patent Office, 18 years after the original application was filed, issued Gould a patent on optical pumping of lasers (no. 4,053,845) (**Figure 5.1**) and Refac immediately notified manufacturers that they would have to pay royalties on optically pumped lasers. Refac asked royalties ranging from 3.5% to 5% that generated more than $1 million per year in licensing fees in the sales of solid-state lasers alone, not to speak of other types of lasers that could have been covered by the patent. In its entire 17-year life, Townes' optical maser patent generated only $1 million in royalties because Bell according to an agreement with the government had agreed to ask only for low-level royalties on its patents.

Therefore, the laser producers resisted this request and Refac, barely a week after the patent was issued, filed a suit against Control Laser Corp, a leader in

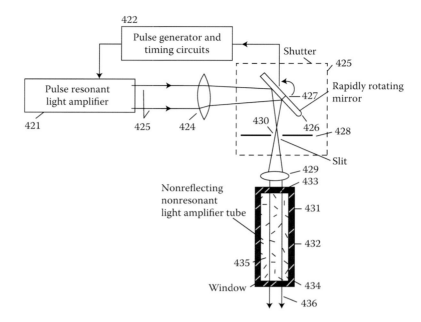

FIGURE 5.1 A design from a Gordon Gould patent.

resisting the patent claims. Seven other laser companies joined Control Laser to fight the Gould patent.

The following year, Gould received a second patent on a broad range of laser applications including machining. News of that patent sent Refac's stock up $10 per share to $34, making Lang's 56% of the stock worth some $40 million. Speculators did not wait for the courts to act on the patents. Gould sold part of his remaining interest to what seemed an unlikely buyer, the Panelrama Corp of Ardmore, PA, a public company that had operated retail building-support stores. The company changed its name to Patlex and bought 20% of Lerner David Littenberg and Samuel, a Westfield, New Jersey, law firm that represented Gould's interests for a 25% stake in the patents.

One of the arguments that the laser companies brought to the court in trying to resist Gould's patent was that the information he gave was insufficient to build a laser. So Gould at Optelecom Inc., in 1981, together with some other researchers and with the financial support of Patlex, employing a commercial sodium lamp manufactured by General Electric for use as a street light to pump an organic dye, rhodamine B, built a laser using only the design information disclosed in his 1959 patent application and other data readily available at that time. This result obviously dealt a blow for the companies that opposed the patent.

The first case to come to trial was against the tiny General Photonics. The company could not afford much defence, and on March 1, 1982, a federal judge ruled it owed royalties on the optical pumping patent. Soon afterward, however, patent opponents took advantage of new regulations that allowed the Patent Office to re-examine claims of patents that had already been issued. They were happy in early 1983, when the Patent Office denied Gould's claims. However, Gould was very stubborn and returned to court and succeeded in reinstating his claims. That proved the decisive victory, and long-delayed infringement cases, finally came to trial. In March 1987, Patlex won its case against Cooper LaserSonics, a Palo Alto, CA, company formed by the merger of several smaller laser companies. In the same month, Lumonics settled three infringement suits out of court. Other companies settled with Patlex.

The final victory came in October 1987 in the long delayed Control Laser case. Days later, Gould received his third patent, on gas-discharge lasers. A fourth, on Brewster angle windows for lasers, followed in 1988.

Soon everyone was talking with Patlex. The fate of the eight companies that initially fought the optical pumping patent shows how Gould's forces won the patent war. Patlex bought the assets of General Photonics after it filed for bankruptcy and acquired Apollo Lasers in a stock exchange with Allied Chemical, which earlier had paid $10 million for the company. Hadron and National Research Group quietly faded away. Financial problems including court costs and licensing fees helped force the sale of Control Laser to Quantronix. Quantronix, in turn, fell victim to a market slump and was brought by Excel. Spectra-Physics has also gone through tough times, including a traumatic sale, but it remains a major player in the laser industry. Only Coherent remained an independent company.

Patlex merged in December 1992 with AutoFinance Group Inc., a financial-service company. Revenue from laser patents was about $7.5 million in the year ending June 30, 1992, and roughly $7 million for the following year. Gould received far more than he would have received if his first American patent had been issued promptly. That patent on optical pumping expired on August 11, 1993. The applications patent expired in July 1996 and the discharge-pumping and Brewster window patents expired in the early 2000s.

To people who want to ask how much Townes' and Gould's ideas had developed from common information, considering that both were at Columbia and knew each other well, one may give an answer by making two considerations. The first is that an idea needs a receptive substrate to fully develop, that is all the general thoughts must be developed before retaining another person's idea and being able to work on it advantageously. In other words, ideas take root only in prepared minds. Even if talking to Townes had given Gould the idea that it is possible to excite atoms by optical pumping, Gould would already have developed the concept of using a population inversion, an optical cavity and so on in order to capitalize on them in his laser project.

Secondly, if we examine how the two men developed the laser idea and specifically how they went about solving one of the principal problems, namely the resonant cavity, we see that the two proposed solutions are typical of their two different personalities. Townes, the inventor of the maser, a microwave expert, started by considering a cube, which is the typical shape for a microwave cavity, in which the radiation travels by reflecting between the walls, and only later on Schawlow's suggestion removed all the walls except two. Gould with an optical education, considered initially a cavity formed by a long tube (1 m) terminated by two plane parallel mirrors (essentially a Fabry–Perot) and then worked out all possible configuration with plane external mirrors, curved mirrors, total reflecting mirrors, and so on.

Gould was above all an inventor who, after having sketched in a notebook a draft of his idea, threw out a series of partially developed suggestions which were formalized in a proposal for a contract to assemble the device. Townes and Schawlow, with the minds of professional physicists, first thought about writing a paper to communicate their idea to the scientific world, not without having first patented it (do not forget they were supported by a commercial firm), and then worked theoretically on the details before dedicating themselves to experimental activity. There is therefore little doubt that also in this story the idea was born and developed independently and contemporarily by these three researchers.

5.5 The Dicke Coherence-Brightened Laser

Among those who have the right to be remembered as having paved the way for the invention of the laser, R H Dicke, deserves a particular mention. He not only developed novel and original ideas for the production of coherent radiation in the microwave and IR spectrum, without making recourse to the feedback concept, but also was among the first to suggest the use of a Fabry–Perot cavity. These two points are different and must be treated separately.

Dicke was a physics professor at Princeton University and a consultant at the RCA Laboratories at Princeton. He was interested in basic physical problems and made important contributions to the study of gravity, but amused himself as well by inventing things. His consultant position at RCA made him fell obliged to translate some of the ideas into patents or projects for the company laboratory. He was interested in methods of narrowing the emission line of the hydrogen atom to study the interaction of the electron with the nucleus and this study led him to examine more, in general, the properties of coherent radiation and to introduce the concept of *superradiance*.

Narrow lines were important to build frequency standards, and RCA had a contract with the Army Signal Corps based on Dicke's idea of coherent emission. Later in 1960, he became interested in gravitation and in 1964 in the Big Bang theory, encouraging his colleague Peebles to calculate the blackbody radiation temperature of the great explosion (the Big Bang).

As a proponent of new coherent sources, Dicke, in the paper *Coherence in Spontaneous Radiation Processes*, published in the January 1954 issue of the *Physical Review*,[2] treated for the first time spontaneous emission of electromagnetic radiation from an atomic (or molecular) system in a correlated state. In this paper, Dicke developed the concept that molecules do not radiate independent of each other. This latter, simplified picture overlooks the fact that all the molecules are interacting with a common radiation field and hence cannot be treated as being independent. By considering a radiating gas as a single quantum mechanical system, energy levels corresponding to certain correlations between individual molecules were described. Using a point-source model, he defined collective energy eigenstates of two-level atoms.

The quantum numbers labeling these Dicke states were the energy quantum number M and the cooperation number J. If N is the total number of atoms in the sample, then $0 < J < N/2$ and $-J < M < J$. Dicke found that the initial radiation rate from a system in such a collective state is proportional to $(J + M)(J - M + 1)$.

A state with $J \sim N/2$ and $M \sim 0$ has the maximum emission rate, proportional to $N/2$. Such a state he called *superradiant*.

Two ways in which a superradiant state may be excited were described. The first is if all molecules are excited; the second is to have the gas in its ground state and irradiate it with a pulse of radiation. This latter excitation method was clearly related to Hahn's photon echo method, and Dicke made several references in his paper to magnetic resonance experiments.

A laser which does not employ mirrors in order to produce feedback amplification, but rather is a source of spontaneous emission of radiation with the emission process taking place coherently, was first discussed briefly by Dicke in a talk before the American Physical Society on January 23, 1953. It was also discussed at the *Fourth Congress of the International Commission on Optics* held in Boston in 1956, and published in the *Journal of the Optical Society of America*.[20]

In 1953, the words "maser" and "laser" having not yet been coined, Dicke called his device an *optical bomb* because he predicted that a laser of this type would be characterized by an unusual short and intense light burst.[3]

A sharp, intense source was also described in principle by Dicke in 1957 in a paper written in collaboration with R B Griffiths.[21] It was based on the use of resonance radiation coherently scattered from a collimated beam of atoms. Dicke observed that if the excitation were carried out with monochromatic radiation, the coherent re-radiation would simply be the coherent forward scattering familiar as a phenomenon of resonance fluorescence.

In the optical region, a rather simple technique for inducing coherent radiation is therefore to excite coherent forward scattering using resonance fluorescence. He also proposed the separation of the coherently scattered light from the exciting radiation either by placing the atomic beam in one arm of a Mach–Zehnder interferometer or by diffraction from several parallel atomic beam pencils. The latter technique was tested experimentally by observing hyperfine structure in light coherently scattered from a beam of sodium atoms. Apparently, this was the first experiment in forward scattering spectroscopy.[22]

Superradiance was observed much later by the MIT group of Feld and co-workers in 1973.[23] It received a large amount of theoretical[24] and experimental[32] attention. Before the Feld work, Abella et al.,[26] in a paper on photon echoes published in the *Physical Review* in 1966, described an experiment in which a dilute ruby crystal was found to emit spontaneously a short pulse of light, the photon echo, at a time τ_s after irradiation by two successive ruby-laser pulses separated by τ_s. In the very beginning of the paper, they said:

> The purpose of the first pulse is to excite a superradiant state exhibiting an oscillating macroscopic electric dipole moment. This dipole moment quickly dephases because of inhomogeneous crystal-field strains, and the atoms then radiate at the normal spontaneous emission rate. The second excitation pulse reverses the dephasing process so that the system rephases at the same rate at which it dephased. When the rephrasing process is complete, the macroscopic electric dipole moment is momentarily reformed, and the crystal emits an intense burst of light, the photon echo.

The second important contribution of R H Dicke comes from the patent *Molecular amplification and generation systems and methods* filed in 1956 and issued in 1958.[4] He was concerned with the exploitation of ammonia transitions for producing emission in the millimeter region, so considered shaping the source of the ammonia molecules into a circle around a microwave cavity and proposed to use this "circular maser" to generate waves in the region from 0.25 to 0.03 mm, which is in the infrared region. To make his maser work in this part of the spectrum, he substituted the microwave cavity with a couple of parallel mirrors, that is, with a Fabry–Perot and in his patent the use of the Fabry–Perot interferometer is described.

5.6 Soviet Research

FIGURE 5.2 Valentin Aleksandrovic Fabrikant (1907–1991). (From S G Lukishova, *JEOS:RP* **5**, 10045 (2010).)

As we have already observed, at the forefront among the proponents of the idea to build an optical generator that exploited the principle of stimulated emission was Valentin Aleksandrovic Fabrikant (1907–1991) who in the Soviet Union made his proposal in the 1940s.[27]

Fabrikant began his scientific career as a student at the Physics and Mathematics Faculty of the University of Moscow in 1929 with G S Landsberg. After taking his degree, he was employed in the Soviet Institute of Electronics where soon he became head of a laboratory. In 1932, his attention was focused on problems of optics and on the properties of electric discharges in gases. He published a series of papers in which he studied the spectral composition and intensity of radiation emitted during an electric discharge in a gas, investigating particularly the collision processes between the excited atoms of the discharge and the electrons and the energy transfers that occurred in these processes.

An atom or a very fast electron may strike another atom and transfer part of its energy which, if sufficient, is able to excite the target atom, raising it to an excited level. This is referred to as a *collision of the first kind*. There also exists a different kind of collision, called *collision of the second kind*, in which an atom which is already in an excited state collides with another atom which is in the ground state and gives it its energy. The result is that the first atom returns to the fundamental state, and the second atom jumps to an excited level. The two atoms need not necessarily be of the same kind; it suffices that the two excited levels have the same energy. If the two atoms are of different kinds, through this mechanism it is possible that atoms of one type, say A, excite atoms of a different type, say B, to an excited level in larger number than that which would occur through thermal collisions of the first kind. In this way, B atoms may show a distribution within the different energy levels that departs from the Maxwell–Boltzmann distribution and this possibility was exactly what eventually interested Fabrikant.

In 1939, he started to examine the possibility of obtaining populations of excited atoms larger in number than predicted by the Boltzmann distribution law and came to show that when radiation travels in a medium in which a population inversion has been realized, one should observe amplification rather than absorption. Then he proposed a way of realizing experimentally an inverted population in a discharge of a mixture of gases using collisions among the discharge atoms. These results were included in his doctorate thesis for full professorship that he defended in 1939 and published in 1940.[28] At this time, Fabrikant's interest was associated with obtaining an experimental proof of the existence of stimulated emission. Curiously enough in the Russian text of his thesis appears a section on "negative absorption." The thesis was later translated into English and reprinted in a book[29] containing both English translations and Russian originals of some important papers reprinted from original publications. However, the section on negative absorption does not appear in the paper note 29. The story is well exposed by Lukishova.[27]

In the 1940 paper, Fabrikant however wrote about possible experiments in a gas discharge for selectively destroying lower lying levels.

World War II interrupted Fabrikant's research which he resumed in 1945 at the Laboratory of New Light Sources of the All-Union Electro-Technical Institute (VEI).

He was considering the problem of negative absorption and on June 26, 1951, together with his former PhD student M M Vudynsky and the senior scientist F A Butaeva, submitted a request for a patent on a new method of amplification of light under the title "Method for amplification of electromagnetic radiation (ultraviolet, visible, infrared and radiowaves."[30]

On July 16, 1951, they submitted an addendum to their application. In September 1951, this application was sent for clearance and only on December 16, 1958, the application was classified and approved. The Russian patent office, in fact, was not willing initially to accept the request. Only later it was accepted.

In the patent was written (from Lukishova[27]).

A method for amplification of electromagnetic radiation is proposed which exploits the induced emission phenomenon analyzed theoretically by A Einstein in 1917. In this method the energy of the amplified radiation is not converted into other forms of energy. The method is suitable for amplification of ultraviolet, visible, infrared and radio frequency waves.

To implement the method a medium with a negative absorption coefficient for the radiation is produced. The radiative flux through such a medium increases in intensity—hence the amplification effect. The gain is given by $\exp(|K|L)$, where K denotes the absorption coefficient and L is the thickness of the layer.

A medium with a negative absorption coefficient results when its particles (for example, atoms or molecules) have a non-equilibrium distribution over energy levels. The concentration of particles in the upper energy state must exceed (with account for statistical weights) that at the lower states. As an example, a gaseous medium filling a suitable vessel is suggested, in which the required non-equilibrium condition is achieved, e.g., by an auxiliary radiation which excites particles to appropriate energy states; or by passing an electric current with a simultaneous use of impurities that selectively depopulate the lower states; or finally by passing a current modulated by using the ion–electron recombination effect to populate the upper state.

What we claim is a method for the amplification of electromagnetic radiation (ultraviolet, visible, infrared or radio wavebands), distinguished by the fact that the amplified radiation is passed through a medium, which by the means of auxiliary radiation or by other means, generates excess concentration, in comparison with the equilibrium concentration of atoms, other particles or their systems, at upper energy levels corresponding to excited states.

The patent was only published in 1959. The title is so general to cover practically anything connected with maser or laser action.

After the application for the patent in 1951, Fabrikant and his students[31] continued experimental work with different materials but without success. Even though they published experimental confirmation of their ideas, this was later disproved.

They experimented with, among other materials, cesium optically pumped with the 3889 Å He line, but they looked at the wrong transition.[32]

In their work at the Optical Institute of the Academy of Sciences, Fabrikant and Butayeva claimed to have obtained light amplification in a discharge of mercury vapor in which the transfer of excitation produced population inversion and negative absorption. However, Sanders et al.,[33] who had a translation of the work by Lengyel,[34] were not able to repeat their work.

Although Fabrikant's work is historically interesting because he developed the concept of the laser, arriving at it from the optical side without passing through the maser phase, it had no influence on the developments of either maser or laser because the work was known only after both devices had been realized. Even in Russia his colleagues probably did not take him very seriously and therefore he did not stimulate anybody even there. Only after the first masers and lasers were built was the Government reminded of him. In 1965, the Soviet Academy of Sciences awarded him with the S I Vavilov gold medal for *the important work on the optics of gas discharges* and he received the prize of the Soviet Socialist Republics for his contribution to the development of luminescent lamps.

Later on, Basov and Prokhorov at the Lebedev Institute in Moscow started research to extend the maser concept to optical wavelengths. Basov started his work on lasers in 1957 with research into physical methods of obtaining nonequilibrium states in semiconductors, as will be discussed later. The year after, A M Prokhorov[35] in a letter to the editor of the *Soviet Journal of Experimental and Theoretical Physics* in June 1958, pointed out the possibility of assembling a molecular generator and amplifier (MAG) for waves of wavelength shorter than a millimeter, that is, in the optical region, by using the rotational transitions of ammonia molecules.

The required population inversion was obtained by letting the molecular beam pass through a suitable quadrupolar separator, as in a maser; the proposed pumping was, substantially, nonoptical.

In his letter, Prokhorov proposed a theoretical example of how to realize the MAG, giving information regarding the negative absorption coefficient of the material for which amplification was obtained, and considering a cavity with mirrors for which calculation was made of the Q factor as a function of wavelength.

In 1959, N G Basov[36] proposed a method for producing light by exciting a homogeneous semiconductor with electrical pulses. In a semiconductor for a sufficient high strength of the electric field, due either to ionization or to tunneling effects, the concentration of nonequilibrium carriers distributed in a large energy band is strongly increased.

To obtain a population inversion and produce light, it is necessary to cut off the applied field rapidly, in a time much shorter than the lifetime of the nonequilibrium carriers. This situation is realized when Gunn domains propagate inside some semiconductors (see Chapter 6). In 1974, Basov obtained the effect in a somewhat different setup.[37]

Notes

1. V A Fabrikant, Doctoral Dissertation, FIAN P N Lebedev Physical Institute, Academy of Sciences, USSR (1939). Transactions of the All-Union Order of Lenin Electrotechnical Institute 41, Electron and Ion devices, pp. 236, 254 (1940).

2. R H Dicke, *Phys. Rev.* **93**, 99 (1954).

3. R H Dicke, *The Coherence Brightened Laser*, in *Quantum Electronics*, edited by P Grivet and N Bloembergen (Dunod: Psris, 1964) p. 35.

4. R H Dicke, Patent *Molecular amplification and generation systems and methods* requested in 1956 and obtained in 1958, US Patent 2,851,652 (9 September 1958).

5. The book was published later: C H Townes and A L Schawlow, *Microwave Spectroscopy* (McGraw-Hill: New York, 1955).

6. *Lasers, Spectroscopy, and New Ideas: A Tribute to A L Schawlow*, W M Yen, M D Levenson and A L Schawlow, editors (Springer: Berlin, 1987).

7. A L Schawlow, *From Maser to Laser* in *Impact of Basic Research on Technology* edited by B Kursunoglu and A Perlmutter (Plenum Press: New York, 1973) p. 113.

8. C H Townes, *Science* **159**, 699 (February 16, 1968).

9. A L Schawlow and C H Townes, *Phys. Rev.* **112**, 1940 (1958).

10. A L Schawlow and C H Townes, US Patent No 2,929,922. The filing date was July 30, 1958. The title *A Medium in which a Condition of Population Inversion Exists*. The patent was issued March 22, 1960.

11. Today it is not a problem to built cavities of the dimension of wavelength in the visible, but that was unimaginable at those times.

12. The very fine fringes used in the Fabry–Perot interferometer were discovered in the year 1893 by M R Boulouch (*J. Phys.* **2**, 316 (1893)). He prepared half silvered glass plates, observed the rings, and using Na light showed that for a given distance between the plates the rings of the D_1 and D_2 lines were separated. Boulouch was high-school teacher at the "Lycèe de Bordeaux" where during the years 1879–1880, Fabry was his colleague before being nominated at the University of Marseille where he developed with his colleague Perot what is today known as the Fabry–Perot interferometer. He never mentioned the priority of Boulouch. See S Tolansky, *Multiple Beam Interferometry* (Oxford, 1948) p. 10.

13. H Z Cummins, I D Abella, O S Heavens, N Knable and C H Townes, in *Advances in Quantum Electronics*, edited by J Singer (Columbia University Press: New York, 1961) p. 12.

14. S Jacobs, G Gould and P Rabinowitz, *Phys. Rev. Lett.* **7**, 415 (1961) and P Rabinowitz, S Jacob and G Gould, *Appl. Opt.* **1**, 513 (1962).

15. A L Schawlow, *First International Quantum Electronics Conference*, September 1959, Columbia (Columbia University Press: New York, 1960) p. 553.

16. The story of Gordon Gould has been written in the form of a tale with Gould's full co-operation by N Taylor, *Laser: The inventor, the Nobel laureate, and the thirty-year patent war* (Kensington Publishing Corp.: New York, 2000).

17. The contents of Gould's application of 1959 were eventually published by the Patent Office (No. 3,388,314) in June 11, 1968. The claims allowed in this patent regards "apparatus for generating radiation of frequencies higher than those of light" and are worthless, because the instrument constructed on these principles will not produce x-rays. However, in this patent are published Gould's ideas. The British patents are G Gould, Brit. Patent Specs. 953721-953727, published April 2, 1964.

18. Gould had used a thallium lamp to produce optically excited states in beams of thallium, G Gould, *Phys. Rev.* **101**, 1828 (1956).

19. The story of Gould's patents made the object of recurrent information on many scientific magazines. The information reported here come mainly from *Laser Focus* (April 1981) p. 6; (April 1981) p. 14; Vol. **24**, n. 5 (May 1988) p. 96; Vol. **30**, n. 12 (December 1994) p. 49; *Photonics Spectra*, Vol. **22**, n. 5 (May 1988) p. 74.

20. R H Dicke, *J. Opt. Soc. Am.* **47**, 527 (1957).

21. R B Griffiths and R H Dicke, *Rev. Sci. Instrum.* **28**, 646 (1957).

22. W Gawlick and G W Series, in *Laser Spectroscopy*, Vol. **4**, edited by H Walther and K W Rothe (Springer: Berlin, 1979) p. 210.

23. N Skribanowitz, I P Herman, J C MacGillivray and M S Feld, *Phys. Rev. Lett.* **30**, 309 (1973).

24. Although not exhaustive, a list of works can be found in: F Haake in *Laser Spectroscopy*, Vol. **4**, edited by H Walther and K W Rothe (Springer: Berlin, 1979) p. 451; R Glauber and F Haake, *Phys. Lett.* **68**, 29 (1978); F Haake, H King, G Schoeder, J Haus, R Glauber and F Hopf, *Phys. Rev. Lett.* **42,** 1740 (1979); F Haake, H King, G Schroeder, J Haus and R Glauber, *Phys. Rev.* **A20**, 2047 (1979); R Bonifacio, P Schwendimann and F Haake, *Phys. Rev.* **A4,** 302 (1971); R Bonifacio

and L A Lugiato, *Phys. Rev.* **A11**, 1507 (1975); *Phys. Rev.* **A12**, 587 (1975); M F H Schuurmans and D Bolder, *Laser Spectroscopy*, Vol. **4**, edited by H Walther and R W Rothe (Springer: Berlin, 1979) p. 459; G Banfi and R Bonifacio, *Phys. Rev. Lett.* **33**, 1259 (1974); *Phys. Rev.* **A22**, 2068 (1975); J C MacGillivray and M S Feld, *Phys. Rev.* **A14**, 1169 (1976).

25. D Polder, M F H Schuurmans and Q H F Vrehen, *J. Opt. Soc. Am.* **68**, 699 (1978); *Phys. Rev.* **A19**, 1192 (1979); H M Gibbs, Q H F Vrehen and H M J Hikspoors, *Phys. Rev. Lett.* **39**, 547 (1977); Q H F Vrehen and M F H Schuurmans, *Phys. Rev. Lett.* **42**, 224 (1979), etc. aggiornare.

26. I D Abella, S R Hartmann and N A Kurnit, *Phys. Rev.* **141**, 391 (1966).

27. See the paper by S G Lukishova, *J. Eur. Opt. Soc. Rapid Pub.* **5**, 10045 (2010) and E I Pogrebyskaya, Valentin Alexandrovich Fabrikant: social-political aspects of the biography of a physicist in *Scientific Society of Soviet Physicists*, 1950–1960 years. Documents, Recollections, Investigations, V P Vizgin and A B Kessenikh, eds. 1, 484 (Russian Academy of Sciences, S I Vavilov Institute for the History of Science and Technology, Russia Christian Academy, St-Petersburg, 2005) in Russian.

28. V A Fabrikant, The emission mechanism of a gas discharge in *Trudy (proceedings) of VEI* (the All-Union Electro-Technical Institute), *Electronic and Ion Devices* **41**, 236 (1940) in Russian.

29. V A Fabrikant, *Selected Works* (MEI Moscow Power Institute Publisher: Moscow, 2007) both in Russian and with English translation.

30. Materials of Soviet Patent application N 576749 (1951–1964), 87 pages (Archive, Agency on Patents and Trademarks, Institute of Industrial Property of Russian Federation Moscow) quoted in Lukishova.

31. V A Fabrikant, M M Vudynskii and F Butayeva, USSR Patent No 123209 submitted June 18, 1951 and published in 1959. Supplemented by USSR Patent No. 148441.

32. F A Butayeva and V A Fabrikant, Investigations in Experimental and Theoretical Physics (A memorial to G S Landsberg) (USSR Acad. Sci. Pubbl.: Moscow, 1959) pp. 62–70.

33. J H Sanders, M J Taylor and C E Webb, *Nature* **113**, 767 (1962); see also a footnote in W R Bennett Jr., *Appl. Opt. Suppl.* **1**, 24 (1962) and S Jacobs, G Gould and P Rabinowitz, *Phys. Rev. Lett.* **7**, 415 (1961).

34. See also B A Lengyel, *Am. J. Phys.* **34**, 903 (1966).

35. A M Prokhorov, *Zh. Eksp. Teor. Fiz.* **34**, 1658 (1958); *Sov. Phys. JETP* **7**, 1140 (1958).

36. N G Basov, B M Vul and Yu M Popov, *Zh. Eksp. Teor. Fiz.* **37**, 587 (1959); *Sov. Phys. JETP* **10**, 416 (1959); V V Bagaev in *Quantum Electronics* edited by P Grivet and N Bloemergen (Dunod: Paris, 1964) p. 1899.

37. N G Basov, A G Molchanov, A S Nasibov, A Z Obidin, A N Pechenov and Yu M Popov, *Sov. Phys. JETP Lett.* **19**, 336 (1974).

6

The First Lasers

6.1 Introduction

Immediately after the publication of the paper by Schawlow and Townes (see Chapter 5), a number of researchers started to think about different systems for the production of inverted populations in the infrared and visible regions. We shall see in the following sections that many different approaches were considered almost simultaneously and independently and that, in the case of semiconductors, considerations of the possibility of producing radiation through stimulated pair recombination even preceded the Schawlow and Townes discussion.

Of course the main lines were influenced by the ideas of these two researches and most people were expecting the first laser action to take place in an excited gas. However, not everybody was working on gases. It so happened that the first working laser was realized in May 1960 at the Hughes Research Laboratories by Theodore Maiman, using ruby as the active material.[1]

6.2 The Ruby Laser

The first laser was realized in May 1960 at the Hughes Research Laboratories, Malibù (Southern California) by Theodore H Maiman, using ruby as the active material.

Theodore H Maiman (1927–2007) was the young head of the quantum electronics division at Hughes. After supporting himself in college by repairing radios and other electrical appliances—an inclination he inherited by his father Abe, an electrical engineer—and then serving in the Navy, Maiman earned a BSc in engineering physics from Colorado University, an MSc in electrical engineering, and a PhD (1955) in physics from Stanford University, where his doctoral research[2] was in microwave spectroscopy under Professor Willis Lamb. Then he became a research scientist at Lockheed Aircraft for a short while, studying communication problems connected with guided missiles.

Later he accepted a position at Hughes Research Laboratories (HRL) in Culver City, California. At Hughes, he went to work in the newly formed Atomic Physics Department, where the principal interest was to generate higher coherent frequencies than were currently available. This was about the time that the ammonia maser came about. Hughes had an intense interest in maser research at that time. At first, however, Maiman worked on a different contract. When he finished this work, he had wanted to work in a fundamental research capacity, but the Army Signal Corps, which sponsored the research, required at that time a practical X-band (i.e., at a wavelength of 3 cm) maser. They did not want any state-of-the-art advances, but simply wanted that maser, and Maiman was asked to head the project. He was not very enthusiastic at first because the project involved a practical device and he was more research-oriented. But then he became more interested and even though they had not demanded any tremendous advances, he decided he could certainly make the maser more practical.

Masers at that time had two serious practical drawbacks. The main difficulty was that a solid-state maser, which is a more useful type, needs to work at very low temperatures. Indeed, liquid helium temperature was needed, that is, only 4 K. The other problem was that the conventional maser used a huge magnet, weighting about 2 tons, to obtain the Zeeman levels needed for the maser action. Inside this magnet was a dewar in which had to be poured liquid nitrogen in order to start lowering the temperature. Inside this dewar, another one was placed which was full of liquid helium. The real maser was a small microwave cavity, with the crystal in its interior, which was positioned in the liquid helium dewar between the pole faces of the magnet (see **Figure 6.1**). The magnet had to create a strong magnetic field within the whole volume occupied by the two dewars and the maser cavity, which justified its great size. The preferred maser material at that time was ruby. Maiman decided that there were a certain things he might be able to do still using ruby. He made a miniature cavity from ruby by cutting it into the shape of a small parallelepiped. Then he painted a highly conductive silver paint over the ruby and put a small hole in it. In this way, the ruby behaved as both the active material and the resonant cavity and so space was saved. He then decided that instead of putting the double dewar inside the monster magnet, he could put a small permanent magnet inside the dewar. It was thought that the magnet would crack and break although it

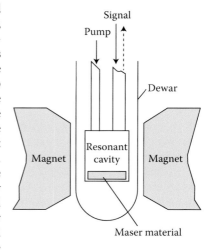

FIGURE 6.1 A typical set-up of a traditional solid-state maser.

actually worked nicely. So the whole thing, magnet, dewars, and everything, was less than 15 kg instead of 2 tons, and it performed technically much better and was much more stable than before.

Later, he made an even smaller maser that weighted less than 2 kg and developed a "hot" maser cooled at liquid nitrogen temperature and then one cooled at only dry ice temperature.[3]

Speculations about laser materials during the first half of 1960 centered around gases, specifically optically excited alkali vapors and noble gases excited in an electrical discharge. Maiman's achievement of the ruby laser came as a surprise, but it was not an accidental discovery. Having worked with ruby as a (microwave) maser material for some time, he wanted to see whether it could be used also in the optical region.

In the beginning, he performed some calculations but was discouraged because Irwin Wieder[4] had published a paper which indicated that the quantum efficiency of ruby (i.e., the number of fluorescence light photons emitted for every absorbed light photon) was only around 1%.

Ruby is a crystal of aluminum oxide (corundum Al_2O_3) with some Cr_2O_3 as an impurity in which some chromium atom loses three of its electrons and becomes a chromium ion that replaces one of the aluminum ions in the lattice. These chromium ions have a series of energy levels in the visible region which provide the material, which otherwise would be a transparent crystal, with a color between pink and dark red according to the quantity of chromium present.

A simplified energy-level diagram for triply ionized chromium in corundum is shown in **Figure 6.2**.[5] When the ruby is irradiated with light at a wavelength of about 5500 Å (green), chromium ions are excited to the $^4F_2$ state

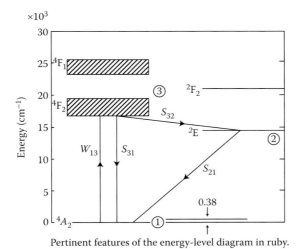

Pertinent features of the energy-level diagram in ruby.

FIGURE 6.2 Energy levels of the chromium ion in ruby. (From T H Maiman, *Phys. Rev. Lett.* **4**, 564 (1960).)

and then quickly lose some of their excitation energy through non-radiative transitions to the ^2E state (this state in reality is made by two very near levels). This state then slowly decays by spontaneously emitting a sharp doublet, the components of which at 300 K are at 6943 and 6929 Å (the R doublet).

Irwin Wieder[4] at Westinghouse Research Laboratories had studied the emission of the R doublet using a tungsten lamp whose light was absorbed in order to excite the two green and blue band of the ruby (the $^4F_2$ and $^4F_1$ bands, respectively). The energy was transferred to the ^2E levels and Wieder calculated that the efficiency of this transfer was about 1%. If that was true, one red photon for every 100 absorbed photons practically ruled out the possibility of using optical pumping to obtain a laser. After having examined other materials, however, Maiman decided to perform more accurate measurements on ruby and studied the spectroscopy of the chromium ion in pink ruby, making an accurate spectroscopy study of fluorescence, received by the *Physical Review Letters* on April 22, 1960, and published in June 1960.[6] He found the efficiency was indeed very high.

In his research, Maiman was helped only by an assistant, Irnee J D'Haenens, had only limited support from Hughes, and held the impression that a strong scepticism existed among his bosses George Birnbaum and Harold Lyons regarding the probability of his success.

The lifetime of the ^2E level was determined by Maiman to be about 5 ms. This rather long permanence time of atoms in the metastable state ^2E, and the successive decay to the fundamental state, is responsible for the fluorescence phenomenon in ruby. In addition, the decay rate S_{32} from the $^4F_2$ to the doublet ^2E was found to be $S_{32} \sim 2 \times 10^7$ s^{-1}.

A measurement of fluorescent quantum efficiency, that is, the number of fluorescent quanta emitted compared to the number absorbed by the crystal from the exciting beam, yielded a value close to unity.

Maiman calculated that population changes could easily be produced in the $^4A_2$ ground state of ruby under suitable optical excitation. He measured them under the influence of 5600 Å radiation. This was done by observing changes in the magnitude of the 11.3 GHz absorption due to zero-field ground-state splitting under illumination with a short light pulse from a flash tube. He also studied the 4100 Å optical absorption ($^4A_2 \rightarrow \, ^4F_1$ transition) resulting from 5600 Å excitation.

When an intense pulse of radiation at 5600 Å was turned on, the 4100 Å radiation passing through the crystal abruptly increased and subsequently decayed in about 5 ms. This result was fairly well explained by the temporary reduction in the ground-state population, and both experiments allowed him to calculate that a population change of about 3% was obtained.

Maiman was working with pink ruby with a concentration of approximately 0.05% by weight of Cr_2O_3 to Al_2O_3.

At this moment, the principal problem was to find a green light source powerful enough to pump the atoms to the upper level. Roughly speaking, a lamp at high temperature emits light as if it is a blackbody. Preliminary calculations had shown that lamps equivalent to a blackbody at 5000 K were needed. Maiman started to make calculations with some commercial mercury lamps but found that the performances were just at the limit of usefulness. He then remembered that pulsed xenon flash lamps have an equivalent temperature of 8000 K. There was no reason which prevented the laser from working with a pulsed regime and indeed in many cases using pulsed sources was attractive.

After carefully examining the catalogs, he found out there were only three lamps that could work, all made by General Electric which produced a very powerful type of helical flash lamp used by professional photographers. They were listed under the names FT503, FT506, and FT624. He choose the smallest of the three lamps, namely the FT506.

In a few months, in the new laboratories in Malibu, California, where he had moved in early 1960, by applying a sufficient and rapid irradiation from the flash lamp, he obtained inversion of population and laser emission between the ^{2}E level and the ground state. Lasing was obtained in May 1960 with a laser signal that was not very strong because the ruby used was a residue of the microwave maser experiments and was of poor optical quality. Maiman then ordered special rubies and his discovery was made public through a press announcement (*New York Times*) on July 7, 1960.

On June 24, he submitted a paper containing his first results that was, in fact, refused by the *Physical Review Letters*[7] for the reason that maser physics was believed to have reached a stage where further advances no longer merited rapid publication. Moreover, people were confident in Schawlow's assertion that R lines in ruby were not suitable for laser action. However, Maiman's paper was soon published in British journals[8,9] instead. He had intended to publish a more complete account of his experiment in the *Journal of Applied Physics* but had to withdraw this because the paper had already been published without his authorization by the English journal *British Communications and Electronics*. In the August 6 issue of *Nature*, he described the experiment simply as follows:

> … a ruby crystal of 1 cm dimensions, coated on two parallel faces with silver, was irradiated by a high-power flash lamp; the emission spectrum obtained under these conditions is shown in [**Figure 6.3(b)**]. These results can be explained on the basis that negative temperature were produced and regenerative amplification ensued.

In the paper in the *British Communications and Electronics*, the shortening of the decay time was also discussed.

When Hughes' public relations photographer took the photographs of the first laser, he found the FT506 lamp was too small and asked Maiman to pose instead with the FT503 flash lamp because it was more photogenic (see **Figure 6.4**). So when the press release was circulated everybody thought that this was the lamp used. There was a run of sales for those lamps and consequently all of the reproductions of the ruby laser made in other labs used the FT503.

A detailed account of the experiment was published the following year in two papers in the *Physical Review*. The first paper[10] contained a full theoretical treatment in which, by using the rate equation formalism, the pumping power for the threshold condition was derived. The general class of materials

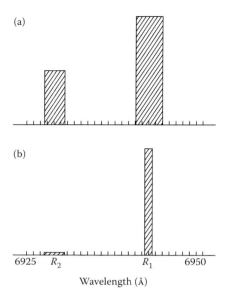

FIGURE 6.3 Histogram of the power emitted by the ruby as a function of frequency. (a) The broad emission fluorescence on the two R lines. (b) The narrow emission on the R_1 line which largely exceeds over the R_2 line. (From T H Maiman, *Nature* **187**, 493 (1960).)

considered was that of fluorescent solids whose emission spectra consist of one or more sharp lines. The excitation mechanism was considered to be by radiation of frequencies which produce absorption into one or more bands. Some of this excitation energy is lost by a combination of spontaneous emission and thermal relaxation to low-lying states; however, if the solid has a relatively high

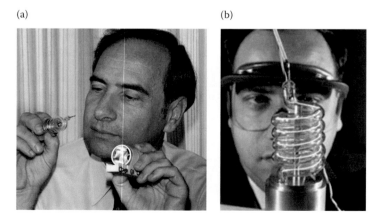

FIGURE 6.4 Maiman with the real ruby laser and lamp (a) and with the big lamp (b).

fluorescent efficiency most of the energy is transferred to the sharp fluorescent levels by means of a non-radiative process. Subsequently, by a combination of spontaneous emission and thermal relaxation, the excited atoms (ions) return either to the ground state or to another low-lying state. The spontaneous emission from these sharp levels is the observed fluorescent radiation. If the exciting radiation is sufficiently intense, it is possible to obtain a population density in one of the fluorescent levels greater than that of the low-lying thermal state. In this situation, spontaneously emitted (fluorescent) photons traveling through the crystal stimulate upper-state atoms to radiate and a net component of induced emission is superposed on the spontaneous emission.

Three-level and four-level schemes were discussed (see **Figure 6.5a** and **b**, which are Figures 1 and 2 of the original paper[10]). One of the features of the solid-state system was the possible use of a broad absorption band for the

(a)

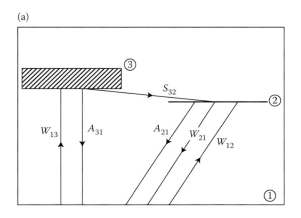

(b)

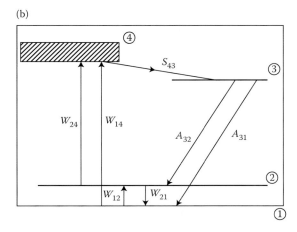

FIGURE 6.5 Optical energy-level diagram (a) for a three-level fluorescence solid and (b) for a four-level system. (From T H Maiman, *Phys. Rev.* **123**, 1145 (1961).)

pump transition. This situation allows a relatively high pumping efficiency to be realized, since most high-power optical sources have very broad spectral distributions in their radiation energy.

This overcame one of the particular problems foreseen by Schawlow and Townes.

In the three-level scheme, the rate equations were written by Maiman as

$$dN_3/dt = W_{13}N_1 - (W_{31} + A_{31} + S_{32})N_3,$$
$$dN_2/dt = W_{12}N_1 - (A_{21} + W_{21})N_2 + S_{32}N_3,$$
$$N_1 + N_2 + N_3 = N_o,$$

where W_{13} is the induced transition probability per unit time for the transition $(1 \rightarrow 3)$ caused by the exciting radiation of frequency ν_{13} and W_{21} is the induced probability $(2 \rightarrow 1)$ due to the presence of radiation of frequency ν_{21}. A_{31} and A_{21} are Einstein spontaneous emission coefficients and S_{32} is the transition probability for the non-radiative process $(3 \rightarrow 2)$, N_1, N_2, and N_3 are, respectively, level population densities, and N_o is the total active ion density in the crystal.

Thermal processes $(2 \rightarrow 1)$ and $(3 \rightarrow 1)$ were assumed to be negligible compared with the corresponding radiative processes, but that the reverse was true for the transition $(3 \rightarrow 2)$.

In the steady state, where the time derivatives are zero, an approximate solution was found to be

$$(N_2 - N_1)/N_o \sim (W_{13} - A_{21}) / (W_{13} + A_{21} + 2W_{12}),$$

provided $A_{31} \ll S_{32}$, $N_3 \ll N_1$, $N_3 \ll N_2$. Therefore, by making $W_{13} > A_{21}$, stimulated emission can, in principle, be obtained. Considerations concerning losses led to the more restrictive condition that $N_2 - N_1$ must be sufficient to overcome circuit losses, that is,

$$(N_2 - N_1)/N_o \sim (1 - r)/\alpha_o l,$$

where r is the reflection coefficient of the silvered end plates, α_o is the normally measured absorption coefficient for the transition $1 \rightarrow 2$ under low-power excitation, and l is the length of the material.

A similar analysis was carried out in the case of four levels. The use of rate equations of the kind written here, which we have already seen in the Bloembergen treatment of the solid-state maser, immediately became of wide and common use in all treatments of lasers.

The next argument to be dealt with in the paper concerned the spectral width. The presence of multimode oscillation was considered and the importance of strains, inhomogeneities, and deviations from single crystallinity—which tend to scatter energy into undesired modes—was recognized. No discussion of coherence properties of the radiation was entered into, nor was there any consideration of the statistics of the radiation which was clearly produced in a non-equilibrium thermodynamic situation. Apparently, nobody was aware of this, and no discussion of this point can be found in any of the papers on lasers published during that time. Attention was focused essentially on the use of rate equations to calculate threshold pumping levels and, later on, spiking behavior, Q-switching properties, etc.

The second paper[11] was a description of the results of spectroscopic and stimulated emission experiments. In particular, spiking behavior and line narrowing were observed. The various ruby samples were divided into two categories.

1. Crystals which exhibited R_1 line narrowing by only about four or five times had a faster but smooth output time decay (compared with the fluorescence), an output beam angle of about 1 rad, and no clear-cut evidence of a threshold excitation. Clearly these were highly strained and bad crystals.

2. Crystals which exhibited pronounced line narrowing of nearly four orders of magnitude, an oscillatory behavior of the output pulse, and a beam angle of about 10^{-2} rad; these crystals were, in particular, characterized by a very clear-cut threshold input energy where this pronounced line and beam narrowing occurred.

The sketch of the laser realized by Maiman is shown in **Figure 6.6** (Figure 7 of the original paper[11]). Maiman wrote[11]

Due to the need for high source intensities to produce stimulated emission in ruby and because of associated heat dissipation problems, these experiments were performed using a pulsed light source .... The material samples were ruby cylinders about 3/8 inch in diameter and 3/4 inch long with the ends flat and parallel within $\lambda/3$ at 6943 Å. The rubies were supported inside the helix of the flash tube, which in turn was enclosed in a polished aluminium cylinder [see our Figure 6.6]; provision was made for forced air cooling. The ruby cylinders were coated with evaporated silver at each end; one end was opaque and the other was either semitransparent or opaque with a small hole in the center.

The population inversion was obtained between the level $_2E$ and the ground level and light emission was amplified by multiple reflection at the end mirrors.

In the first laser, the line R_1 was active (**Figure 6.7**). The energy for the flash tube was obtained by discharging a 1350 µF capacitor bank. The input energy was varied by changing the charging potential. Threshold was obtained at energies between 0.7 and 1.0 J according to the way the terminal faces of the rod were prepared.

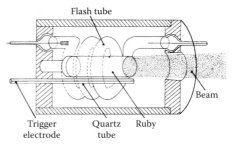

Apparatus used by Maiman for the first ruby laser.

FIGURE 6.6 The original Maiman device. (From T H Maiman et al., *Phys. Rev.* **123**, 1151 (1961).)

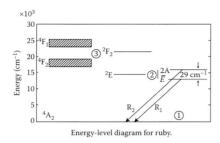

Energy-level diagram for ruby.

FIGURE 6.7 Energy levels in ruby. (From T H Maiman et al., *Phys. Rev.* **123**, 1151 (1961).)

The ruby laser system was patented by Maiman (patent no. 3,353,115 issued in November 14, 1967).

At the end of 9 months of work to make the ruby laser, Maiman had spent only $50,000. Immediately after building his laser, he left Hughes and founded in 1962 a company, the Korad Corporation, which became a market leader in the construction of high-power ruby lasers. In 1968, he sold Korad to Union Carbide and founded a venture capital firm called Maiman Associated and in 1972 the Laser Video Corporation. In 1976, he joined TRW Electronics in California where he was vice president of R&D section, the position from which he retired in 1984. In the same year (1984), he was inducted into the National Inventors Hall of Fame. He took great pride in the laser's medical applications and was inducted into the Royal College of Surgeons of England as the only non-physician member of the society.

In his later years, Maiman served as an adjunct professor at Simon Fraser University in Vancouver, Canada. He died in Vancouver, British Columbia, on May 5, 2007.

The day after Maiman's announcement that ruby was successful, many people disbelieved it. Many people who had not seen the English papers by Maiman read instead a paper Schawlow and other people from Bell had published in the *Physical Review Letters*,[12] where they duplicated Maiman results, and thought that Bell Laboratories were the first to build the laser. This misunderstanding came about because the proposal for a laser was made at Bell by Schawlow and Townes, who everybody knew were working on its practical implementation, while Hughes in California was completely extraneous to this research and isolated from the principal teams based on the East Coast.

Moreover, Bell's people claimed that Maiman had really not achieved laser action. A controversy issued based essentially on the assertion Maiman had not declared the spiking behavior of the laser in his first paper. Echo of it can be still found in a series of letters to *Physics Today* in 2010.[13] The Bell's point of view—in addition to Arthur Schawlow's account of the work done at Bell in the summer of 1960[14]—has been recently resumed by some of the survivors.[15]

Taking his clue from the misleading *Times* picture, Nelson at Bell in a group including Schawlow ordered a General Electric FT524, a powerful helical xenon flashtube powered by a 400-μF capacitor bank up to 4 kV and duplicated Maiman's laser. In August 1960 with a long pink ruby rod the duplicated laser worked and the results were published in the October issue of the *Physical Review Letters*.[12]

The principal characteristics of the laser beam, its coherence, directionality, and collimations were studied in this paper. In the *Physical Review* paper by Maiman[11] which was published in August 1961, references are made to this and other papers showing the frantic activity in the field which immediately started after Maiman's announcement in June 1960.

The group with Schawlow, in the paper sent to the *Physical Review Letters*,[16] verified the narrowing of the emission line, finding a linewidth of 0.2 cm⁻¹ in lasing conditions when compared to 6 cm⁻¹ in normal emission.

They also verified the sudden variation of the directional distribution of the emitted light when the threshold condition for excitation was overcome, and the presence of relaxation oscillations (**Figure 6.8**). At that time, a similar behavior was known to occur at microwave frequencies in ruby masers[17] and was attributed by Statz and de Mars[18] to time-dependent interaction between the inverted population of the electron-spin systems of the paramagnetic substance and the resonant cavity.

In the case of ruby, one possible explanation given in the Schawlow et al. paper[12] was that the stimulated

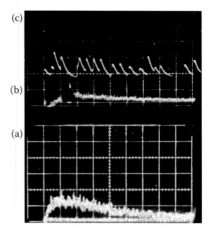

FIGURE 6.8 The spiking behavior of a free-operating ruby laser observed by the Schawlow group. (a) Decay of fluorescence for a low excitation level. (b) Lasing action at a higher excitation level when the trace, after 500 μs, breaks. (c) An enlargement of 100 μs of the lasing trace, showing spikes. (From R J Collins et al., *Phys. Rev. Lett.* **5**, 303 (1960).)

emission, once it sets in, proceeds at a rate greater than that at which atoms are being excited to the ²E state. When this occurs, the stimulated emission may drive the inverted population below the level at which the process first sets in, so that, when the stimulated emission is finally quenched, some definite time interval will be required for the negative population to be restored.

In this paper, there is also the first investigation of spatial coherence of the laser (first-order coherence), Schawlow et al. opened a rectangular aperture 50 × 150 μm in size in a heavily silvered coating of one end of the ruby rod and looked for diffraction. The image was found to consist of a Fraunhofer diffraction pattern for a rectangular aperture illuminated by wave fronts which were approximately plane and approximately coherent. This pattern disappeared when the excitation was reduced below the threshold. Another interesting result of this experiment was that the laser emission was taking place from small active regions on the face of the rods, each about 500 μm in diameter. Later on, this filamentary structure of the emission was intensively studied. The investigators also observed that a decrease in the excitation threshold of 30% could be obtained by lowering the working temperature.

Further studies were made of the first-order spatial and temporal coherence of the emission,[19] and mode structure was observed by Evtuhov and Neeland[20]

(**Figure 6.9**). Longitudinal modes were studied through their beats by Siegman.[21] Direct studies of longitudinal modes were made using a high-resolution, high-dispersion grating spectrometer both by Duncan et al.[22] and Ciftan et al.[23]

Schawlow[24] had originally proposed that the so-called N lines in ruby be used,[25] since these transitions have a somewhat lower state above the ground state and can therefore be depopulated at low temperature. Shortly after the announcement of laser action at the R_1 transition, he and Devlin[26] and also Wieder and Sarles[27] reported the observation of laser action at 7009 and 7041 Å, which are two of the strongest N lines. Emission on the R_2 line was obtained in 1962 by McClung, Schwarz, Hellwarth, and Meyers.[28]

The transverse mode emission

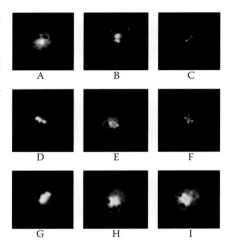

FIGURE 6.9 Typical patterns on the end of a ruby laser. (From V Evtuhov and J K Neeland, *Quantum Electronics*, Vol. 3, edited by P Grivet and N Bloembergen (Columbia University Press: New York, 1964), p. 1405).)

was strongly influenced by the optical quality of the ruby.[29] The original ruby laser material was grown by the Verneuil flame fusion technique,[30] a process long used in both Europe and the United States for gems and bearings, which had been developed in 1904. The Czochralski process was applied later.[31] Improvement of ruby laser characteristics was obtained by using better crystals and more efficient light pumping schemes.[32]

A continuously operating ruby laser was developed by Nelson and Boyle[33] in 1962 by using an arc lamp and a trumpet-shaped laser rod with a special arrangement of mirrors to image the lamp upon the end of the ruby.

Abella, a student of Townes who was working on the cesium laser project, was asked by Townes to replicate Maiman's achievement what he did in 2 months together with Cummins.[34] Then Abella, Hartmann, and Kurnit did the first experiment that observed photon echoes in ruby.[35]

In the early days of research, it had already been shown that by focusing ruby laser light it was possible to vaporize any kind of materials. Very soon people began to drill holes in razor blades by focusing ruby lasers onto them and someone suggested that the laser output could be measured in Gillettes!

Schawlow became a well-trained demonstrator of the spectral properties of laser light, aiming the ruby beam onto a blue balloon contained in a larger red one. The red ruby light was not absorbed by the red balloon, which was unaffected, but made the blue balloon to explode because it absorbed the light. The experiment was perfected by building a small, portable ruby laser into the housing from a toy ray-gun.[36]

The first demonstration was given at the meeting of the American Association for the Advancement of Science at Cleveland in December 1963, and Schawlow still used it to show how this exercise eventually led him to obtain a patent for a laser eraser, which demonstrates the more amusing paths to discovery!

6.3 The Four-Level Laser

A few months after the realization of the ruby laser, a four-level laser at low threshold was obtained by Sorokin and Stevenson[37] by using trivalent uranium in calcium fluoride.

In the four-level laser, the population inversion is more easily obtained (Figure 6.5b). In this case, the critical population inversion between levels 3 and 2 can be obtained simply by maintaining a population in level 2 equal to the theoretical minimum.

In September 1959, Townes organized a conference on "*Quantum Electronics—Resonance Phenomena*" during which, even though no laser had been built, most of the informal discussion centered on lasers.

Peter Sorokin and Mirek J Stevenson of the IBM Thomas J Watson Research Center attended the conference and became enthusiasts of the laser concept. The Watson Center had been created in 1956 and offered all comforts to its researchers in pleasant countryside close to the cultural resources of New York.

The director of the physics section of the center, William V Smith, after reading the paper by Schawlow and Townes, suggested that his microwave spectroscopy group, among whom were Sorokin and Stevenson, should redirect their effort toward lasers since he believed they would benefit IBM and help establish the reputation of the new laboratory.

Peter Sorokin was the son of a sociology professor at Harvard University where he studied physics, receiving his PhD in 1958 with a dissertation on nuclear magnetic resonance under the supervision of Bloembergen. The young man[38] had planned to go into theoretical solid-state physics. In his second year of graduate work, together with a friend, he registered for a reading programme on nuclear magnetic resonance given by Bloembergen, which they thought would be an easy course. At the time Bloembergen was not a very polished lecturer, and the two friends let everything go over their heads. At the end of the course, however, the professor announced he wanted a term paper from each student, to which Sorokin and his friend responded unsatisfactory. Bloembergen commented "These papers don't say anything about what I was teaching". So Sorokin spent part of the summer trying to understand nuclear magnetic resonance, and then wrote another essay which this time Bloembergen accepted. By that time he felt he had invested so much time in the subject, which actually seemed interesting, that he may as well sign up to take a thesis with Bloembergen. First Bloembergen assigned him a theoretical problem, and for a year he sat at a desk with a pad of paper. Finally he came back to the professor and said: "The divergent part cancel, and all you are left are terms that are very hard to evaluate, but they are finite". So Bloembergen looked at him and said: "Well Peter, I think you had better do experiments", and assigned Sorokin a thesis in which he had to make nuclear magnetic resonance measurements on cesium atoms. However, the cesium resonance appeared to have long relaxation times, which hindered the measurements. Another year went by and Sorokin became very discouraged; then, a young scientist came to the lab as a post-doctoral researcher and built a resonance system based on the cross-coil approach of Bloch's group at Stanford. Sorokin understood immediately that this was the right way to perform the measurement and built a similar apparatus and finished his thesis.

After his thesis, Sorokin was hired by IBM to work on microwave resonance in solids. When the paper by Schawlow and Townes appeared, his boss suggested he study the possibility of building a laser. Along with Mirek Stevenson, who had obtained his PhD with Townes a couple of years before and who had been hired about the same time of Sorokin, he decided to focus on this new problem. Therefore, after the September 1959 Conference, they plunged into this work. They wanted to build a laser which emitted continuously using lamps having a power of the order of watts. Sorokin thought the principal problem was the pumping. To increase efficiency, it was essential to decrease the losses and he considered eliminating the mirrors of the Fabry–Perot cavity and substituting them with two total reflection prisms and choose calcium fluoride as a host material. After examination of the scientific publications on the subject, he found that the Russian P P Feofilov had studied light emission from uranium or samarium ions in calcium fluoride. Uranium has a fluorescence emission at about 2.5 μm. The uranium or samarium ions in the calcium fluoride substitute a calcium ion in the crystal and have energy levels similar to those of chromium in ruby with the only difference (shown in **Figure 6.10**) being that there is one additional level and therefore light emission may occur between a level populated by the decay of one band to an intermediate level which, if one works at low temperature, is practically unpopulated because thermal agitation is not able to populate it from the ground state (four levels). Calcium fluoride doped with uranium has moreover a strong absorption band in the visible and could be pumped with a high pressure xenon arc lamp. The system needed, however, to be cooled at a low temperature.

The two researchers while receiving the crystal heard about Maiman's result with ruby and immediately decided to forget about their starting idea of making a resonant cavity using total reflection, cut their calcium fluoride crystal into the shape of cylinders, silvered the two extremities, and in November obtained emission.

Emission was at 2.5 μm in a crystal of CaF_2 with 0.05% molar concentration of uranium. Later, with 0.1% concentration, they observed emission at 2.6 μm.[39]

Considerable confusion existed regarding the origin of these lines, although an extensive literature on optical properties of ions in solids existed at that time[40] and most of the spectral information needed to develop these lasers still had to be obtained while the research was in progress. A group of investigators at Bell Telephone Laboratories undertook a detailed study of the energy levels

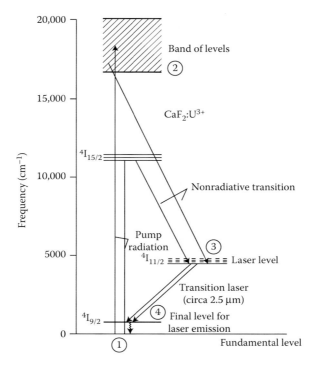

FIGURE 6.10 Energy levels of three times ionized uranium atoms in a crystal of calcium fluoride (CaF_2:U^{3+}). The transition is obtained by pumping from the ground level (1) to the band (2). (The light absorbed by level $^4I_{15/2}$ can be neglected.) Electrons decay from band (2) to levels (3) and the laser transition at about 2.5 μm occurs between (3) and (4).

of U in CaF_2. By means of paramagnetic resonance techniques they showed, after some uncertainty,[41] that the U ion may be located in CaF_2 in tetragonal,[42] trigonal, and orthorhombic sites. Emission at 2.51 and 2.44 μm was assigned to U^{3+} in tetragonal sites. Orthorhombic U^{3+} is responsible for emission at 2.61, 2.24, and 2.57 μm, and emission at 2.24 μm was observed in crystals containing 90% of U^{4+} ions in trigonal sites.[43]

The experiments of Sorokin and Stevenson were performed at liquid helium temperature; later Bostick and O'Connor[44] reported operation at 77 K, and Miles[45] at up to 300 K. In the early experiments flash excitation was used, and regular relaxation oscillations, as in ruby, were observed. Continuous operation was obtained by Boyd et al.[41] at 77 K, with the arrangement shown in **Figure 6.11**.

Meanwhile, in 1961, Sorokin and Stevenson[46] obtained stimulated emission in divalent Sm in CaF_2 at a wavelength of 7083 Å. In the same year, Johnson and Nassau[47] reported stimulated emission of Nd^{3+} in $CaWO_4$ at 1.06 μm, and a few months later they reported continuous operation at room temperature using the same equipment as shown in Figure 6.11.[48] In the first experiments,

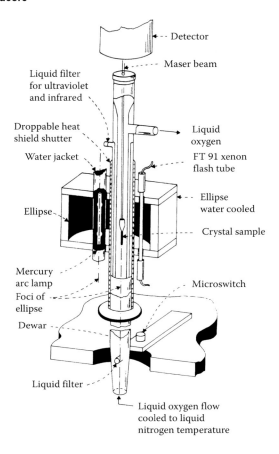

FIGURE 6.11 Experimental set-up used at Bell Telephone Laboratories. (From L F Johnson et al., *Phys. Rev.* **126**, 1406 (1962).)

operation was obtained for a few seconds, but when the problem of charge compensation was understood and the number of available levels was reduced by the addition of Na to the crystal, continuous operation became possible for periods up to an hour.[49]

The next lasing materials were Tm and Ho,[50] still in $CaWO_4$.

6.4 The Neodymium Laser

The other solid-state laser realized in 1961 and one of the most important till today is the neodymium glass laser. In 1959–1960, the American Optical Company also entered laser research.[51] The company's original focus was optical instruments and ophthalmic products. It had also strong capabilities in glassmaking and glassworking. During the 1950s, the company decided to expand its product lines and it therefore initiated research projects in new area such as military electro-optics and fiber optics. Elias Snitzer (1925–2012), a PhD from the University of Chicago who had his university career hampered because he had been heavily involved in left-wing politics as a student at Chicago, was hired into the research group in the early 1959, and began with research on the propagation of electromagnetic waves through optical fibers. For the company, this work would strengthen its already considerable patent production in fiber optics and bolster its image in the field within the general scientific community. Snitzer saw that there were connections between his optical fiber research and laser work. Since a glass fiber may sustain electromagnetic modes, it could be converted into a laser resonator if one placed mirrors at its ends. This prospect was the more interesting because of the doubts within the scientific community about whether the Fabry–Perot resonator that was being used with gaseous media would work. The glass itself could become a lasing material by doping it with suitable substances like samarium or ytterbium that could possess the required levels for incoherent light excitation sent through the sides or the end of the fiber. Snitzer considered he would be able to concentrate even more pumping light into the fiber by covering it with a thick layer of glass with a slightly different refractive index.[52]

Early in 1960, Snitzer with two more colleagues began examining a series of glass fibers doped with ions that had fluorescent lines in the visible range. Glass was an unusual choice. All materials under study were gases or crystals. After Maiman's successful laser, Snitzer also tried with ruby fibers. Until then he had used high-pressure mercury lamps which emitted light in a continuous way. Now he gave up and brought flash lamps. The group examined 200 fibers. At the end of 1960, Snitzer's two co-workers had transferred to an Air Force classified project aimed at creating a solar-powdered laser transmitter. Snitzer continued alone and decided to shift from visible to infrared. This decision meant changing the doping materials. In the infrared, rare earths like neodymium, praseodymium, holmium, erbium, and thallium could be used. Snitzer decided also to give up with fibers and concentrate on a simple doped glass rod. In October 1961, he succeeded to obtaining the laser effect from a neodymium-doped glass rod.[53]

Snitzer used laser rods consisting of inner core of doped crown glass 6 mm in diameter and 45 cm long coated, except at its ends, with a layer of undoped glass of lower refractive index to give total internal reflection. A 45-cm long xenon flash tube was used for pumping in a close configuration. The flash tube and the rod were together wrapped in silver foil which served as a reflector.

He also observed laser action of Nd^{3+} in barium glass with Nd concentration varying from 1.4×10^{10} to 2.16×10^{20} cm^{-3}. Stimulated emission was produced in thin Nd-doped glass fibers of 0.03 cm and also 32 μm in diameter. The length of the fibers varied from 1 to 7.5 cm. These fibers were coated with an ordinary glass of similar index of refraction.

The neodymium ions exhibit narrow spectra both when they are incorporated in crystals and when in amorphous materials like glass. There are several advantages in using glass as a laser material. The preparation methods are well established and making glass is certainly easier than growing crystals. Besides, the homogeneity that can be obtained with glasses is much better

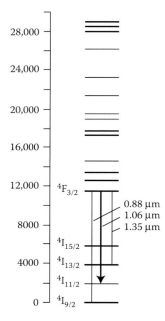

than with crystals and therefore much larger samples can be produced. Moreover, glasses doped with rare earth ions had already been in production for many years for their applications as photographs lenses.

The levels of neodymium in glass are shown in **Figure 6.12**. The $^4F_{3/2}$ level is the fluorescent one and laser transitions occur between this level and the $^4I_{13/2}$, $^4I_{11/2}$, and $^4I_{9/2}$ levels at wavelengths of 1.35, 1.06, and 0.88 μm, respectively. Excitation is obtained by optically pumping from the fundamental level to levels above the $^4F_{3/2}$ state. There are three levels which absorb infrared radiation, levels which absorb in the yellow region around 5800 Å and other levels that absorb mostly in the ultraviolet. In Figure 6.12, the absorbing levels higher than the $^4F_{3/2}$ level are shown as bold lines. From these levels, the excited atoms decay with nonradiative transitions to the $^4F_{3/2}$ level from which the laser emission begin.

The lasing levels all belong to transitions of the inner electrons on the 4f levels. These electrons are screened by eight outer electrons ($5s^2$ and $5p^6$) and are therefore weakly influenced by the crystal field, which explains why the emission wavelength is mostly insensitive to the kind of host material.

FIGURE 6.12 Energy levels of neodymium ions (Nd^{3+}) dispersed in a barium "crown" glass.

Lasers using neodymium in glass are important first of all because they represent an example of a solid material different from a synthetic crystal, also because certain glasses doped with neodymium have great output energies per unit volume, and finally because the versatility of the glass matrix allows lasers to be shaped into rods or fibers.

Later on, in 1966, Snitzer discussed the properties of glass lasers doped with Nd, Yb, Er, and Ho.[54]

As we said previously, the laser effect was obtained also by L F Johnson and K Nassau[55] at Bell Telephone Laboratories with Nd^{+3} in $CaWO_4$ crystal. Later, Johnson, G D Boyd, K Nassau, and R R Soden[56] obtained quasi-continuous operation at 1.06 μm with neodymium doping into a crystal of calcium tungstate cooled at liquid nitrogen temperature. The same authors also described other crystals doped with Nd in continuous operation.[57]

In December 1961, ARPA (Advanced Research Project Agency by 1959 became the agency for the support of basic research into advanced military technology) organized a scientific committee who gave high priority to research on ruby and glass lasers. The following year, Snitzer obtained emission from a fiber of 32 μm diameter. Today, rare earth ions incorporated into glass fibers make excellent amplifiers for signals sent along optical fibers and used for telecommunications.

The trivalent ion of neodymium has been incorporated in a great number of different hosts. One of those is a lattice of $Y_3Al_5O_{12}$, which is commonly represented by YAG acronym for yttrium aluminum garnet. The neodymium energy levels are essentially the same, irrespective of the host matrix. The neodymium YAG laser is used both in continuous and pulsed operation. This laser was made at Bell Laboratories by Joseph E Geusic, who in 1962 with E D Scovil started to write a review paper on masers and lasers[58] in which an analogy was discussed that Scovil had made between an optically pumped laser and a thermodynamic heat pump. The analogy yielded to certain criteria for selecting laser materials which guided Geusic to select about 40 crystals among which was YAG. The trouble with this material was that there were no sufficiently long crystals. Geusic asked Bell colleagues, among whom were LeGrand G van Uitert, how to obtain such crystals. The two men at first obtained from an outside company a sample on which they performed encouraging measurements. Then van Uitert succeeded in growing a long enough crystal for good optical quality, and Geusic and his technician Horatio M Marcos had the laser working[59] in a continuous way showing that it required a power about one-fifth of that used for calcium tungstate with neodymium.

Geusic and van Uitert started a collaboration with the Linde division of Union Carbide to develop longer crystals of better optical quality and demonstrated all the advantages of this laser that constitutes a valid alternative to other power lasers (ruby and CO_2). This laser is a typical example of how multidisciplinary collaboration, characteristic of large American laboratories, permitted in a couple of years the development of a new laser with exceptional performances.

In 1962, the Bell Laboratories group obtained emission from Nd^{3+} in a number of different host crystals.[60]

Shortly after Maiman's laser, investigators began dreaming of directly converting sunlight into laser radiation. This result was quickly realized using Nd YAG crystal obtaining 1 W of cw laser output at room temperature.[61] A previous result had been obtained in 1963 at RCA Labs, with an improbable sun-pumped continuous optical maser in $CaF_2:Dy^{2+}$ at liquid neon temperature $(27°K)$[62] emitting at 2.36 μm.[63]

6.5 The Gas Laser

Although gas lasers were not operational before the realization of the ruby laser, gaseous systems had been considered since the beginning. The use of a gas discharge for obtaining population inversion was proposed by Fabrikant[64] as early as 1939. He suggested the use of a buffer gas to shorten the relaxation time of the lower laser level through collisions of the second kind. Also A Ferkhman and S Frish claimed to have realized population inversion[65] way back in 1936.

After Schawlow and Townes[66] had considered selective optical pumping in gases, proposals for obtaining, through selective excitation, the inversion of the upper laser level in a one-gas system were made in 1959 by Sanders[67] and

Javan,[68] who both suggested producing the required population inversion by using an electric discharge in a gas.

John H Sanders—an experimental physicist on leave from the Clarendon Laboratory, Oxford, England—had been invited to Bell from January to September 1959 by Rudolph Kompfner, an associate director of the division on communication science, to try to realize an infrared maser. With less than one year of time for the research, Sanders decided to excite pure helium in a discharge tube inside a Fabry–Perot cavity and to attempt to obtain the laser effect through a process of trial and error by varying the discharge parameters that he had no time to evaluate theoretically. The maximum distance at which the Fabry–Perot mirrors could be put and still remain aligned parallel to each other so as to realize the cavity was 15 cm, and Sanders had to use discharge tubes no longer than this which was later judged by Javan a fundamental limitation, due to the very low gain.

In a very short letter to the *Physical Review Letters*, Sanders observed that it was difficult to obtain a sufficient number of excited atoms with a flash lamp and suggested the use of excitation produced by electron collisions. Such an excitation could easily be produced in an electrical discharge in a gas or a vapor. Population inversion could be produced if excited states with a long lifetime existed in the active material and there were states present at lower energies with a short lifetime. He observed, too, that in an electrical discharge many processes exist which "… may disturb the relative population of such levels from the value implied by these lifetime values, and it may be possible to choose conditions which maintain a high population in the upper state."

In 1950, W E Lamb[69] had already noted that electron collisions in a gas discharge can create an inversion of population, as we have seen in Chapter 2.

The very next paper in the same issue of the *Physical Review Letters* was written by A Javan,[68] who too had considered these problems.

Ali Javan[70] was a native of Iran who received his PhD with Townes in 1954 for microwave spectroscopy work and remained in Townes' group for four years working on microwave spectroscopy and masers. After his thesis, while Townes was on sabbatical leave in Paris and Tokyo, Javan became more involved with masers and had an idea for a three-level maser before Scovil, Feher, and Seidel from Bell Laboratories published the first experimental work on the subject. He found a method to achieve amplification without population inversion using a Raman effect in a three-level system (see Chapter 9).

In April 1958, while looking for a position at Bell Laboratories, he spoke with Art Schawlow who told him about lasers. In August 1958, he was hired at Bell but before starting work on lasers had to settle a problem between RCA and Bell because he had sold RCA the patent on the three-level Raman maser and RCA was contesting Bell's application on three-level masers. Luckily, RCA decided that the maser amplifier was not commercially viable and the possible litigation was dropped down.

So eventually Javan could dedicate himself to the idea of making laser, which he was thinking of building using gases, publishing the proposal in the paper on the *Physical Review Letters*. He had decided to use a gas as an active medium because he believed the simplicity of gases would simplify the study.

Javan realized that any single physical process tends to produce a Boltzmann distribution; therefore, a medium with population inversion can be produced in a steady-state process only as a result of the competition of several physical processes proceeding at different rates. After having discussed pure gases, he then focused attention on certain kinds of gas mixtures.

The following exchange processes were considered:

1. electron collision of the first kind in which an atom gains energy from an electron;

2. electron collision of the second kind in which an excited atom loses energy to an electron;

3. spontaneous emission of radiation from an excited atom;

4. absorption of radiation by an atom;

5. stimulated emission of radiation by an atom.

Javan used θ_{ij} to indicate the lifetime of the transition of an atom from level i to level j, when the atom is subjected only to collisions with electrons of a given density in equilibrium among themselves at the absolute temperature T. If no other processes than these collisions were to take place, the rate of change of the number of atoms at level i would be given by

$$dN_i/dt = \sum_j [(N_j/\theta_{ji}) - (N_i/\theta_{ij})]. \tag{6.1}$$

If we assume that thermodynamic equilibrium is established at temperature T, the number of atoms in each level will be stationary. Then

$$\sum_j [(N_j^*/\theta_{ji}) - (N_i^*/\theta_{ij})] = 0, \tag{6.2}$$

where N^* is the stationary value of N.

The principle of detailed balance now requires that the exchange between each pair of energy levels should balance out. This means that not only must the sum in Equation 6.2 vanish, but that

$$(N_j^*/\theta_{ji}) - (N_i^*/\theta_{ij}) = 0 \tag{6.3}$$

must hold for every i and j. Therefore,

$$\theta_{ij}/\theta_{ji} = N_i^*/N_j^* = (g_i/g_j)\exp[-(E_i - E_j)/kT] \tag{6.4}$$

and Equation 6.4 is always valid.

Now let us consider the simplest case of two levels, with a spontaneous radiative transition, with a lifetime τ_2 from level 2 to 1; the rate equation is

$$dN_2/dt = (N_1/\theta_{12}) - (N_2/\theta_{21}) - (N_2/\tau_2). \tag{6.5}$$

In the stationary state,

$$N_2/N_1 = (1/\theta_{12})/[(1/\theta_{21}) + (1/\tau_2)]. \qquad (6.6)$$

When the radiative process is comparatively fast, so that $\tau_2 \ll \theta_{21}$, we have

$$N_2/N_1 \sim \tau_2/\theta_{12} = (g_2\tau_2/g_1\theta_{21})\exp[-(E_2 - E_1)/kT]. \qquad (6.7)$$

The ratio τ_2/θ_{21} is a measure of the departure from the Boltzmann equilibrium distribution.

To some extent, this factor is under the control of the experimenter, since τ_2 is fixed: but $1/\theta_{21}$ is proportional to the electron density in the discharge. However, N_2/N_1 cannot be increased arbitrarily by increasing the electron density, because the validity of Equation 6.7 is predicted on the assumption $\tau_2/\theta_{21} \ll 1$.

However, if a third level 3 is introduced above level 2, and we assume no direct interaction between levels 2 and 3, we find, under similar assumptions,

$$N_3/N_1 \sim (g_3/g_1)(\tau_3/\theta_{31})\exp[-(E_3 - E_1)/kT]. \qquad (6.8)$$

In the case of equal multiplicities, we conclude from Equations 6.7 and 6.8 that

$$N_3/N_2 \sim (\tau_3/\tau_2)(\theta_{21}/\theta_{31})\exp\left[-(E_3 - E_2)/kT\right]. \qquad (6.9)$$

This equation was given in the paper by A Javan, without the detailed discussion we have presented here. Javan argued that, even though the exponential factor is always less than 1, the factor in front of the exponential may be sufficiently larger than 1 for certain values of the parameters to cause $N_3/N_1 > 1$.

Javan considered that the 3^1D and 2^1P levels of He have a lifetime ratio $\tau_3/\tau_2 =\times 35$ and an estimate of θ_{31}/θ_{21} gave 15.

Equation 6.9 indicates then that a negative temperature is possible. Unfortunately, the collision cross-section for excitation of 3^1D is relatively small. At a pressure of about 5×10^{-3} mmHg, the relatively high fraction of 1% of the total number of atoms would have to be ionized to give 10^9 excited atoms per unit volume needed to overcome resonator losses and exceed the threshold conditions for oscillation.

Javan also considered Ne, and, to prevent an unfavorable population of the excited levels and suggested the introduction of a very small amount of a quenching gas, such as argon, to decrease the lifetimes of the longer-lived excited atoms.

An alternative scheme was finally suggested, which used the transfer of excitation between excited states of two different atoms in a gaseous mixture. Javan wrote:

> Consider a long-lived state of an atom (such as a metastable state). This state can be populated appreciably at moderate electron densities.

If an excited state of a second atom happens to lie very close in energy to that of the level of the first atom, a large cross section is expected to exist for an inelastic collision resulting in a transfer of excitation from the metastable state to the excited state of the other atom and vice versa.

Due to the non-adiabatic nature of the process of collision, the levels of the second atom which differ in energy considerably from that of the metastable level of the first atom do not show appreciable cross sections for transfer of excitations.

A few other considerations then followed and mixtures of Kr–Hg and He–Ne were proposed.

The paper was received at the *Physical Review Letters* on June 3, 1959. The ideas contained therein were also presented at the *Quantum Electronics Conference*.[71] Javan knew Ladenburg's work quite well, as is shown by a quotation he made in the *Physical Review Letters* paper of the *Review of Modern Physics* paper by Ladenburg.[72]

A few months later, Javan with a few other colleagues at Bell succeeded in realizing a gaseous laser in a He–Ne mixture.

Javan worked in strict contact with William R Bennett Jr. (1930–2008), a spectroscopist at Yale University who had been his friend at Columbia[73] (see **Figure 6.13**).

The two men worked until late at night for a full year. In the spring of 1959, Javan asked Donald R Herriott (1928–2007), a specialist at Bell on optical apparatus later to become president of the *Optical Society of America*, to help in the designing of the discharge tube. The structure was designed (see **Figure 6.14**) with the mirrors inside the discharge tube providing a special mounting with micrometric screws which facilitated the alignment.

In September 1959, Bennett left Yale to join Bell and together with Javan started an intensive and meticulous programme to calculate and measure the spectroscopic properties of helium–neon mixtures under various conditions in order to determine the factors governing population inversion. They found that in the best conditions one may expect only a very low gain, of the order of 1.5%. Such a low gain made absolutely necessary to minimize losses and to employ mirrors with the highest possible reflectivity, which were made with a multilayer dielectric deposition with a maximum reflection at the chosen wavelength that was 1.15 μm.

In 1960, Javan, Bennett, and Herriott finally tested the laser. Initially, they attempted to generate an electric discharge in a quartz tube that contained the gas mixture by using a powerful magnetron, but the tube melted. The apparatus had to be remade and after

FIGURE 6.13 William Bennett and Ali Javan at work on their He–Ne laser. (From *Phys. Rev. Focus* **26**, 24 (2010).)

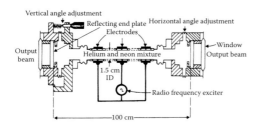

FIGURE 6.14 Diagram of the He–Ne laser. (From D R Herriott, *J. Opt. Soc. Am.* **52**, 31 (1962).)

other setbacks, on December 12, 1960, they realized the discharge in a new tube and tried to align the mirrors to obtain the lasing effect but without success. Then in the afternoon, Herriott saw the signal:

I was casually turning the micrometer on one of the mirror adjustment when a signal suddenly appeared on the oscilloscope. We adjusted the monochromator and found the signal peaked at 1.153 microns, one of the expected wavelengths.

The first continuous gas laser was born.

As usual during the research a number of people did not believe to it and told Javan the gas discharges were too chaotic to be handled. Javan was spending a lot of money and was lucky enough to have the positive result before Bell decided to stop the efforts. At the end Bell admitted to having spent two million dollars on this research and even though probably this figure is perhaps exaggerated, certainly the project required considerable financial effort.

Following the example of Hughes, Bell also gave a public demonstration of the helium–neon laser on December 14, 1960, and in order to demonstrate the importance it could have for telecommunications, a transmission of a telephone conversation was organized utilizing as a carrier the laser beam that was modulated via a Kerr cell by the telephone signal.

The He–Ne gas laser was presented to the press for the first time on January 31, 1961, by A Javan, W R Bennett Jr., and D R Herriott. The paper describing the laser was received by the *Physical Review Letters* on December 30, 1960, and published in the February 1961 issue.[74]

The laser was realized with a radiofrequency discharge in a mixture of He–Ne. Five different wavelengths in the infrared were active, giving continuous operation. Population inversion was achieved between several Ne levels by means of excitation transfer from the metastable He (2^3S) to the 2s levels of neon (see **Figure 6.15**).[75] Although the $2s^2$ and $2s^4$ levels of neon may radiate to the Ne ground state, in the limit of complete resonance trapping, their lifetimes are determined primarily by radiative decay to the 2p levels. Under these conditions, the lifetimes of all of the 2s levels are about one order of magnitude longer than those of the corresponding 2p states. (The decay of the 2p levels is due to their radiative decay to the 1s levels). Thus, population inversions may be obtained on each of the 30 allowed 2s → 2p transitions.

The continuous-wave oscillation was observed in a discharge containing 0.1 mmHg of Ne and 1 mmHg of He in a quartz tube with an inside diameter of 1.5 cm and length 80 cm terminated with 13-layer evaporated-dielectric-film plane mirrors placed inside the tube (Figure 6.13 shows A Javan and W Bennett near the laser).

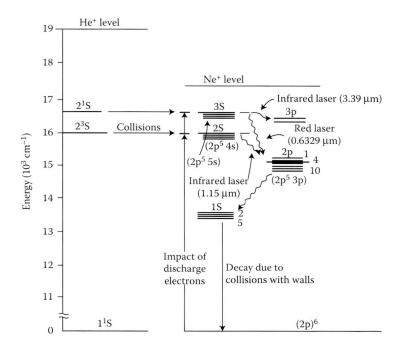

FIGURE 6.15 Energy levels of helium and neon. The helium levels are designated by the Russell–Saunders notation and the neon levels by Paschen notation. The laser transitions in the infrared (2.8–4.0 µm, 1.1–1.5 µm) and visible (0.59–0.73 µm) are shown.

Oscillations at 11,180, 11,530, 11,600, 11,990, and 12,070 Å were observed. The linewidth of the 11530 Å line (15 mW—the strongest one) was measured using the beating technique in a photomultiplier. The measured linewidth was in the range of 10–80 kHz and the angular spread of the beam was less than one minute of arc. Evidence of modes was obtained. The emitted beam had a diameter of 0.45 in. (about 1 cm).

A complete account of the experimental realization was given by Donald R Herriott at the Optical Society of America *Spring Meeting* at Pittsburg, March 1961 and at the *Second International Conference on Quantum Electronics*, Berkeley, March 23–25, 1961, and published a few months later in the *Journal of the Optical Society of America*.[76]

FIGURE 6.16 First realization and mounting of the He–Ne laser. (From D R Herriott, *J. Opt. Soc. Am.* **52**, 31 (1962).)

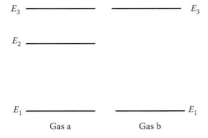

FIGURE 6.17 Energy-level diagram of the two-gas system used by Basov.

Figure 6.16 (Figure 1 of that paper) shows the first realization and its mounting. The paper also studied several properties of the laser and the importance of this laser for understanding of many lasing properties (*hole burning, mode patterns*, etc.) became immediately apparent.

Visible output at 6328 Å was obtained in 1962,[77] by changing the reflectivity of the external mirrors[78] and later operation at 3.39 μm in the IR was achieved.[79] The pertinent levels are also indicated in Figure 6.15. The 3.39-μm line shares a common upper state with the 6328 Å line. This means that they compete. Moreover, the 3.39-μm line has a much greater gain than that on the 6328 Å line, and in tubes a meter or more in length the infrared emission is strongly superradiant. This is the reason why the gain on the 6328 Å line does not increase beyond a certain limit when the length of the discharge tube is increased.

Much later, an emission in the green was also obtained.

The kinetics of the processes of interaction between the atoms of the two gases was developed by Basov and Krokhin.[80] The two gases were distinguished by the letters a and b. The atoms of the working gas "a" had three relevant levels; the atoms of the auxiliary gas "b" had two relevant levels, and the upper level of the two gases was coincident or nearly so (**Figure 6.17**).

The rate of change of the number of atoms of gas "a" on the upper level E_3 was written as

$$dN_3^a/dt = N_1^a[(1/\theta_{13}) + (1/t_{ba})] - N_3^a(1/\theta_3) + (1/t_{ab}) + (1/\tau_3)], \qquad \textbf{(6.10)}$$

where $1/t_{ba}$ is the rate per ground-state atom of gas "a" of resonant transfer of energy in collisions with atoms of gas "b" on level 3; $1/t_{ab}$ is the rate per excited atom of gas "a", for the inverse process; $1/\theta_{13}$ is the rate of excitation per atom from level E_1 to level E_3 by electron collisions of the first kind and $1/\theta_3$ is the total transition rate per atom of gas from level E_3 resulting from collisions of the second kind with electrons; and $1/\tau_3$ is the total transition rate from level E_3 arising from the remaining relaxation process. Excitation from level 2 to level 3 was neglected in this analysis because a term describing the process is proportional to N_2^a, and hence much smaller than the leading terms in Equation 6.10.

The rate of change of the number of atoms N_2^a on level E_2 was written as

$$dN_2^a/dt = (N_1^a/\theta_{12}) + N_3^a[(1/\tau_{32}) + (1/\theta_{32})] - N_2^a[(1/\theta_{21}) + (1/\tau_2)] \qquad \textbf{(6.11)}$$

where $1/\theta_{12}$ is the rate of excitation by electrons, $1/\theta_{21}$ is the rate for the inverse process, $1/\theta_{32}$ is the transition rate from level E_3 to level E_2 arising from electron collisions, and the τ's indicate radiative lifetimes.

In the stationary case, $dN/dt = 0$ and the condition for the existence of a population inversion between levels E_3 and E_2 is

$$N_3^a/N_2^a > 1.$$

According to collision theory, $t_{ba}/t_{ab} = N_1^b/N_3^b$ and assuming a Maxwellian distribution of electrons, or

$$\theta_{31}/\theta_{13} = \exp[-(E_3 - E_1)/kT],$$

the previous inequality can be written as

$$(\theta_{13}/t_{ba})\left\{1 - \left(N_1^b/N_3^b\right)\exp(-E_2/kT) + \theta_{21}\left[(1/\tau_{21}) - (1/\tau_{32}) - (1/\theta_{32})\right]\right\}$$
$$> \left\{\theta_{31}\left[(1/\theta_3) + (1/\tau_3)\right]\exp\left[(E_3 - E_2)/kT\right] - 1 + \theta_{21}\left[(1/\tau_{32}) + (1/\theta_{32}) - (1/\tau_{21})\right]\right\},$$

$$(6.12)$$

where the ratio N_1^b/N_3^b is to be considered an external parameter, that is, the effective excitation temperature for atoms of the gas "b" will be considered as given.

The quantities θ_{31}/θ_3 and θ_{21}/θ_{32} do not depend on the density of the electrons in the discharge, but are determined by the cross-sections of the corresponding processes and by the temperature of the electrons.

Inequality 6.12 has a different meaning in different regions of the variables $\theta_{21}[(1/\tau_{21}) - (1/\tau_{32})]$ and θ_{31}/τ_3. The range of these variables is divided into four regions by two curves. The equations of the curves are obtained by setting the expression in square brackets and the right-hand side equal to zero. In a first approximation, $1/\theta_{32}$ may be neglected in comparison with $1/\tau_{32}$ and one may take $\theta_{31}/\theta_3 \sim 1$; the curves then become straight lines of equations (**Figure 6.18**)

$$\theta_{21}[(1/\tau_{21}) - (1/\tau_{32})] = N_1^b/N_3^b\exp(-E_2/kT) - 1, \qquad \textbf{(6.13)}$$

$$\theta_{31}/\tau_3 = \left\{\theta_{21}\left[(1/\tau_{21}) - (1/\tau_{32})\right] + 1 - \exp\left[(E_3 - E_2)/kT\right]\right\}/\exp\left[(E_3 - E_2)/kT\right].$$

$$\textbf{(6.14)}$$

The four regions in Figure 6.18 therefore correspond to:

Region I, where the formation of a population inversion in gas "a" is obtained even in the absence of gas "b";

Region II, where population inversion occurs because of the presence of gas "b";

Region III, where gas "b" impedes the formation of a population inversion; and

Region IV, where the occurrence of population inversion is impossible.

Although this analysis did not in fact predict any possible laser system, it proved valuable in giving a fuller understanding of why certain gas lasers work.

Experimental research into the gas laser was, at the time these calculations came out, most intense. In his paper, Javan[81] presented his first experimental

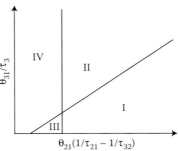

FIGURE 6.18 The Basov–Krokhin diagram. (From Basov and Krokhin, *Appl. Opt.* **1**, 213 (1962).)

realization of the He–Ne laser and mentioned concomitant research by Ablekov et al.,[82] writing:

> Recently some evidence for the presence of inverted populations in a Hg–Zn mixture has been reported …

In their paper, Ablekov et al. reported the observation of negative temperature or optical gain on the $4^1D_2 \rightarrow 4^1P°_1$ transition of Zn at 6562 Å in a DC discharge containing Hg and Zn. Ablekov also claimed to have observed line narrowing.[83] These results have never been repeated and no laser action has subsequently been obtained. However, these experiments show how the idea of utilizing a discharge in a two-component gas was common to several research groups independently.

During the next year or so, more than 40 different transitions were obtained from the visible (0.63 μm) to the medium infrared (12 μm) by using 10 different gaseous systems and at least four different excitation mechanisms.[84]

6.6 The Cesium Laser

The year 1961 witnessed the realization of two more lasers on which people had worked since the very beginning of the laser concept. One of them was the cesium laser. Townes and Schawlow, after writing their paper, had agreed that Townes would attempt to build a laser with potassium vapors, both because the calculations had shown it could work but also because potassium is a simple monoatomic gas with well-known properties and he desired a system whose properties could be analyzed in detail. "My style of physics"—he said later[85]—"has always been to think through a problem and theoretically analyse it, and then do an experiment which has to work. If it doesn't work at first, you make it work. You analyse and duplicate the theoretical conditions in the laboratory until you beat the problem into submission."

His preliminary calculations had revealed that a potassium laser would have a highly monochromatic output that would have been very useful for special applications. It would also have had drawbacks: low efficiency (about 0.1%) and a power output of fractions of a milliwatt.

While Townes concentrated on potassium vapor, Schawlow at Bell was studying ruby, concluding, however, that the lines later used by Maiman to build the first laser were not suitable. Townes, after asking for and obtaining financial support from the Air Force Office of Scientific Research, recruited two graduate students to work on the project: Herman Z Cummins (1933–2010) and Isaac D Abella (born 1934).

However, the project encountered a series of problems. The potassium vapors were darkening the glass tube of the discharge and chemically attacking the seals. Often the gas blew up the distillation apparatus; at other times, it plucked up impurities during distillation such that, later, potassium ions that had been excited would lose energy. At the end of 1959, Townes asked Oliver S Heavens, a British expert in the physics of dielectric layer mirrors, to come to help in the research and had decided to shift to a laser which used cesium vapors instead of potassium, pumping them with a helium lamp.

One of the narrow absorption lines of the cesium atom has exactly the same energy as one of the narrow helium lines. It is therefore possible to use the light emitted at this wavelength (389 nm) from a helium lamp to pump selectively a cesium level which populates more than the lower levels. So population inversion is obtained (**Figure 6.19**).

They used an elliptical reflecting cylinder and tried sapphire tubing and sapphire windows to avoid cesium darkening in glass. By June 1960, they had measured gain but no laser oscillation. Heavens attended the June spectroscopy meeting in Rochester, New York, and gave an overly optimistic progress report of the cesium work.

After the announcement by Maiman, Townes shifted Abella to work on ruby, while Cummins continued on the cesium project. The cesium laser was made to work at TRG between the end of 1961 and the first months of 1962 by Paul Rabinowitz, Stephen F Jacob, and Gould[86] and emitted radiation at 3.20 and 7.18 µm. It was one of the lasers made possible using confocal mirrors. Gould had mentioned in his patent application that it was a feasible but not promising laser. The research at TRG had started with potassium, but after a seminar by Oliver Heavens who had spoken of cesium as a better candidate, TRG also diverted to study this material and, pushed by the necessity to show the government that the million dollars they had received had been well spent, eventually succeeded in obtaining population inversion in March 1961 and oscillation early in 1962.

The laser was more a curiosity than a practical source of radiation because today there are other more easily constructed lasers which emit in the same wavelength range, and in addition cesium vapor is noxious.

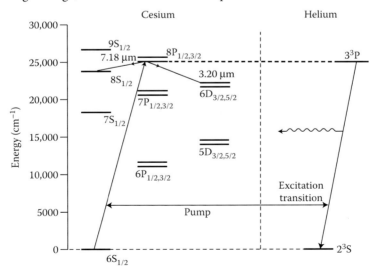

FIGURE 6.19 Energy levels of cesium and helium.

6.7 Laser Cavities

At this point, the role of resonant cavities in laser design and evolution should be considered because of the strong connection between the development of resonators and of the different kinds of lasers which took place at this time, after the first ruby and He–Ne lasers.

We have already noted that the use of a Fabry–Perot-type cavity was independently proposed by Basov and Prokhorov, by Dicke, by G Gould, and by Schawlow and Townes. In their paper,[66] Schawlow and Townes understood immediately that this was a crucial point in laser design and discussed their choice of a Fabry–Perot cavity at length. It is remarkable also that the first lasers and almost all present working lasers use the same basic principle, although a number of different configurations have been introduced. Indeed, the original proposal of using plane parallel mirrors was employed by Maiman and Javan. However, plane mirrors are difficult to align and the silver or aluminum coating for good reflective surfaces, as first used, is also critical.

Immediately after the first laser achievements, resonator characteristics started to be improved. The first improvement was the use of a dielectric coating, which decreases losses, improves the quality of the mirror, and also allows an initial selection of the wavelength to be amplified in the active material. Javan, in his first He–Ne laser, where a very high reflectivity was needed due to the small gain of the medium, used 13-layer evaporated dielectric films on fused silica plates flat to $\lambda/100$. The layers were alternate films of zinc sulfide and magnesium fluoride which gave a maximum reflectance of $98.9 \pm 0.2\%$, transmittance of 0.3%, and scattering and absorption losses of 0.8%.[73,87]

The two Fabry–Perot plates of the cavity were mounted within the gas chamber (Figure 6.14) to eliminate the losses that would be caused by unwanted reflections at windows.

Most technical complications occurring in the construction of this first laser could be avoided by simply putting reflectors outside. William W Rigrod, Herwig Kogelnik, Donald R Herriott, and D J Brangaccio[88] realized the first laser with external mirrors using the strong 1.1530 μm transition in the He–Ne mixture. The laser beam was allowed to pass through flat mirrors whose surface normals were placed at the Brewster angle with respect to the beam axis (**Figure 6.20**). This arrangement eliminates Fresnel refection loss for radiation polarized in the plane of incidence and prevents oscillations in the polarization normal to this plane.

The resultant losses in window transmission come entirely from scattering by imperfections or absorption in the glass and, with reasonable care, may be kept below about 0.5%. The tolerances (about ±3%) in setting the window at the Brewster

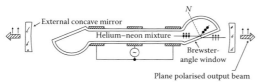

FIGURE 6.20 The external mirrors confocal cavity used by Rigrod et al. using Brewster angle windows. (From W W Rigrod et al., *J. Appl. Phys.* **33**, 743 (1962).)

angle are quite coarse and easily met. Another innovation in this set-up was the use of spherical mirrors of equal radius with coincident foci (*confocal system*). This system allows rather low-quality windows (~1/4λ) to be used and the criterion for mirror alignment is less severe than the plane-parallel case by about two orders of magnitude in the plate angles.

The confocal geometry had already been suggested by Connes to ameliorate interferometer performances.[89] It was also suggested for laser cavities on the basis of theoretical studies on diffraction losses by A Gardner Fox and Tingle Li (1931–2012),[90] Series,[91] and Lewis.[92] Considering a confocal resonator and a plane-parallel resonator each of spacing d and equal Q, the energy distribution in the former is more concentrated on the axis. Therefore, the confocal resonator has a smaller effective mode volume. The volume of active material will thus be less for the confocal than for the plane-parallel resonator. Therefore, the confocal resonator requires less pump power than the plane-parallel resonator.

The approximate theory of the plane-mirror resonator, as developed by Schawlow and Townes,[66] required a more detailed analysis for diffraction losses. It was known that modes exist in microwave cavities, but there was no definite proof that existed in a Fabry–Perot that is very different from the closed box of a traditional microwave cavity. The first satisfactory approach to the problem was due to Fox and Li[90,92] who investigated the effects of diffraction on the electromagnetic field in a Fabry–Perot interferometer in free space and proved modes indeed exist. Their approach was to consider a propagating wave which is reflected back and forth by two parallel, plane mirrors. They observed that this is equivalent to the case of a transmission medium comprising a series of collinear, identical apertures cut into parallel and equally spaced black partitions of infinite extent. They assumed initially an arbitrary initial field distribution at the first mirror and proceeded to compute the field produced at the second mirror as a result of the first transit. The newly calculated field distribution was then used to compute the field produced at the first mirror as a result of the second transit. The computation was repeated over and over again for subsequent successive transits, and the authors inquired whether, after many transits, the relative field distribution approached a steady state, if there were more than one steady solutions and what were the losses associated with these solutions. The self-reproducing distribution of phase and relative amplitude over the aperture that they found may be regarded as the proper resonant mode of the cavity.

This calculation, based on Huyghen's principle, was rather cumbersome: in the end, a computer was used to give numerical solutions for rectangular-plane, circular-plane, and confocal-spherical or parabolic mirrors.

A parameter which was useful in the calculation was the quantity $N = a^2/\lambda L$ (*Fresnel number*), defined for circular mirrors of a radius a at a distance L from each other. From diffraction theory, when $L \ll a^2/\lambda$, the center of one mirror is in the near field of the other regarded as an aperture, and the field in the central region of the second mirror may be calculated from the field on the first by means of geometrical optics. When, on the other hand, $L > a^2/\lambda$, the Fresnel zones appear and geometrical optics is not adequate for the calculations.

A sample of the distribution of amplitude and phase of the dominant mode (TEM_{oo}) obtained by Fox and Li is shown in **Figure 6.21** for a pair of circular-plane mirrors for $N = 2$, 5, and 10. The undulations on the curves are related to the number of Fresnel zones.

The electric-field configurations of modes were also derived and were depicted for square and circular mirrors (**Figure 6.22**).

Using the same method, Gary D Boyd, James Gordon, and Herwig Kogelnik[93] investigated the modes in a resonator formed by two spherical reflectors. They reached many interesting conclusions.

One was the equality of the resolving power of the Fabry–Perot cavity with its Q, within the small loss approximation; another was the nowadays well-known Gaussian field distribution of the field in the TEM_{oo} mode. They showed that the confocal resonator has several advantages. The diffraction losses are orders of magnitude less than for the plane-parallel resonator. The optical alignment of the two reflectors is not critical. On the basis of an analysis in terms of geometrical optics, Boyd and Kogelnik,[93] following a suggestion published later by Fox and Li,[94] were able to divide resonators into two classes: stable resonators and unstable resonators. Resonators are stable if

$$0 < [(d/b_1) - 1][(d/b_2) - 1] < 1, \qquad (6.15)$$

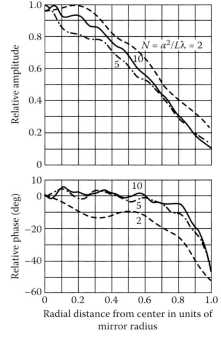

FIGURE 6.21 Relative amplitude and phase distributions of the dominant (TEMoo) mode for circular plane mirrors. (From A G Fox and T Li, *Bell Syst. Tech. J.* **40**, 453 (1961).)

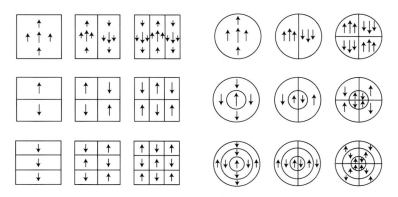

FIGURE 6.22   Field configuration of normal modes for square and circular mirrors. (From A G Fox and T Li, *Bell Syst. Tech. J.* **40**, 453 (1961).)

where the two mirrors of radii b_1 and b_2 are at a distance d apart and unstable whenever this condition is not fulfilled. A bundle of light rays launched in a stable system is periodically refocused as it travels back and forth between the two mirrors, whereas in an unstable system it is dispersed more and more.

A stability diagram for the various resonator geometries was given by Boyd and Kogelnik in the form shown in **Figure 6.23**. According to Equation 6.15, the boundary lines between the stable and unstable regions are two straight lines given by

$$d/b_1 = 1, \quad d/b_2 = 1 \tag{6.16}$$

and a hyperbola which satisfies $d = b_1 + b_2$.

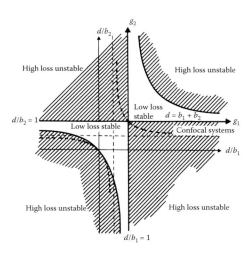

FIGURE 6.23   Two-dimensional diagram of stable and unstable regions. (From G D Boyd and H Kogelnik, *Bell Syst. Tech. J.* **41**, 1347 (1962).)

For confocal systems, that is, systems with coinciding reflector foci, we have $2d = b_1 + b_2$, which may be written as

$$[(d/b_1) - 1/2][(d/b_2) - 1/2] = 1/4. \tag{6.17}$$

These systems are represented in Figure 6.23 by points on another hyperbola and fall within the high-loss region. Boyd and Kogelnik also showed that the stable and unstable regions can be predicted from the theory of optical modes and that one must expect relatively high diffraction losses for unstable systems and relatively low diffraction losses for stable systems. This fact was also well demonstrated by Fox and Li.[94]

These relations were also shown to derive from the stability of a sequence of equally spaced lenses of equal focal length.[95] Later on, the same relation was derived using simple geometrical optics and the transport matrix method.[96]

The results obtained by Fox and Li and Boyd et al. were immediately studied and verified by a number of other researchers[97] using various analytical techniques and different kinds of mirrors or configurations. **Figure 6.24** shows several possible cavity configurations which were studied.

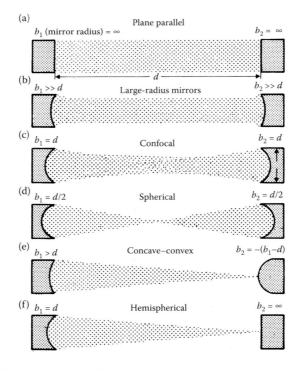

FIGURE 6.24 Resonator configurations giving uniphase wavefronts (the intracavity radiation pattern is shown in light grey). (From A L Bloom, *Spectra Phys. Laser Tech. Bull.* 2 August (1963).)

The importance of the study of resonant cavities in promoting the development of gas lasers cannot be underestimated. Until this theoretical work was done, the gas laser was at best a marginally operating device whose oscillation depended upon almost impossible tolerances in the end mirror adjustments. Theoretical studies on curved mirror resonators showed that resonators could be devised that were relatively insensitive to mirror adjustment and whose intrinsic losses could be lower than those of a plane-parallel resonator, allowing observation of laser action in media with much lower gain than was at that time thought possible. The plane-parallel resonator has since almost dropped out of existence for practical laser work, and all discoveries of new gas laser transitions have been performed with curved mirrors.

One of the predictions of the theoretical models concerned the field and theoretical intensity distribution of mode patterns. An approximate description of the transverse (x,y) field distribution and the resonant frequencies of the modes of stable resonators was given in several works.[97,98] The modes are distinguished by their mode number (m, n, and q for rectangular geometries, and p, l, and q for cylindrical geometries). The mode number q measures the number of field zeros of the standing wave pattern along the z-axis. In a rectangular geometry, the transverse mode numbers m and n measure the field modes in the x- and y-directions. For rectangular mirrors, the field distribution of a TEM_{mnq} mode is given by

$$E(x,y) = E_o H_m[(x/w)2^{1/2}]H_n[(y/w)2^{1/2}]\exp[-(x^2 + y^2)/w^2], \qquad \textbf{(6.18)}$$

where E_o is a constant amplitude factor, $H_n(x)$ is the Hermite polynomial of the nth order, and the parameter w is the beam radius or spot size. This measures the beam width of the mode of lowest transverse order ($m = n = 0$), the fundamental. According to Equation 6.18, the field distribution of this mode is described by a Gaussian profile. A similar formula was obtained for circular mirrors. Formulae for the beam radius were derived by Boyd et al.[93]

Resonant frequencies were also derived. The frequency spacing v_o between the resonances of two modes with the same transverse mode numbers m and n (or p and l) and neighboring longitudinal-mode numbers q and $q + 1$ was found to be given by

$$v_o = c/2d.$$

For a confocal system,[93] the resonant frequencies v were further given exactly by

$$v = v_o[(q + 1) + 1/2(m + n + 1)],$$

and the same for cylindrical symmetry with ($m + n + 1$) replaced by ($2p + l + 1$).

For mirrors of unequal curvature, Boyd and Kogelnik[93] derived the approximate expression

$$v = v_o\{(q + 1) + (1/\pi)(m + n + 1)\cos^{-1}[(1 - d/R_1)(1 - d/R_2)]^{1/2}\},$$

where R_1 and R_2 are the radii of curvature of the two mirrors.

Masers and Lasers

The standing wave pattern of a mode inside the resonator is equivalent to the one of two traveling waves which propagate in opposite directions along the z-axis. If a mirror is partially transmitting, the wave incident upon it propagates along the optic axis outside the resonator, thus maintaining the relative field distribution and changing the beam radius, which thus becomes a function of z. Also the phase $R(z)$ changes. The study of the way these quantities change was started by Ammon Yariv and Gordon[99] and others[100] and is well known nowadays under the name of *Gaussian optics*.

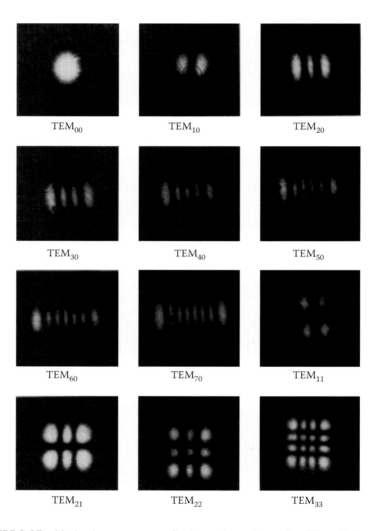

FIGURE 6.25 Modes in a concave-reflector rectangular cavity. (From H Kogelnik and W W Rigrod, *Proc. IRE* **50**, 220 (1962).)

The mode patterns predicted were easily verifiable quantities. Mode patterns of pure rectangular modes of a concave-mirror interferometer were obtained by Kogelnik and Rigrod[101] (**Figure 6.25**) and later by Rigrod[102] for pure circular modes of a concave-plane mirror interferometer (**Figure 6.26**). Mode patterns in ruby were obtained by Evtuhov and Neeland.[20]

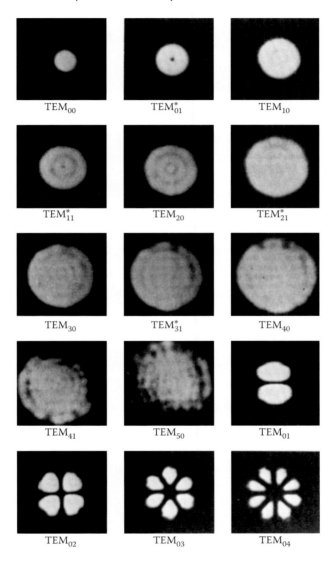

FIGURE 6.26 Asymmetrical modes of a concave-mirror interferometer. An asterisk designates two degenerate modes combining in space and phase graduated to form a composite circular-symmetric mode. (From W W Rigrod, *Appl. Phys. Lett.* **2**, 51 (1963).)

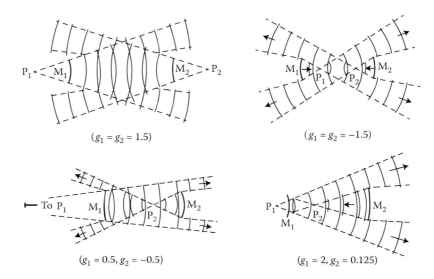

$(g_1 = g_2 = 1.5)$

$(g_1 = g_2 = -1.5)$

$(g_1 = 0.5, g_2 = -0.5)$

$(g_1 = 2, g_2 = 0.125)$

FIGURE 6.27 Some typical unstable mode patterns illustrating the mode behavior in different unstable regions of Figure 6.23. M_1 and M_2 are the cavity mirrors. (From A E Siegman, *Proc IEEE* **53**, 277 (1965).)

Attention was also given to unstable resonators by Siegman[103] in 1965, using a simplified geometrical optics analysis. He also pointed out that unstable cavities may be useful in some cases, and gave the following reasons:

1. Unstable resonators can have large-mode volume even in very short resonators

2. The unstable configuration is readily adapted to adjustable diffraction output coupling

3. Unstable resonators have very substantial discrimination against higher-order transverse modes

The unstable resonator has been recognized as being almost ideal for transverse-mode control in TEA lasers, enabling maximum energy extraction in a collimated beam with hard-to-damage reflecting optics.[104]

More extensive analyses were done by several groups.[105] Unstable resonators correspond to the shaded regions of Figure 6.23. **Figure 6.27** shows examples of unstable resonators.

Aside from the calculated structures shown in Figures 6.24 and 6.27, a number of other configurations were investigated.[106] Probably, the most interesting class is formed by resonators made using corner-cube reflectors instead of plane or curved mirrors, as illustrated in **Figure 6.28**. Peck[107] and Murty[108] already used cube-corner prisms in Michelson interferometers to relax alignment tolerances. Peck[109] then showed theoretically that a Fabry–Perot

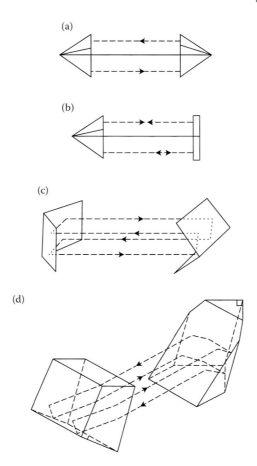

FIGURE 6.28 Several types of resonators using corner or cube reflectors to form the cavity: (a) a pair of corner reflectors cavity (from E R Peck, *J. Opt. Soc. Am.* **52**, 253 (1962)); (b) a corner and flat reflector cavity (from E R Peck, *J. Opt. Soc. Am.* **52**, 253 (1962)); (c) crossed 90° corner reflector cavity (from G Gould et al., *J. Opt. Soc. Am.* **52**, 452 (1962)); (d) crossed roof prism optical cavity (from G Gould et al., *J. Opt. Soc. Am.* **52**, 452 (1962).)

interferometer with cube-corner reflectors replacing one or both flats possesses six eigen polarizations (i.e., polarizations which are reproduced after traversing the prisms). Gould et al.[110] considered a cavity consisting of two crossed 90° roof reflectors, replacing the Fabry–Perot mirrors and showed that it has only two plane eigen polarizations, still maintaining relaxed alignment tolerances. They also considered the variants roof-reflector versus flat, and the use of entrance faces at or near Brewster's angle.

The reflection losses in corner reflectors are virtually eliminated by employing total internal reflection in a prism.[111] A ruby with one end flat (covered with dielectric coating) and the other cut with a 90° dihedral angle like a roof top so as to be effectively 100% reflective was considered by Dayhoff.[112] External glass roof prisms were also used to end ruby rods.[113]

Finally, coupling techniques using the phenomenon of frustrated total internal reflection were proposed by several authors.[111-114] References to Russian work may be found in Anan'ev.[115]

Another interesting structure was the one in which light traveled in a closed path, as in a ring. These ring lasers,[116] as they were called, have several interesting properties which found application, for example, in making gyroscopes.[117]

A particular kind of modes are the so-called *whispering gallery modes* characteristic of ring and spherical cavities. These modes were first described by Lord Rayleigh[118] to explain an acoustic phenomenon occurring in the St. Paul's Cathedral in London. In the Cathedral, ghostly whispers are sometimes heard in any point around the circumference of the gallery—a circular platform running around the base of the dome's interior from which visitors can get up close to the dome's decoration—when someone is speaking inside. Rayleigh demonstrated that the sound is carried by waves that travel around the circumference clinging the walls.

In the electromagnetic domain, their existence in dielectric spheres was already discussed by G Mie[119] and P Debye.[120] Richtmyer[121] proposed to create high-Q resonators using whispering gallery modes. They were later used in microwave theory.[122] The first observation in a laser contest was made by C G B Garrett et al.[123] using Sm:CaF$_2$. However, they did not attract much attention until the 1990s when they suddenly became the object of wide studies.

The radiation of a whispering gallery mode propagates within a narrow layer adjacent to the wall surface of the cavity. They can be understood as closed circular beams supported by total internal reflection from the boundaries of the resonator. The simpler geometry of the resonator is either a ring or a cylinder or a sphere. When the reflecting boundary has high index contrast, and the radius of curvature exceeds several wavelengths, the radiation losses, similar to bending losses of a waveguide, become very small, and the Q-factor becomes limited only by material attenuation and scattering caused by geometric imperfections and may be very high.

They were discussed by several authors.[124] Lasers using whispering modes were constructed.[125]

A method which had escaped the Fabry–Perot logic was studied by E Snitzer who, in two papers written in 1960,[126] considered the use of fibers as dielectric waveguides to provide a resonant structure for an optical maser. Being an expert in propagation in optical fibers, he immediately appreciated the possibility of having a mode selection and a stronger mode coupling in a fiber which was small enough to support only one or two modes. In his analysis, he considered a fiber terminated with partially reflecting ends and observed that laser action should be possible at much lower power levels than in an ordinary Fabry–Perot cavity.

At the same time in which the study of the mode configuration of the electromagnetic field in an empty cavity was made, some attempt was done to understand the behavior of a cavity filled with an active medium.

An analysis by A Kastler[127] pointed out that a Fabry–Perot cavity filled with atoms, even in the case when the system works below the threshold of laser action, shows remarkable properties. In the case of external illumination, the local light intensity in the stationary waves inside the interferometer can be much higher than the intensity of the incident light beam. When atoms are emitting, he showed that narrow fringes of very strong intensity are obtained. He also considered two cases of non-uniform distribution of emitting atoms. The first occurred when the emitting atoms were located in an atomic beam, showing that Doppler broadening can be suppressed; and the second occurred if atoms had a lamellar distribution between the plates. This last case was a precursor of present distributed-feedback structures.[128]

6.8 Further Progress in Gaseous Lasers

In consequence of the notable advances in the understanding of the properties of an active medium in a resonant cavity, researchers were able to optimize design geometry in their endeavors to construct new lasers.[129]

It could be said with safety that today everything can lase; consequently, it would be impossible to trace out the development of all the hundreds of different possible lasers in existence, so we shall confine ourselves here and in the following few sections to a few comments pertaining to some of the most well-known lasers, beginning here with gaseous systems.

It is customary to classify lasers obtained in a gaseous medium according to the following general scheme:

1. Neutral atom lasers

2. Ion lasers

3. Molecular lasers

4. Excimer lasers

6.8.1 Neutral Atom Lasers

A typical representative of the neutral atomic gas laser is the He–Ne laser already described. Other neutral atom lasers are atomic vapor lasers which include the Cu, Pb, and Tl systems. They are characterized by the features that the upper laser level is optically connected to the atom ground state and must therefore have the opposite parity to the ground state. The lower laser level is a metastable level of fairly low excitation energy. One of the attractive features of this class of lasers is their high efficiency.[130]

The first of these lasers was found by Fowles and Silfvast[131] in 1965 at the University of Utah using lead, which emitted at 723 nm. They discovered a new class of metal vapor lasers trying several materials and obtaining laser action at 492.4 nm vaporizing zinc, then Cd which lased at 441.6 nm, and in 1967 considered He–Cd.

The laser action in lead occurs at a single wavelength, 7229 Å, which corresponds to the transition $6s^2 6p7s(^3P^\circ_1) \rightarrow 6s^2 6p^2(^1D_2)$ in the neutral lead atom (**Figure 6.29**). Lead was vaporized heating the metal in an oven at a temperature in the range 800–900°C and then excited in a discharge tube.

The best studied example of this type of laser is the copper vapor laser which uses vapors of copper as the lasing medium in a three-level system (**Figure 6.30**). It was first discovered by Walter, Pilch, Solimene, Gould and Bennett[132] at TRG and is described by Walter.[133] Lasing is obtained in the $^2P_{1/2} \rightarrow ^2D_{5/2}$ and $^2P_{1/2} \rightarrow ^2D_{1/2}$ transitions of neutral copper atom at 511 and 578 nm, respectively. Metallic copper was vaporized at high temperature (1400°C) and excited in a longitudinal pulsed discharge.

The relative values of the corresponding cross-sections are such that the rate of electron impact excitation of the P states is greater than that to the D states; thus, the P states are preferentially excited by electron impact. At the high temperature used ($T = 1500°C$), the only effective decay of the P states is to the D states with two main lines at 510 nm (green) from $4p^2P_{3/2}$ to $4s^2\ ^2D_{5/2}$ and 578 nm (yellow) from $4p^2P_{1/2}$ to $4s^2\ ^2D_{3/2}$.

The pulse width is typically from 5 to 60 ns and peak power from 50 to 5000 kW has been obtained. The average power can range from 25 W to more than 2 kW.

A major advance occurred in 1972 when the Soviets obtained high power outputs, exceeding 15 W. A A Petrash et al.[134] at Lebedev Institute in Moscow showed that, by repetitive pulsing the discharge at a high repetition rate, the device becomes self-heating, obviating the need for an external oven. I Smilanski[135] at the Nuclear Research Centre Negev in Israel showed that it was possible to scale the output power with the tube diameter. Through the 1970s and 1980s, many laboratories were developing copper laser systems for dye laser pumping. In the United States, the interest came from the Lawrence Livermore

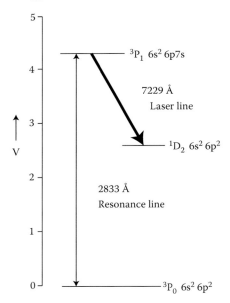

FIGURE 6.29 Energy-level diagram for the lead laser. (From G R Fowles and W T Silfvast, *Appl. Phys. Lett.* **6**, 236 (1965).)

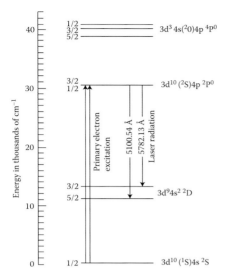

FIGURE 6.30 Energy-level diagram for the copper laser. (From W T Walter et al., *IEEE J. Quantum Electr.* **QE-2**, 474 (1966). © 1966 IEEE. With permission.)

National Lab which was using these lasers to pump dye lasers for uranium enrichment.[136]

6.8.2 Ion Gas Lasers

Ionized gas lasers use excited ions obtained in an electric discharge. The excitation energy of an ionized atom is much larger than in the case of a neutral atom and the spacing between energy levels is greater; therefore, the emission of these lasers is usually in the visible or near ultraviolet.[137]

They also require a much higher current density for the discharge than is required by neutral gases. This is because the lasing level is populated through two or more successive collisions, the first one being to produce the ion and the other to excite it. As a consequence, these lasers have a high plasma temperature in the order of several thousands of degrees. The first laser action in gaseous ions was obtained by William Earl Bell, one of the founders of Spectra Physics, toward the end of 1963. He observed two visible and two infrared transitions between levels of singly ionized mercury.[138] This first ion laser was excited in a pulsed mode by discharging capacitors charged to high voltage into a discharge tube filled with a 500:1 mixture of He:Hg at a pressure of 0.5 Torr. The laser was interesting in that it gave visible wavelength operation ($\lambda = 5678$ Å), high peak power (up to 40 W), and exhibited high gain (more than 0.8 dB cm^{-1}). The possibility of obtaining cw laser action in an ionic system of this kind also seemed likely, as the length of the laser pulse could be extended by increasing the length of the discharge excitation pulse.

Early in 1964 a report appeared in the magazine *Electronics*[139] of a laser transition observed at a wavelength of 5225 Å in a mercury–argon mixture, and laser action was subsequently reported.[140] The transition was not assigned to either argon or mercury but it may have been the earliest observation of a noble gas ion laser transition, as noted by Bridges,[141] who reported the observation of 10 laser transitions in the green and blue portions of the visible spectrum between levels of singly ionized argon. These transitions were excited in the pulse mode using either pure argon or mixtures of argon with helium or neon buffer gas. Bridges wanted to study the effect of noble gases especially helium in the He–Hg ion laser,[142] so substituted He with Ne and found the system still works, then tried with Ar. At first the system did not work, so he pumped Ar out and re-introduced He to check the mirrors' alignment.

To his surprise, the helium–mercury laser had now also a new line at 4880 Å in addition to the red and green lines from mercury. That line was produced by ionized Ar. It was February 14, 1964.

Independently, and almost simultaneously, similar results were reported by Convert et al.[143] and Bennett et al.[144] Bennett et al. obtained superradiance and quasi-cw oscillation on several lines and proposed an excitation mechanism through electron impact excitation. Some of the observed transitions exhibited very high gain (several dB m⁻¹) and produced large peak powers (several hundred watts).

Bridges[145] also obtained pulsed ion laser operation in xenon and krypton and the gain in Ar at 4880 Å was the highest yet demonstrated for any line in the visible region.

Almost immediately afterwards, Eugene I Gordon et al.[146] reported continuous laser action in many of the previously reported pulsed transitions in singly ionized argon, krypton, and xenon. They used reflecting prisms (see **Figure 6.31**) as described by White[147] to discriminate between the various wavelengths and obtained single-wavelength operation. Other authors extended the number of ion laser transitions into the infrared with mercury,[148] and with the rare gases into the visible,[149] the ultraviolet,[150] and the infrared.[151]

Gerritsen and Goedertier[152] reported the first laser transition between levels of a doubly ionized atom: mercury.

Using singly ionized Cu, cw operation has been obtained at ultraviolet wavelengths.[153]

Longitudinal discharge lasers[154] and transverse-discharge copper vapor lasers were reported.[155] To lower the operation temperature, one may obtain the metal from the vapor phase of a compound, such as a halide, which has appreciable vapor pressure at much lower temperatures (450–500°C only for CuI). Laser action in copper halide vapors was first demonstrated in 1973 by Liu et al.[156] and Chen et al.[157] Transversely excited CuI laser was made by Piper.[158]

The discovery of new ion laser transitions had, by the middle of 1965, covered 11 elements with 230 reported transitions.[159] Ten years later, in 1975, ion laser oscillation had been observed in 32 elements, up to the fifth stage of ionization in some ion transitions.[160]

The most important laser of this category is the Ar⁺ laser. Due to the high power output attainable in the cw mode of operation and to the convenient wavelength region spanned by the oscillation frequencies (0.45–0.52 μm), the Ar⁺ laser soon became one of the most important lasers available. It can be operated in a pure

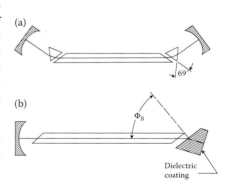

FIGURE 6.31 (a) Typical high-gain tube utilizing two Brewster-angle prisms and (b) half-confocal dispersive cavity. (From A D White, *Appl. Opt.* **3**, 431 (1964).)

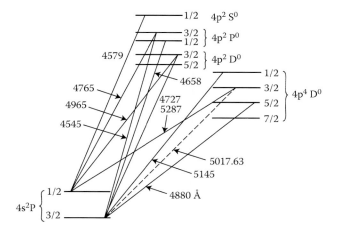

FIGURE 6.32 Energy levels of Ar$^+$ and nine laser transitions. (From W B Bridges, *Appl Phys. Lett.* **4**, 128 (1964).)

Ar discharge that contains no other gases. The excitation mechanism is discussed in detail in Note 160. The pertinent energy-level scheme was first given by Bridges[141] who also identified most of the observed transitions and is given in **Figure 6.32**. The more prominent transitions are at 4880 and 5145 Å. In the paper, the first line was erroneously assigned to the transition $4p^4D°_{1/2} \rightarrow 4s^2P_{3/2}$ instead of $4p^4D°_{5/2} \rightarrow 4s^2P_{3/2}$. This line has been corrected in Figure 6.32.

The early commercial argon ion lasers were unreliable and short-lived. Their problems are discussed in the book by Lisa Bromberg.[51]

6.8.3 Molecular Lasers

Molecular lasers use transitions between vibrational and rotational levels of a molecule. Two possible general schemes are feasible. In the first one, transitions between vibrational states of the same electronic level (fundamental) are used. In the second one, vibrational states of different electronic states are used. A typical exponent of the first class is the CO_2 laser, and of the second is the N_2 laser. Both were discovered in 1963.

Legay and Barchewitz in 1963[161] observed strong infrared emission from CO_2 when it was mixed with vibrationally excited nitrogen, and they interpreted this as an indication of collisional energy transfer. Subsequently, Legay and Legay-Sommaire[162] suggested on theoretical grounds the possibility of obtaining laser action on the rotation–vibration bands of gases excited by active nitrogen and mentioned specifically the CO_2 10.4 μm transition.

At about the same time, Patel et al.[163] observed laser action in pure CO_2 at 10.6 μm. This discovery came at a time when the search for new laser transitions from ionic and molecular species was near its peak and this particular laser did not attract any more attention than the hundreds of other laser transitions reported at that time. A few months later, Patel[164] published a detailed account of his results. Using a 5-m tube, he had obtained outputs of 1-mW cw with DC excitation and 10-mW peak with 1-μs excitation pulses. He also presented an interpretation of previously reported results.[165] A short while after, he observed laser action in a CO_2–N_2 mixture[166] and realized that this was a much more efficient system. In this paper, he discussed the transfer excitation mechanism and the lasing lines which are shown in **Figure 6.33**. The selective excitation of the CO_2 molecule from its ground state to the 00°1 state[167] takes place during a two-body collision involving a CO_2 ground-state molecule and a vibrationally excited N_2 molecule also in its ground electronic state. The first vibrationally excited level (v = 1) of N_2 is only 18 cm^{-1} distant in energy from the 00°1 vibrational level of CO_2.

The experimental set-up is shown in **Figure 6.34**. It is a continuous-flow system with nitrogen flowing through a high-frequency discharge. The active nitrogen thus formed was mixed with CO_2 in an interacting region between the mirrors of a Fabry–Perot resonator; an output of over 1 mW was observed with a 20-cm long interaction region on the strongest line at 10.6 μm (10.5915 μm). Other lines from 10.5322 to 10.6537 μm were observed, all belonging to the P branch from P(14) to P(16). Further work by Patel[168] and Legay-Sommaire[169] led to the direct excitation of the gas within the laser resonator and the attainment of 12 W from a 2-m tube containing a mixture of CO_2 and air.[168] Howe[170] in the same year investigated the effect of various molecular gases on the laser output; he observed oscillation on R-branch lines as well as P-branch lines and noted that the flow rate of the mixture affected the gain. Moeller and Rigden[171] reported 10 W m^{-1} from a sealed tube containing a CO_2–He mixture and also mentioned, with no details, that a flowing mixture of CO_2, N_2, and He gave good results, but the simultaneous report by Patel et al.[172] gave a

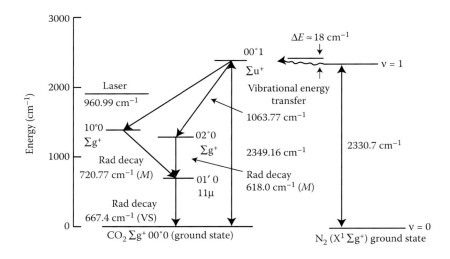

FIGURE 6.33 Energy-level diagram showing pertinent levels of CO_2 and N_2. (From C K N Patel, *Phys. Rev. Lett.* **13**, 617 (1964).)

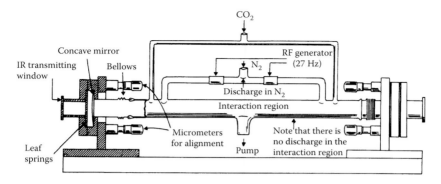

FIGURE 6.34 The experimental apparatus used by Patel in 1964 for laser action in N_2–CO_2 (the drawing is not to scale). (From C K N Patel, *Phys. Rev. Lett.* **13**, 617 (1964).)

detailed account of a water-cooled tube using a flowing CO_2–N_2–He mixture which produced over 50 W m^{-1}. In 1966, Q-switching techniques were applied to the CO_2 and N_2O molecular gas laser systems.[173] This paper marks the end of the discovery phase in the history of the CO_2 laser. In 1968, Hill[174] obtained 5 J, 200 kW pulses from CO_2 gas at 10.6 μm at pressures as high as 60 Torr using an ultrahigh-voltage axially pumped discharge. This improvement of nearly three orders of magnitude over previous results was a direct consequence of the use of high-voltage (10^6 V), fast rise time (<5 μs) electrical pulses to excite the CO_2 laser at somewhat elevated pressure (60 Torr). Later on, to meet the requirement of a short discharge time and to lessen the requirement for extremely high applied voltages, pulsed transverse excitation on the CO_2–N_2–He was used.[175] These transversely excited atmospheric-pressure lasers are now called TEA lasers. The technique was immediately applied to several systems[176] and gives high power.

By the mid-1960s, it had already been realized that an electron beam might be used both to improve the performance of existing lasers[177] and to extend laser action into the UV portion of the spectrum.[178] The importance of these ideas was not realized until about 1970 or 1971, when the electron-beam-pumped molecular Xe excimer laser[179] and the electron-beam-controlled discharge atmospheric-pressure CO_2 laser appeared.[180]

These developments were quickly followed in 1971 by the first electron-beam initiation of an HF chemical laser[181] and extension of laser action in CO_2 to 25 atm[182] using an electron-beam-controlled discharge. A good and interesting historical survey of the development of pulsed molecular lasers up to 1974 is given by O R Wood II.[183]

Finally, although not strictly connected with the CO_2 laser are gas dynamic lasers. First proposals were by Basov and Oraevskii[184] and other authors.[185] In these lasers, the rapid adiabatic expansion of a gas is able to create population inversion between vibrational levels of molecules in a gas mixture which is initially in equilibrium at a very high temperature obtained simply by heating.

Three characteristic times are of importance in this case: the relaxation time τ_S of the upper level, the relaxation time τ_L of the lower level, and the time needed to reach equilibrium τ_E. If $\tau_E \sim \tau_L \ll \tau_S$ during expansion, the population of the lower level is able to follow the temperature and pressure variations of the gas and so remains in equilibrium with it. However, the population of the upper level is not able to follow the fast variations of conditions of the gas and remain "frozen" to the equilibrium value at the initial temperature before expansion.

To fulfill these requirements, it is necessary to expand the gas through a supersonic nozzle. Gas dynamic lasers usually operate with a mixture of gases (usually CO_2, N_2, and H_2O or He).

N_2 is representative of the class of molecular lasers which use transitions between vibrational states of different electronic states. An initial proposal for laser action in this system was made as early as 1960 by Houtermans.[186] Matthias and Parker[187] were the first to observe laser action in N_2 in 1963, using a peak-pulse voltage and current of about 40 kV and 90 A, respectively, in 2 Torr pressure of nitrogen. They obtained laser action in a number of transitions which have been shown to belong to the first positive system ($B^3\Pi_g - A^3\Sigma^+$) of nitrogen in the range of wavelengths from about 7500 to 12,500 Å.

Later, Heard[188] obtained laser action in a pulsed nitrogen discharge in the near-UV region of the spectrum. The laser transitions extended from about 3000 to 4000 Å and belonged to the second positive system $C^3\Pi_u^+ - B^3\Pi_g$ (electronic transitions). The lower laser level here is the upper laser level for the laser action in the near IR. The strongest transition is at 3371.3 Å. In 1965, Leonard[189] observed 200 kW pulses at 3371 Å in N_2 gas at 20 Torr pressure using pulsed transverse excitation, and similar results were also obtained by Gerry.[190] The nitrogen laser is used to pump tunable dye lasers covering the near-ultraviolet and visible regions and is very simple to construct. The discharge in this laser runs transversally across the width of the active volume rather than along its length and the laser is working in a large range of N_2 pressures from a few Torr to one atmosphere.

6.8.4 Excimer Lasers

The last type of gas laser to be discussed here is the excimer laser. These lasers are an important workhorse for numerous industrial and material processing applications in the ultraviolet and are also the key component in corneal corrective surgery. Excimers are formed by the interaction between two atoms or molecules A and B, one of which is electronically excited

$$A + B^* \rightarrow (AB)^*.$$

The bound molecule $(AB)^*$ may then decay radiatively to the ground state and dissociate

$$(AB)^* \rightarrow A + B + h\nu.$$

The ground state may be either repulsive or sufficiently weakly bound so as to be unstable at normal temperatures. The term *excimer*, first introduced in 1960,[191] should properly denote an excited homopolar molecule, for example, Xe^*_2, whereas for an electronically excited heteropolar complex, for example, KrF^* the term *exciplet* should be used. Excimer lasers are capable of efficient generation of high-power pulses of radiation at ultraviolet and vacuum ultraviolet wavelengths in a spectral region which has no laser transitions. The potential of excimers as laser media was first pointed out by Houtermans[186] in 1960, before the first successful operation of the ruby laser by Maiman. However, the first demonstration of this type of laser action had to wait until 1970, when Basov et al.,[179] following an earlier proposal,[192] used liquid xenon pumped with a pulsed high-energy electron beam. At high current densities of the order of 3060 A cm^{-2}, narrowing of the luminescence line at 1760 Å with a half-width reaching 20 Å was observed.

In 1972, the first gaseous excimer using electron-beam excitation of high-pressure Xe gas was reported by Koehler et al.[193] at 173 nm. The first high-power output was subsequently demonstrated by Ault et al.[194] and Hughes et al.[195] These last researchers also obtained laser action in Ar, and laser emission from a krypton excimer was obtained in the same year by Hoff et al.[196]

In 1974, Golde and Thrush[197] at Cambridge University discovered that rare-gas monohalide molecules could be created in an excited state. Velazco and Setser[198] at Kansas State University discovered several other types of rare-gas halide molecules and suggested their use for making lasers. In May 1975, S K Searles and G A Hart[199] of the Naval Research Laboratory in Washington, DC, made the first laser of this kind using xenon bromine (XeBr) emitting at 282 nm. A number of other lasers[200] were produced: KrF at 249 nm, XeF at 354 nm, and XeCl at 308 nm.

6.9 The Liquid Laser: Dye and Chelate Lasers

The earliest published works suggesting that organic materials could be used as active media for laser appear to be those of Brock[201] and Rautian and Sobel'mann,[202] who in 1961 proposed that triplet-state phosphorescence could serve as the basis for an organic laser.

In 1964 Stockman, Mallory, and Tittel[203] discussed a laser process based upon singlet-state fluorescence and Stockman[204] described early results in the experimental efforts to realize a dye laser using the dye perylene excited by a fast, powerful flash lamp. His efforts, were, however, without success: it is clear in retrospect that the 1-μs flash lamp pulses which he used were too long for the production of laser action in perylene.

The first unambiguously successful attempt to produce stimulated emission from organic molecules was reported in 1966 by Sorokin and coworkers who used a giant ruby laser to excite solutions of the dyes chloro-aluminum pthalocianine (CAP)[205] and, a few months later, 3,3′-diethylthiadicarbocyanine (DTTC) iodide in an optical cavity.[206]

If most of the lasers we have considered up to now arose as a result of a highly coordinated effort, and required to employ advanced technologies, explaining in some way why they were all developed in the United States, the case of organic dyes (simple dyes) can be considered to be different. The first laser of this kind arose by chance, thanks to the technique of Q-switching (Chapter 7). At IBM, Peter Sorokin and John Lankard showed in 1966 that a simple and convenient material to use for this effect could be a group of organic dyes called metallic phthalocyanines dissolved in organic liquids. Phthalocyanines are complexes with metal ions at their centers. The two researchers asked a colleague of theirs, John Luzzi, to synthesize some and Luzzi, rather than making a gram which would have been the norm, made them a whole pound. So Sorokin placed these phthalocyanines right in the cavity of a ruby laser and switched on the laser. A single powerful pulse, 20-ns long, was the immediate result.

Whilst trying to better understand what was happening, Sorokin thought that these materials could also be used for other experiments and focused on two of them. In one experiment, he wanted to stimulate a Raman scattering. The other experiment was aimed at verifying if the dyes, pumped with the light from a ruby laser, were able to give the laser effect by themselves.

Sorokin decided to start with the first experiment, sending light from a ruby laser onto the sample. When the spectrum of the emitted light was examined, it was evident that the second experiment had succeeded. Placing the dye sample between two mirrors, Sorokin and Lankard obtained a powerful laser beam at 7555 Å. They tried with all the dyes they could find in the store and discovered the effect was quite general. One afternoon, Sorokin went down the aisle at his lab asking colleagues "What colour do you want?", so many were the wavelengths that could be obtained by changing the dye. One thing they did not notice was that the new laser could be tuned, that is, it emitted at a wavelength which could be varied over a reasonable range using the same material.[38,207]

A little later, and independently, the chemist Fritz P Schafer, then at Marburg University in Germany, while studying the saturation characteristics of certain organic dyes of the cyanine family, obtained the same effect. He was studying the light emitted by a dye pumped with a powerful Q-switched ruby laser with the help of a spectroscope, when his student Volze, by examining highly concentrated solutions, obtained signals thousands of times stronger than expected. Soon the two researchers understood they were witnessing laser emission, and together with Schmidt, then a doctorate student, photographed the spectra at different concentrations obtaining the first proof that a laser had been built which was tunable in wavelength over 600 Å by changing the concentration of the dyes or the reflectivity of the mirrors of the resonant cavity. The effect was soon confirmed by Spaeth and Bortfield[208] and by Schaefer et al.[209] using several cyanine dyes with structures similar to DTTC; and later by Stepanov et al.[210] and extended to a dozen different dyes of the cyanine family. A surge of interest resulted, as happens in these cases, and worldwide research soon identified thousands of dyes showing the laser effect.

Later still the Sorokin group suggested the use of flash lamp pumping[211] and obtained a working laser of this type;[212] Schmidt and Schaefer[213] followed.

Visible coherent emission was reported by three different research groups almost simultaneously.[214]

One of the most attractive properties of dye lasers is their tunability. In contrast to most other laser media, the emission spectra of fluorescent dyes are broad, permitting the lasing wavelength to be tuned to any chosen value within a fairly broad range. Furthermore, the number of fluorescent dyes is very large and compounds may be selected for emission in any given region of the optical spectrum.

Laser mechanism in dyes can be explained with reference to **Figure 6.35**. State S_0 is the ground state. S_1, S_2, T_1, and T_2 are excited electronic states. State S are singlets, that is, the spin of the excited electron is anti-parallel to the spin of the remaining molecule. States T are triplets in which the two spins are parallel. Singlet-to-triplet or triplet-to-singlet transitions are forbidden at first order. Transitions between singlet states or between triplet states give rise to intense absorption of light and to fluorescence. The characteristic color of organic dyes is due to the absorption from S_0 to S_1. All the states are then split into vibrational and rotational levels as shown in Figure 6.35.

Excitation is through the absorption of pumping light which raises the energy of molecules from the fundamental singlet state up to a vibrational level in the first excited singlet state.

This is followed by a very fast nonradiative decay to the bottom of the S_1 levels. Laser emission takes place between this level and a high vibrational level of the fundamental singlet state with lifetime τ_S. A nonradiative decay from this level to the fundamental state terminates the process.

The vibrational levels of the fundamental state, if sufficiently high in energy, can normally be found unoccupied. It is then possible to have enough population inversion for laser action between the states of the excited singlet and some of the vibrational levels higher than the fundamental state. This condition poses a link between the wavelength of the pumping light, the absorption coefficient of the material at that wavelength, and the wavelength at which the laser emission takes place.[215]

There is also some probability, W_{ST}, which depends upon the structure of the molecule, that a molecule in the excited singlet level relaxes to a triplet state. Since this transition is forbidden, its rate is usually much smaller than τ_S^{-1}. The lifetime τ_T^f or decay of T_1 to the ground state is relatively long. Therefore, the crossing between the singlet and the triplet is harmful

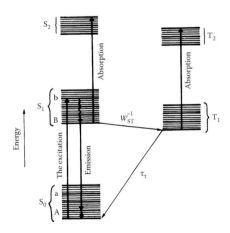

FIGURE 6.35 Diagram of the energy levels of an organic dye molecule. The heavy horizontal lines represent vibrational states and the lighter lines represent rotational fine structure.

for the laser for two reasons. Firstly, the accumulation of molecules in the triplet state lowers the concentration of molecules available to participate in the laser process. Secondly, and more seriously, the absorption of molecules due to a $T_1 \rightarrow T_2$ transition is very strong. If the wavelength region of this absorption coincides with that of the laser emission, an accumulation of molecules in T_1 will increase the laser losses. For this reason, many organic-dye lasers only operate on a pulsed basis.

Eventually in 1969, Snavely and Schaefer[216] obtained indication of a possible continuous emission, which was obtained the following year at the Eastman-Kodak Research Laboratory by Peterson, Tuccio, and Snavely[217] who used an argon laser to pump a solution of rhodamine 6G dissolved in water.

These lasers allowed a long pursued dream to be realized: to obtain a laser that was easily tunable over a wide range of frequencies. The dyes are also interesting from other reasons. They can be used dissolved in solids or liquids with easily controllable concentrations. The liquid solutions are particularly convenient. Cooling is obtained by recirculating the solution. Moreover, the liquid is not damaged and the cost of obtaining the active material is very low, in contrast to solid-state lasers, for example. By choosing a suitable dye among thousands available and the wavelength of the pump radiation, it is possible to cover all the visible spectrum up to the near infrared. A laser tunable to a desired wavelength was finally born!

In mid-1967, Bernard H Soffer and Brill B McFarland[218] substituted a diffraction grating for one of the mirrors of a laser-pumped dye laser and obtained efficient spectral narrowing and continuous tunability over a range of more than 400 Å, simply by tuning the position of the grating. They further reported the use of solid solutions of dyes in plastic as a practical dye laser material.

Dye lasers, due to their broad band emission, can be operated on a pulsed regime which may be shorter than a picosecond and play a fundamental role in short pulse generation (Chapter 8).

The first narrow linewidth tunable laser oscillator was made by Haensch.[219] Synchronous tuning in grazing incidence grating cavities[220] was demonstrated in 1981.[221]

Solid-state dye lasers were introduced by Soffer and McFarland[222] and Peterson and Snavely.[223] In the 1990s, new and improved gain media were introduced.[224]

An amusing experiment was performed in 1970 by Schawlow and Theodor Hansch who built the first *edible laser*.[225] The story is recollected by Hansch[226] who remembers that at some time Schawlow was convinced that anything would lase if you hit it hard enough, and wished to give a demonstration using one of the colorful gelatine desserts popular with children. They used 12 different flavors of Knox Jell-O preparing the flavor according to the instructions by adding water to the content of one bag and tried to make it lase using a nitrogen laser. They saw the strong fluorescence but did not obtain laser action. Schawlow at the end savoured the material as a snack in his office. The experiment was repeated several days, with all 12 flavors without success and Schawlow had 12 snacks. At the end, he pointed out that sodium fluorescein is almost non-toxic and the two researchers mixed up some clear gelatine with a small amount of this dye obtaining laser action, but Schawlow no longer insisted on eating it.

Another class of liquid lasers consists of organic-rare-earth compounds based upon the fluorescence of rare-earth ions in liquid solvents. The earliest suggestions of the possible use of rare earths in organic compounds for laser action were made in 1962 by Whan and Crosby[227] and Schimitschek and Schwartz.[228] The first laser of this kind, using a class of metallo-organic compound called rare-earth chelate, was reported the following year by Lempicki and Samelson[229] and involved the use of an alcohol solution of the tetrakis chelate, europium benzoylacetonate at 140 K. In these molecules, a beta-diketone chelate cage surrounds the rare-earth ion and absorbs light from the pump. This absorbed energy is then transferred to the rare-earth ion which is actually responsible for the lasing process. The following year, they obtained operation at room temperature.[230]

Chelate lasers have had, however, no great application.

6.10 The Chemical Laser

In a chemical laser, population inversion and laser output are produced directly by a chemical reaction. Although chemical lasers were discussed theoretically very early in 1960, the first realization was not made until 1965.

On June 8, 1960, the Canadian chemist J Polanyi, in the section on spectroscopy of a conference of the Royal Society of Canada, proposed an infrared and visible analogue of the maser based on chemical reactions.[231] The point was that chemical reaction products can have negative temperature; so amplification of infrared and visible radiation is possible. He named a device of this type *iraser* or *vaser*. After the conference, the paper was sent to the *Physical Review Letters*, but the editor rejected it for the reason that lasers were exclusively an engineering topic not of scientific interest. It was subsequently published in the *Journal of Chemical Physics*.[232]

One of the reactions proposed in this work, the one with HCl, was later used in the first chemical laser by Kasper and Pimentel[233] in 1965. Polanyi and his staff had already studied excited states by means of infrared chemiluminescence,[234] and it came quite naturally to them to think about vibrational states as the source of inverted population, and chemical reactions as one way of creating the population inversion. Lasers based on chemiluminescence were discussed at length during the interval 1960–1965.[235] Photodissociation phenomena for laser excitation was proposed also by Rautian and Sobel'man in the ex-Soviet Union in 1961.[236]

Although chemical lasers were not yet realized, two conferences were dedicated to them: a session on chemical reactions as a possible source of induced radiation in the *First Symposium on the Elementary Processes of High-energy Chemistry*, Moscow, March 18–22, 1963 and the *First Conference on Chemical Lasers*, San Diego, September 9–11, 1964.

The first chemical laser was built in 1965 by Kasper and Pimentel as said who demonstrated operation using HCl molecules emitting at 3.7 μm. The excited HCl was generated in a Cl_2–H_2 mixture under pulsed photodissociation of Cl_2 using a flash lamp. This achievement was followed by a iodine laser in the ex-Soviet Union.[237]

During the next few years, the principal workers in the field were chemical kineticists, who found the chemical laser was a useful tool for studying the vibrational distribution of reaction products. The photodissociation laser in which a UV flash lamp is used to initiate the chemical laser reaction was used in all of the chemical laser kinetic work during the period 1965–1969. Although it is not a chemical laser in the sense that chemical energy is converted into laser emission, it is a laser of considerable importance.

The most highly developed laser of this kind is the iodine laser, for which the typical reactions are

$$h\nu_{UV} + CF_3I \rightarrow CF_3 + I^\star(^2P_{1/2})$$

$$I^\star(^2P_{1/2}) \rightarrow I\,(^2P_{3/2}) + h\nu_{laser}.$$

The UV radiation in the band from 2500 to 2900 Å is provided by a flash lamp. Laser emission occurs at 1.315 μm. Typical gases used are CF_3I, CH_3I, C_2H_5I, etc.

In 1967, T Deutsch[238] first used an electric discharge to initiate the chemical laser reaction. He also carried out extensive spectroscopic measurements.

The real growth in the field of the chemical lasers did not occur until 1969, when there were three major developments:

1. Extensive work with electrical initiated chemical lasers, as reported by Tal'roze, Basov, and other Soviet scientists.[239]

2. The first true cw operation of a chemical laser by Spencer[240] who followed previous work by Airey and McKay;[241] in Spencer work, the chemical reagents were mixed rapidly in a supersonic flow stream.

3. The first purely chemical laser by Cool and Stevens in which bottled gases were mixed directly, without any auxiliary source of reaction initiation, and produced laser output.[242]

Later on, pulsed high-energy electron beams[243] came into operation and Zharov[244] showed that laser output energy was about twice the energy put into the system by the electron beam. A review of the state of the art in 1973 was made by Cool.[245] One of the first realization[242] is shown in **Figure 6.36**.

Iodine lasers became popular because with them it is possible to generate coherent-light pulses of more than 1000 J energy at several tens of microsecond duration with a 100 GW power per nanosecond. The laser is scalable. In the 1970s, two large installations existed: one at Max-Planck Institute for Plasma Physics in Garching (1500 J with 15 μs pulses), and the other at the Lebedev Physical Institute in Moscow (750 J with 60 μs pulses).[246]

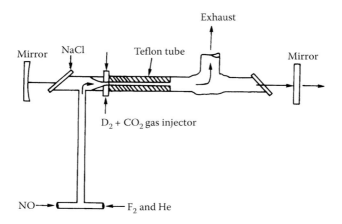

FIGURE 6.36 The continuous all-chemical laser developed by T A Cool and R R Stephens. (From T A Cool and R R Stephens, *Appl. Phys. Lett.* **16**, 55 (1970).)

6.11 The Semiconductor Laser

Semiconductors were considered at an early stage to offer a good means with which to obtain stimulated emission.

A patent was filed on April 22, 1957 and published later, September 20, 1960[247] by Y Watanabe and J Nishizawa in Japan in which recombination radiation produced by injection of free carriers in a semiconductor was considered. The patent title was *semiconductor maser* and, as a specific example, recombination radiation at about 4 μm in tellurium was considered. Naively, the authors considered the semiconductor in a resonant cavity of the kind used in the microwave region (**Figure 6.37**); but the concept of using an injection of carriers and their recombination radiation were sound. A few months later, in 1958, P Agrain (1924–2002) from France, in a speech at Brussels during an international conference on solid-state physics in electronics and telecommunications,[248] presented the idea on the extension of maser action to the field of optical frequencies using semiconductors. Unfortunately, the paper was not published in the proceedings of the conference.

Bernard and Duraffourg[249] remember he had shown that population inversion between two localized levels is not required in the case of emission of two bosons, one stimulated and the other thermalized. In semiconductors such as germanium and silicon, the valence and conduction bands are displaced relative to each other in energy–momentum space (**Figure 6.38**). Therefore, in the band-to-band transitions, the absorption or emission of a phonon is necessary for conservation of momentum. The longest wavelength emission corresponds to the electron transition from the conduction band minimum to the valence band maximum with the emission of a phonon. The reverse of the above process is a simultaneous absorption of the photon and phonon.

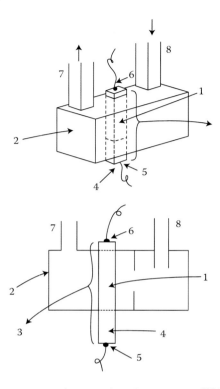

FIGURE 6.37   Japanese patent for a semiconductor maser.[225] Tellurium is used as a semiconductor (3) in the resonant cavity (2). Holes from region (4) are injected into the semiconductor region (1) and recombine with free electrons, resulting in a radiation of wavelength of 4 μm. Electrodes (5) and (6) provide electrical contacts. Output (7) and input (8) terminals are indicated for the amplified radiation.

If the sample temperature is sufficiently small and the phonons necessary for the inverse transition are absent, absorption of the long wavelengths under consideration will be small. Therefore, in this case, one can expect that with a slight increase in the carrier concentration compared with the equilibrium concentration, a state with negative temperature relative to the transition under consideration might arise.

In the proceedings of the Brussels conference quoted in Note 248, after the summary of the Agrain paper, at the end of p. 1766, B Lax observes that the original proposal of an injection semiconductor maser had already been made by John von Neumann in a private communication to John Bardeen in 1954.

In the collected works by von Neumann[250] dated September 16, 1953, reviewed by John Bardeen, the following notes on photon disequilibrium amplification scheme appear:

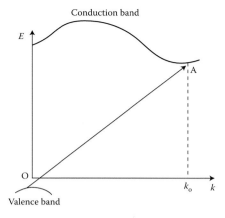

FIGURE 6.38 Band structure in germanium. If holes are created in the valence band (in O) and electrons in the conduction band (in A), they cannot recombine directly without the help of a phonon of wavevector k_0.

The possibility of making a light amplifier by use of stimulated emission in a semiconductor is considered. By various methods, for example by injection of minority carriers from a p–n junction, it is possible to upset the equilibrium concentrations of electrons in the conduction band and holes in the valence band. Recombination of the excess carriers may occur primarily by radiation, with an electron dropping from the conduction to the valence band and the energy emitted as a photon with an energy slightly greater than the energy gap. The rate of radiation may be enhanced by incident radiation of the same frequency in such a way as to make an amplifier. The basic principle was used later by Townes and by Bloembergen in the MASER (microwave amplification by stimulated emission of radiation), although not with recombination radiation in a semiconductor.

Calculations are made for an ideal semiconductor with effective masses in both valence and conduction bands equal to the ordinary electron mass. It is concluded that a very large concentration of excess carriers is required ($\sim 10^{21}$ cm^{-3}). No method is given for obtaining and maintaining large excess carrier concentrations in this concentration range. It is concluded that the large amount of heat released by the radiation might be carried away without requiring an excessive rise in temperature.

Again in 1959, Kromer[251] proposed the use of negative effective mass of carriers to amplify electromagnetic radiation. This proposal was criticized by various authors[252] who showed that a necessary condition for amplification is the formation of states at negative temperature. Zeiger[253] also considered laser action in semiconductors.

In the Soviet Union, a group at the Lebedev Institute of Physics (FIAN) of the Academy of Sciences of the USSR headed by Basov had been discussing the possibility of obtaining states with negative temperature in semiconductors since 1959.[254] In the first of these works, the use of the impurity ionization mechanism in a semiconductor at low temperatures using an electric pulse is proposed. After a sufficiently rapid removal of the electric field, and given that the temperature of the crystal lattice is low, it is possible to have negative temperature due to the difference between the recombination τ and slowing down time of charge carriers in the bands, (t_s), if $t_s \ll \tau$. Later on, the slowing-down time t_s was calculated[255] and found to be in the order of 10^{-10} s in Ge and Si, much shorter than the lifetime. This mechanism led, several years later,[256] to laser action.

In 1960, a different system was considered in which use was made of the radiation emitted in the electron–hole recombination between bands whose minima are not facing.[254,257] Semiconductors such as germanium and silicon which present indirect transitions were considered in depth. This mechanism is the same as in the two-boson laser considered by Agrain.

In 1961, the Soviet group proposed a different method, which eventually proved successful, utilizing p–n junctions in degenerate semiconductors.[257]

Historically, electroluminescence was observed by Round as early as 1907, when a current was passed through a silicon carbide detector.[258] Similar results were obtained by Lossev in 1923,[259] who made a systematic study of the effect and published a series of papers on the subject.[260]

The theory of the effect investigated by Lossev was given by Lehovec, Accardo, and Jangochian,[261] who proposed a model based on the band structure diagram of a p–n junction. In their model, electrons are injected across a forward-biased p–n junction and combine with holes in the p-region of the junction. In this recombination process, the energy lost by the electron is emitted as a photon. In 1961, the group at Lebevdev Institute also considered experimentally excitation of semiconductors. Excitation with a strong electric field was considered in InSb[262]; only indirect demonstration of the formation of negative temperature states was obtained. Later, a study of negative photoconductivity in some samples of silicon was performed.[263]

At the same time, at the 1st and 2nd *Quantum Electronics Conference*, where semiconductor lasers were discussed in depth, B Lax[264] suggested the use of transitions between energy levels generated in strong magnetic fields (*Landau levels*) or transitions between levels of some impurities to obtain generation in the submillimeter region or in the IR. However, the very short relaxation times of these levels coupled with the presence of absorption transitions as higher levels are approached, created difficulties in generation.

Later, Popov[265] showed that due to the shortening of the lattice relaxation time with current, it is practically impossible to obtain a negative temperature state.

The possibility of obtaining stimulated emission in semiconductors by transitions between conduction and valence bands was discussed in a complete and exhaustive manner in 1961 by Bernard and Duraffourg.[266] By using the concept of Fermi quasi-levels, they obtained a fundamental relation which must be fulfilled to enable a laser effect in semiconductors. This relation was also derived independently by the Basov group.[267]

Bernard and Duraffourg considered that, in the one-particle approximation, an electronic state in a solid may be represented by a Bloch wave defined over the whole crystal. They considered two such states: one in the valence band with a wave vector k_i and energy $E_v(k_i)$, the other in the conduction band with a wave vector k_j and energy $E_c(k_j)$.

If the crystal is not in equilibrium, under certain conditions, the occupation probability of any state of the conduction band is given by

$$f_c = [1 + \exp(E(k) - F_c)/kT]^{-1},$$

where F_c is the "quasi-Fermi level" for the electrons in the conduction band. In the same way for holes in the valence band, one may introduce a "quasi-Fermi level" F_v. At equilibrium, $F_c = F_v = F_0$. These "quasi-Fermi levels" are useful for the description of carriers in a p–n junction.

Let us now consider the case in which a state $E_v(k_i)$ of the valence band is connected to a state $E_c(k_j)$ of the conduction band by a direct radiative transition. Let W_v^c be the probability per unit time of such a process. In a radiation field containing a density of photons $P(v)$ of energy hv, the number N_a of quanta absorbed per unit time is

$$N_a = A W_v^c f_v(k_i) \left[1 - f_c(k_j) \right] P(v).$$

The number N_e of quanta emitted per unit time by stimulated emission is

$$N_e = A W_c^v \left[1 - f_v(k_i) \right] f_c(k_j) P(v),$$

where the proportionality coefficient A includes the density of states of the valence band and of the conduction band. The necessary condition for amplification is now

$$N_e > N_a,$$

or

$$f_c(k_j)[1 - f_v(k_i)] > f_v(k_i)[1 - f_c(k_j)],$$

if it is considered that $W_v^c = W_c^v$. This condition is equivalent to

$$\exp(F_c - F_v)/kT > \exp[E_c(k_j) - E_v(k_i)]/kT.$$

Since

$$E_c(k_j) - E_v(k_i) = hv,$$

the above condition simply reduces to

$$F_c - F_v > hv. \tag{6.19}$$

In this case, a population inversion is not necessary for amplification. This distinct difference from the other kinds of lasers we have considered is due to the continuous distribution of electron levels when compared with the discrete distribution present in atoms or molecules.

Relation 6.19 refers to an energetic condition which allows photons of energy hv fulfilling it only to be emitted, not absorbed.

Bernard and Duraffourg also considered some materials where condition 6.19 was likely to be fulfilled and in which radiative recombination was large, and suggested among others the III–V compounds GaAs and GaSb.

After the publication of the paper of these two authors, many groups started active research. In January 1962, Nasledov et al.[267] in Leningrad reported that the linewidth of the radiation emitted from GaAs diodes at 77 K narrowed slightly at high current densities (1.5×10^3 A cm^{-2}). They suggested that this might be a sign of stimulated emission.

In the United States, some group had started experimental work at about the same time with their effort also focused on developing the first semiconductor laser device. The groups were at IBM, RCA, the Lincoln Laboratory at MIT and General Electric in two different laboratories in Schenectady and Syracuse.[268] The competition to be the first to obtain laser action became a frantic race, briefly described here.

At the Watson Research Center of IBM, Rolf W Landauer (1927–1999) in December 1961 formed a small group to examine the problem in a systematic way. William P Dumke,[269] also from IBM even if not a direct collaborator of Landauer, showed that simple semiconductors like silicon and germanium, which are the most commonly used for electronics applications, would not be suitable, due to their band configuration, and suggested the use of more complex semiconductors from the structural point of view, like gallium arsenide, in which the minimum energy of the conduction band just coincides with the maximum of the valence band (*direct gap semiconductors*). IBM was well placed to consider GaAs because it had already started a programme to study it for electronics applications.

Compound semiconductors, especially GaAs, were also being studied at General Telephone and Electronics Laboratories (GT&E) at Bayside, New York. Here Summer Mayburg headed a small group which studied which kinds of devices could be made with GaAs. In March 1962, at an American Physical Society meeting, Mayburg[270] presented as a post-deadline paper, the work titled *Efficient electroluminescence with GaAs diodes at 77 K*, and on the invitation of Landauer, who had first met in their college days, visited the IBM laboratory. The work presented by Mayburg reported results obtained by Harry Lockwood and San-Mei Ku, who had shown that under certain circumstances practically every charge injected through a p–n junction resulted in a photon.

Jacques I Pankove at RCA had spent one year in Paris during 1956–1957, working with Agrain. Coming back from France he had started research but without much support because the head, William Webster, considered that the commercial income from semiconductor laser sales would be very low. In January 1962, at the American Physical Society conference, Pankove[271] announced that, together with his colleague M J Massoulie, he had obtained recombination radiation from gallium arsenide junctions. Mayburg acknowledged he could be beaten by RCA and redoubled his efforts.

At IBM, after the Mayburg seminar, the theoretician Gordon Lasher started to study how to make a cavity for a semiconductor laser and at the same time in the nearby IBM laboratory in Yorktown Heights in New York state, Marshall I Nathan started thinking how to build a gallium arsenide laser.

At the MIT Lincoln Laboratory, a group headed by Robert H Rediker[272] had been studying GaAs diodes since 1958. In mid-1961, they started to measure recombination radiation as a means of investigating diode characteristics and found the same strong luminescence already observed by GE and RCA.

In July 1962, Mayburg's results were discussed at a *Solid State Device Research Conference* at the University of New Hampshire, and R J Keyes and

T M Quist[273] of the Lincoln Laboratory announced they had built gallium arsenide diodes with a luminescence efficiency that they estimated to be 85%. Earlier, Pankove[271] had pointed out that a GaAs p–n junction would prove an efficient source of infrared radiation.

The Lincoln Labs used the electroluminescence emitted by a diode to transmit a television channel, and the news was reported in the *New York Times*.

At this point, a fourth group entered the race. Robert N Hall[274] from General Electric at Schenectady had attended the 1962 conference in New Hampshire and was struck by the presented results. The high efficiency with which light was emitted by GaAs p–n junctions impressed him greatly, and returning by train he started to make calculations of the possibility of building a laser and thought about how to obtain a Fabry–Perot cavity. The idea was to make a p–n junction, cut it, and polish it. Hall had been an amateur astronomer and at school had built a telescope; he knew how to polish optical components. Nowadays cavities are made by cleaving the crystal in the right directions, but at that time he was not familiar with such techniques. After having approached some of his colleagues, he obtained permission from his boss to start work on this project. The principal difficulty was to make a junction in GaAs which, to satisfy the Bernard and Duraffourg conditions, needed to be heavily doped. A second difficulty was to cut and polish the structure and make the cut sides highly parallel to each other. It was then necessary to send a very high current through the junction to inject a sufficiently high number of electrons, and this current had to be pulsed for a very short time to avoid melting of the structure. To prevent the temperature increasing too much, it was necessary to cool to liquid nitrogen temperature (77 K).

Even though Hall had been the last to enter the competition, he was the first, if only by a short while, to obtain in September 1962 the first laser diode.[275]

Bernard visited Hall's laboratory several times, discussing the possibility of semiconductor lasers. During one visit he happened to appear just after Hall's group had got the first one going, but before they had submitted the paper describing it. The achievement was therefore kept secret to avoid being beaten. Hall had to discuss the problem of how to make laser without being able to tell Bernard that he had one working in the next room.

The New Hampshire conference also inspired N Holonyak,[276] an expert of gallium arsenide in General Electric's Syracuse laboratory. When the first diode was operated, nearly simultaneously several groups announced laser action in p–n junctions in GaAs. In all cases, GaAs was cooled at 77 K and pumped with current pulses of high intensity and short duration (a few microseconds). The second laser was announced on October 4 by a group with Marshall I Nathan of IBM at Yorktown Heights[277] and a third on October 23 at Lincoln Laboratory of MIT, directed by T M Quist.[278] At the General Electric Laboratory at Syracuse, Holonyak also succeeded in making a semiconductor laser, as described in a paper submitted on October 17.[279] Nick Holonyak came from a strong background in semiconductor physics. He was the first Bardeen's graduate student and during a period of work at Bell Labs had built with J Moll the first pnp switch in 1954.

All these lasers were made with a gallium arsenide junction cooled at liquid nitrogen temperature and pumped with current pulses of high intensity and duration of a few microseconds.

Hall's device[275] was a cube of 0.4 mm edge with the junction lying in a horizontal plane through the center. The front and back faces were polished parallel to each other and perpendicular to the plane of the junction, creating a cavity. In many studies, it had been speculated that the mirrors would be parallel to the junction. Hall's laser, as well as all other working lasers, comprised an arrangement of the Fabry–Perot mirrors so that the radiation would bounce back and forth in the junction plane, that is the region where the injected carriers recombine to emit light. This gave a relatively long path for amplification (**Figure 6.39**). Current was applied in the form of pulses of 5–20 μs duration, with the diode polarized in the forward direction and immersed in liquid nitrogen. The threshold for laser action was found about 8500 A cm^{-2}. Radiation patterns and narrowing of the emission line were studied. Below threshold the spectral width was 125 Å, which constricted suddenly to 15 Å at threshold.

Nathan[277] used a somewhat different system. He started out with GaAs junctions made by diffusing Zn into GaAs doped with Te. These diodes were then banded on to a Au-plated kovar washer and the junction was etched to approximately 1×10^{-4} cm^2. In this case, no cavity was made. Threshold for laser emission at 77 K was given in the range 10^4–10^5 A cm^{-2}.

Quist[278] used a mesa structure with an area of 1.4×0.6 mm^2, the short sides being polished optically flat and nearly parallel. At 77 K threshold was found at approximately 10^4 A cm^{-2}; this was decreased by a factor 15 at 4.2 K (**Figure 6.40**).

Spectral measurements were also made which confirmed the narrowing of the emission line.

Holonyak and Bevacqua[279] used forward-biased Ga (As$_{1-x}$P$_x$) p–n junctions instead. The diodes were rectangular parallelepipedons, or cubes, with two opposite, parallel sides carefully polished so as to give a resonant cavity. Using this compound, the authors were able to obtain emission in the region 6000–7000 Å[280] instead of about 8400 Å as obtained using simple GaAs.

In Russia, a short while after the construction of the American lasers, V S Bagarev, N G Basov, B M Vul, B D Kopylovskii, O N Krokhin,

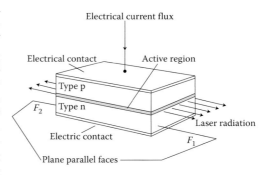

FIGURE 6.39 Layout of a pn junction semiconductor laser of the simplest type. Laser radiation is emitted in the thin active region between the p and n zones and is reflected back and forth by the two plane parallel facets F_1, F_2 which act as mirrors.

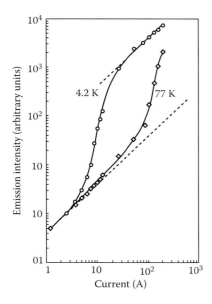

FIGURE 6.40 Variation of radiation intensity with current in a GaAs diode at two different temperatures. (From T M Quist et al., *Appl. Phys. Lett.* **1**, 91 (1962).)

Yu M Popov, A P Shotov, and others built a laser diode at FIAN. The results were discussed at the third Conference on Quantum Electronics in Paris in 1963.

Almost at the same time as these first result, McWhorter, Zeiger, and Lax[281] produced the first theoretical treatment of mode confinement in a p–n junction.[282]

Stimulated light emission from InP was obtained by Weiser and Levitt.[283] The emission was at low temperature 77 or 4.2 K at around 9050 Å.

In a semiconductor, laser transitions take place in a p–n junction between occupied electronic states in the conduction band and empty electronic states in the valence band. A notable difference between these lasers and the other lasers is that here the transitions are between states within which there is a distribution of energy and not between levels of defined energy.

Figure 6.41 shows the distribution of the electron states in a doubly degenerate semiconductor. The small dots indicate the energy position of the allowed electron states as a function of the propagation constant k of the electron for both the conduction and valence bands. The position of energy bands and electron occupation along a junction, without any bias applied, are

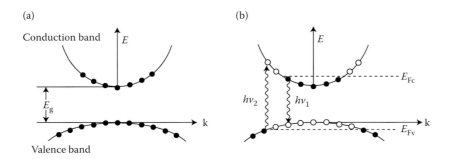

FIGURE 6.41 Position of the energy levels near the energy gap in a direct semiconductor (such as GaAs) as a function of the electron propagation constant k; (a) equilibrium situation; (b) a doubly degenerate semiconductor.

shown in **Figure 6.42a**. The situation for laser action (**Figure 6.42b**) can exist only under nonthermal equilibrium conditions and can be depicted using two quasi-Fermi levels, E_{Fc} and E_{Fv}, in the two bands.

At equilibrium, the Fermi energy has the same value along the sample (Figure 6.42a).

If a forward-bias voltage, V_{appl} is applied (Figure 6.42b), the Fermi level in the n region is raised by a quantity eV_{appl} with respect to that in the p region. There now exists a narrow region, the active region of the device, which contains both electrons and holes and is doubly degenerate. Electromagnetic radiation of frequency υ such that

$$E_g/h < \upsilon < (E_{Fc} - E_{Fv})/h,$$

which is propagating in this region, is amplified.

Many of these first injection lasers were, typically, rectangular parallelepipedons or trapezoids made by cutting chips of material and polishing two parallel ends of the chip. The plane of the p–n junction was perpendicular to the polished ends of the parallelepipedon so as to form a cavity corresponding to a small Fabry–Perot interferometer.

Bond et al.[284] were the first to make use of cleaving along parallel crystal planes. These lasers, which use a single semiconductor, are usually referred to as *homostructure lasers*. Homostructure junction lasers had very high threshold current densities for lasing at room temperature (>50,000 A cm⁻²) as a result of which continuous operation was not achieved until 1967.[285] They usually operated at very short pulses (≤1 μs) and low duty cycles (<0.1%).

The development of semiconductor lasers has been slow owing to various reasons. It was in fact necessary to develop a new technology for manipulating semiconductor materials, considering that silicon, for which there already existed a well-established technology, could not be used. There were also problems connected with the high currents needed to obtain laser action that limited the operation to short pulses and required the use of very low temperatures, resulting in a low efficiency of the device.

A notable step forward for the solution of these problems was made in 1969 with the introduction of heterostructures. Already in 1963, Kroemer[286] had suggested the use of heterojunctions, in which a layer of a semiconductor with a relatively narrow energy gap is sandwiched between two layers of a wider energy-gap semiconductor. A similar suggestion was made in the

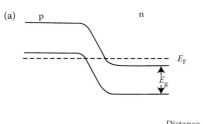

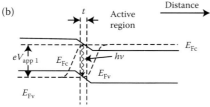

FIGURE 6.42 Degenerate p–n junction; (a) at no applied voltage; (b) polarized in the forward direction. Electrons and holes recombine in the junction region.

same period by Alferov and Kazarinov at the Ioffe Physico-Technical Institute in Leningrad but the two researchers did not publish it although received an inventor certificate in the USSR in which they mentioned a number of benefits of using double heterostructures.[287] In 1966, Alverov et al.[288] predicted several properties and advantages of the system.

Six years elapsed until Hayashi and Panish[289] of Bell Laboratories and Kressel and Nelson[290] at RCA developed the first heterostructure lasers in which the active region was surrounded on one side by a cladding layer of a higher band gap semiconductor. At the same time, Alverov et al.[291] developed the more complicated multilayer structures which are nowadays called double-heterostructure lasers in which the active region is surrounded on both sides by layers of higher band gap semiconductor. These efforts were awarded in 2000, with the Nobel Prize in Physics to Zhores Alverov (1930–2002) and Herbert Kroemer (1928–) both "for developing semiconductor heterostructures used in high-speed and opto-electronics" and Jack Kilby (1923–) "for his part in the invention of the integrated circuit."

In a heterostructure laser, the simple p–n junction is replaced by multiple layers of semiconductors of different composition (**Figure 6.43**). The active region is then reduced in thickness and the total current required for laser operation decreases notably, with a corresponding decrease in the heating effect. It followed that it was no longer necessary to cool and one could achieve a working laser even at room temperature.[292] The use of a heterostructure, however, requires careful matching of the lattice constants of the two semiconductors.

Two factors are largely responsible for the transformation of semiconductor lasers from laboratory devices working only at very low temperature to practical optoelectronic devices able to work continuously at room temperature. One is the exceptional and lucky similarity of the lattice constants of aluminum arsenide (AlAs) and gallium arsenide (GaAs) that allow heterostructures

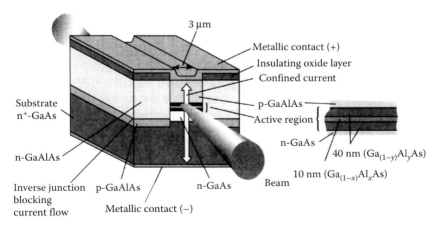

FIGURE 6.43 Buried heterostructure laser diode. (From *Laser Focus World*, May 1992, p. 127.)

to be made by layers of different composition of the compound gallium and aluminum arsenide ($Al_xGa_{1-x}As$). The second is the presence of many important optoelectronic applications for which semiconductor lasers are particularly suited given their properties: small dimensions (a few cubic millimeters), high efficiency (sometimes more than 50%), pumping obtained directly from the current and longer life compared with other lasers. The fact that the laser is pumped directly by the current allows direct modulation of its output, simply by modulating the current. This characteristic is ideal for information transmission systems.

In the double heterostrusture (DH) laser, the active region is sandwiched by semiconductors of larger band gap. The resulting band discontinuities confine the injected carriers in the active region and the refractive index difference between the materials creates a waveguide forcing light generated by stimulated emission to propagate parallel to the layers. The optical-mode confinement significantly reduces the internal loss that would otherwise occur in the absence of index guiding due to the spreading of the optical mode in the lossy regions.[293] Already in 1969, double-heterostructore GaAs lasers were constructed[291] with a room-temperature value of threshold current $J_{th} \sim 5$ kA/cm^2. This value was reduced to about 1.6 kA/cm^2 in 1970[294] and to 0.5 kA/cm^2 by 1975 with AlGaAs lasers.[295]

The next important step occurred with the invention of the quantum well (QW) laser patented by Dingle and Henry in 1976.[296] Their calculations predicted the advantage of using quantum wells as the active layer in semiconductor lasers: the carrier confinement and nature of the electronic density of states should result in more efficient devices operating at lower threshold currents than ordinary semiconductor lasers. In addition, the use of a quantum well with discrete transition energy levels dependent on the quantum well thickness provides a means of tuning the resulting wavelength of the material. Semiconductor heterostructure laser incorporating quantum wells in its active medium should therefore result superior over the conventional double heterostructure laser.

In these new devices, the active region is not a single GaAs layer but rather consists of several ultrathin (~0.01 μm) layers composed alternatively of GaAs and AlGaAs materials. The use of QWs further reduces the threshold and allows some control over the wavelength by changing the QW thickness.[297] If the dimensions of the quantum well in all the three directions become comparable with the wavelength of carriers the system acquires full quantum properties and the resulting quasi-zero-dimensional structures are known as *quantum dots*.

The energy levels in a single quantum dot (QD) are discrete, similar in atoms below the ionization threshold. The corresponding wave functions are spatially localized within the QD, but extend over many periods of the crystal lattice. A quantum dot contains a small finite number (in the order of 1–100) of conduction band electrons, valence band holes, or excitons. QDs, such as colloidal semiconductor nanocrystals, can be as small as 2–10 nm, corresponding to 10–50 atoms in diameter and a total of 100–100,000 atoms inside.

Self-assembled QDs are typically between 10 and 50 nm in size. Quantum dots defined by lithographically patterned gate electrodes, or by etching on two-dimensional electron gates in semiconductor heterostructures can have lateral dimensions exceeding 100 nm.

The confined electrons move in the semiconductor host crystal, whose band structure plays an important role for all quantum dot properties. Typical energy scales, for example, are in the order of 10 eV in atoms, but only 1 meV in QDs. In contrast to atoms, the energy spectrum of a QD can be engineered by controlling the geometrical size, shape, and the strength of the confinement potential. QDs provide an ultimate limit of size quantization in solids. The density of states is a delta function[298] and is responsible for many properties of the system.

Although the first realizations of heterostructures were of poor quality, the situation was reversed by 1981 through the use of new material growth capabilities (molecular beam epitaxy) and optimization of the heterostructure laser design by Tsang.[299]

Greater benefits were predicted for lasers with embedded QD active layers. Arakawa and Sakaki[300] predicted in the early 1980s that quantum dot lasers should exhibit performances that are less temperature-dependent than existing semiconductor lasers and that will in particular not degrade at elevated temperatures. Other benefits of QD active layers include further reduction in threshold current and a more efficient laser operation.[301]

The first semiconductor dots based on II–VI microcrystals in glass matrix were proposed and demonstrated by Ekimov and Onushchenko[302] but had practically no application because of poor quality and in an insulating matrix. QDs in a semiconductor matrix were fabricated later by L Goldstein et al.[303]

The growth of nonplanar substrates led to the realization of quantum wire heterostructures and to the first demonstration of lasing in one-dimensional system.[304]

Initially, the most widely followed approach to forming QDs was through electron beam lithography of suitably small feature patters (~300 Å) and subsequent dry-etch transfer of dots into the substrate material.[305] The lateral dimensions of the dots are usually considerably larger than the thickness of the well, leading to pancake-shaped dots, with the consequence that the quantization energy from lateral confinement is usually small compared to the one from the parent quantum well itself. Another technique takes advantage of the fact that a layer of a material having a lattice constant different from that of the substrate after some thickness is deposited can spontaneously transform to an array of three-dimensional islands[306] which may be small (10 nm) and have similar size and shape and form dense arrays. Incorporating of single-sheet array of InGaAs–GaAs QDs inserted in a double-heterostructure GaAs–AlGaAs waveguide structure allowed in 1993 realization of photopumped lasing via QD states both at low and at room temperature[307] and injection[308] in self-organized QDs was demonstrated.

Soon after the first realization of lasers based on self-organized QDs in 1993–1994, the threshold current density was dramatically decreased.[309]

Then self-assembled QDs exhibited higher quantum efficiency. The Stranski–Krastanow (SK) crystal growth has emerged as the most mature growth technique to prepare self-organized QDs on planar substrates of high uniformity and large density. The first demonstration of a QD laser with high threshold density at low temperature was reported by Ledentsov and colleagues in 1994.[310]

The demonstration of an injection laser operated in a continuous-wave mode at room temperature, at 1 μm from SK dots came in 1997.[311] The following year, the achievement of room temperature lasing at 1.3 μm from InAs/GaAs SK dots first in a pulsed injection mode[312] and then in continuous-wave mode in 1999[313] was reported. The progresses in realizing QD lasers based on self-organized structures was described by Bimberg et al.[314]

Semiconductor heterostructures with self-organized QDs exhibit the properties expected for zero-dimensional systems. When used as active layer in the injection laser, these advantages help to strong increase in material gain, to improve temperature stability of the threshold current, to improve dynamic properties, and to tune the operation frequency via the change in density of states.

Coming back to traditional diode lasers, long-wavelength semiconductor lasers in the range 1.1–1.6 μm are of considerable interest for optical fiber communications and the research showed that the combination InGaP–InP was the most suitable.[315]

After the successful operation of near-infrared diodes, great efforts were devoted in decreasing the emission wavelength.

The first laser based on CdS, excited by fast electrons and operating at 0.5 μm, was devised in 1964 by Basov et al.[316]

Gallium nitride diode lasers emitting blue light were invented by Shuji Nakamura[317] (b. 1954) who has been awarded the 2014 Physics Nobel Prize together with I Asaki and H Amano.

Nakamura demonstrated efficient room-temperature blue-laser emission from indium gallium nitride-based diodes in 1995, when he was working at Japan firm Nichia Chemical Industries. After his achievement, which was obtained notwithstanding the scepticism of his boss, Nakamura came to the University of California Santa Barbara (UCSB). The story of this accomplishment is recollected by him in a book.[318] Green laser diodes proved difficult to be built due to the characteristics of the quantum wells that serve as their gain region. By mastering the epitaxy of high indium containing quantum wells and using optimized waveguides, they have been built based on the system (Al,In)GaN.[319] This laser is emitted at 405 nm. The gap between 400 and 600 nm was overcome[320] when three companies, Osram, Nichia, and Sumitomo, simultaneously demonstrated the first green laser diodes in the range 515–530 nm.[321] Blue InGaN-based laser diodes with an emission wavelength of 450 were reported by S Nakamura in 2000.[322] In a completely different geometry, highly efficient blue and true green laser diodes were built by Raring.[323] The field is very active with the goal of producing a green cw-laser diode for applications such as laser projection and television.

6.12 Vertical-Cavity Surface-Emitting Laser Diodes (VCSELs)

Vertical-cavity surface-emitting laser diodes (VCSELs) constitute a new class of light sources with a number of remarkable properties. In contrast to waveguide-based edge-emitting laser diodes, these microlasers are extremely short of just about 1–3 μm effective resonator length, the emission occurs in a circular beam perpendicular to the semiconductor surface, and two-dimensional arrays of laser diodes can easily be formed. Single transverse and longitudinal mode oscillation and alignment-tolerant and efficient coupling into a single-mode fiber make the devices ideally suited for local area fiber networks and optical interconnect systems for microelectronic circuits.

These lasers with vertical cavity and emission perpendicular to the surface of the active layer were suggested by Iga et al. in 1979.[324] Pulsed and cw room-temperature operation were achieved in 1984[325] and 1988,[326] respectively. A major breakthrough occurred when Jewell demonstrated arrays of 1 mA threshold devices.[327] In the 1990s, the advance in VCEL design and growth was tremendous, resulting in devices comparable with and even surpassing the conventional edge emitting lasers in threshold current, efficiencies, and speed of modulation.

The simplest method to achieve optical and carrier lateral confinement is to etch a pillar or a post. The first monolithic VCSELs were air-post ones.[328]

The first surface-emitting laser (VCSEL) operating via QDs was assembled by Schur et al.[329] Tunable VCSELs are discussed by Chang-Hasnain.[330]

6.13 Quantum Cascade Lasers

A new generation of semiconductor lasers was introduced by Federico Capasso in 1994.[331] These are semiconductor injection lasers based on intersubband transitions in a multiple-quantum-well (QW) heterostructure, designed by means of bandstructure engineering and grown by molecular beam epitaxy.

That light amplification is possible in intersubband transitions, that is, transitions between quantized energy states within one energy band of a semiconductor, was first predicted by Kazarinov and Suris.[332] The first inter-subband laser was demonstrated in 1994 by Faist, Capasso et al.

The laser makes use of the discrete electronic states arising from quantum confinement, normal to the layers, in nanometer-thick semiconductor heterostructures grown by molecular beam epitaxy (see **Figure 6.44**). Parallel to the layers, these states have plane wave-like energy dispersion. Electrons make radiative transitions to a lower subband (e.g., from $n = 3$ to $n = 2$). Figure 6.44 shows two active regions: the quantum well and barrier regions that support the electronic states, between which the laser transition is taking place, and—schematically—the intermediate injector region.

Under an appropriate applied bias electrons tunnel from the injector region into energy level 3, the upper laser state, of the active region. Electrons scatter

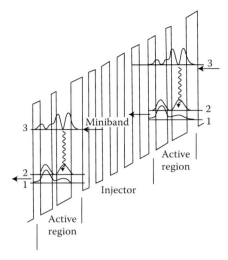

from this level into both lower-lying levels 2 and 1 very rapidly emitting LO phonons. The upper-state lifetime τ_3 is in the order of a picosecond and the scattering time τ_2 between levels 2, the lower laser state, and level 1 is in the order of 0.2 ps. With $\tau_3 \gg \tau_2$ population inversion between the laser levels 3 and 2 may occur. Electrons need only be supplied fast enough into level 3, by tunneling from the preceding injector region, and need to exit from levels 2 and 1 into the following downstream injector region at a high rate by tunnelling. Inside this injector region and brought about by the externally applied electric field, electrons gain again in energy (relative to the band bottom) and are injected into the following downstream active region. Typical quantum cascade lasers have a cascade containing 20–30 active regions alternate with injector regions.

FIGURE 6.44 Conduction band energy diagram of two portions of the active region with the intermediate injector section and the moduli squared of the wavefunctions involved in the laser transitions (1, 2, 3) of the quantum cascade laser. Electrons are injected through a suitable barrier into the $n = 3$ energy level of the active region. The laser transition is indicated by the wavy arrows and the electron flow by the straight arrows. A voltage of appropriate polarity gives the conduction band diagram the staircase shape shown in the figure. (From C Gmachi et al., *Rep. Prog. Phys.* **64**, 1533 (2001). With permission.)

In conclusion, electrons streaming down a potential staircase sequentially emit photons at the steps. The steps consist of coupled quantum wells in which population inversion between discrete conduction band excited states is achieved by control of tunneling.

The first laser realized by Faist et al. was designed in InGaAs/AlInAs on InP alternating 25 undoped coupled-well active regions with compositionally graded layers. The graded regions consisted on an AlInAs–GaInAs superlattice with constant period, shorter than the electron thermal de Broglie wavelength and emitted at 4.3 μm with peak powers in excess of 8 mW in pulsed operation at a low temperature (around 10 K). The threshold current density was $J_{th} = 15$ kA/cm^2.

The intersubband nature of the optical transition has several key advantages. First, the emission wavelength is primarily a function of the QW thickness. This allows choosing well-understood and reliable semiconductors for the generation of light in a wavelength range unrelated to the material's energy band gap. Second, a cascade process in which multiple—often several tens of—photons are generated per electron becomes feasible, as the electron remains inside the conduction band throughout its traversal of the active region. This cascading process is behind the intrinsic high-power capabilities of the lasers. Finally, intersubband transitions are characterized through an ultrafast carrier dynamics and the absence of the line width enhancement factor.

Earlier proposals were made a few years before.[333] For a detailed discussion on this, see Capasso.[334]

Already a year after its invention, the laser operated in cw at cryogenic temperatures[335] and in pulsed mode at room temperature.[336] More recent advances are reported in C Gmachi et al.[337]

6.14 The Free-Electron Laser

Free-electron lasers arise out of the undulator proposed by Motz in 1951 (Chapter 4), although the first actual proposal that an electron beam could be used to produce coherent light was made much later. They are inherently able to produce coherent tunable radiation anywhere in the spectrum between submillimeter and ultraviolet up to x-rays. Moreover, they are potential sources of very high power, since no damage can occur to the lasing medium, as it happens with liquid or solid-state lasers. The basic operating principle comes from the phenomenon of the scattering of photons from electrons in the presence of radiation, that is, stimulated Compton scattering.

In 1933, Kapitza and Dirac[338] predicted stimulated Compton scattering and proposed an experiment to observe the effect with non-relativistic electrons scattered by standing waves. Much later, Pantel et al.[339] considered the generation of short-wavelength radiation by stimulated scattering of an electromagnetic pump wave from a relativistic electron beam.

It was, however, only in 1971 that Madey,[340] combining all these concepts, analyzed the radiation emitted by a relativistic electron beam moving through a periodic, transverse, DC magnetic field. The free relativistic electrons "see" the virtual quanta representing, in the Weizsacker–Williams approximation, the static, transverse, periodic magnetic field as long-wavelength photons and scatter them into real short-wavelength photons by means of stimulated Compton scattering.

The strict connection with the Motz undulator described in Chapter 4 is evident.

In 1972, Madey and coworkers began to build a laser based on these principles. After overcoming innumerable difficulties, they eventually succeeded in demonstrating stimulated amplification of radiation in 1976[341] and laser oscillation the following year.[342]

The experimental setup for the demonstration of the amplification of infrared radiation by relativistic free electrons in a constant, spatially periodic, transverse magnetic field is shown in **Figure 6.45**.[341]

The periodic magnetic field was generated by a superconducting right-hand double helix with a period of 3.2 cm and a length of 5.2 m. The helix consisted of two bifilar helical coils with current flowing in each coil in opposite directions wound around a 10.2-mm internal diameter evacuated copper tube which enclosed the interaction region. The field due to the helix was transverse

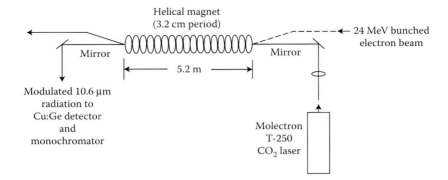

FIGURE 6.45 Experimental set-up used by L R Elias et al. in 1976 to demonstrate amplification of radiation by a free electron beam. The electron beam was magnetically deflected around the optical components on the axis of the helical magnet. (From L R Elias et al., *Phys. Rev. Lett.* **36**, 717 (1976).)

and rotated in the plane normal to the axis with the period of winding. The electron beam was obtained from the superconducting linear accelerator of the Hansen High Energy Physics Laboratory of Stanford University at an energy of 24 MeV. The electron beam and an infrared beam from a CO_2 laser ($\lambda = 10.6$ μm) were steered so as to pass through the magnet on the axis.

The experiment was performed by sending the electron beam through the periodic field and measuring the gain and absorption coefficients for the 10.6 μm radiation sent through the field parallel to the electron beam axis. A gain of 7% per pass was obtained at an electron current of 70 mA.

Nearly 2 years later, oscillation at 3.4 μm using a 43 MeV electron beam was demonstrated.[342] In this case, the same superconducting helix as previously was used which generated a periodic magnetic field of 2.4 kG. The radiation was amplified in the space between a pair of mirrors (**Figure 6.46**). An average power output of 0.36 W was obtained with a peak power of 7 kW.

The optical cavity was 12.7 m long, this length being chosen so that the back-and-forth travel time of the light pulse in the cavity be equal to the time separation between electron bunches from the line. The initial signal was the

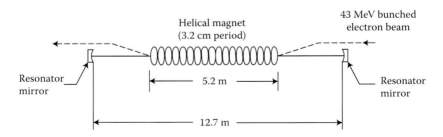

FIGURE 6.46 Schematic setup for the FEL oscillator. (From D A G Deacon et al., *Phys. Rev. Lett.* **38**, 892 (1977). With permission.)

undulator radiation generated by the beam itself and the system can be called a self-amplified spontaneous emission (SASE) amplifier.

Several theoretical works appeared on the subject meanwhile. Just to mention two of them, in 1971, R P Palmer[343] developed a classical picture of how energy would be efficiently transferred between free electromagnetic waves and free, pre-bunched relativistic electrons passing through a static helical magnet, and discussed the possibility of self-induced bunching. Colson[344] developed the classical theory in the small gain approximation.

FELs, notwithstanding their large size, cost, and complexity, offer operation over a large wavelength range, capability of very high average power, and a more favorable scaling for the gain at short wavelengths. After the first achievements of Madey group, a large activity started and some 30 free-electron lasers have been built around the world. Most of these lasers work in the infrared and far-infrared regime, with may be 10 producing visible and ultraviolet light.

Initially the main research emphasis was the development of high average power, MW level, infrared systems. In continuous mode, an average power of 1.7 kW was produced at a wavelength of 3 μm[345] and power up to 2 kW in the infrared with pulses of 1 ms.[346]

Two different but complementary approaches were theoretically developed relying on stimulated Compton and Raman scattering, respectively. The Compton scattering regime occurs when the electron current is sufficiently small that the beam space-charge potential is smaller than the ponderomotive potential; the first realizations by Elias et al. and by Deacon et al. were of this type. The Raman scattering regime occurs when the electron current is large. Raman FELs were developed in a number of laboratories.[347]

In the late 1970s and early 1980s, the FEL theory was developed in the high gain regime. In the small-signal case theory, the electric field is kept constant during the interaction and at the undulator exit the electron energy loss is given to the radiation field so that the theory is not self-consistent. The high gain case included the evolution of the electromagnetic field during the interaction.[348] The theory showed the self-bunching of the electron beam.

The physics and technical status of FELs are covered by several review articles.[349] A good summary with an historical excursus has been traced by C Pellegrini.[350]

The amplification medium of a free-electron laser is made up of unbound (free) electrons that are forced to move in a strong magnetic field. Bunches of free electrons are first generated in an electron gun and travel in the undulator emitting radiation. The photons are initially incoherent and concentrated over a narrow range of wavelengths, with the peak wavelength depending on the energy of the electrons and the properties of the magnetic field. After leaving the undulator, the electron beam is finally separated from the light beam by a bending magnet. The incoherent light can be made coherent by using an optical resonator with mirrors at both ends of the undulator. The photon pulses then pass repeatedly through the undulator and meet up on each round trip with an electron bunch.

Self-amplified spontaneous emission (SASE) may be used. SASE is based on the fact that while the synchrotron radiation emitted by the electrons moves through the undulator at the speed of light, the electrons themselves actually travel slightly slower. The electron therefore lag a little behind their emitted radiation, which can catch up with—and interact with—earlier electrons. The interaction will either accelerate or decelerate the electrons depending on their exact position and the phase of the light wave with which they interact.

The net result is that the light wave pushes the electrons into smaller so-called microbunches, which are separated by a distance corresponding to the wavelength of the undulator magnetic field. Several electrons now start to emit light in tandem, producing light of a higher intensity. This light then sorts the electrons into tighter and tougher bunches and causes them to radiate in phase. As a result, the radiation power rises exponentially with distance along the undulator, until it finally saturates.

For this process to work electrons, beams of extremely high quality are needed.

Since the electrons in a free-electron laser are not bound to atoms, the wavelength of the laser is not limited to specific atomic transitions. It can therefore be tuned by adjusting the energy of the accelerator or the magnetic field strength and modulation frequency of the undulator. In Chapter 9, we discuss how FELs may be used to obtain coherent radiation in the x-ray region.

A different kind of geometry can be used. In storage rings, high-density electron beams can circulate for periods of up to several days. Originally developed by high-energy physicists for studying electron–positron collisions, new machines were constructed all over the world dedicated to the production of synchrotron radiation. These machines are optimized to be photon sources as bright as possible, which means with the largest possible electron density. On a straight section of such a storage ring, it is possible to insert an undulator. Some of these storage-ring FELs are operating around the world, generating in the UV region.

6.15 Does the Laser Exist in Nature?

The answer is yes! A laser emission at a wavelength of about 10 µm was identified in the Mars and Venus atmospheres in 1981 by Michael J Mumma[351] of the Laboratory for Extraterrestrial Physics of the NASA Goddard Space Flight Centre. The emission had been already observed by some of Townes' students[352] in 1976, but was confirmed to originate from a natural laser only in 1981.

The population inversion of carbon dioxide that forms the great part of the atmosphere of these planets is produced by sunlight and therefore occurs only in the illuminated hemisphere. The mechanism is exactly similar to that of CO_2 lasers that are built on Earth and emit at 10 µm. The emission lines in the atmosphere of the two planets are about 100 million times more intense than could be obtained if the gas was in thermodynamic equilibrium conditions at atmospheric temperature. A part of the observed light is amplified radiation by the inverted medium. If two mirrors could be placed in an orbit around the planets, oscillations could be obtained just as on Earth. An independent confirmation of the 10-μm laser action was provided by theoretical studies of Gordiets and Panchenko[353] and Stepanova and Shved.[354]

The emission lines have been useful to measure the temperature and the winds on Mars[355] and Venus.[356]

Masers in space have been found for many years, as we saw, and there is no reason why one could exclude the existence of lasers. The process, however, because it requires greater photon energies, is relatively more difficult. Early in 1995, a team of astronomers detected amplified infrared light coming from a hydrogen disc whirling around a young star of the Cygnus constellation, 4000 light years away from us. The intensity of the emission at one wavelength compared with its neighbors shows that one is in the presence of stimulated emission (**Figure 6.47**).

Previous observations of the star, called MWC349, had already revealed in 1994 an intense maser emission from its disc, at wavelengths of 850 and 450 μm, produced by hydrogen.[357] The study of the processes according to which this emission occurred suggested that also emission at shorter wavelengths could be present in disc regions nearer to the star. Therefore, Vladimir Strelnitski[358] of the National Air Space Museum of Washington, Edwin Ericson and Michael Haas of the Ames SETI Institute of Mountain View in California, placed an infrared telescope on an aeroplane, let it fly at a height of 12,500 m to avoid the absorption of infrared radiation by our atmosphere, and observed a line at 169 μm that is six times more intense than one would expect by an emission in thermal equilibrium. This line is produced by the hydrogen atoms that have been ionized by intense ultraviolet light coming from the star or through more complex processes that occur in the disc. When

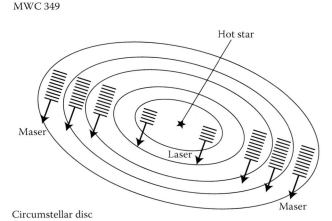

FIGURE 6.47 The natural laser in the star MWC349. The laser emission occurs in the region of the hydrogen disc that are near to the central star, while the maser emission occurs in the far region. The radiation is emitted in the figure plane and reaches Earth which by chance lies in the same plane.

the free electrons recombine with the ions most of the emission occurs spontaneously, but it is possible also to have stimulated emission. The same process produces maser emission in other parts of the disc, but in the central parts one observes laser emission, partly because the hydrogen there is more dense and partly because the ultraviolet pumping light is more intense. By chance the disc is oriented almost edge-on so that some of the light aims at the Earth and therefore the emitted laser beam can be received. The disc is a region where one believes that planets could form and the observed emission comes from one of this "nursery of planets" that is at the same distance from the star as the distance between the Earth and the Sun. Laser emission may therefore help our understanding of the state of the gases in the disc. The wavelength of 169 μm, in the medium infrared, is in a region that is really between submillimeter waves and the region typically assigned to optics. One may thus speak of a laser at huge wavelength or of a maser at short wavelength, according to how the situation is interpreted.

Other emissions in the near infrared were discussed.[359]

Lasers in the ultraviolet region exist too. An emission in this region has been identified using the Hubble space telescope, from a gaseous cloud near to the star η-Carinae.[360] These observations and much more was discussed in a review paper by Johansson and Letokhov.[361]

We may therefore conclude that also celestial lasers, like masers, are objects already existing in the cosmos and therefore it could be questioned if masers and lasers have not been invented but discovered.

Notes

1. Maiman has written his recollection: T Maiman, *The Laser Odyssey* (Laser Press: Blaine, WA, 2000); see also interview by James Cavuoto, *Laser & Applications*, May 1985, p. 85.
2. The thesis was presented at the 1955 *Annual meeting of the American Phys. Soc.* T H Maiman and W E Lamb Jr., *Phys. Rev.* **98**. 1194 (1955) and later published in full W E Lamb Jr. and T H Maiman, *Phys. Rev.* **105**, 573 (1957).
3. Operation of a ruby maser at 60 K was reported by C R Ditchfield and P A Forrester, *Phys. Rev. Lett.* **1**, 448 (1958) and, independently, by T H Maiman, *Quantum Electronics*, edited by C H Townes (Columbia University Press: New York, 1960), p. 234. Operation at 195 K was reported by T H Maiman, *J. Appl. Phys.* **31**, 222 (1960).
4. I Wieder, *Phys. Rev. Lett.* **3**, 468 (1959).
5. The spectroscopy of ruby had already been discussed by S Sugano and Y Tanabe, *J. Phys. Soc. Jpn.* **13**, 880 (1958). The two components R_1 and R_2 of the transition $^4A_2 \to {}^2E$ were discussed by O Deutschbein, *Ann. Phys. Lpz* **14**, 712 (1932) and *Z. Phys.* **77**, 489 (1932). The notation of levels comes from group theory. When an ion is located in a crystal lattice, the electric and magnetic fields prevailing at the site of the ion may exert a profound influence on the energy-level structure of the ion. The number of components into which a multiple level may be split is determined by group theory; it involves the irreducible representation of the symmetry group applicable to the crystal in question. The split levels are designated by the appropriate symbol of group theory with the spectroscopic symbol of the original

level suppressed or abbreviated. See, for example, D S McClure in *Solid State Phys.* 9, 399 (Academic: New York, 1959). In the case of ruby, the ground state of free Cr^{3+} in the crystal is split into three levels denoted by $^4F_1$, $^4F_2$, and $^4A_2$ with multiplicity 12, 12, and 4, respectively. The symbols F_1, F_2, and A_2 refer to the matrix representations of the octahedral group and are not to be related to the values of the orbital angular momentum. The upper index 4 is the only reminder that the three levels in question arise from the ground state of the free Cr^{3+} ion which had orbital angular momentum $L = 3$, and spin $S = 3/2$. The next lowest group of states of the free Cr^{3+} ion had $L = 4$, $S = 1/2$. In the crystal it splits into four sublevels, which are designated by the symbols $^2A_1$, $^2F_1$, $^2F_2$, and ^2E with multiplicities 2, 6, 6, and 4, respectively. The levels of interest in our discussion are the ones shown in Figure 6.1.

6. T H Maiman, *Phys. Rev. Lett.* **4**, 564 (1960).
7. This circumstance is quoted in B A Lengyel, *Am. J. Phys.* **34**, 903 (1966) and in Maiman's book.
8. T H Maiman, *Nature* **187**, 493 (1960); sentences reproduced by permission Copyright © 1960 McMillan Journals Limited.
9. T H Maiman, *British Commun. Electron.* **7**, 674 (1960).
10. T H Maiman, *Phys. Rev.* **123**, 1145 (1961).
11. T H Maiman, R H Hoskins, I J D'Haenens, C K Asawa and V Evtuhov, *Phys. Rev.* **123**, 1151 (1961).
12. R J Collins, D F Nelson, A L Schawlow, W Bond, C G Garrett and W Kaiser, *Phys. Rev. Lett.* **5**, 303 (1960).
13. *Phys. Today* **63**, 8 (May 2010).
14. A L Schawlow in *Impact of Basic Research on Technology*, edited by B Kursunoglu, A Perlmutter (Plenum Press: New York, 1973), p. 113.
15. D F Nelson, R J Collis and W Kaiser, *Phys. Today* **63**, 40 (January 2010).
16. R J Collins, D F Nelson, A L Schawlow, W Bond, C G Garrett and W Kaiser, *Phys. Rev. Lett.* **5**, 303 (1960).
17. C Kikuchi, J Lambe, G Makhov and R W Terhune, *J. Appl. Phys.* **30,** 1061 (1959).
18. H Statz and G de Mars in *Quantum Electronics,* vol. 1, edited by C H Townes (Columbia Unversity Press: New York, 1960), p 530.
19. P Kisliuk and D J Walsh, *Appl. Opt.* **1**, 45 (1962); D F Nelson and R J Collins, *J. Appl. Phys.* **32**, 739 (1961); M Hercher, *Appl. Opt.* **1**, 665 (162); J I Masters and G B Parrent, *Proc. IRE* **50**, 230 (1962); M D Galanin, A M Leontovich and Z A Chizhikova, *Sov. Phys. JETP* **43**, 347 (1962); D A Berkley and G J Wolga, *Phys. Re. Lett.* **9**, 479 (1962); M S Lipsett and L Mandl, *Nature* **199**, 553 (1963); M S Lipsett and L Mandel, *Proc. Third Int. Conf. Quantum Electronics* (1963), vol. 3, p. 1271; I D Abella and C H Townes, *Nature* **192**, 957 (1961); G R Hanes and B P Stoicheff, *Nature* **195**, 587 (1962); T P Hughes, *Nature* **195**, 325 (1962); M Ciftan, A Krutchkoff and S Koosekanani, *Proc. IRE* **50**, 84 (1962); J Borie and A Orzag, *Comptes Rendus* **255**, 874 (1962).
20. V Evtuhov and J K Neeland, *Appl. Opt.* **1**, 517 (1962); these observations were reported in a more complete way in *Quantum Electronics* edited by P Grivet and N Bloembergen (Columbia University Press: New York, 1964), vol. 3, p. 1405 and *IEEE J. Quantum Electron.* **QE-1**, 7 (1965).
21. A E Siegman, Microwave phototubes and light demodulators, in *Northeast Electronics Research and Engineering Meeting (NEREM) Record* (Boston, MA, 1961); B J McMyrtry and A E Siegman, *Appl. Opt.* **1**, 51 (1962).
22. R C Duncan Jr., Z J Kiss and J P Wittke, *J. Appl. Phys.* **33**, 2568 (1962).
23. M Ciftan, A Krutchkoff and S Koozekanani, *Proc. IRE 50*, **84** (1962).

24. A L Schawlow in *Quantum Electronics,* vol. 1, edited by C H Townes (Columbia University Press: New York, 1960), p. 553.

25. N lines in ruby were studied by several researchers (C E Mendenhall and R W Wood, *Phil. Mag.* **30**, 316 (1915); O Deutschbein, *Ann. Phys. Lpz* **14**, 712 (1932); ibid. **14**, 729 (1932); ibid. **20**, 828 (1932)). Deutschbein labeled them as N for Nebenlinien. A L Schawlow, D L Wood and A MM Clogston (*Phys. Rev. Lett.* **3,** 271 (1959)) had shown that they disappear at low concentration of Cr, but appear strongly in emission at concentrations in the order of 0.5% (red ruby). The energy-level structure of Cr in the red ruby is different from that of pink ruby because of the exchange interaction between the strongly magnetic Cr ions. Fluorescence in red ruby occurs at 7009 (N_2) and 7041 (N_1)Å.

26. A L Schawlow and G E Devlin, *Phys. Rev. Lett.* **6**, 96 (1961).

27. I Wieder and L R Sarles, *Phys. Rev. Lett.* **6**, 95 (1961).

28. F J McClung, S E Schwarz, R W Hellwarth and F J Meyers, *Bull. Am. Phys. Soc.* **6**, 511 (1961); *J Appl. Phys.* **33**, 3139 (1962).

29. M Hercher (*Appl. Opt.* **1**, 665 (1962)) studied the effects of optical path length variations with a Twyman–Green interferometer. Also Evtuhov and Neeland (Note 20) used this interferometer to study the emission of ruby. A deterioration of the near-field pattern with decrease in optical quality was observed by C M Stickley, *Appl. Opt.* **2**, 855 (1963) and J W Grove, at a meeting of the American Optical Society, Spring 1963, studied the effect of scattering centres on laser-mode formation.

30. R L Hutcheson, *IRE Int. Conv. Rec.* **10**, 147 (1962).

31. Linde Crystal Product Bulletin F-2330 (Linde Co: East Chicago, IL, 1965); see R D Olt, *Appl. Opt.* **1**, 25 (1962).

32. The first ruby lasers were pumped with the rod surrounded by a helical xenon flash lamp, possibly with a reflector over the outside. A right-elliptic cylinder with a lamp along one focus and the laser crystal along the other was used by M Ciftan, CC F Luck, C G Shafer and H Statz, *Proc. IRE* **49**, 960 (1961). A modification with several intersecting ellipses with a common focus containing the crystal was later employed by C Bowness, D Missio and T Rogala, *Proc. IRE* **50**, 1704 (1962). Rougher and simpler arrangements used the rod placed adjacent to one or more linear flash lamps surrounded by aluminium foil, magnesium oxide, or other highly reflecting material (P A Miles and H E Edgerton, *J. Appl. Phys.* **32**, 740 (1961)). Other dispositions were used by P H Keck, J J Redman, C E White and R E Dekinder Jr. (*Appl. Opt.* **2**, 827 (1963)), D F Nelson and W S Boyle (*Appl. Opt.* **1**, 181 (1962)) and D Roers (*Appl. Opt.* **3**, 259 (1964)). The efficiency of laser cavities was calculated by O Svelto, *Appl. Opt.* **1**, 745 (1962).

33. D F Nelson and W S Boyle, *Appl. Opt.* **1**, 181 (1962).

34. I D Abella and H Z Cummins, *J. Appl. Phys.* **32**, 1177 (1961); I D Abella, *Phys. Rev. Lett.* **9**, 453 (1962).

35. N A Kurnit, I D Abella and S R Hartmann, *Phys. Rev. Lett.* **13**, 567 (1964).

36. This and much more is recounted in a pleasant paper by P Daukantas, *OPN* **22**, 22 (May 2011).

37. P P Sorokin and M J Stevenson, *Phys. Rev. Lett.* **5**, 557 (1960).

38. The story is told by Sorokin in an interview by J Hecht (in *Lasers & Applications*, **4**, 53 March 1985) later published in the book J Hecht, *Laser Pioneers* (Academic Press, New York, 1992), p. 139.

39. P P Sorokin and M J Stevenson, *Adv. Quantum Electronics*, edied by J R Singer (Columbia University Press: New York, 1961), p. 65.

40. See, for example, P Pringsheim, *Fluorescence and Phosphorescence* (Wiley: New York, 1949); D S McClure, *Solid State Phys.* **9**, 400 (1959); W A Runciman, *Rep.*

Prog. Phys. **21**, 30 (1958). The first spectroscopic studies of CaF_2:U^{3+} were those of L N Galkin and P P Feofilov, *Dokl. Akad. Nauk USSR* **114**, 745 (1957) (in Russian) Transl. in *Sov. Phys. Doklady* **2**, 255 (1957) and L N Galkin and P P Feofilov, *Opt. Spectrosc.* **1**, 492 (1959).

41. G D Boyd, R J Collins, S P S Porto, A Yariv and W A Hargreaves, *Phys. Rev. Lett.* **8**, 269 (1962); S P S Porto and A Yariv, *J. Appl. Phys.* **33**, 1620 (1962).
42. This result was already known; see B Bleaney, P M Lewellyn and D A Jones, *Proc. Phys. Soc.* **B69**, 858 (1956).
43. S P S Porto and A Yariv, *Quantum Electronics*, edited by P Grivett and N Bloembergen (Dunod: Paris, 1964), p. 717.
44. H A Bostick and J R O'Connor, *Proc. IRE* **50**, 219 (1962).
45. P A Miles, Discussion in *Advances in Quantum Electronics*, edited by J R Singer (Columbia University Press: New York, 1961), p. 76.
46. P P Sorokin and M J Stevenson, *IBM J. Res. Dev.* **5**, 56 (1961). A US patent on a four-level laser base on this laser was assigned to P P Sorokin on 1/11/1966 no. 3.229.306.
47. L F Johnson and K Nassau, *Proc. IRE* **49**, 1704 (1961). Also this laser was patented by the authors US patent no. 3.225.306 issued December 21, 1965.
48. L F Johnson, G D Boyd, K Nassau and R R Soden, *Proc. IRE* **50**, 213 (1962) and *Phys. Rev.* **126**, 1406 (1962).
49. P H Keck, J J Redman, C E White and D E Bowen, *Appl. Opt.* **2**, 833 (1963).
50. L F Johnson, G D Boyd and K Nassau, *Proc. IRE* **50**, 86, 87 (1962).
51. See Lisa Bromberg, *The Laser in America: 1950–1970* (The MIT Press: Cambridge, MA, 1991), p. 90.
52. E Snitzer, *J. Appl. Phys.* **32**, 36 (1961).
53. E Snitzer *Phys. Rev. Lett.* **7**, 444 (1961).
54. E Snitzer, *Appl. Opt.* **5**, 1487 (1966).
55. L F Johnson and K Nassau, *Proc. IRE* **49**, 1704 (1961).
56. L F Johnson, G D Boyd, K Nassau and R R Soden, *Proc. IRE* **50**, 213 (1962).
57. L F Johnson, G D Boyd, K Nassau and R R Soden, *Phys. Rev.* **126**, 1406 (1967).
58. The paper was later published: J E Geusic and H E D Scovil, *Rep. Progr. Phys.* **27**, 241 (1964).
59. J E Geusic, H M Marcos and L G Van Uitert, *Appl. Phys. Lett.* **4**, 182 (1964).
60. L F Johnson, *J. Appl. Phys.* **33**, 756 (1962); L F Johnson and R R Soden, *J. Appl. Phys.* **33**, 757 (1962); a resumè of these researches is in Johnson, *J. Appl. Phys.* **34**, 897 (1963).
61. C G Young, *Appl. Opt.* **5**, 993 (1996). See also N A Kozlov, A A Mak and B M Sedov, *Sov. J. Opt. Technol.* **33**, 549 (1966).
62. Z J Kiss, H R Lewis and R C Duncan Jr., *Appl. Phys. Lett.* **2**, 93 (1963).
63. Laser action in CaF_2:Dy^{2+} was reported by Z J Kiss and R C Duncan Jr., *Proc. IRE* **50**, 1531 (1962).
64. V A Fabrikant, Doctoral Dissertation, FIAN P N Lebedev Physical Institute, Academy of Sciences, USSR (1939).
65. A Ferkhman and S Frish, *Phys. Z. Sowjet.* **9**, 466 (1936), reported by N G Basov and O N Krokhin, *Appl. Opt.* **1**, 213 (1962).
66. A L Schawlow and Townes, *Phys. Rev.* **112**, 1940 (1958).
67. J H Sanders, *Phys. Rev. Lett.* **3**, 86 (1959).
68. A Javan, *Phys. Rev. Lett.* **3**, 87 (1959).
69. W E Lamb and R C Retherford, *Phys. Rev.* **79**, 549 (1950).
70. See *Lasers and Applications*, **4**, 49 (October 1945), interview by J Hecht, later published in the book J Hecht, *Laser Pioneers* (Academic Press, New York, 1992), p. 155.

71. A Javan, in *Quantum Electronics*, edited by C H Townes (Columbia University Press: New York, 1960), p. 564.

72. R Ladenburg, *Rev. Mod. Phys.* **5**, 243 (1933).

73. W R Bennett, *IEEE J. Sel. Top. Quantum Electron.* **6**, 869 (2000).

74. A Javan, W R Bennett Jr. and D R Herriott, *Phys. Rev. Lett.* **6**, 106 (1961). W R Bennett and A Javan had a US patent for this laser no. 3,149,290 issued September 15, 1964.

75. The level notation in Figure 6.15 is the Russell-Sanders notation for helium and the Paschen notation for neon. In the Russell-Sanders coupling scheme, the number of the total angular momentum is indicated with a capital letter as follows: S for $L = 0$, P for $L = 1$, D for $L = 2$, F for $L = 3$, G for $L = 4$, H for $L = 5$, I for $L = 6$, K for $L = 7$. If necessary, the value of the total electron spin added to the angular momentum is indicated as a small subscript on the right. As a superscript at the left is the multiplicity of the state or the number of possible values of the vector sum of an orbital angular momentum and a spin angular momentum. So, for example, the ground state of helium which is $1s^2$ with the two electron spins antiparallel to each other, is indicated as 1^1S. The first excited state 1s2s is 2^3S if the spins are parallel and 2^1S if the spins are antiparallel. Coupling of the outer electron of an excited noble gas to the core electrons does not follow the Russell-Sanders rules. Paschen notation is a system of shorthand symbols. The ground state of neon is $2p^6$ which in Paschen notation is simply indicated as 1p. The excited configurations are $2p^53s$ (in Paschen notation 1s), $2p^53p$ (in Paschen notation 2p), $2p^54s$ (in Paschen notation 2s), $2p^55s$ (3s), and so on.

76. D R Herriott, *J. Opt. Soc. Am.* **52**, 31 (1962); D R Herriott in *Advances in Quantum Electronics,* edited by J R Singer (Columbia University Press: New York, 1961), p. 44.

77. A recollection of the discovery of this laser is A D White, *OpN* **22**, 35 (Oct 2011).

78. A D White and J D Rigden, *Proc. IRE* **50**, 1697 (1962).

79. W R Bennett Jr., *Appl. Opt. Suppl.* **1**, 24 (1962).

80. N G Basov and O N Krokhin, *JETP* **39**, 1777 (1960) (in Russian); *Sov. Phys. JETP* **12**, 1240 (1961); N G Basov, O N Krokhin and Yu M Popov, *Usp. Fiz. Nauk* **72**, 1961 (1960) (in Russian); N G Basov and O N Krokhin, *Appl. Opt.* **1**, 213 (1962).

81. A Javan, W R Bennett Jr. and D R Herriott, *Phys. Rev. Lett.* **6**, 106 (1961); W R Bennett and A Javan had a US patent for this laser no. 3,149,290 issued September 15, 1964.

82. V K Ablekov, M S Persin and I L Fabelinski, *Zh. Eksp. Teor. Fiz.* **39**, 892 (1960) (in Russian); *Sov. Phys. JETP* **12**, 618 (1960).

83. V K Ablekov, *Sov. Phys. JETP* **15**, 513 (1962).

84. P Rabinowitz, S Jacobs and G Gould, *Appl. Opt.* **1**, 513 (1962); WR Bennett Jr., W L Faust, R A MacFarlane and C K N Patel, *Phys. Rev. Lett.* **8**, 470 (1962); A D White and J D Rigden, *Proc. IRE* **80**, 1697 (1972); C K N Patel, W R Bennett Jr., W L Faust and R A MacFarlane, *Phys. Rev. Lett.* **9**, 102 (1962); R A McFarlane, C K N Patel, W R Bennett Jr. and W L Faust, *Proc. IRE* **50**, 2111 (1962).

85. Interview with C Breck Hitz in Laser & Applications, January 1985, p. 67.

86. P Rabinowitz, S Jacobs and G Gould, *Appl. Opt.* **1**, 513 (1962).

87. The first use of dielectric mirrors for a ruby laser seems to have been by McClung et al. (Note 28) to select emission on the R_2 line against the one on the R_1; Sorokin and Stevenson (Note 37) had already used them in $CaF_2:U^{3+}$ and $CaF_2:Sm^{2+}$.

88. W W Rigrod, H Kogelnik, D J Brangaccio and D R Herriott, *J. Appl. Phys.* **33**, 743 (1962).

89. P Connes, *Revue d'Optique* **35**, 37 (1956); *J. Phys. Radium* **19**, 262 (1958).

90. A G Fox and T Li, *Proc. IRE* **48**, 1904 (1960).

91. Quoted in G D Boyd and J P Gordon, *Bell Syst. Tech. J.* **40**, 489 (1961).

92. A G Fox and T Li, *Bell Syst. Tech. J.* **40**, 453 (1961); see also *Advances in Quantum Electronics,* edited by J Singer (Columbia University Press: New York, 1961), p. 308.

93. G D Boyd and J P Gordon, *Bell Syst. Tech. J.* **40**, 489 (1961); this paper immediately followed the one of Fox and Li (Note 92). See also G D Boyd in *Advances in Quantum Electronics,* edited by J Singer (Columbia University Press: New York, 1961), p. 318; G D Boyd in *Quantum Electronics,* edited by P Grivet and N Bloembergen (Columbia University Press: New York, 1964), p. 1193; G D Boyd and H Kogelnik, *Bell Syst. Tech. J.* **41**, 1347 (1962).

94. A G Fox and T Li, *Proc. IRE* **51**, 80 (1963).

95. G Goubau and F Schwering, *IRE Trans. Antennas Propag.* (May 1961) **AP-9**, p. 256 (May 1961); J R Pierce, *Proc. Natl Acad. Sci.* **47**, 1808 (1961); F Schwering, *Arch. Electr. Uber.* **15**, 555 (1961).

96. M Bertolotti, *Nuovo Cim.* **26**, 401 (1962).

97. G Toraldo di Francia in *Quantum Electronics and Coherent Light,* edited by P A Miles (Academic Press: New York, 1964), p. 53; C L Tang, *Appl. Opt.* **1**, 768 (1962); J M Burch in *Quantum Electronics,* edited by P Grivet and N Bloembergen (Columbia University Press: New York, 1964), p. 1187; M Pouthier, ibid. p. 1253; A G Fox and T Li, ibid. p. 1263; J Kotik and M C Newstein, *J. Appl. Phys.* **32**, 178 (1962); R F Soohoo, *Proc. IEEE* **51**, 70 (1963); S R Barone, *J. Appl. Phys.* **34**, 831 (1963); J B Beyer and E H Scheibe, *IRE Trans. Antennas Propag.* **10**, 349 (1962); A G Fox and T Li, *Proc. IEEE* **51**, 80 (1963); W Culshaw, *IRE Trans.* **MTT 10**, 331 (1962); H Kogelni and T Li, *Proc. IEEE* **54**, 131–132 (1966); L A Vainshtein, *JETP* **17**, 709 (1963); *Sov. Phys. Tech. Phys.* **9**, 157 (1964); J P Gordon and H Kogelnik, *Bell Syst. Tech. J.* **43**, 2873 (1964); C Y She and H Heffner, *Appl. Opt.* **3**, 703 (1964).

98. I A Veinshtein, *JETP* **18**, 471 (1964); *Sov. Phys. Tech. Phys.* **9**, 166 (1964); P O Clark, *Proc. EEE* **53**, 36 (1965); *J Appl. Phys.* **36**, 66 (1965); W A Specht, *J. Appl. Phys.* **36**, 1306 (1965).

99. A Yariv and J P Gordon, *Proc. IEEE* **51**, 4 (1963).

100. For example, N Kambe, *Proc. IEEE* **52**, 327 (1964); K Miyamoto, *J. Opt. Soc. Am.* **54**, 989 (1964); S A Collins, *Appl. Opt.* **3**, 1263 (1964); S A Collins and D T M Davis, *Appl. Opt.* **3**, 1314 (1964); T Li, *Appl. Opt.* **3**, 1315 (1964); J P Gordon, *Bell Syst. Tech. J.* **43**, 1826 (1964).

101. H Kogelnik and W W Rigrod, *Proc. IRE* **50**, 220 (1962).

102. W W Rigrod, *Appl. Phys. Lett.* **2**, 51 (1963).

103. A E Siegman, *Proc. IEEE* **53**, 277 (1965).

104. A E Siegman, *Laser Focus* **7**, 42 (May 1971).

105. A E Siegman and R W Arrathoon, *IEEE J. Quant. Electron.* **QE-3**, 156 (1967), and many others. For a general review, see A E Siegman, *Appl. Opt.* **13**, 353 (1974).

106. For example, R J Collins and J Giordmaine in *Quantum Eòlectronics III,* vol. 2, edited by P Grivet and N Bloembergen (Columbia Univesity Press: New York, 1964), p. 1239, considered a ruby crystal in the shape of a rectangular parallelepipedon. P A Kleiman and P P Kisliuk, *Bell Syst. Tech. J.* **41**, 453 (1962) considered a double mirror ended interferometer which allowed discrimination against unwanted modes.

107. E R Peck, *J. Opt. Soc. Am.* **38**, 66 (1948); ibid. **38**, 1015 (1948); ibid. **47**, 250 (1957).

108. M V R K Murty, *J. Opt. Soc. Am.* **50**, 7 (1960); ibid. **50**, 83 (1960).

109. E R Peck, *J. Opt. Soc. Am.* **52**, 253 (1962).

110. G Gould, S Jacob, P Rabinowitz and T Shultz, *Appl. Opt.* **1**, 533 (1962); see also P Rabnowitz, S F Jacobos, T Schultz and G Gould, *J. Opt. Soc. Am.* **52**, 452 (1962).

111. L Bergstein, W Kahn and C Shulman, *Proc. IRE* **50**, 1833 (1962).

112. E S Dayhoff, *Proc. 10th Colloqium Spectroscopicum International*, 1962, p. 421.

113. M Bertolotti, L Muzi and D Sette, *Nuovo Cim.* **26**, 401 (1962).

114. H A Daw, *J. Opt. Soc. Am.* **53**, 915 (1963); I N Court and F K Willisen, *Appl. Opt.* **3**, 719 (1964); D F Holshouser, *Quantum Elecronics*, vol. 3, edited by P Grivet and N Bloembergen (Columbia University Press: New York, 1964), p. 1453, suggested the use of conical ends for total internal reflection.

115. Yu Anan'ev, *Laser Resonators and the Beam Divergence Problem* (Adam Hilger: Bristol, 1992).

116. A H Rosenthal, *J. Opt. Soc. Am.* **52**, 1143 (1962).

117. W M Macek and D T M Davis Jr., *Appl. Phys. Lett.* **2**, 67 (1963).

118. Lord Rayleigh, Theory of Sound, vol. II, 1st edition (MacMillan: London, 1878), see also Lord Rayleigh, Phil. Mag. 20, 1001 (1910); ibid. **27**, 100 (1914).

119. G Mie, *Ann Phys.* **25**, 377 (1908).

120. P Debye, *Ann. Phys.* **30**, 57 (1909).

121. R D Richtmyer, *J. Appl. Phys.* **10**, 391 (1939).

122. M Gastine et al. *IEEE Trans. Microw. Theor. Tech.* **MTT-15**, 694 (1967); P Affolter and B Eliasson, *IEEE Trans. Microw. Theor. Tech.* **MTT-21**, 573 (1973).

123. C G B Garrett, W Kaiser and W L Bond, *Phys. Rev.* **124**, 1807 (1961).

124. A N Oraevski, *Quantum Electron.* **32**, 377 (2002); See also V B Braginsky, M L Gorodetsky and V S Ilchenko, *Phys. Lett.* **A137**, 393 (1989); A Matsko and V S Ilchenko, *IEEE J. St. Quantum Electron.* **12**, 3 (2006).

125. See, for example, H B Lin et al., *Opt. Lett.* **11**, 614 (1986); S L McCall et al., *Appl. Phys. Lett.* **60**, 289 (1992); M L Gorodetsky and V S Ilchenko, *J. Opt. Soc. Am.* **B16**, 147 (1999).

126. E Snitzer, *J. Opt. Soc. Am.* **51**, 491 (1961); *J. Appl. Phys.* **32**, 36 (1961).

127. A Kastler, *Appl. Opt.* **1**, 17 (1962).

128. The major distinction between a distributed feedback laser and a conventional laser is that the distributed feedback laser does not use cavity mirrors. Instead, feedback is provided via Bragg scattering from spatially periodic perturbation of the optical parameters of the active medium. Such parameters can be the refractive index of the laser medium as in the first realization by H Kogelnik and C V Shank, *Appl. Phys. Lett.* **18**, 152 (1971), see chapter 7.

129. See, for example, G. Gould, *Appl. Opt.* **4**, 59 (1965); W R Bennett Jr., *Appl. Opt.* **4**, 3 (1965).

130. G G Petrash, *Sov. Phys. Usp.* **14**, 747 (1972); Yu A Babeiko et al., *Sov. J Quantum Electron.* **9**, 651 (1979).

131. G R Fowles and W T Silfvast, *Appl. Phys. Lett.* **6**, 236 (1965).

132. W T Walter, M Pilch, N Solimene, G Gould, and W R Bennett, *Bull. Am. Phys. Soc.* **11**, 113 (1966); The results are also reported in W T Walter et al., *IEEE J. Quantum Electron.* **QE-2**, 474 (1966).

133. W T Walter, N Solimene, M Piltch and G Gould, *IEEE J. Quantum Electron.* **QE-2**, 474 (1966); W T Walter, *IEEE J. Quantum Electron.* **QE-4**, 355 (1968); G R Russell, N M Nerheim and T J Pivirotto, *Appl. Phys. Lett.* **21**, 565 (1972).

134. A A Isaev, M A Kazaryan and G G Petrash, *JETP Lett.* **16**, 27 (1972).

135. I Smilanski, *Opt. Comm.* **25**, 79 (1978).

136. I L Bass et al., *Appl. Opt.* **31**, 6993 (1992).

137. For a discussion on this point, see A L Bloom, *Appl. Opt.* **5**, 1502 (1966).

138. W E Bell, *Appl. Phys. Lett.* **4**, 34 (1964).

139. Electronics p. 17, January 24, 1964.
140. H G Heard, G Makhov and J Peterson, *Proc. IEEE* **52**, 414 (1964).
141. W B Bridges, *Appl. Phys. Lett.* **4**, 128 (1964).
142. More on the discovery of the argon laser may be found in Lisa Bromberg, *The Laser in America,* pp. 163–173. The story is also told by W B Bridge, *IEEE J. Sel. Topics Quantum Electron.* **6**, 885 (2000).
143. G Convert, M Armand and P Martinot-Lagarde, *Comptes Rendus* **258**, 4467 (1964); **258**, 3259 (1964).
144. W R Bennett Jr., J W Knutson Jr., G N Mercer and N J L Detch, *Appl. Phys. Lett.* **4**, 180 (1964); see also Lisa Bromberg, *The Laser in America.*
145. W B Bridges, *Proc. IEEE* **52**, 843 (1964).
146. E I Gordon and E F Labuda, *Bell Syst. Tech. J.* **43**, 1827 (1964); E I Gordon, E F Labuda and W B Bridges, *Appl. Phys. Lett.* **4**, 178 (1964).
147. A D White, *Appl. Opt.* **3**, 431 (1964).
148. A L Bloom, W E Bell and F O Lopez, *Phys. Rev.* **135**, A578 (1964).
149. P Laures, L Dana and C Frapard, *Comptes Rendus* **258**, 6363 (1964); **259**, 745 (1964).
150. L Dana and P Laures, *Proc. IEEE* **53**, 78 (1965).
151. F A Horrigan, S H Koozekanani and R A Paananen, *Appl. Phys. Lett.* **6**, 41 (1965).
152. H J Gerritsen and P V Goedertier, *J. Appl. Phys.* **35**, 3060 (1964).
153. J McNell et al., *Appl. Phys. Lett.* **28**, 207 (1976); K Jain and S Newton, *Appl. Phys.* **B26**, 434 (1981).
154. A A Isaev et al., *JETP Lett.* **16**, 27 (1972); C M Ferrar, *IEEE J. Quantum Electron.* **QE-10**, 655 (1974).
155. S Aleksandrov et al., *Sov. J. Quantum Electron.* **5**, 1132 (1975).
156. C S Liu, E W Sucov and L A Weaver, *Appl. Phys. Lett.* **23**, 92 (1973).
157. C J Chen, N M Nerheim and G R Russell, *Appl. Phys. Lett.* **23**, 514 (1973).
158. J A Piper and C E Webb, *J. Phys. D: Appl. Phys.* **6**, 400 (1973).
159. W B Bridgs and A N Chester, *IEEE J. Quantum Electron.* **QE-1**, 66 (1965).
160. C C Davis and T A King in *Advances in Quantum Elecronics,* vol. 3 (Academic Press: New York, 1975), p. 169.
161. F Legay and P Barchewitz, *Comptes Rendus* **256**, 5304 (1963).
162. F Legay and N Legay-Sommaire, *Comptes Rendus* **259**, 99 (1964).
163. C K N Patel, W L Faust and R A McFarlane, *Bull. Am. Phys. Soc.* **9**, 500 (1964).
164. C K N Patel, *Phys. Rev.* **136**, A1187 (1964).
165. C K N Patel, *Phy. Rev. Lett,* **12**, 588 (1964).
166. C K N Patel, *Phys. Rev. Lett.* **13**, 617 (1964).
167. The fundamental modes of vibration of the CO_2 molecule are three: symmetric stretching mode (with frequency v_1); bending mode (v_2); and asymmetric stretching mode (v_3). The label nm^lp means, n quanta of frequency v_1, m of frequency v_2, and p of frequency v_3. The superscript l gives the angular momentum of the vibration about the axis of the molecule in units of $\hbar$.
168. C K N Patel, *Appl. Phys. Lett.* **7**, 15 (1965).
169. N Legay-Sommaire, L Henry and F Legay, *Comptes Rendus* **260**, 3339 (1965).
170. J A Howe, *Appl. Phys. Lett.* **7**, 21 (1965).
171. G Moller and J D Rigden, *Appl. Phys. Lett.* **7**, 274 (1965).
172. C K N Patel, P K Tuin and J H McFee, *Appl. Phys. Lett.* **7**, 290 (1965).
173. M A Kovacs, G W Flynn and A Javan, *Appl. Phys. Lett.* **8**, 62 (1966). A somewhat unusual technique was proposed in the same year by T J Bridges (*Appl. Phys. Lett.* **9**, 174(1966)) by using a moving mirror technique.
174. A E Hill, *Appl. Phys. Lett.* **12**, 324 (1968).

175. R Dumanchin and J Rocca-Serra, *Comptes Rendus* **269**, 916 (1969); A J Beaulieu, *Appl. Phys. Lett.* **16**, 504 (1970).

176. O R Wood, E G Burkhardt, M A Pollack and T J Bridges, *Appl. Phys. Lett.* **18**, 261 (1971).

177. W R Bennett Jr., *Appl. Opt.* **2**, 3 (1965); P K Tien, C MacNair and H L Hodges, *Phys. Rev. Lett.* **12**, 30 (1964); Yu V Tkach, Ya B Fainberg, L I Bulotin, Ya Ya Bessarab and N P Gadetsuii, *JETP Lett.* **6**, 371 (1967).

178. N G Basov, *IEEE J. Quantum Electron.* **QE-2**, 354 (1966); A G Molchanov, I A Polacktrova and Yu M Popov, *Sov. Phys. Solid State* **9**, 2655 (1968).

179. N G Basov, V A Danilychev, Yu M Popov and D D Khodkevich, *JETP Lett.* **12**, 329 (1970).

180. N G Basov, *JETP Lett.* **14**, 285 (1971); *Sov. J. Quantum Electron.* **1**, 306 (1971); J D Daugherty, E R Pugh and D H Douglas-Hamilton, *Bull. Am. Phys. Soc.* **17**, 399 (1972); C A Fenstermacher, M J Nutter, W T Leland and K Boyer, *Appl. Phys. Lett.* **20**, 56 (1972); C A Fenstermacher, M J Nutter, J P Rink and K Boyer, *Bull. Am. Phys. Soc.* **16**, 42 (1971).

181. D W Gregg, *Chem. Phys. Lett.* **8**, 609 (1971); Y I Pan, C E Turner Jr., K J Pettipiece, *Chem. Phys. Lett.* **10**, 577 (1971).

182. N G Basov et al., *JETP Lett.* **14**, 285 (1971).

183. O R Wood II, *Proc. IEEE* **62**, 355 (1974).

184. N G Basov and A N Oraevskii, *JETP* **17**, 1171 (1963).

185. N G Basov et al., *Sov. Phys. Tech. Phys.* **13**, 1630 (1969); E T Gerry, *Appl. Phys. Lett.* **7**, 6 (1965).

186. F G Houtermans, *Helv. Phys. Acta* **33**, 933 (1960).

187. L E S Matthias and J T Parker, *Appl. Phys. Lett.* **3**, 16 (1963).

188. H G Heard, *Nature* **200**, 667 (1963); *Bull. Am. Phys. Soc.* **9**, 65 (1964).

189. D A Leonard, *Appl. Phys. Lett.* **7**, 4 (1965).

190. E T Gerry, *Appl. Phys. Lett.* **7**, 6 (1965).

191. B Stevens and E Hutton, *Nature* **186**, 1045 (1960).

192. N G Basov, *IEEE J. Quantum Electron.* **QE-2**, 354 (1966).

193. H A Koehler, H A Ferderber, D L Redhead and P J Ebert, *Appl. Phys. Lett.* **21**, 198 (1972).

194. E R Ault et al., *IEEE J. Quantum Electron.* **9**, 1031 (1972).

195. W M Hughes, J Shannon, A Kolb, E Ault and M Bhaumik, *Appl. Phys. Lett.* **23**, 385 (1973).

196. P W Hoff, J C Swingle and C K Rhodes, *Appl. Phys. Lett.* **23**, 245 (1973).

197. M F Golde and B A Thrush, *Chem. Phys. Lett.* **29**, 486 (1974).

198. J E Velazco and D W Setser, *J. Chem. Phys.* **62**, 1990 (1975).

199. S K Searles and G A Hart, *Appl. Phys. Lett.* **27**, 243 (1975).

200. J J Ewing and C A Brau, *Appl. Phys. Lett.* **27**, 350 and 435 (1975).

201. E G Brok, P Czavinsky, H Hormats, H C Neddeman, D Stirpe and F Unterleitner, *J. Chem. Phys.* **35**, 759 (1961).

202. S G Rautian and I I Sobel'mann, *Opt. Spectrosc.* **10**, 134 (1961); *Opt. Spectrosc.* **10**, 65 (1961).

203. D L Stockman, W R Mallory and F K Tittel, *Proc. IEEE* **52**, 318 (1964).

204. D L Stockman, *Proc. 1964 ONR Conf. on Organic Lasers.*

205. P P Sorokin and J R Lankard, *IBM J. Res. Dev.* **10**, 162 (1966).

206. P P Sorokin, W H Culver, E C Hammond and J R Lankard, *IBM J. Res. Dev.* **10**, 401 (1966).

207. Lisa Bromberg, *The Laser in America,* p. 187.

208. M L Spaeth and D P Bortfield, *Appl. Phys. Lett.* **9**, 179 (1966).

209. F P Schaefer, W Schmidt and J Volze, *Appl. Phys. Lett.* **9**, 306 (1966).

210. B I Stepanov, A N Rubinov and V A Mostovsikov, *JETP Lett.* **5**, 117 (1967).

211. P P Sorokin, J R Lankard, E C Hammond and V L Moruzzi, *IBM J. Res. Dev.* **11**, 130 (1967).

212. P P Sorokin and J R Lankard, *IBM J. Res. Dev.* **11**, 148 (1967).

213. W Schmidt and F P Schaefer, *Z. Naturf.* **229**, 1563 (1967).

214. F P Schaefer, W Schmidt and K Marth, *Phys. Lett.* **24A**, 280 (1967); P P Sorokin, J R Lankard, E C Hammond and V L Moruzzi, *IBM J. Res. Dev.* **11**, 130 (1967); B B McFarland, *Appl. Phys. Lett.* **10**, 208 (1967).

215. B I Stepanov and A N Rubinov, *Sov. Phys. Usp.* **11**, 304 (1968).

216. B B Snavely and F P Schaefer, *Phys. Lett.* **28A**, 728 (1969).

217. O G Peterson, S A Tuccio and B B Snavely, *Appl. Phys. Lett.* **17**, 245 (1970).

218. B H Soffer and B B McFarland, *Appl. Phys. Lett.* **10**, 266 (1967).

219. T W Haensch, *Appl. Opt.* **11**, 895 (1972).

220. I Shoshan, N N Danon, U P Oppenheim, *J. Appl. Phys.* **48**, 4495 (1977); M G Littman and H J Metcalf, *Appl. Opt.* **17**, 2224 (1978); S Saikan, *Appl. Phys.* **17**, 41 (1978).

221. K Liu and M G Littman, *Opt. Lett.* **6**, 117 (1981).

222. B H Soffer and B B MacFarland, *Appl. Phys. Lett.* **10**, 266 (1967).

223. O G Peterson and B B Snavely, *Appl. Phys. Lett.* **12**, 238 (1968).

224. J C Altman et al., *IEEE Photon. Technol. Lett.* **3**, 189 (1991); A Maslyukov et al., *Appl. Opt.* **34**, 1516 (1995); A Costela et al., *Laser Chem.* **18**, 63 (1998); *Chem. Phys. Lett.* **369**, 656 (2003); F J Duarte *Appl. Opt.* **33**, 3857 (1994); **38**, 6347 (1999); M Rifani et al., *J. Am. Chem. Soc.* **117**, 7572 (1995).

225. T W Hansch et al., *IEEE J. Quantum Electron.* **QE-7**, 45 (1971).

226. T W Hansch, Optics & Photonics News, February 2005, p. 14.

227. R E Whan and G A Crosby, *J. Mol. Spectrosc.* **8**, 315 (1962).

228. E J Schimitschek and E G K Schwartz, *Nature* **196**, 832 (1962).

229. A Lempicki and H Samelson, *Phys. Lett.* **1**, 133 (1963); *Proc. Symp. on Optical Masers* (Polytechnic Press: Brooklyn, New York (1963), p. 347.

230. H Samelson, A Lempicki, C Brecher and V Brophy, *Appl. Phys. Lett.* **5**, 173 (1964).

231. J C Polanyi, *Proc. R. Soc. V Canada* **54**, 25 (1960).

232. J C Polanyi, *J. Chem. Phys.* **34**, 347 (1961).

233. J V V Kasper and G C Pimentel, *Phys. Rev. Lett.* **14**, 352 (1965).

234. See, for example, J C Polanyi, *J. Chem. Phys.* **31**, 1338 (1959); J K Cashion and J C Polanyi, *Proc. R. Soc.* **A258**, 529 (1960); J K Cashion nd J C PPolanyi, *J. Chem. Phys.* **29**, 455 (1958).

235. N G Basov, A N Oraevskii, *JETP* **44**, 1742 (1963) (in Russian); G Karl and J C Polanyi, *Discuss. Faraday Soc.* **33**, 93 (1962); A N Oraevskii, *JETP* **45**, 177 (1963) (in Russian); VV L Tal'roe, *Kinetica i Katalize* **1**, 11 (1964) (in Russian); A N Oraevskii, *JETP* **48**, 1150 (1965) (in Russian); R A Young, *J. Chem. Phys.* **40**, 1848 (1964); and finally *Appl. Opt. Supl.* **2**: *Chemical Lasers* 1965 which gives an impressive look at the research in the field up to the moment of the first realization.

236. S G Rautian and I I Sobel'man, *Zurn. Eksp. Teor. Fiz.* **41**, 2018 (1961).

237. T L Andreeva et al., *Zurn. Eksp. Teor. Fiz.* **49**, 1408 (1965).

238. T F Deutsch, *Appl. Phys. Lett.* **10**, 234 (1967).

239. N G Basov, E PP Markin, A I Nikitin and A N Oraevskii in *Proc. Symp. Chemical Lasers*, St. Louis (1969); O M Batovskii, G K Vasil'ev, E F Makarov and V L Tal'roze, *JETP Lett.* **9**, 200 (1969); N G Basov, L V Kulakov, E P Markin, A I Nikitin and A N Oraevskii, *JETP Lett.* **9**, 375 (1969); see also *Int. Symp. Chemical Lasers*, Moscow, USSR, 2- September 4, 1969.

240. D J Spencer, T A Jacobs, J H Mirels and R W F Gross, *Int. J. Chem. Kinet.* **1**, 493 (1969), Addendum ibid. **2**, 337 (1970); D J Spencer, H Mirels, T A Jacobs and R W F Gross, *Appl. Phys. Lett.* **16**, 235 (1970).

241. J R Airey and S F McKay, *Appl. Phys. Lett.* **15**, 401 (1969). They obtained cw operation for 1.8 ms, using a shock-tube geometry.

242. T A Cool and R R Stephens, *Appl. Phys. Lett.* **16**, 55 (1970); *J. Chem. Phys.* **51**, 5175 (1969); T A Cool, R R Stephens and T J Falk, *Int. J. Chem. Kinet.* **1**, 495 (1969).

243. Y I Pan, C E Turner Jr. and K J Pettipiece, *Chem. Phys. Lett.* **10**, 577 (1971).

244. V F Zharov,V K Malinovskii, Yu S Neganov and G M Chumak, *JETP Lett.* **16**, 154 (1972).

245. T A Cool, *IEEE J. Quantum Electron.* **QE-9**, 72 (1973).

246. A V Antonov et al., *Sov. Journ. Quantum Electron.* **5**, 123 (1975).

247. Japanese Patent no. 273217, September 20, 1960, Yasushi Watanabe, Sendai and Jun-ichi- Nishizawa, Sendai, Appl. No. 32-9899 filed April 22, 1957: Semiconductor maser.

248. P Agrain, *Congrès International sur le Physique de l'Etat Solide et ses Applications à l'Electronique et aux Tèlècommunications* Bruxelles 1958, quoted by P Agrain in *Quantum Electronics* edited by P Grivet and N Bloembergen (Dunod: Paris, 1964), p. 1761; see also M G A Bernard and G Duraffourg in *Quantum Electronics* edited by P Grivet and N Bloemrgen (Dunod: Paris, 1964), p. 1849.

249. M G A Bernard and G Duraffourg in *Quantum Electronics,* edited by P Grivet and N Bloemrgen (Dunod: Paris, 1964), p. 1849.

250. J von Neumann, *Collected Works,* vol. 5 (Pergamon: London, 1963), p. 420. See also *IEEE J. Quantum Electr.* **QE-23**, 659 (1987).

251. N Kromer, *Proc. IRE* **47**, 397 (1959).

252. N G Basov, O N Krokhin and Yu M Popov, *Zh. Eksp. Teor. Fiz.* **38**, 1001 (1960); D C Matthis and M J Stevenson, *Phys. Rev. Lett.* **3**, 18 (1959); P Kaus, *Phys. Rev. Lett.* **3**, 20 (1959).

253. H J Zeiger, in *Quarterly Progress Report on Solid State Research, Lincoln Laboratory, MIT* (October 15, 1959), pp. 41–53, AD 33 1991.

254. N G Basov, B M Vul and Yu M Popov, *Zh. Eksp. Teor. Fiz.* **37**, 587 (1959) (in Russian); *Sov. Phys. JEPT* **10**, 416 (1960); N G Basov, O N Krokhin and Yu M Popov, *Zh. Eksp. Teor. Fiz.* **39**, 1001 (1960); ibid. 1486 (1960) (in Russian), *Sov. Phys. JETP* **12**, 1033 (1961); *Zh. Eksp. Teor. Fiz.* **40**, 1203 (1961) (in Russian); *Sov. Phys. JETP* **13**, 845 (1961); in *Advances in Quantum Electronics,* edited by J R Singer (Columbia University Press: New York, 1961), p. 496; N G Basov, in *Quantum Electronics,* edited by P Grivet and N Bloembergen (Dunod: Paris, 1964), p. 1769.

255. O N Krokhin and Yu M Popov, *Sov. Phys. JETP* **38**, 1589 (1960).

256. The mechanism led to laser action in CdS by F H Nicoll, *Appl. Phys. Lett.* **23**, 465 (1973) and later by the Lebedev group in several semiconductors, under the name of *semiconductor streamer lasers*; see N G Basov et al., *JETP* **70**, 1751 (1976); N G Basov, A G Molchanov, A S Nasibov, A Z Obidin, A N Pechenov and Yu M Popov, *IEEE J. Quantum Electron.* **QE-13**, 699 (1977); *JETP* **19**, 336 (1972); *JETP Lett.* **19**, 336 (1974); *IEEE J. Quantum Electron.* **QE-10**, 794 (1974).

257. N G Basov, O N Krokhin and Yu M Popov, *JETP* **10**, 1879 (1961).

258. T Round, *Electrical World* 309 (1907).

259. O V Lossev, *Telegraphia i Telefonia* **18**, 61 (1923).

260. O V Lossev, *Phil. Mag.* **6**, 1024 (1928).

261. K Lehovec, C A Accardo and E Jamgochian, *Phys. Rev.* **83**, 603 (1951).

262. N G Basov, B D Osipov and A N Khvoshchev, *Zh. Eksp. Teor. Fiz.* **40**, 1882 (1961); B D Osipov and A N Khvoshchev, ibid. **43**, 1179 (1962).
263. N G Basov, O N Krokhin, L M Lisitsyn, E P Markin and B D Osipov, *Zh. Eksp. Teor. Fiz.* **41**, 988 (1961). A historical perspective is presented in Yu M Popov, *Proc. PN Lebedev Physics Institute*, vol. 31 (Consultant Bureau, New York, 1968).
264. B Lax in *Quantum Electronics,* edited by C H Townes (Columbia University Press: New York, 1960), p. 428, and *Advances in Quatum Electronics,* edited by J Singer (Columbia Universiry Press: New York, 1961), p. 465.
265. Yu M Popov, *Fiz. Tverd Tela* **5**, 1170 (1963).
266. M G A Bernard and G Duraffourg, *Phys. Status Solidi* **1**, 669 (1961).
267. D N Nasledov, A A Rogachev, S M Ryvkin and B V Tsarenkov, *Sov. Phys. Solid State* **4**, 782 (1962).
268. Recollections from the principal workers from these centres can be found in a special number of the *IEEE J. Quantum Electron.* **QE-23** (1987); see also R D Dupuis, *OPn* April 2004, p. 30.
269. P W Dumke, *Phys. Rev.* **127**, 1559 (1962).
270. As reported by R N Hall, *IEEE Trans. Electron. Devices* **WD-23**, 700 (1976).
271. J I Pankove and M J Massoulie, *J. Electrochem. Soc.* **109**, 67C (1962); see also J I Pankove, *Phys. Rev. Lett.* **9**, 283 (1962); J I Panove and J E Berkeyheiser, *Proc. IRE* **50**, 1976 (1962).
272. R H Rediker, *IEEE J. Quantum Electron.* **QE-23**, 692 (1987).
273. R J Keyes and T M Quist, see *Proc. IRE* **50**, 1822 (1962).
274. R N Hall, *IEEE J. Quantum Electron.* **QE-23**, 674 (1987).
275. R N Hall, G E Fenner, J O Kingsley, T J Soltys and R O Carlson, *Phys. Rev. Lett.* **9**, 366 (1962), paper received on September 24.
276. N Holoniak, *IEEE J. Quantum Electron.* **QE-23**, 684 (1987); *IEEE J. Sel. Top. Quant. Electron.* **6**, 1190 (2000).
277. M I Nathan, W P Dumke, G Burns, F H Dill Jr. and G Lasher, *Appl. Phys. Lett.* **1**, 62 (1962).
278. R J Keyes and T M Quist, *Proc. IRE* **50**, 1822 (1962); T M Quist, R H Rediker, R J Keyes, W E Krag, B Lax, A L McWhorter and H J Zeiger, *Appl. Phys. Lett.* **1**, 91 (1962).
279. N Holonyak Jr. and S F Bevacqua, *Appl. Phys. Lett.* **1**, 82 (1962).
280. N Holonyak Jr., S F Bevacqua, C V Bielan and S J Lubowski, *Appl. Phys. Lett.* **3**, 47 (1963).
281. A L McWhorter, H J Zeiger and B Lax, *J. Appl. Phys.* **34**, 125 (1963), paper received October 23, 1962.
282. Other treatments followed immediately; see, for example, A Yariv and R C C Leite, *Appl. Phys. Lett.* **2**, 55 (1963); A L McWhorter, *Solid State Electron.* **6**, 417 (1963); G J Lasher, *IBM J.* **7**, 58 (1963); T M Quist, R J Keyes, W E Krag, B Lax, A L McWhorter, R H Rediker and H J Zeiger, *Quantum Electronics,* edited by P Grivet and N Bloembergen (Dunod: Paris, 1964), p. 1833.
283. K Weiser and R S Levitt, *Appl. Phys. Lett.* **2**, 178 (1963).
284. W L Bond, B G Cohen, R C C Leite and A Yariv, *Appl. Phys. Lett.* **2**, 57 (1963).
285. This was achieved by J C Dyment and L A D'Asaro, *Appl. Phys. Lett.* **11**, 292 (1967) who, using an adequate heat sink, obtained cw operation at temperatures up to 205 K.
286. H Kroemer, *Proc. IEEE* **51**, 1782 (1963).
287. Zh I Alverov and R F Kazarinov quoted in Zh I Alverov et al., *Sov. Phys. Solid State* **9**, 208 (1967). For more details, see Z Alferov, *IEEE J. Sel. Topics Quantum Electron.* **6**, 832 (2000).

288. Z Alferov,V B Khalfin and R F Kazarinov, *Sov. Phys. Solid State* **8**, 2480 (1967); Z Alferov, *Sov. Phys. Semicond.* **1**, 358 (1967).

289. I Hayashi, M B Panish and P W Foy, *IEEE J. Quantum Electron.* **QE-5**, 211 (1969); M B Panish, I Hayashi and S Sumski, ibid. **QE-5**, 210 (1969); I Hayashi and M B Panish, *J. Appl. Phys.* **41**, 150 (1970); see also H C Casey Jr. and M B Panish, *Heterostructure Lasers* (Academic Press: New York, 1978), part I, chap 1 where a good historical introduction to semiconductor lasers is provided.

290. H Kresseland and H Nelson, *RCA Rev.* **30**, 106 (1969).

291. Zh I Alverov, V M Andreev, E L Portnoi and M K Trukn, *Sov. Phys. Semicond.* **3**, 1107 (1970).

292. Operation in cw at room temperature was obtained by I Hayashi, M B Panish, P W Foy and S Sumski, *Appl. Phys. Lett.* **17**, 109 (1970) and Zh I Alferov, V M Andreev, D Z Garbuzov, Yu V Zhilyaev, E P Morozov, E L Portnoi and V G Trofim, *Sov. Phys. Semicond.* **4**, 1573 (1971).

293. H Kressel and H Nelson, *RCA Rev.* **30**, 106 (1969).

294. I Hayashi, M B Panish and P W Foy, *IEEE J. Quantum Electr.* **QE-5**, 211 (1969).

295. M Ettenberg, *Appl. Phys. Lett.* **27**, 652 (1975).

296. R Dingle and C H Henry, *Quantum Effects in Heterostructure Lasers,* U.S. Patent No. 3982207, 1976.

297. R Dingle, W Wiegmann and C H Henry, *Phys. Rev. Lett.* **33**, 827 (1974); W T Tsang, C Weisbuch, R C Miller and R Dinge, *Appl. Phys. Lett.* **35**, 673 (1979).

298. J Y Marzin et al., *Phys. Rev. Lett.* **73**, 716 (1994); M Grundmann et al., *Phys. Rev. Lett.* **74**, 4043 (1995).

299. W T Tsang, *Appl. Phys. Lett.* **39**, 786 (1981); **40**, 217 (1982).

300. Y Arakawa and H Sakaki, *Appl. Phys. Lett.* **40**, 939 (1982).

301. M Asada, Y Miyamoto and Y Suematsu, *IEEE J. Quantum Electron.* **QE-22**, 1915 (1986).

302. A I Ekimov and A A Onushchenko, *JETP Lett.* **34**, 345 (1981).

303. L Goldstein et al., *Appl. Phys. Lett.* **47**, 1099 (1985).

304. E Kapon, D M Hwang and R Bhat, *Phys. Rev. Lett.* **63**, 430 (1989).

305. H Hirayama et al., *Electron. Lett.* **30**, 142 (1994).

306. J M Moisson et al., *Appl. Phys. Lett.* **64**, 196 (1994); D Lenard et al., *Appl. Phys. Lett.* **63**, 3203 (1993).

307. N N Ledentsov et al., *Semiconductors* **28**, 832 (1994).

308. N Kirstaedter et al., *Electron. Lett.* **30**, 1416 (1994).

309. N N Ledentsov et al., *Phys. Rev.* **B54**, 8743 (1996).

310. N Kirstaedter, N N Ledentsov, M Grundman et al., *Electron. Lett.* **30**, 1416 (1994).

311. V M Ustinov et al., *J. Cryst. Growth* **175/176**, 689 (1997).

312. D L Huffaker et al., *Appl. Phys. Lett.* **73**, 2564 (1998).

313. G Park et al., *Appl. Phys. Lett.* **75**, 3267 (1999).

314. D Bimberg et al., *Thin Solid Films* **367**, 235 (2000); see also E Kapon, *Quantum Wire and Quantum Dot Lasers*, in *Semiconductor Lasers*, edited by E Kaplon (Academic Press: San Diego, 1999) Chap 4, p. 291.

315. A P Bogatov et al., *Sov. Phys. Semicond.* **9**, 1282 (1975); J J Hsieh et al., *Appl. Phys. Lett.* **28**, 709 (1976); K Oe et al., *Jpn. J. Appl. Phys.* **16**, 1273 (1977); N Kobayashi and Y Horikoshi, *Jpn. J. Appl. Phys.* **18**, 1005 (1979); S Akiba et al., *Electron. Lett.* **15**, 606 (1979); G D Henshall and P D Greene, *Electron. Lett.* **15**, 621 (1979); H T Kawaguchi et al., *Electron. Lett.* **15**, 669 (1969); Kaminov et al., *Electron. Lett.* **15**, 763 (1979); S Arai et al., *Jpn. J. Appl. Phys.* **18**, 2333 (1979); D Botez et al., *Proc. IEEE* **68**, 689 (1980).

316. N G Basov, O V Bogdankevich and A G Devyatkov, *Sov. Phys. JEPT* **20**, 1067 (1965).

317. S Nakamura et al., *Jpn. J. Appl. Phys.* Part 2 **34**, L1332 (1995); T Mukai, H Narimatsu and S Nakamura, *Jpn. J. Appl. Phys.* Part 2 **37**, L479 (1998); S Nakamura et al., *Appl. Phys. Lett.* **69**, 4056 (1996); S Nakamura, *IEEE J. Select. Topics Quantum Electron.* **3**, 435 (1997); S Nakamura and G Fasol, *The Blue Laser Diode* (Springer: Berlin, 1997); see also I Asaki and H Amano, *J. Electrochem. Soc.* **141**, 2266 (1994). S Nakamura and S Chichibu, *Introduction to Nitride Semiconductor Blue Lasers and Light Emitting Diodes* (CRC Press: London, 2000).

318. S Nakamura, *The Blue Laser Diode* (Springer Verlag: Berlin, 2000).

319. See, for example, S Nakamura, *Science* **281**, 956 (1998).

320. S Nakamura et al., *Jpn. J. Appl. Phys.* Part 2, **35** L74 (1996); K Itaya et al., *Jpn. J. Appl. Phys.* Part 2 **35**, L1315 (1996); S Nakamura et al., *Appl. Phys. Lett.* **69**, 4056 (1996); G E Bulman et al., *Electron. Lett.* **33**, 1556 (1997); K Kuramata et al., *Jpn. J. Appl. Phys.* Part 2 **36**, L1130 (1997); M Kneissl et al., *Appl. Phys. Lett.* **72**, 1539 (1998); F Nakamura et al., *J. Cryst. Growth* **189/190**, 841 (1998); H Katoh et al., *Jpn. J. Appl. Phys.* Part 2 **37**, L444 (1998); S Nakamura et al., *Jpn. J. Appl. Phys.* Part 2 **37**, L627 (1998); S Nakamura et al., *Jpn. J. Appl. Phys.* Part 2 **38**, L226 (1999).

321. Y Enya et al., *Appl. Phys. Express* **2**, 082101 (2009); A Avramescu et al., *Appl. Phys. Express* **3**, 061003 (2010); P S Hsu et al., *Appl. Phys. Express* **3**, 052702 (2010); K Kojima et al., *Opt. Express* **15**, 7730 (2007); S Nagahama, T Yanamoto, M Sano and T Mukai, *Phys. Stat. Sol.* (a) **190**, 235 (2002); K Kojima et al., *Opt. Express* **15**, 7730 (2007); see also H Ohta, S P DenBaars and S Nakamura, *J. Opt. Soc. Am.* **B27**, B45 (2010).

322. S Nakamura et al., *Appl. Phys. Lett.* **76**, 22 (2000); S Nakamura et al., *J. Jpn. Soc. Appl. Phys.* **34**, L1332 (1995); T Mukai, H Narimatsu and S Nakamura, *J. Jpn. Soc. Appl. Phys.* **37**, 479 (1998).

323. J W Raring et al., *Appl. Phys. Express* **3**, 112101 (2010).

324. H Soda, K Iga, C Kitahara and Y Suematshu, *Jpn. J. Appl. Phys.* **18**, 2329 (1979).

325. K Iga, S Ishikawa, C Ohkouchi and T Nishimura, *Appl. Phys. Lett.* **45**, 348 (1984).

326. F Koyama, S Kinoshita and K Iga, *Appl. Phys. Lett.* **55**, 221 (1989).

327. J L Jewell et al., *Electron. Lett.* **25**, 1123 (1989).

328. J L Jewell et al. *Electron. Lett.* **25**, 1123 (1989).

329. R Schur et al., *Jpn J. Appl. Phys.* **35**, L357 (1997); H Saito et al., *Appl. Phys. Lett.* **69**, 3140 (1996).

330. C J Chang-Hasnain, *IEEE J. Sel. Topics Quantum Electron.* **6**, 978 (2000).

331. J Faist, F Capasso, D L Sivco, C Sirtori, A L Hutchinson, A Y Cho, *Science* **264**, 553 (1994).

332. R F Kazarinov and R A Suris, *Sov. Phys. Semicond.* **5**, 207 (1971).

333. A Kastalsky, V J Goldman, J Abeles, *Appl. Phys. Lett.* **59**, 2636 (1991); J P Lohr, J Singh, R K Mains, G I Haddad, ibid 2070; S I Borenstain and J Katz ibid **55**, 654 (1989).

334. F Capasso, K Mohammed and A Y Cho, *IEEE J. Quantum Electron.* **22**, 1853 (1986).

335. J Faist et al., *Appl. Phys. Lett.* **67**, 3057 (1995); **66**, 538 (1995).

336. J Faist et al., *Appl.Phys. Lett.* **68**, 3680 (1996).

337. C Gmachi, F Capasso, D I Sivco, A Y Cho, *Rep. Progr. Phys.* **64**, 1533 (2001).

338. P L Kapitza and P A M Dirac, *Proc. Camb. Phys. Soc.* **29**, 297 (1933).

339. R H Pantel, G Soncini, H E Petroff, *IEEE J. Quantum Electr.* **QE-4**, 905 (1968).

340. J M J Madey, *J. Appl. Phys.* **42**, 1906 (1971).

341. L R Elias, W M Fairbank, J M J Madey, G J Ramian, H A Schwettman and T I Smith, *Phys. Rev. Lett.* **36**, 717 (1976); *Phys. Today,* February 1976, p. 17.

342. D A G Deacon, L R Elias, J M J Madey, G J Ramian, H A Schwettman and T I Smith, *Phys. Rev. Lett.* **38**, 892 (1977); *Sci. Am.* June 1977, p. 63.

343. R V Palmer, *J. Appl. Phys.* **43**, 3014 (1972).
344. W B Colson, *Phys. Lett.* **A64**, 190 (1977).
345. G R Neil et al., *Phys. Rev. Lett.* **84**, 662 (2000).
346. E J Minehara et al., *Nucl. Instrum. Methods* **A445**, 183 (2000).
347. V L Granatstein et al., *Appl. Phys. Lett.* **30**, 384 (1977); T J Orzechowski et al., *Phys. Rev. Lett.* **54**, 889 (1985); D A Kirkpatrik et al., *Phys. Fluids* **B1**, 1511 (1989).
348. N M Kroll and W A McMullin, *Phys. Rev.* **A17**, 300 (1978); P Sprangle and R A Smith, *Phys. Rev.* **A21**, 293 (1980); A M Kondradenko and E L Saldin, *Part. Accel.* **10**, 207 (1980); A Gover and P Sprangle, *IEEE J. Quantum Elecron.*, **QE-17**, 1196 (1981); G Dattoli, J C Gallardo, A Renieri and F Romanelli, *IEEE J. Quantum Electron.* **QE-17**, 1371 (1981); R Bonifacio, F Casagrande and G Casati, *Opt. Comm.* **40**, 219 (1982); R Bonifacio, C Pellegrini and L Narducci, *Opt. Comm.* **50**, 373 (1984); P Sprangle, C M Tang and C W Roberson, *Nucl. Instr. Meth.* **A239**, 1 (1985); E Jerby and A Gover, *IEEE J. Quantum Electron.* **QE-21**, 1041 (1985); K Kim *Nucl. Instr. Meth.* **A285**, 43 (1986); *Phys. Rev. Lett.* **57**, 1871 (1986); J M Wang and L H Yu, *Nucl. Instr. Meth.* **A250**, 484 (1986); R Bonifacio, F Casagrande and C Pellegrini, *Opt. Comm.* **61**, 55 (1987).
349. W B Colson et al., *Physics Today*, January 2002, 35; H P Freund and P O'Shea, *Science* **292**, 1853 (2001); H P Freund and G R Neil, *Proc. IEEE* **87**, 782 (1999).
350. C Pellegrini, *Eur. Phys. J. H* **37**, 659 (2012).
351. M J Mumma et al., *Science* **212**, 45 (1987); see also D Deming et al., *Icarus* **55**, 347 (1983) and M J Mumma in *Astrophysical Masers*, edited by A W Clegg and G E Nedoluha (Springer Verlag: Berlin 1993), p. 455.
352. M Johnson et al., *Astrophys. J.* **208**, L145 (1976).
353. B F Gordiets and V Ia Panchenko, *Space Res.* (Russian) **21**, 929 (1983).
354. G I Stepanova and G M Shved, *Sov. Astron. Lett.* **11**, 162 (1985).
355. M J Mumma in *Astrophysics Masers* (Note 351).
356. J J Goldstein et al., *Icarus* **94**, 45 (1991).
357. C Thum et al., *Astron. Astrophys.* **283**, 582 (1994).
358. V Strelnitski et al., *Science* **272**, 1459 (1996).
359. S J Messenger and V Strelnitski, *Mon. Not. Royal Astron. Soc.* **404**, 1545 (2010); S Johansson and V S Letokhov, *A&A*, **378**, 266 (2001).
360. K Davidson and S Johansson in *Photonics Spectra*, Feb. 1996.
361. S Johansson and V S Letokhov, *New Astronomy Rev.* **51**, 443 (2007).

7

Laser Properties and Progresses in Novel Lasers

7.1 Introduction

In the first experiments on lasers, some of their principal properties were observed. In the case of solid-state lasers, one of the most important characteristics was spiking. The emission consisted in thousand short pulses of a microsecond duration separated by 1-µs or so one from the other, during about 1-ms emission. The result was that the power emitted in each spike was in the order of kW even if a total 1 J of energy was emitted, and the emission occurred randomly in time.[1]

The first experiments on He–Ne gas laser revealed its mode structure. The resonators that are used in lasers are inherently multimode devices. The resonant modes that can exist in such devices may be classified as longitudinal and transverse modes. The longitudinal mode order is determined by the number of field variations along the axis of the cavity, while the transverse mode order is determined by the number of field variations in the plane of the mirrors. For each longitudinal mode order, there exist a set of transverse modes. The number of modes that can partake in the oscillations of a laser is dependent on the geometry and the losses of the resonator, the width of the atomic resonance of the active material, and the degree of population inversion. Practically, a laser will oscillate in several modes simultaneously unless special steps are taken to suppress the unwanted ones. The result is an emission on a number of different very closely spaced frequencies. Unfortunately, the number of modes operating at each time was varying randomly which resulted in a fluctuating power emission.

The research focused immediately on the tentative to dominate these effects and correct them so to have more stable and reliable emissions. We start this chapter considering first these efforts.

7.2 *Q*-Switching

In the case of pulsed lasers, a very important improvement came with the understanding that a suitable time control of the quality factor (the so-called *Q* value) of the cavity could result in a very large increase in the power output. It was so that, in 1961, Hellwarth[2] proposed the principle of *Q*-switching whose first realization was done in 1962.

A typical pulse of a solid-state laser (ruby) consists of 1-ms long burst of spikes (see Figure 6.8). Robert W Hellwarth[3] wrote a theoretical paper on spiking from which he drew later the idea of *Q*-switching. He observed that, if a method could be found preventing the laser to oscillate until the exciting flash is over, a population inversion greatly exceeding the usual threshold value could be established. If laser action is allowed at this moment, all available energy is concentrated into one giant pulse of extremely short duration.

The way to obtain this was simply to take losses high, to prevent laser action, and to lower them suddenly, at the moment in which the device had to operate. Because losses in a cavity are connected to the quality factor *Q* of the cavity, the method was coherently named *Q*-control or *Q-switching*.

Gordon Gould had discussed the method to control the *Q* of a laser cavity already in 1958 as we mentioned in Chapter 5.

Hellwarth[4] suggested to use a Kerr cell together with a polarizer to switch open or close the ray path in the cavity (see **Figure 7.1**).

The technique of *Q*-switching was practically demonstrated by F J McClung and R W Hellwarth[5] in 1962.

A ruby rod with its ends polished flat and parallel was aligned between two separately mounted flat and parallel reflectors with its *c*-axis cut at 60° to the rod axis as shown in Figure 7.1. Between one of the external mirrors and the ruby rod, a nitrobenzene Kerr cell was placed so to act as a switch blocking the light between the external mirrors when a suitable electric field was applied. To produce the giant pulse, the Kerr cell was activated and then the pump flash lamp was fired and at a suitable time the cell was suddenly turned off. The resulting pulse is shown in **Figure 7.2**.

There is a great variety of modes by which the cavity can be controlled that may be substantially divided into two categories: *active methods* and *passive methods*. In active methods, *Q* is controlled by a variable attenuator that may be a mechanical device such as a shutter, chopper wheel or spinning mirror/prism, or a modulator such as an acousto-optic or an electro-optic device, such as a Pockels or a Kerr cell.[6]

The use of a rotating disc was made by Collins and Kisliuk.[7] They constructed a ruby laser with external mirrors and placed a chopper wheel between the ruby

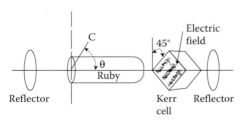

FIGURE 7.1 The general setup for *Q*-switching with a Kerr cell. (From F J McClung and R W Hellwarth, *J. Appl. Phys.* **33**, 828 (1962).)

and one mirror (**Figure 7.3**). A timing pulse was generated when the open sector in the chopper wheel which rotated at 10,000 rpm passed the auxiliary lamp and the detector. Reasonable control was obtained. The use of a mechanical chopper was in fact not the best choice and very often more than a single pulse is obtained.

A mechanical system may be made also by using a rotating Porro prism as a cavity retro-reflector. This requires rotational rates of approximately 24,000 rpm to operate in the fast Q regime where occurrence of secondary pulses is reduced or eliminated.

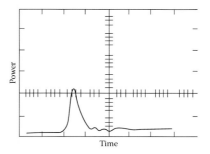

FIGURE 7.2 Output from the laser after switching the shutter. The time calibration is 0.2 µs/cm. (From F J McClung and R W Hellwarth, *J. Appl. Phys.* **33**, 828 (1962).)

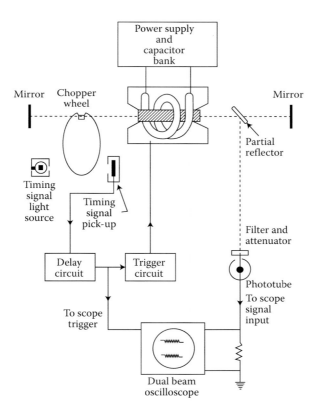

FIGURE 7.3 Schematic diagram of the arrangement by Collins and Kisliuk. (From R J Collins and P Kisliuk, *J. Appl. Phys.* **33**, 2009 (1962).)

A rotating prism was used by Basov[8] for producing short pulses in a neodymium glass laser. The roof prism was rotated at 30,000 rpm. The emission consisted of two pulses. De Maria et al.[9] considered instead use of an ultrasonic shutter.

Other methods may be classified as *passive Q-switching* and employ a saturable absorber. Use of a photosensitive liquid was discussed by Masters et al.[10] Passive Q-switching has been obtained employing, for example, an ion-doped crystal like Cr:YAG[11] which is used for Nd:YAG[11] or Nd:YVO$_4$ lasers,[12] or a bleached dye or a passive semiconductor.[13] In these cases, a material with high absorption at the laser wavelength is put in the cavity and prevents laser oscillation until the population inversion reaches a value exceeding the combined optical losses inside the cavity. When a bleaching material is used, the radiation increase in the cavity produces a self-transparency of the material.

A dye is used as a saturable absorber. Considering it as a two-level system, the energy difference between the two levels must correspond to that of the laser transition. For example, in the case of ruby laser, vanadium phthalocyanine in nitrobenzene was used. The concentration is adjusted in such a way that oscillation can just take place with the dye cell in the cavity. The laser pumps the dye molecules into the upper level, where they remain for a short time. During this time, the molecules in the lower level are few, the dye is relatively transparent to laser light, and thus the Q-switching has been effected.

Q-switching allowed the production of giant pulses with a duration of about 10^{-8}–10^{-9} s and powers up to 10^8 W.

Q-switching techniques were also applied successfully to a variety of lasers as, for example, CO$_2$ and N$_2$O molecular gas laser systems.[14] The theory of Q-switching was treated by several authors.[15]

Q-switching and cavity dumping of Nd:YAG lasers was described by D Maydan and R B Chesler.[16] Cavity *dumping* is one way of obtaining output from a continuously pumped laser. The scheme differs from purely continuous operation in that the output coupling strength varies in time. The authors obtained the variable output coupling with an acousto-optic modulator as shown in **Figure 7.4**.

A fused silica acousto-optic modulator is used as a switch to drive the laser beam out of the cavity. In the case of Q-switching, the modulator deflects out the beam keeping the cavity-Q low until inversion has reached the maximum level, then closes the cavity, and after a few nanoseconds deflects the resulting radiation out. For cavity dumping, the modulator switches the light in and out from the cavity. The cavity is never kept below threshold conditions as in the Q-switched mode of operation and the modulator is used to dump the beam out of the cavity after a time delay. The result may be a shorter output pulse than regular Q-switching and may increase the repetition rate.

7.3 Modes in the He–Ne Laser

The study of the behavior of the He–Ne laser led immediately to discover a series of important nonlinear phenomena in its operation.

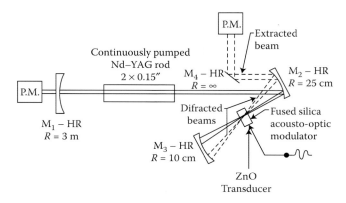

FIGURE 7.4 The intracavity modulation system. (From Maydan and Chesler, *J. Appl. Phys.* **42**, 1031 (1971).)

One of the first of these effects was *mode pulling*. In the Fabry–Perot laser cavity, a number of axial modes separated by $\Delta v = c/2L$ (L is the length of the cavity, v and c have the usual meaning) resonate within the Doppler broadened linewidth.[17] Studying the helium–neon laser, Bennett[18] found that some nonlinear frequency-dependent pulling mechanism exists which makes the spacing between adjacent resonant modes different from $c/2L$. This phenomenon was dubbed *mode pulling*.

The following year, W E Lamb Jr. and other researchers[19] proposed a semiclassical theoretical description of the laser (see Section 7.5).

The theory predicted the existence of a dip in the gain-frequency curve of a laser (*hole burning*). Hole burning was physically explained by Bennett[20] considering that in the case of inhomogeneous broadening, such as Doppler broadening, irradiation at a particular frequency causes an absorption decrease in a range of the order of the homogeneous width centered around the pumping frequency while the rest of the band remains unaffected, producing a hole in the spectral distribution of the emitted light (*hole burning*).

McFarlane et al.[21] failed to observe the dip in a helium–neon laser with natural occurring Ne but were able to find it when isotopically enriched Ne was used. Natural neon is a mixture of two isotopes Ne^{20} and Ne^{22} which have slightly different lasing frequency which masked the effect. The dip was clearly observed by Szoke and Javan[22] and is shown in **Figure 7.5**.

This effect gave rise to a very important result. If two traveling waves in a gas have the same frequency ω but opposite **k** vectors, the wave with $+k$ produces a saturation dip in the Doppler profile for molecules for which the relation $\omega = \omega_a + k\beta$ ($\beta = v/c$ and ω_a atomic resonance frequency) is satisfied. The other beam saturates molecules with $\omega = \omega_a - k\beta$. If ω is scanned across the Doppler profile, the probe registers a saturation dip exactly at the center (hole burning). At optical frequencies, the effect was first demonstrated as a dip in the output of a He–Ne laser and is known as the *Lamb dip*.[23]

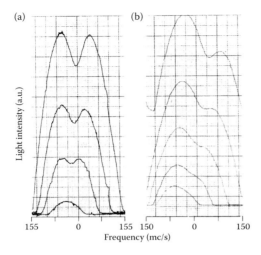

(a)   (b)

Light intensity (a.u.)

155   0   155   150   0   150

Frequency (mc/s)

FIGURE 7.5 Power output versus the oscillation frequency at 1.15 μm at various excitation levels. (a) A laser with Ne^{20} isotope and (b) with normal isotope abundance. (From A Szoke and A Javan, *Phys. Rev.* **10**, 521 (1963). With permission.)

A monochromatic light beam resonantly excites only a small group of atoms or molecules under the inhomogeneously broadened profile and induces in them a population change. This group of atoms or molecules, marked by the population change, can then be selectively studied by either absorption or luminescence. The discovery of this effect allowed a new spectroscopic method to be created: *saturation spectroscopy*. The effect of inhomogeneous broadening is suppressed with this technique.

Christian Bordè[24] in Paris and independently Hansch[25] developed the new technique of saturation spectroscopy.

7.4 Mode Locking

A different technique of controlling laser emission, allowing to have a train of regular pulses, was named *mode locking*. Mode locking is one of the most important ways in which lasers are operated. It makes possible to have emission in the form of a regular train of pulses and to generate pulses as short as 10^{-12} s, as we will show in Chapter 8.

It takes advantage of an apparent drawback in the laser operation. In general, the laser modes in a cavity will oscillate independently (the individual phases of the waves in each mode are not fixed), which gives rise to output power fluctuations. In mode locking, each mode is made to operate with a fixed phase between itself and the other modes. A constructive interference is realized which gives intense burst of light.

There are two methods for obtaining mode locking. One uses a periodic modulation of losses using an optical modulator, typically using the acousto-optic or electro-optic effect, while the other uses a passive phase locking, using a saturable absorber. The final results are very similar.

The basic equations which describe the effect of an internal phase or loss perturbation on a laser oscillation were given by several authors[26] and are today in any textbook.[27]

A simple and simplified way to understand the effect is the following. In a multimode laser, the electric field of the nth mode can be written as

$$E_n = A \exp i(\omega_n t + \delta_n), \tag{7.1}$$

where ω_n is the angular frequency of the nth mode and δ_n its relative phase.

The total field can be written as

$$E_i = A \sum_{n=0}^{N-1} \exp i(\omega_n t + \delta_n),$$ (7.2)

where N is the total number of modes. If the modes are uncoupled, the total intensity is found by adding the intensities of the modes

$$I = NA^2.$$ (7.3)

If modes are coupled so that they have the same relative phase δ,

$$\delta_n = \delta$$ (7.4)

things behave differently. Because the separation between modes is

$$\Delta\omega = \pi c/L$$ (7.5)

where L is the cavity length, we may label the angular frequency of the highest mode by ω, and the frequency of the nth mode by

$$\omega_n = \omega - n\Delta\omega$$ (7.6)

so that the field can be written as

$$E(t) = A\sum\exp[i(\omega - n\Delta\omega)t + \delta] = A \exp i(\omega t + \delta)\sum\exp(-in\varphi),$$ (7.7)

where

$$\varphi = \Delta\omega t = \pi c t/L.$$ (7.8)

The intensity is given by the square of the amplitude and therefore

$$I = A^2 \sin^2(N\varphi/2)/\sin^2(\varphi/2).$$ (7.9)

The maximum intensity is

$$I_{max} = N^2 A^2.$$ (7.10)

Thus, the output of a mode-locked laser is N times the power of the same laser with modes uncoupled.

From Equation 7.9, the pulse maxima occur at those times at which the denominator vanishes, that is, when

$$\varphi/2 = \pi c t^*/2L = m\pi,$$ (7.11)

where m is an integer. This time is therefore

$$t^* = 2mL/c,$$ (7.12)

and two successive pulses are separated by a time

$$\Delta t = 2L/c,$$ (7.13)

that is, by one round trip transit time.

First indications of mode locking appear in the work of Gürs and Müller[28] on ruby laser and of Statz and Tang[29] on He–Ne lasers.

In 1963, S E Harris[30] at Stanford considered the possibility of applying internal time-varying perturbations to modulate phase and obtain mode control and stability (mode locking). The following year, this stabilization was obtained in two ways.

In the July 1, 1964 issue of the *Applied Physics Letters*, Hargrove, Fork, and Pollack[31] of Bell Telephone Labs wrote

> Stabilization (locking) of the amplitude and frequency of the mode in a 0.633 μm He–Ne laser was achieved by internally modulating the laser at a synchronous frequency (i.e. at the reciprocal round-trip travel time, a range of frequencies nominally given by $c/2L$): In a multimode gaseous laser, saturation and dispersion effects generally result in a spectrum of longitudinal mode difference-frequencies near $c/2L$. In typical environments, and especially for very closely spaced modes, the amplitudes and phases of the individual modes fluctuate randomly. As a consequence, the laser output is modulated with a fluctuating amplitude at a frequency approximately equal to the frequency spacing between adjacent longitudinal modes. DiDomenico[32] has independently predicted that modulating the internal losses of a laser at a frequency equal to some multiple of the axial mode spacing causes all axial modes to couple with a well defined amplitude and phase.

The experimental arrangement used an acoustic modulator and obtained synchronous modulation around 56 MHz. When locking was attained, the fundamental signal at around 56 MHz changed from erratic to constant amplitude. Random and systematic amplitude fluctuations were virtually eliminated.

In November 1964, Harris and Targ[33] reported in *Applied Physics Letters*:

> … the operation of a He–Ne laser in a manner such that all of the laser modes oscillate with FM phases and nearly Bessel function amplitudes, thereby comprising the sidebands of a frequency-modulated signal. The resulting laser oscillation frequency is, in effect, swept over the entire Doppler line-width at a sweep frequency which is approximately that of the axial mode spacing. This type of FM oscillation is induced by an intra-cavity phase perturbation which is driven at a frequency which is approximately but not exactly the axial mode spacing.

The phase perturbation was obtained via the electro-optic effect with a KDP crystal.

Following that paper, in the same issue of the journal, Harris with McDuff[34] published in a letter a first-order theory of the oscillation using Lamb's theory.

The first paper clearly identifying the mechanism was written in 1964 by DiDomenico[35] from Bell Labs who, following a suggestion by E I Gordon, showed theoretically that mode locking could be obtained by internal loss modulation at the resonator mode-spacing frequency.

Lamb[36] also described how the nonlinear properties of the laser medium could cause the modes of a laser to lock with equal frequency spacing. These papers were followed by the ones of Hargrove et al.,[31] and Yariv[37] who made independently similar theoretical predictions.

Hargrove et al. achieved mode locking by internal loss modulation inside the resonator. This is called *active* mode locking. The analytic theory of active mode locking was later established by Siegman and Kuizenga.[38]

Mocker and Collins[39] showed that the saturable dye used in ruby laser to Q-switch the laser could also be used to mode lock. This was the first example of *passive* mode locking.[40] The saturable absorber is bleached by the radiation in the resonator, closes the cavity, and allows the emission of radiation. This emission stops when the gain medium is depleted, and the process starts all over again. The two authors observed that the Q-switched pulse broke up into a train of very short pulses separated by the roundtrip time. The analytic theory of passive mode locking was developed by New.[41]

Harris[42] reviewed the stabilization and modulation of laser oscillators describing both phase and amplitude modulation techniques.

Passive mode locking of dye lasers was observed first by Smidt and Schaefer[43] by using a flash-lamp pumped rhodamine 6G laser with DODCI as an absorber. This work was followed by the one of Ippen, Shank, and Dienes[44] who generated the first cw saturable absorber mode locking using a saturable dye in a dye laser. Shortly thereafter, this led to the production of pulses of sub-picosecond duration[45] (see Section 8.10).

The complex historical steps toward mode locking are well discussed, for example, in papers by Smith[46] and Haus.[47]

New techniques of mode locking were developed in the following years. The passive mode-locked lasers are the simplest and cheaper devices, producing a train of stable tunable subpicosecond pulses in a wide spectral region. There are also hybrid mode-locking techniques embodying some of the advantages of both active and passive schemes. The most successful technique of passive mode locking is based on semiconductor saturable absorber mirrors (SESAMs).[48] The more the mirror reflects light, the more intense the light is. A SESAM normally consists of a semiconductor saturable absorber that is integrated into a Bragg mirror structure, which has alternating layers of two different optical materials with optical thickness corresponding to one-quarter of the wavelength of light for which the mirror is designed. When the light intensity increases, the semiconductor absorbs light and electrons are excited from the valence to the conduction band. Under strong excitation, the absorption is saturated because possible initial states in the valence band are depleted while the final states are partially occupied, the semiconductor becomes transparent and therefore the reflectivity of the system varies with the intensity of light incident upon it.

A self-mode-locking technique, termed Kerr lens mode locking[49] (KLM),[50] will be described in Chapter 8. In this case, the rod of laser material acts as a lens for high-intensity light due to the optical Kerr effect. Increasing the light intensity in the material, the focusing from the Kerr lens restrict the beam until the beam's diameter becomes narrow enough that its linear diffraction is large enough to balance out the Kerr effect and all the light can pass through a small diaphragm put on the axis of the laser. The system then acts as a fast saturable absorber.

Mode-locked lasers were initially limited to output power in the order of 1 W. In the late 1990s, a generation of mode-locked high power lasers was born.

The first attempts toward mode-locked lasers with multiwatt average power output were based on Nd:YAG[51] obtaining average power around 10 W with 16-ps pulses.

Passively mode-locked Yb:YAG using the active material in the shape of a disk[52] was used by U Keller et al.[53] obtaining 16 W average power in 730-fs pulses. With improved technique, higher powers were obtained[54] as, for example, 60 W in femtosecond pulses.[55]

7.5 Lamb's Theory

Willis Lamb Jr. (1913–2008) was born in Los Angeles, California, in 1913. He received a bachelor of science in chemistry from the University of California, Berkeley, in 1934 and a PhD in physics in 1938 under the guidance of Robert Oppenheimer (1904–1967). He was at Columbia from 1938 to 1951. The years 1951–1956 were spent at Stanford and 1956–1962 at Oxford. Then he was in Yale from 1962 to 1974 and finally at the Optical Sciences Center, University of Arizona.

In 1943, he started working in the Columbia Radiation Laboratory at Columbia University on magnetrons. The important work with Rertherford on the fine structure of hydrogen was made between 1946 and 1951. He started work on his model for a gas laser around 1961, in Oxford, considering single and multimode oscillation. For the single-mode oscillation, his equations gave the threshold population inversion for oscillation. The steady-state solutions gave the operating field amplitude and frequency as a function of pumping, Doppler frequency, resonator tuning, and Q. The laser frequency varied with the amount of detuning of the cavity resonance from the atomic transition frequency (*frequency pulling*) and with the excess of population inversion above the threshold value (*frequency pushing*). The presence of an intensity dip when the cavity was tuned to resonance came to him as a surprise. The paper *Theory of an Optical Maser* was published only in 1964. The first public description of the work was given at the *Third International Symposium on Quantum Electronics* in Paris, during February 11–15, 1963. The theory was fully exposed the same year to the International School Enrico Fermi, in Varenna, Italy.

In the latter part of his career, he paid increasing attention to the field of quantum measurements. He died in 2008.[56]

Lamb[57] considered a high-Q multimode cavity in which there is a given classical electromagnetic field acting on a material medium which consists of a collection of atoms described by the laws of quantum mechanics.[58] The effect of the electromagnetic field on the atoms in the cavity is to produce a macroscopic electric polarization $P(r,t)$ of the medium. This acts as a source for the electromagnetic field in accordance with Maxwell equations. The conditions that the field produced should be equal to the field assumed (self-consistency) determine the amplitude and frequencies of the possible oscillations. The calculation included nonlinear effects, so that phenomena of frequency pulling and pushing, mode competition, frequency locking, etc., could be described.

The theory was developed considering that only two atomic states a and b contributed to the maser oscillation. Assuming to have n normal modes in the cavity with eigenfunctions

$$U_n(z) = \sin K_n z, \tag{7.14}$$

where n is a large integer and z is the axial coordinate, in the presence of a given polarization $P(z,t)$, quasi-stationary forced oscillations of the electric field can be expanded in normal mode eigenfunctions

$$E(z,t) = \sum_n A_n(t)U_n(z). \tag{7.15}$$

The amplitudes $A_n(t)$ obey a differential equation of a forced, damped simple harmonic oscillator

$$d^2A_n/dt^2 + (\sigma/\varepsilon_o)(dA_n/dt) + \Omega n^2 An = -(1/\varepsilon_o)(d^2P_n(t)/dt^2), \tag{7.16}$$

in which $P_n(t)$ is the space Fourier component of $P(z,t)$

$$P_n(t) = (2/L)\int_0^L dz\ P(z,t)\sin K_n z. \tag{7.17}$$

Since $P_n(t)$ will be very nearly monochromatic at an optical frequency υ, its second derivative in Equation 7.16 is replaced by $-\upsilon^2 P_n$. One then chooses the conductivity σ so to give the desired Q_n of the nth mode, that is,

$$\sigma = \varepsilon_o\upsilon/Q_n. \tag{7.18}$$

So $A_n(t)$ obeys

$$d^2A_n/dt^2 + (\upsilon/Q_n)(dA_n/dt) + \Omega_n^2 A_n = (\upsilon^2/\varepsilon_o)P_n. \tag{7.19}$$

Considering a gas laser in which the separation of the principal modes is 150 MHz which is much larger than the cavity mode bandwidth $\upsilon/Q \sim 1$ MHz, one may neglect time Fourier components of $A_n(t)$ and $P_n(t)$ which are far from the cavity resonance frequency Ω_n and write

$$A_n(t) = E_n(t)\ \cos(\upsilon_n t + \varphi_n(t)), \tag{7.20}$$

and

$$P_n(t) = C_n(t)\cos[\upsilon_n t + \varphi_n(t)] + S_n(t)\sin[\upsilon_n t + \varphi_n(t)], \tag{7.21}$$

where the amplitudes $E_n(t)$ and phases $\varphi_n(t)$, as well as the in phase and quadrature coefficients $C_n(t)$ and $S_n(t)$, are slowly varying functions of t which, together with the frequencies υ_n, are still to be determined. Putting Equations 7.20 and 7.21 into Equation 7.19 retaining only the first derivatives of E and φ, equating the coefficients of cos and sin separately to zero, and neglecting all terms involving $\upsilon_n \dfrac{dE}{dt}Q_n, \dfrac{d\varphi_n}{dt}, \dfrac{dE_n}{dt}$, and $\upsilon_n \dfrac{d\varphi_n}{dt}\dfrac{E_n}{qN}$ and recognizing that $\upsilon_n + \varphi_n$ is very close to Ω_n one finds the self-consistent equations

$$(\upsilon_n + d\varphi_n/dt - \Omega_n)E_n = -(1/2)(\upsilon/\varepsilon_o)C_n, \qquad (7.22)$$

and

$$dE_n/dt + (1/2)(\upsilon/Q_n)E_n = -(1/2)(\upsilon/\varepsilon_o)S_n, \qquad (7.23)$$

which will give the amplitude, frequencies, and phase of the radiation once the polarization state of the medium is known in terms of $E_n(t)$.

The second equation describes the effect of the damping and the active medium on the mode amplitude. If the in-quadrature component of the polarization is zero, the amplitude is exponentially damped out, as for a passive, lossy cavity. The in-quadrature component of the polarization represents the gain introduced by the active medium, which overcomes the cavity losses, and allows oscillation to occur. The first equation describes the part the in-phase component of the polarization plays in altering the frequency of the field from that associated with the passive cavity and therefore describes frequency pushing and pulling effects, etc.

This can be seen considering the case when the polarization is a linear function of the electric field, thus

$$C_n = \varepsilon_o E_n \chi' \qquad (7.24)$$

$$S_n = \varepsilon_o E_n \chi''. \qquad (7.25)$$

On substituting into Equations 7.22 and 7.23, one obtains

$$dE_n/dt = -(1/2)\upsilon(1/Q_n + \chi_n'')E_n \qquad (7.26)$$

$$\upsilon_n + d\varphi_n/dt - \Omega_n = -(1/2)\upsilon\chi_n'. \qquad (7.27)$$

The second equation indicates that the oscillation frequency differs from the eigenvalue for the mode, Ω_n, by a "pulling term" $(-\upsilon\chi_n')/2$ due to the presence of the dielectric. From the first equation, it can be seen that if $\chi_n'' > 0$, the dielectric adds to the damping already present. On the other hand, if $\chi_n'' < (-Q_n)^{-1}$, the equation describes an exponential built-up of the oscillation, and it is at this point that higher-order terms (since in general the polarization is not a linear function of the field) need to be considered to take into account the saturation behavior of the medium.

To proceed further, it is necessary to relate the macroscopic polarization to the atomic properties of the active medium. This is done by Lamb utilizing the density matrix formalism.

Under the approximation of a constant population inversion, the macroscopic polarization turns out to be a linear function of the electric field.

The field amplitude results to obey the following equation:

$$dE_n/dt = \{-\upsilon/2Q_n + \Pi\upsilon N\gamma/2\varepsilon_o\hbar[(\upsilon_n - \upsilon_{ab})^2 + \gamma^2)]\}, \qquad (7.28)$$

where Π takes account of the potential function describing the interaction within the atom itself, γ is the inverse of the lifetime of the transition between the two states a and b, N is the population inversion, and $\nu_{ab} = (E_a - E_b)/h$ being $E_{a,b}$ the energy of the level a and b.

If the amplitude is to increase with time, rather than being exponentially damped by the losses of the cavity, then Equation 7.28 shows that

$$\Pi N\gamma/\varepsilon_o h[(\nu_n - \nu_{ab})^2 + \gamma^2] > 1/Q_n. \tag{7.29}$$

The threshold for laser oscillation is obtained by an equality sign in Equation 7.29. From the above expression, it can be seen that in order to achieve a low value for the threshold population inversion associated with a particular active medium, the transition probability (which is proportional to Π) for the laser transition must be large, while the damping constant, γ, must be small (i.e., the homogeneous linewidth of the laser transition must be narrow).

For a population inversion in excess of the threshold inversion, Equation 7.29 predicts that the amplitude of the mode builds up exponentially, and without limit in this approximation. In practice, of course, this does not happen, for as the field builds up in amplitude, the population of the upper level decreases (due to stimulated emission), while that of the lower level increases, and it is just such effects that were neglected in the first-order theory. Without assuming a constant population inversion, the macroscopic polarization becomes a nonlinear function of electric field and this leads to predict a saturation value for the amplitude of the electric field.

Considering now Equation 7.27 for the phase one obtains

$$(\nu_n - \Omega_n) \approx (1/2)(\nu/\varepsilon_o)(\Pi/h)(N(\nu_{ab} - \nu_n)/[(\nu_n - \nu_{ab})^2 + \gamma^2]. \tag{7.30}$$

At threshold Equation 7.30 becomes

$$\nu_n - \Omega_n \approx \nu(\nu_{ab} - \nu_n)/2Q_n\gamma. \tag{7.31}$$

The oscillation frequency of the mode involves the oscillation frequency of the passive cavity (Ω_n) and the central frequency of the atomic transition (ν_{ab}) and is pulled toward the center of the natural linewidth by an amount proportional to the detuning of the cavity, ($\Omega_n - \nu_{ab}$), from this central frequency.

Lamb then treated the nonlinear theory taking into account the influence of the stimulated emission on the population of the laser levels. Using the new polarization, the amplitude of the field at steady state ($dE/dt = 0$) is found

$$E_n^2 = qh^2\gamma_a\gamma_b/\Pi[N/N_T - 1 - (\nu_{ab} - \nu_n)^2/\gamma^2], \tag{7.32}$$

where q is a numerical factor, γ_a and γ_b are the natural linewidth of states a and b, respectively, $\gamma = (\gamma_a + \gamma_b)/2$, and N_T is the inversion population at threshold. Saturation is now clearly present.

When the mode coincides with the line center, Equation 7.32 reduces to

$$E_n^2 = qh^2\gamma_a\gamma_b/\Pi[N/N_T - 1]. \tag{7.33}$$

Equation 7.31 continue to be valid. Lamb then describes the case of moving atoms, and multimode operation, finding that in general the curve displaying

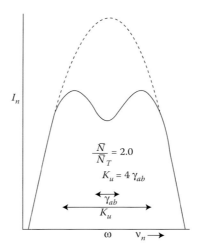

FIGURE 7.6 Relative intensity of oscillation as a function of detuning from line center. The solid curve theory is drawn for $N = 2N_T$ and $Ku = \pm(v_n - v_{ab}) = 4\gamma$. The dotted curve indicates the Doppler gain profile of the line. (From W E Lamb Jr., *Phys. Rev.* **134**, A1429 (1964). With permission.)

intensity of oscillation as a function of cavity detuning from the line center will have a flattened peak at the line center, and for a sufficiently large relative excitation may exhibit a central dip lying between two maxima (*Lamb dip*) (see **Figure 7.6** taken from Lamb paper).

Generalization of Lamb's theory[59] was made by a number of authors. Alternative theoretical treatments were done by Haken.[60]

One of the effects of saturation is that it leads to mode competition, that is, the reduction of gain available to a given mode by saturation due to some other mode. The theory was developed by Gordon et al.[61] who also performed the experiment.

7.6 Mode Selection and Frequency and Amplitude Stabilization

An enormous effort was devoted to the establishment of a technology for the optimization and control of laser radiation.

Mode volume and alignment tolerances of mirrors and the active medium are important. Mode selection and stabilization were the object of active studies.

For applications such as optical communication, it is desirable from the standpoint of noise, coherence, spectral purity, etc., to suppress all but one mode in a laser. Therefore, mode selection schemes were studied. By using a long, thin configuration (small mirrors and large mirror separation), it is possible to suppress all but the dominant transverse mode.[62] Also, by operating just above the oscillation threshold[63] or by using a short resonator,[64] it is possible to restrict the oscillations to a single longitudinal mode.

In general, mode selection involves the introduction of loss to the resonator in some prescribed manner. An understanding of cavity properties enabled substantial improvement to be made in ameliorating the quality of laser output and in giving hints for mode selection.[65] Kleiman and Kisliuk[66] were one of the first examples of mode selection. Several propositions and geometries were used.[67] The problem of mode deformation due to nonuniformities in the gain produced by nonlinearities in the medium was thoroughly studied by Statz and Tang.[68] In laser oscillators, the gain is strongly depressed by saturation in regions where the amplitude of the mode is large, thus giving a nonuniform gain distribution in the cavity. By numerical calculations, the two researchers showed that even with nonuniform gain distribution in the transverse direction, the lower-order optical modes as originally calculated represent the electromagnetic field in the cavity rather accurately. Goldsborough[69] had studied from an experimental point of view the beat frequencies between the various cavity modes in a concave mirror optical resonator finding good agreement with theory.

Light diaphragm,[70] Gauss selector, Fabry–Perot interference modulators, unstable resonators, orthogonal prism cavity, half-homocentric cavity, non-uniform reflecting resonators, and cat's eye cavity[71] are some of the many methods that have been used over the years to control beam pattern. A prism reflector was described by Giordmaine and Kaiser.[72] A review of first efforts is given by P W Smith.[73]

In lasers having gain at several wavelengths, operation on one of the transitions only could be obtained with a suitable prism as shown by Bloom.[74] The prism was designed so that the radiation meets it at the Brewster angle to reduce insertion losses. Another realization was by Zoot.[75] Also mirrors with a varying spectral reflectance were used.[76] A somehow different system employed a thin film with layers all proportionally graded in thickness in one direction so that the wavelength of the narrow band transmission peak varies across the surface. The filter was inserted in the cavity at the Brewster angle and selection was obtained translating the film.[77]

The causes of frequency variations were soon recognized and suitable structures were designed to minimize them. The frequency stability of a single-frequency laser is determined entirely by the stability and length of its resonator. The process by which the electromagnetic radiation attains equilibrium in a laser cavity was studied deriving the linewidth under the mirror motion and considering in detail the effect of thermal vibration.[78] Bennett et al.[79] reported a frequency stability of one part in 10^{10} over 8-h periods on the 3.39-μm He–Ne transition by using a modulation of the discharge. Complete isolation[80] of the laser from external perturbations gave stabilities better than 1 part in 10^{13} and reseattability of 1 part in 10^9. Bell's people were very active. Cavity stabilization was discussed by Smith[81] and intensity fluctuations in the output of laser oscillators were investigated theoretically by McCumber.[82] A detailed review of the first methods used to stabilize a single frequency laser was given by White.[83] Even the use of a magnetic field was considered.[84] Targ, Osterink, and French[85] described a system based on minimizing the amplitude of the AM component of the output at the frequency equal to the intermode spacing. Systems based on competition between two longitudinal resonator modes were described by R L Fork[86] and by P W Smith[87] for ring lasers. The use of an axial weak magnetic field was described by Tomlinson and Fork.[88]

There were many methods of stabilizing helium–neon lasers.[89] A vibrating interferometer mirror to sense a zero slope point of the laser power versus frequency profile was used by Rowley and Wilson.[90] Laser stabilization using the laser line profile was obtained by White, Gordon, and Labuda.[91]

A good discussion was made by A L Bloom in one of the first books on gas lasers.[92]

The Lamb dip has been used to stabilize lasers. After the initial demonstration of the dip, it was quickly realized that it could be used to define the output frequency of a single-frequency laser to a higher degree of accuracy than could be obtained merely by attempting to center on a relatively broad Doppler curve.[93] An updating of the methods of wavelength stabilization was made by Baird and Hanes in 1974[94] in an excellent paper which provides a clear summary of the situation.

Later the locking to the resonance of Fabry–Perot cavities was demonstrated by Salomon et al.,[95] and Young et al.[96] reported a linewidth of 0.6 Hz for averaging times up to 32 s using a high finesse Fabry–Perot cavity.

7.7 Tunable Solid-State Lasers

At the turn of 1980s, a number of solid-state tunable new lasers were developed which had notable impact on the following research.

The most important of these lasers are alexandrite, titanium–sapphire, and color center lasers; color center lasers, however, operate typically at cryogenic temperature.

Alexandrite, chromium-doped chrysoberyl, is a crystal of $BeAl_2O_4$ in which Cr^{3+} ions replace some of the Al^{3+} ions.[97] The laser emits around 760 nm with a bandwidth of around 100 nm. This laser may be considered the archetype of what is now a large class of solid-state lasers, usually referred as *tunable solid-state lasers*. The emission wavelength of these lasers can in fact be tuned over a wide spectral bandwidth. In this class is included Ti:sapphire ($Ti:Al_2O_3$) that is the most widely used tunable solid-state laser[98] with an emission band between 660 and 1180 nm with a peak at 790 nm.

The electronic structure of the Ti^{3+} ion is a closed shell plus a single 3d electron. Because the ionic radius of the titanium ion is 26% larger than the aluminum one it replaces, a strong local distortion is induced which creates a strong local electric field. This means that the absorption band is abnormally wide in the bluish-green part of the spectrum. The absorption at these visible wavelengths excites electrons from a ^2T_g ground ion state to a ^2E_g excited level, which then splits into two sublevels which are 50 nm apart. For operation, several watts of pump power are needed because the upper-state lifetime of Ti is very short at 3.2 μs and the saturation power, the incident optical power required to achieve significant saturation of an absorber, is very high. Originally Ti:sapphire lasers in cw were pumped using argon ion lasers at 514 nm, which were powerful but bulky, inefficient, and expensive to run. In pulsed operation, frequency-doubled Nd:YAG or Nd:YLF were used as well as flash lamps.

A better pumping is through frequency-doubled neodymium-doped yttrium orthovanadate (Nd:YVO$_4$) laser to provide 5 W of power at 532 nm. The Nd:YVO$_4$ emits at 914, 1064, and 1342 nm.[99]

An additional good property of Ti:sapphire laser is that sapphire has excellent thermal conductivity which relieves thermal effects at high power and intensities.

Other large band lasers are: Cr:LiSAF (Cr^{3+}:LiSrAlF$_6$) and Cr:LiCAF (Cr^{3+}:LiCaAlF$_6$)[100] which emit at 850 nm (with a bandwidth 780–1010 nm) and 780 nm (bandwidth 720–840 nm), respectively. Other materials are Co:MgF$_2$, Cr^{4+}:YAG, Cr^{4+}:Fosterite (Mg$_2$SiO$_4$), and many others.

In this category can also be included color center lasers that operate in the near-infrared wavelength range from 0.8 to about 4 μm with up to 100 mW average output power. The first lasers of this type operated at cryogenic temperature. Today, they are continuously tunable over wide spectral ranges.

The hosting material is a halide of Na, K, Na, or Li that is x-ray irradiated to provide defects which act as a quantum well for the charges. These F centers, as they are called, have quantum energy levels which provide the laser transition in a four-level scheme.

The color center lasers were introduced by L F Mollenauer at Bell Labs. in 1974.[101] He was very active in the study of this type of lasers and examined several kinds of defects in a number of alkali halides[102] obtaining also room-temperature operation.[103] A number of other researchers were interested in color center lasers.[104]

Mollenauer[105] reported cw laser action with F$_2^+$ centers in NaF with emission continuously tunable from 0.885 to 1.00 μm, pumping with a krypton ion laser at 0.7525 μm with a threshold at about 40 mW. With KCl laser, action was obtained in a band near 1.69 μm pumped with a Nd:YAG at 1.34 μm.

The F$_2^+$ center consists of a single electron shared by two halide anion vacancies adjacent to each other along a [110]-axis. The emission can be modeled as that from H$_2^+$ molecular ion. The center can be created by x-ray or electron beam irradiation and annealing.

A LiF:F$_2^-$ color-center laser operating in the spectral region around 1.14 μm and the visible was described.[106]

7.8 Distributed Feedback Lasers

Conventional semiconductor lasers in general emit on many longitudinal modes. Methods to have emission in single longitudinal mode were studied extensively in the 1980s. One way of improving mode selectivity is to make the feedback frequency-dependent so that the cavity loss is different for different longitudinal modes. Two mechanisms have been found useful in this respect and are the *distributed feedback* and the *coupled-cavity* mechanism.

A distributed feedback laser (DFB) is a type of laser diode, quantum cascade laser, or optical fiber laser where the active region of the device is periodically structured as a diffraction grating. The structure builds a one-dimensional interference grating (Bragg scattering) and the grating provides optical feedback for the laser. The device does not utilize a conventional mirror cavity, but provides feedback via backward Bragg scattering from the periodic perturbation of the refractive index and/or the gain of the laser medium itself.[107] This is a feedback mechanism distributed over the whole length of the periodic structures. Another important feature of DFB is the high spectral selectivity which originates in the Bragg effect. This allows selection of longitudinal modes even for broad gain profile of the laser medium.

The theory was developed by Kogelnik and Shank,[108] using coupled-wave theory.

Distributed feedback lasers are very compact and have a mechanical stability which is intrinsic to integrated optical devices. In addition, the grating nature of the device provides a filter mechanism which restricts the oscillation to a narrow spectral range.

A DFB structure can be produced, for example, by introducing a spatial variation of the refractive index n such as

$$n(z) = n + n_1 \cos Kz, \tag{7.34}$$

where z is the coordinate along the optic axis and $K = 2\pi/\Lambda$. Here Λ is the period of the spatial modulation. A DFB structure of this kind will oscillate in the vicinity of a wavelength λ_o given by the Bragg condition

$$\lambda_o/2n = \Lambda, \tag{7.35}$$

Kogelnik and Shank (1971) derived expressions for the threshold and the spectral width of stimulated emission in such periodic structures from a simple coupled-wave analysis. The field E in the device is of the form

$$E = R(z)\exp\frac{-iKz}{2} + S(z)\exp\frac{iKz}{2}, \tag{7.36}$$

consisting of two counter-running waves with the complex amplitudes R and S. These waves grow in the presence of gain and they feed energy to each other due to the spatial modulation of n. The boundary conditions for the wave amplitudes are

$$R(-L/2) = S(L/2) = 0, \tag{7.37}$$

where L is the length of the DFB laser. At the endpoints of the device, a wave starts with zero amplitude receiving its initial energy through feedback from the other wave.

The start oscillation condition is derived in the approximation of large gain G for

$$n_1 = (\lambda_o/L)[\ln G/\pi (G)^{1/2}]. \tag{7.38}$$

When the gain G exceeds the threshold value at center frequency by a factor 2, the threshold is exceeded over a spectral bandwidth $\Delta\lambda$ approximately given by

$$\Delta\lambda/\lambda_o = (\lambda_o/4\pi nL)\ln G. \tag{7.39}$$

Altering the temperature of the device causes the pitch of the grating to change due to the dependence of refractive index on temperature. A change in the refractive index alters the wavelength selection of the grating structure and thus the wavelength of the laser output, producing a tunable laser.

DFB was first observed in dye lasers in 1971 by Shank et al.[109] It was an efficient and tunable device. DFB was achieved by pumping the dye (Rhodamine 6G in ethanol) with the fringes formed by the interference of two coherent beams. They were obtained by passing the second harmonic of a single-mode ruby laser through a beam splitter. The laser could be tuned between 0.57 and 0.64 μm wavelength by variation of the interference angle of the coherent pumping beams. Single-mode operation was obtained with less than 0.01 Å linewidth.

The theory and experiment of Kogelnik initiated a number of theoretical papers discussing various aspects of DFB and coupled wave theory.[110] First solid-state and thin-film DFB lasers were operated.[111]

7.9 Optical Amplifiers

In principle, any laser-active gain medium operated just below laser threshold may act as an amplifier. Such amplifiers are commonly used to produce high-power laser systems. Special types such as regenerative amplifiers and chirped pulse amplifiers are used to amplify ultrashort pulses.

One of the first laser amplifier was considered by Geusic and Scovil[112] who discussed the basic principles which are necessary for realizing a nonreciprocal optical amplifier, reporting the operation of a pulsed unidirectional traveling-wave laser, using ruby. The device consisted of a succession of amplifying sections, each separated by nonreciprocal elements so that the power is easily transmitted in one direction but strongly attenuated in the reverse direction. It could also be used as an image amplifier.

Great importance have doped fiber amplifiers. The method of placing rare-earth ions in the core of an optical fiber as an amplifier medium was first demonstrated in 1964 by Koester and Snitzer.[113] They observed 40 dB of gain at 1.06 μm in a flash-lamp-pumped neodymium-doped fiber 1 m in length. Amplification of light in the wavelength region of minimum loss for a silica-based optical fiber (1.5 μm) using transitions of the erbium ion was demonstrated more than 20 years later.[114] The first optical fibers with low transmission loss had appeared[115] in the early 1970s, but after a brief interest in laser-source applications in the mid-1970s, rare-earth-doped fibers fell into oblivion to resurrect in the early 1980s, based now on the effect of stimulated Raman effect.

Later, two methods have gained great popularity: Er-doped fibers (EDF) and semiconductor optical amplifier (SOA).

In the 1980s, Michel Digonnet and Herbert Shaw from Stanford University applied for a number of patents in the United States for a fiber-optic amplifier.[116] The amplifier was obtained with two optical fibers interacting through evanescent field coupling, one doped with a suitable material which will lase at the signal frequency and the other to supply the pump power (**Figure 7.7**).

In this type of rare-earth-doped fiber amplifiers, the signal and pump are introduced into the doped fiber and the signal is amplified via stimulated emission provided by downward electron transitions in the pumped ions. Active ions can be Er^{3+}, Pr^{3+}, Tm^{3+}, Nd^{3+}, Yb^{3+}, and Ho^{3+}, operating over a broad range of wavelengths, principally in the near-infrared. The amplifier consists of a length of monomode fiber, the core of which is doped with the suitably chosen rare earth during fabrication. When this special fiber is excited optically by a comparatively strong pump source, it provides gain for a weak incident signal in a wavelength range longer than the pump wavelength. The gain occurs via the process of stimulated emission. The wavelength range in which gain is obtained depends among other things on the particular rare earth that is used. Of particular interest for optical communications applications are fiber amplifiers doped with erbium ions which can exhibit gain in the range 1.51–1.60 µm.

Digonnet and Gaeta[117] performed a theoretical analysis of optical fiber laser amplifiers and oscillators considering, in particular, Nd-doped glasses, dye solutions, and Nd:YAG.

The erbium-doped fiber amplifier (EDFA) was first demonstrated 2 years later by two groups from the University of Southampton[118] and AT&Bell Labs,[119] respectively.

The group led by David N Payne at the University of Southampton, England,[120] showed amplification of light in the wavelength region of minimum loss for a silica-based optical fiber (1.5 µm) using transitions of the erbium ion.[121] Nearly 20 years had passed before the right way was found. The EDFA was contemporarily demonstrated also by a group at AT&Bell Labs[122] which demonstrated traveling-wave amplification at 1.53 µm at room temperature in an Er^{3+}-doped single-mode fiber using a 514.5 nm pump source.

The amplifier was improved and in 1989 very good results were reported.[123] In Japan,[124] InGaAsP diodes were used to pump at 1.48 µm, and diode-pumped EDFAs rapidly became very popular. Only a few milliwatts of pump power is enough to generate a gain of thousands that is intrinsically insensitive to light polarization. The gain is stable over a 100°C temperature range and EDFA can operate in a regime of minimum spontaneous emission noise. Erbium-doped fiber amplifiers (EDFAs) are widely used in optical fiber communication systems.[125]

The use of the cladding-pumping technique to pump single-mode silica-based fiber lasers has proved to be successful for a variety of dopants like Nd,[126] Yb,[127] and Er,[128] Tm[129] (emission at 2 µm) and ZBLAN (ZrF_4-BaF_2-LaF_3-AlF_3-NaF) fibers doped with Ho[130] or Er.[131] EDFA advances were resumed by Desurvire.[132]

Semiconductor optical amplifiers (SOAs) are amplifiers which use a semiconductor to provide the gain medium. These amplifiers have structures similar to Fabry–Perot laser diodes but with antireflection elements at the end faces

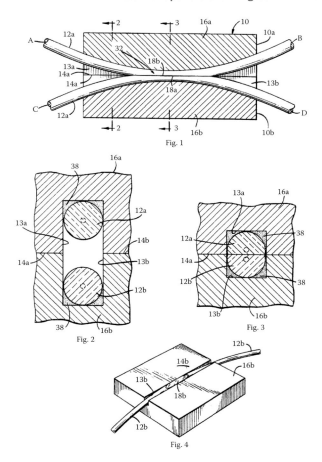

FIGURE 7.7 The optical fiber bidirectional amplifier includes a pair of small-diameter optical fibers, arranged in a side-by-side configuration, the first fiber providing a pumping source and the second fiber doped with a material which will lase at the frequency of the signal to be amplified. (From the Digonnet's patent US4515431 published May 7, 1985.)

or can operated in a traveling-wave way.[133] They are typically made from group III–V semiconductors such a GaAs/AlGaAs, InP/InGaAs, InP/InGaAsP, and InP/InAlGaAs.

Initial optical amplifier studies were carried out on GaAs homostructure devices in the mid-1960s.[134] Later, extensive work on AlGaAs laser amplifiers was carried out in the 1980s.[135] InGaAsP was used with gain centered at 1.3 or 1.5 μm.[136]

The semiconductor optical amplifier is of small size and electrically pumped. It can be potentially less expensive than EDFA and can be integrated with semiconductor lasers, modulators, etc. However, the performance is still not comparable with the EDFA. The SOA has higher noise, lower gain, moderate polarization dependence, and high nonlinearity with fast transient time.

Another way to amplify optical signals in a fiber is through stimulated Raman scattering (we will come back to this in Chapter 8). The amplification effect is achieved by a nonlinear interaction between the signal and a pump laser within the optical fiber. In Raman amplification, the signal frequencies which can be amplified are determined by the pump frequency and the Raman gain curve for the fiber material. Raman amplifiers had a period of great use during the 1980s.[137] Nakamura et al.[138] showed that amplification at 1.5 μm can be obtained using the second Stokes line in fused-silica fiber excited by a 1.32 μm Nd:YAG laser.

For some time it was the most popular amplification method.[139] The most attractive feature of fiber Raman amplifiers is their very wide bandwidth and their ability to provide distributed amplification within the transmission fiber.

7.10 Diode Pumped Solid-State Lasers

As early as 1963, laser diode pumped solid-state lasers were proposed.[140] A first realization of a diode pumped laser was a diode laser-pumped uranium-doped CaF_2 by Keyes and Quist[141] in 1964. However, this laser operated at cryogenic temperature. It was not until 1972 that Nd:YAG was pumped at room temperature,[142] and it was only in 1978 that Scifres et al.[143] demonstrated a 1 W continuous wave laser diode bar that allowed diode laser pumping easy.

The fundamental advantage of diode pumping versus arc-lamp-pumping roots in the spectral composition of the pump. A traditional Nd:YAG laser, for example, can be pumped by an arc lamp. These lamps emit radiation across a broad spectrum. Most of this emission is wasted, since Nd only absorbs light at narrow frequency bands. In comparison, a diode laser can be tuned to emit at a specific wavelength. Thus, through proper selection and temperature tuning of the diode laser, it is possible to set the center frequency of the diode laser around 808 nm that is the optimum frequency at which the majority of the output of the diode laser is absorbed by the Nd:YAG material.

In the early 1980s, the introduction of room-temperature GaAlAs diodes producing tens of milliwatts made diode-pumped solid-state lasers commercially feasible.

The evolution of diode-pumped solid-state lasers and their application was described by A Leuzinger.[144]

Notes

1. Some tentative was made to obtain a control of the spikes. See, for example, B I Davis and D V Keller, *Appl. Phys. Lett.* **5**, 80 (1964); C L Tang, H Statz and G de Mars, *Appl. Phys. Lett.* **2**, 222 (1963).
2. R W Hellwarth, *Advances in Quantum Electronics*, edited by J R Singer (Columbia University Press: New York, 1961), p. 334.

3. R W Hellwarth, *Phys. Rev. Lett.* **6**, 9 (1961).

4. R W Hellwarth in *Advances in Quantum Electronics*, edited by J R Singer (Columbia University Press: New York, 1961), p. 334.

5. F J McClung and R W Hellwarth, *J. Appl. Phys.* **33**, 828 (1962); *Proc. IEE* **51**, 46 (1963).

6. For a general discussion of the different methods, see, for example, W Koechner, *Solid-State Laser Engineering* (Springer: New York, 2006).

7. R J Collins and P Kisliuk, *J. Appl. Phys.* **33**, 2009 (1962); see also N G Basov, V S Zuev and P G Krjukov, *Appl. Opt.* **1**, 1 (1962).

8. N G Basov, V S Zuev and Yu V Senat-Ski, *J. Exp. Theoret. Phys.* **48**, 1562 (1965) (in Russian); translated in *Soviet Phys. JETP* **21**, 1047 (1965).

9. A J DeMaria, R Gagosz and G Barnard, *J. Appl. Phys.* **34**, 453 (1963).

10. J I Masters, J Ward and E Hartouni, *Rev. Sci. Instrum.* **34**, 365 (1963); M A Kovacs, G W Flynn and A Javan, *Appl. Phys. Lett.* **8**, 62 (1966).

11. I V Klimov, I A Shcherbakov and V B Tsvetkov, *Laser Phys.* **8**, 232 (1998).

12. Y Bai et al., *Appl. Opt.* **36**, 2468 (1997).

13. L W Braverman, *Appl. Phys. Lett.* **27**, 602 (1975).

14. M A Kovacs, G W Flynn and A Javan, *Appl. Phys. Lett.* **8**, 62 (1966); T J Bridges, *Appl. Phys. Lett.* **9**, 174 (1966).

15. A A Vuylsteke, *J. Appl. Phys.* **34**, 1615 (1963); W G Wagner and B A Lengyel, *J. Appl. Phys.* **34**, 2040 (1963); A Szabo and R A Stein, *J. Appl. Phys.* **36**, 1562 (1965); L E Erickson and A Szabo, *J. Appl. Phys.* **37**, 4953 (1966).

16. R B Chesler and D Maydan, *J. Appl. Phys.* **42**, 1028 (1971); D Maydan and R B Chesler, *J. Appl. Phys.* **42**, 1031 (1971).

17. D R Herriott, *J. Opt. Soc. Am.* **52**, 31 (1962).

18. W R Bennett Jr., *Phys. Rev.* **126**, 580 (1962).

19. W E Lamb Jr., *Int. Sch. E. Fermi, course XXXI* (1963), p. 78; W E Lamb Jr., *Phys. Rev.* **134**, A1429 (1964); H Haken and H Sauermann, *Z. Phys.* **173**, 261 (1963); H Haken and H Sauermann, *Z. Phys.* **176**, 47 (1963); A Szoke and A Javan, *Phys. Rev. Lett.* **10**, 521 (1963); R A McFarlane, W E Bennett, W E Lamb, *Appl. Phys. Lett.* **2**, 189 (1963).

20. W R Bennett Jr., *Phys. Rev.* **126**, 580 (1962).

21. R A McFarlane, W E Bennett and W E Lamb, *Appl. Phys. Lett.* **2**, 189 (1963).

22. A Szoke and A Javan, *Phys. Rev. Lett.* **10**, 521 (1963).

23. A Szoke and A Javan, *Phys. Rev. Lett.* **10**, 521 (1963); R A McFarlane, W E Bennett, W E Lamb, *Appl. Phys. Lett.* **2**, 189 (1963); W E Lamb Jr., *Phys. Rev.* **134**, A1429 (1964).

24. C Bordé, *CR Acad. Sci. Paris* **271**, 371 (1970).

25. T W Hansch et al., *Nature* **235**, 63 (1972).

26. E I Gordon and J D Rigden, *Bell Syst. Tech. J.* **42**, 155 (1963); A Yariv, *J. Appl. Phys.* **36**, 388 (1965); *IEEE J. Quantum Electron.* **QE-2**, 30 (1966); S E Harris and O P McDuff, *Appl. Phys. Lett.* **5**, 205 (1964); *IEEE J. Quantum Electron.* **QE-1**, 245 (1965); M H Crowell, *IEEE J. Quantum Electron.* **QE-1**, 12 (1965).

27. One for all is A E Siegman, *Lasers* (University Science Books: Stanford, 1986).

28. K Guers and R Mueller, *Phys. Lett.* **5**, 179 (1963); K Guers, *Quantum Electronics III*, edited by P Grivet and N Bloembergen (Columbia University Press: New York, 1964), p. 1113.

29. H Statz and C L Tang, *Quantum Electronics*, edited by P Grivet and N Bloembergen, p. 469.

30. The proposal was made by S E Harris in October 1963 at Stanford University under Contract AF 33 (657)-11144 as quoted in S E Harris and R Targ, *Appl. Phys. Lett.* **5**, 202 (1964).

31. L E Hargrove, R L Fork and M A Pollack, *Appl. Phys. Lett.* **5**, 4 (1964).
32. M Di Domenico, *J. Appl. Phys.* **35**, 2870 (1964).
33. S E Harris and R Targ, *Appl. Phys. Lett.* **5**, 202 (1964).
34. S E Harris and O P McDuff, *Appl. Phys. Lett.* **5**, 205 (1964).
35. M Di Domenico, *J. Appl. Phys.* **35**, 2870 (1964).
36. W E Lamb Jr., *Int. Sch. E.Fermi, course XXXI* (1963), p. 78.
37. A Yariv, *J. Appl. Phys.* **36**, 388 (1964).
38. D I Kuizenga and A E Siegman, *IEEE J. Quant. Electron.* **QE-6**, 803 (1970).
39. H W Mocker and R J Collins, *Appl. Phys. Lett.* **7**, 270 (1965).
40. H A Haus, *J. Appl. Phys.* **46**, 3049 (1975).
41. G H C New, *IEEE Quant. Electron.* **QE-10**, 115 (1974); see also *Proc. IEEE* **67**, 380, 115 (1979). In this paper are references also to the previous author's papers.
42. S E Harris, *Proc. IEEE,* **54**, 1401 (1966).
43. W Schmidt and F P Schaefer, *Phys. Lett.* **26A**, 558 (1968).
44. E P Ippen, C V Shank and A Dienes, *Appl. Phys. Lett.* **21**, 348 (1972).
45. C V Shank and E P Ippen, *Appl. Phys. Lett.* **24**, 373 (1974).
46. P W Smith, *Proc. IEEE* **58**, 1342 (1970).
47. H A Haus, *IEEE J. Sel. Topics Quant. Electron.* **6**, 1173 (2000).
48. U Keller et al., *Opt. Lett.* **17**, 505 (1992); *IEEE J. Sel. Top. Quant. Elecron.* **2**, 435 (1996).
49. U Keller, *Opt. Lett.* **16**, 1024 (1991).
50. D E Spence, P N Kean and W Sibbett, *Opt. Lett.* **16**, 42 (1991).
51. C Hoenhinger et al., *J. Opt. Soc. Am.* **B16**, 46 (1999).
52. A Giesen et al., *Appl. Phys.* **B58**, 363 (1994).
53. J Aus der Au et al., *Opt. Lett.* **25**, 859 (2000).
54. R Paschotta et al., *Appl. Phys.* **B72**, 267 (2001); F Brunner et al., *Opt. Lett.* **27**, 1162 (2002); see also R Paschotta and U Keller, *Optics and Photonics News*, May 2003, p. 51.
55. E Innerhofer et al.*, Opt. Lett.* **28**, 367 (2003). An updating on SESAM and Kerr-lens mode-locking with the development up to 2003 is in U Keller, *Nature* **424**, 831 (2003).
56. A biographical memoir by L Cohen, M Scully and R Scully has been published by the National Academy of Sciences, Washington, DC, in 2009.
57. A recollection of his activity in the laser field was given by W Lamb Jr., *IEEE J. Quant. Electron.* **QE-20**, 551 (1984).
58. W E Lamb Jr., *Phys. Rev.* **134**, A1429 (1964).
59. C V Heer and R D Graft, *Phys. Rev.* **140**, A1088 (1965); G Durand, *IEEE J. Quant. Electron.* **QE-2**, 448 (1966); M Sargent III, W E Lamb Jr. and R L Fiore, *Phys. Rev.* **164**, 436, 450 (1967); M I D'Yakonov and A S Fridrikhov, *Sov. Phys. Uspekhi* **9**, 837 (1967); W van Haeringen, *Phys. Rev.* **158**, 256 (1967); S Stenholm and W E Lamb Jr., *Phys. Rev.* **181**, 618 (1969).
60. H Haken, *Z. Phys.* **181**, 96 (1964); *Phys. Rev. Lett.* **13**, 329 (1964); *Z. Phys.* **190**, 327 (1966); see also H Haken, *Light*, Vol. 2 (North Holland: Amsterdam, 1985).
61. E I Gordon, A D White and J D Rigden, *Proc. Symp. on Opt. Masers*, New York, 1963 (Polytechnic Press: New York, 1963), p. 309.
62. W W Rigrod, H Kogelnik, D J Brangaccio and D R Herriott, *J. Appl. Phys.* **33**, 743 (1962).
63. A Javan, E A Ballik and W L Bond, *J. Opt. Soc. Am.* **52**, 96 (1962).
64. J Haisma and H DeLang, *Phys. Lett.* **3**, 240 (1963).
65. See, for example, T Li, *Bell Syst. Tech. J.* **42**, 2609 (1963).
66. D A Kleiman and P P Kisliuk, *Bell Syst. Tech. J.* **41**, 453 (1962).

67. H Kogelnik and C K N Patel, *Proc. IRE* **50**, 2365 (1962); S A Collins and G R White, *Appl. Opt.* **2**, 448 (1963); J M Burch, *J. Opt. Soc. Am.* **52**, 602 (1962); J A Baker and C W Peters, *Appl. Opt.* **1**, 674 (1962); J G Skinner and J E Geusic, *J. Opt. Soc. Am.* **52**, 1319 (1962).

68. H Statz and C L Tang, *J. Appl. Phys.* **36**, 1816 (1965).

69. J P Goldsborough, *Appl. Opt.* **3**, 267 (1964).

70. A L Mikaelian, Yu G Turkov and V G Savel'ev, *Sov. Phys. JETP Lett.* **6**, 161 (1967).

71. Z Xu et al., *Opt. Comm.* **261**, 118 (2006).

72. J A Giordmaine and W Kaiser, *J. Appl. Phys.* **35**, 3446 (1964).

73. P W Smith, *Lasers,* Vol. 4, edited by A K Levine and A J De Maria (Marcel Dekker: New York, 1976), p. 74.

74. A L Bloom, *Appl. Phys. Lett.* **2**, 101 (1963).

75. R M Zoot, *Appl. Opt.* **5**, 349 (1966).

76. F J McClung, S E Schwarz and F J Meyers, *J. Appl. Phys.* **33**, 3139 (1962); E Snitzer, *Appl. Opt.* **5**, 121 (1966).

77. G R Hanes and J A Dobrowolski, *Appl. Opt.* **8**, 482 (1969).

78. M Bertolotti, D Sette and F Wanderling, *Nuovo Cim.* **48**, 301 (1967).

79. W R Bennett Jr. et al., *Appl. Phys. Lett.* **5**, 56 (1964).

80. T S Jaseja, A Javan and C H Townes, *Phys. Rev. Lett.* **10**, 165 (1963).

81. P W Smith, *IEEE J. Quant. Electron.* **QE-1**, 343 (1965).

82. D E McCumber, *IEEE J. Quant. Electron.* **QE-2**, 219 (1966).

83. A D White, *IEEE J. Quant. Electron.* **QE-1**, 349 (1965).

84. T G Polanyi et al., *IEEE J. Quant. Electron.* **QE-2**, 178 (1966).

85. R Targ, L M Osterink and J M French, *Proc. IEEE* **55**, 1185 (1967).

86. R L Fork (US Patent 3,395,365).

87. P W Smith, *IEEE J. Quant. Electron.* **QE-4**, 485 (1968).

88. W J Tomlinson and R L Fork, *Appl. Opt.* **8**, 121 (1969).

89. W R Bennett et al., *Appl. Phys. Lett.* **5**, 56 (1964); A D White, *IEEE J. Quant. Electron.* **QE-1**, 349 (1965); F Spieweck, *Z. Naturforsch.* **22a**, 2007 (1967); R Balhorn et al., *Appl. Opt.* **11**, 742 (1972); G Birnbaum, *Proc. IEEE* **55**, 1015 (1967).

90. W R C Rowley and D C Wilson, *Nature* **200**, 745 (1963).

91. A D White, E I Gordon and E F Labuda, *Appl. Phys. Lett.* **5**, 97 (1964).

92. A L Bloom, *Gas Lasers* (John Wiley & Sons: New York, 1968).

93. V S Letokhov and B D Pavlik, *Sov. J. Quant. Electron.,* **6**, 32 (1976).

94. K M Baird and G R Hanes, *Rep. Prog. Phys.* **37**, 927 (1974).

95. C Salomon, D Hils and J L Hall, *J. Opt. Soc. Am.* **B5**, 1576 (1988).

96. B C Young et al., *Phys. Rev. Lett.* **82**, 3799 (1999).

97. J C Walling et al., *IEEE J. Quant. Electron.* **QE-16**, 1302 (1980).

98. P F Moulton, *J. Opt. Soc. Am.* **B3**, 125 (1986); P Albers et al., *J. Opt. Soc. Am.* **B3**, 134 (1986).

99. R A Fields, M Birnbaum and C L Fincher, *Appl. Phys. Lett.* **51**, 1885 (1987).

100. S A Payne et al., *J. Appl. Phys.* **66**, 1051 (1989); *IEEE J. Quant. Electron.* **QE-24**, 2243 (1988).

101. L F Mollenauer and D H Olson, *Appl. Phys. Lett.* **24**, 386 (1974).

102. See, for example, L F Mollenauer, *Opt. Lett.* **1**, 164 (1977); L F Mollenauer, D M Bloom and A M DelGaudio, *Opt. Lett.* **3**, 48 (1978); L F Mollenauer, *Opt. Lett.* **5**, 188 (1978); **6**, 342 (1981).

103. L F Mollenauer, *Opt. Lett.* **5**, 188 (1980).

104. K P Koch, G Liftin and H Welling, *Opt. Lett.* **4**, 387 (1979); I Schneider and M J Marrone, *Opt. Lett.* **4**, 390 (1979); I Schneider and C L Marquardt, *Opt. Lett.* **5**, 214

(1980); G Litfin and R Beigana, *J. Phys. E* **11**, 984 (1978): G Litfin, R Beigana and H Welling, *Appl. Phys. Lett.* **31**, 381 (1977).

105. L F Mollenauer, *Opt. Lett.* **1**, 164 (1977).
106. T T Basiev, P G Zverev, V V Fedorov and S B Mirov, *Appl. Opt.* **36**, 2515 (1997).
107. C V Shank, J E Bjorkholm and H Kogelnik, *Appl. Phys. Lett.* **18**, 395 (1971); H Kogelnik and C V Shank, *Appl. Phys. Lett.* **18**, 152 (1971).
108. H Kogelnik and C V Shank, *J. Appl. Phys.* **43**, 2327 (1972).
109. H Kogelnik and C V Shank, *Appl. Phys. Lett.* **18**, 152 (1971); C V Shank and E P Ippen, *Appl Phys. Lett.* **24**, 373 (1974); I P Kaminow, H P Weber and Chandross, *Appl. Phys. Lett.* **18**, 497 (1971).
110. S R Chinn, *IEEE J. Quant. Electron.* **QE-9**, 574 (1973); R E De Wames and W F Hall, *Appl. Phys. Lett.* **23**, 28 (1973); S Wang, *J Appl. Phys.* **44**, 767 (1973); *Opt. Comm.* **10**, 149 (1974); *IEEE J. Quant. Electron.* **QE-10**, 413 (1974); *Appl. Phys. Lett.* **26**, 89 (1975); S Wang and W T Tsan, *J. Appl. Phys.* **45**, 3978 (1974); S Wang, R F Cordero and Ch Tsang, *J. Appl. Phys.* **45**, 3975 (1974); A Yariv, *IEEE J. Quant. Electron.* **QE-9**, 919 (1973); A Yariv and A Gover, *Appl. Phys. Lett.* **26**, 537 (1975); M A Nakamura et al., *Appl. Phys. Lett.* **22**, 515 (1973); **23**, 224 (1973); C V Shank, R V Schmidt and B I Miller, *Appl. Phys. Lett.* **25**, 200 (1974); S Wang, *J Appl. Phys.* **44**, 767 (1973); A Yariv, *IEEE J. Quant. Electron.* **QE-9**, 919 (1973); for more information, see, for example, G P Agrawal and N K Dutta, *Semiconductor Lasers* (Kluwer Academic Publishers: Boston, 1993), Chapters 7 and 8 for coupled-cavity semiconductor lasers.
111. J E Bjrkholm, T P Sosnowski and C V Shank, *Appl. Phys. Lett.* **22**, 132 (1973); M Nakamura et al., *Appl. Phys. Lett.* **22**, 515 (1973); **23**, 224 (1973); D P Schinke et al., *Appl. Phys. Lett.* **21**, 494 (1972).
112. J E Geusic and H E D Scovil, *Bell Syst. Tech. J.* **41**, 1371 (1962).
113. C J Koester and E Snitzer, *Appl. Opt.* **3**, 1182 (1964).
114. R J Mears, L Reekie, S B Poole and D N Payne, *Electron. Lett.* **22**, 159 (1986); E Desurvive, L Simpson and P C Becker, *Opt. Lett.* **12**, 11 (1987).
115. C J Koester and E Snitzer, *Appl. Opt.* **3**, 1182 (1964); J Stone and C A Burrus, *Appl. Opt.* **13**, 1256 (1974); N Periasamy and Z Bor, *Opt. Comm.* **39**, 298 (1981).
116. M J F Digonnet, R A Lacy and H J Shaw, US Patent US4553238 filed Sept. 30, 1983; M Chodorow, M J F Digonnet and H J Shaw, US 4515431 filed Aug 11, 1982.
117. M J F Digonnet and C J Gaeta, *Appl. Opt.* **24**, 333 (1985).
118. R J Mears, L Reekie, I M Jaucey and D N Payne, *Electron. Lett.* **23**, 1026 (1987).
119. E Desurvire, J Simpson and P C Becker, *Opt. Lett.* **12**, 888 (1987).
120. R J Mears, L Reekie, I M Jaucey and D N Payne, *Electron. Lett.* **23**, 1026 (1987).
121. R J Mears, l Reekie, S B Poole and D N Payne, *Electron. Lett.* **22**, 159 (1986).
122. E Desurvire, J Simpson and P C Becker, *Opt. Lett.* **12**, 888 (1987).
123. E Desurvire and J R Simpson, *J. Lightwave Tech.* **7**, 835 (1989); E Desurvire et al., *Opt. Lett.* **14**, 1266 (1989).
124. M Nakazawa, Y Kimura and K Suzuki, *Appl. Phys. Lett.* **54**, 295 (1989).
125. A number of books treat the subject. See, for example, E Desurvire, D Bayard, B Desthieux and S Bigo, *Erbium-doped Fiber Amplifiers: Device and System Development* (Wiley, 2002).
126. H Po et al., *Electron. Lett.* **29**, 1500 (1993).
127. H M Pask et al., *Electron Lett.* **30**, 863 (1994); Y Jeong et al., *Opt. Express* **12**, 6088 (2004).
128. J D Minelly et al., *IEEE J. Photon. Technol. Lett.* **5**, 301 (1993).
129. S D Jackson and T A King, *Opt. Lett.* **23**, 1462 (1998).
130. X Zhu et al., *Opt. Lett.* **37**, 4185 (2012).

131. C. Wei et al., *Opt. Lett.* **37**, 3849 (2012).

132. E Desurvire, *Physics Today*, January 1994, 20.

133. J C Simon, *J. Lightwave Technol.* **LT-5**, 1286 (1987).

134. M J Coupland, K G Mambleton and C Hilsum, *Phys. Lett.* **7**, 231 (1963); J W Crowe and R M Craig Jr., *Appl. Phys. Lett.* **4**, 57 (1964); W F Kosnocky and R H Cornely, *IEEE J. Quant. Electron.* **QE-4**, 225 (1968).

135. M Nakamura and S Tsuji, *IEEE J. Quant. Electron.* **QE-17**, 994 (1981); J C Simon, *J. Lightwave Technol.* **LT-5**, 1286 (1987); T Saitoh and T Mukai, *J. Lightwave Technol,* **LT-6**, 1656 (1988); M J O'Mahony, *J. Lightwave Technol.* **LT-6**, 531 (1988).

136. W B Joyce and B C DeLosch Jr., *Appl. Opt.* **23**, 4187 (1984); N K Dutta et al., *J. Appl. Phys.* **67**, 3943 (1990).

137. G A Koepf et al., *Electron. Lett.* **18**, 942 (1982); E Desurvire et al., *Electron. Lett.* **19**, 751 (1983); Y Aoki et al., *Electron. Lett.* **19**, 620 (1982); **21**, 191 (1985); *Appl. Opt.* **25**, 1056 (1986); K Nakamura et al., *IEEE J. Lightwave Technol.* **2**, 379 (1984); M Nakasaka, *Appl. Phys. Lett.* **46**, 628 (1985).

138. K Nakamura et al., *IEEE J. Lightwave Technol.* **2**, 379 (1984).

139. G A Koepf et al., *Electron. Lett.* **18**, 942 (1982); E Desurvire et al., *Electron. Lett.* **19**, 751 (1983); Y Aoki et al., *Electron. Lett.* **19**, 620 (1982); **21**, 191 (1985); *Appl. Opt.* **25**, 1056 (1986); K Nakamura et al., *IEEE J. Lightwave Technol.* **2**, 379 (1984); M Nakasaka, *Appl. Phys. Lett.* **46**, 628 (1985).

140. R Newman, *J. Appl. Phys.* **34**, 437 (1963).

141. R J Keyes and T M Quist, *Appl. Phys. Lett.* **4**, 50 (1964).

142. H G Danielmeyer and F W Ostermayer, *J. Appl. Phys.* **43**, 2911 (1972).

143. D R Scifres, R D Burnham and W Streifer, *Appl. Phys. Lett.* **33**, 1015 (1978).

144. A Leuzinger, *Optics & Photonic News*, May 1999, p. 37.

8

Nonlinear Optics

8.1 Introduction

Nonlinear optics is concerned with the behavior of light–matter interaction when the material response becomes a nonlinear function of the applied electromagnetic field.

The main consequences of a linear response are that the optical properties of a material, such as refractive index and absorption, are independent of the light intensity, the principle of superposition is valid, the frequency of light is never altered, and light beams never interact between them.

The discovery of nonlinear optics demonstrated that the dielectric constant and magnetic permeability can be a function of the field strengths; it can so happen that the refractive index may depend on light intensity, the principle of superposition is violated, the frequency of light is altered as it passes through a nonlinear medium, and light beams can interact between them through the nonlinear properties of the medium.

An example of a nonlinear behavior in the quasi-static domain is the nonlinear permeability of ferromagnetic materials.

To be more specific and restricting for the moment to the electric case, the electric polarization **P** that usually is written as

$$\mathbf{P} = \varepsilon_o \chi : \mathbf{E}, \tag{8.1}$$

where ε_o is vacuum permittivity, χ is called electric susceptibility and usually is a tensor of rank 2 with no dimension, and **E** is the electric field vector, in the general case takes the form

$$\mathbf{P} = f(\mathbf{E}), \tag{8.2}$$

where $f(\mathbf{E})$ is a function depending in the considered material. In many cases, Equation 8.2 may be expanded as a function of the different powers of E, as

$$P = \varepsilon_o \chi E + \chi^{(2)} : E \cdot E + \chi^{(3)} : E \cdot E \cdot E + \cdots, \tag{8.3}$$

where $\chi^{(2)}$, $\chi^{(3)}$, and so on are called *susceptibility of second, third, ... order* and are tensors.

In general, a similar expression could hold for the magnetic vector **M** and both in the expansion of P or M may appear mixed terms (i.e., terms with E and B in both the expansions of P or M).

In the early 1900s, Lorentz[1] calculated the linear susceptibility appearing in Equation 8.1 with the electron model as a harmonic oscillator as we discussed in Chapter 2. If he had introduced some anharmonicity, as Nick Bloembergen did later, he could have developed the field of nonlinear optics 50 years in advance.

8.2 The Prehistory of Nonlinear Optics

Consider for a moment the linear case and write the dipole moment per unit volume or *polarization P* as

$$P_i = P_i^{\circ} + \varepsilon_o \chi_{ij} E_j, \tag{8.4}$$

where P_i° is a possible pre-existing natural polarization and χ_{ij} are the components of the tensor of *linear susceptibility*. In the nonlinear case, we may write

$$P_i = P_i^{\circ} + \varepsilon_o \chi_{ij} E_j + \chi_{ijk}^{(2)} E_j E_k + \chi_{ijkl}^{(3)} E_j E_k E_l + \cdots \tag{8.5}$$

where the successive terms with increasing power orders in the electric field describe the *second-*, *third-*, and so on order *nonlinear effects* and the coefficients $\chi^{(2)}{}_{ijk}$, $\chi^{(3)}{}_{ijkl}$, and so on are the corresponding tensor elements of the corresponding nonlinear *susceptibilities*.

Using this definition, we may see that the *quadratic electro-optic effect*, discovered in 1875 by the Rev. John Kerr (1824–1907)[2] of the Free Church Training College in Glasgow, United Kingdom, and the *linear electro-optic effect,* discovered by Fredrich Carl Alwin Pockels (1865–1913) at the University of Gottingen in 1893,[3] are both a manifestation of nonlinear optics.

In fact, we may roughly write for the Kerr effect that a change in relative dielectric constant is produced by applying a quasi-static electric field E_o squared, or

$$\Delta\varepsilon_r = rE_o^2, \tag{8.6}$$

where **r** is a characteristic parameter of the material. Usually Equation 8.6 is a tensor equation but we do not need this refined presentation for our considerations. Writing Equation 8.4 in the simplified form

$$\mathbf{P} = \varepsilon_o \chi E, \tag{8.7}$$

the relative dielectric constant may be defined as

$$\varepsilon_r = 1 + \chi. \tag{8.8}$$

If we consider the third-order term in the polarization, producing still an effect at the same frequency as the linear term, it is

$$\mathbf{P} = \varepsilon_o \chi E + \chi^{(3)} |E_o|^2 \, E = \varepsilon_o [\chi + (\chi^{(3)}/\varepsilon_o)|E_o|^2]E = \varepsilon_o \chi' E, \qquad (8.9)$$

with

$$\chi' = (\chi + \chi^{(3)}/\varepsilon_o)|E_o|^2 \qquad (8.10)$$

or

$$-1 + \varepsilon_r + \Delta\varepsilon_r = -1 + \varepsilon_r + \chi^{(3)}/\varepsilon_o|E_o|^2, \qquad (8.11)$$

which in the static case gives

$$\Delta\varepsilon_r = (\chi^{(3)}/\varepsilon_o)E_o^2 = rE_o^2 \qquad (8.12)$$

with

$$\mathbf{r} = \chi^{(3)}/\varepsilon_o. \qquad (8.13)$$

So we see that the Kerr effect is due to the presence of a third-order nonlinearity.

In a similar way, it can be shown that the Pockels effect is the manifestation of a second-order nonlinearity.

Also *two-photon absorption* was already discussed at the beginning of the XX century, although only as a theoretical possibility; in fact, Einstein[4] in his famous paper on the photoelectric effect in 1905 mentioned the possibility of processes involving more than one photon.

Later Maria Goeppert-Mayer (1906–1972) in Göttingen, in her doctoral dissertation "Uber Elementarakte mit zwei Quantenspringen" published in 1931 developed the quantum theory of two-photon absorption.[5] Maria was awarded the Physics Nobel Prize in 1963 with Hans Jensen (1907–1973) and Eugene P Wigner (1902–1995) for their work on nuclear shell theory.

Curiously enough, the term *nonlinear optics* appeared for the first time in two papers by E Schroedinger (1887–1961)[6] in which the term was used discussing a nonlinear Maxwell equation introduced by M Born (1882–1970)[7] and M Born and L Infeld.[8] M Born, frustrated by the divergences of the pre-renormalization quantum electrodynamics, in the early 1930s, inaugurated an alternative nonlinear Maxwell equation which was resumed by Schroedinger in a series of papers to "escape" from the "infinities" which were in the quantum conceptions of photons, electrons, etc., at the time.

Multiphoton processes were also considered by G Breit and E Teller[9] and J Wheeler,[10] discussing the simultaneous emission of two photons in the transition from one discrete quantum level to another one in the case of the transition $2s \rightarrow 1s$, in hydrogen.

In this pre-historical excursus, we may mention also, A T Forrester[11] who, in the 1940s, suggested creating electromagnetic waves from 1 mm down to about 0.05 mm by mixing light of different frequencies, and in the 1950s, succeeded in demonstrating this combination process with incoherent light.[12]

The way to obtain the mixing was, however, based on the use of nonlinear devices (photoelectric cells) and not of material's nonlinearities.

8.3 The First Experiments

The nonlinear properties in the optical region were demonstrated for the first time in 1961 by Franken et al.[13] with an experiment of second harmonic generation, before a theory of nonlinear processes was made. Sending the red light of a ruby laser which produced approximately 3 J of 6.943 Å light in 1 ms pulse onto a crystal of quartz, they were able to observe a tiny production of ultraviolet light.

Peter Franken (1928–1999) (**Figure 8.1**) finished his PhD in experimental atomic physics at Columbia University in 1952 under the Nobel laureate Polykarp Kusch (1911–1993). Franken then spent several years at Stanford before coming to the University of Michigan in 1956 where he stayed until accepting the Directorship of the Optical Science Center at the University of Arizona in 1973. By 1963, Franken was interested also in *quark* theory.

Peter had generous spirit with an infectious *joie de physique* and was famous for his humor.

In an interview to the Bohr Library of AIP,[14] he recalls he attended one of the first conferences on lasers in Pittsburgh (organized by OSA) and Schawlow was speaking of ruby laser applications in "eye surgery, communication and communication and eye surgery …," so he started wandering if something else could be done and calculated a huge electric field was produced by the ruby laser beam and realized a second harmonic could be generated. He spoke of his idea to Gaby Weinreich, a solid-state experimentalist and theorist, asked the senior experimentalist in spectroscopy Wilbur Peters to join in the experiment and decided to use fused quartz as a nonlinear material, but Weinreich draw attention they had to use a crystal that lacked a center of inversion. So with a commercial laser rented from the firm Trion and with Alan Hill, at that time a senior student who performed the hard work, the system, a very simple setup (see **Figure 8.2**) but which needed a perfect alignment because they had no phase matching and the second harmonic had very low power, was assembled and … it worked!

FIGURE 8.1 Peter Franken (1928–1999).

The story is that when Franken published the paper describing the experiment on second harmonic production, the proofreader of the *Physical Review Letters* corrected the photo of the prism output, suppressing the tiny spot of the second harmonic believing it was just a defect. You may see this in **Figure 8.3** in which under the arrow you see … nothing!

The four researchers worked at the experiment 4 or 5 months, did not keep their work secret, and received many visitors—among others Willis Lamb and people from Bell Labs—who all were skeptical. At that time, all people were

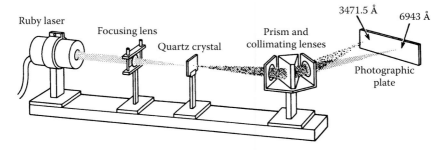

FIGURE 8.2 Sketch of the second harmonic experiment by Franken et al. (From A Yariv, *Quantum Electronics* (J. Wiley & Sons, New York, 1975).)

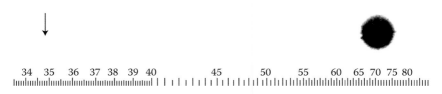

FIGURE 8.3 This is Figure 1 of Franken paper in which the big spot of the ruby laser light is shown on the right side. An arrow at the left side is pointing to the position where the tiny spot of the blue light of the produced second harmonic had to appear, but nothing is visible! (From P A Franken et al., *Phys. Rev. Lett.* **7**, 118 (1961). With permission.)

thinking in terms of photons and anybody knew you cannot change the frequency of a photon. In the 1960s that was a revolutionary concept. It was not appreciated that the change is via the quantized levels of matter.

Multiphoton absorption was also first observed in 1961 by Werner Kaiser and Garrett[15] in a sample of $CaF_2:Eu^{2+}$ under strong illumination with a ruby laser pulse and is shown

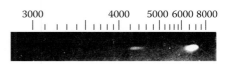

FIGURE 8.4 Photographic plate showing the laser spot at 6943 Å and the blue emission of a $CaF_2:Eu^{2+}$ crystal under strong illumination with light from a ruby laser. (From W Kaiser and C G B Garrett, *Phys. Rev. Lett.* **7**, 229 (1961). With permission.)

in **Figure 8.4**. The overexposed bright spot at 6943 Å results from the incident laser light. The light which is observed around 4250 Å is characteristic for the blue fluorescence of $CaF_2:Eu^{3+}$.

The dependence of the fluorescence intensity on the square of the ruby intensity, as predicted for two-photon absorption, was experimentally verified (see **Figure 8.5**).

In the 1970s, S J R Sheppard and R Kompfner[16] suggested to build a nonlinear microscope using two-photon absorption and eventually the microscope was built in 1990 by W Denk, J H Strickler, and W W Webb.[17]

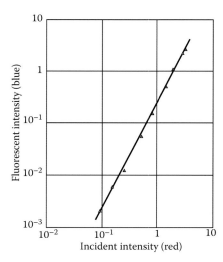

FIGURE 8.5 Blue fluorescent intensity versus ruby laser intensity. (From W Kaiser and C G B Garrett, *Phys. Rev. Lett.* **7**, 229 (1961). With permission.)

8.4 Nonlinear Optics

Other processes in which three or more quanta are involved were studied in 1959–1962[18] but we may say that nonlinear optics entered the stage when Bloembergen, immediately after the first experiments described previously, in 1962, published his famous paper giving the sound basis for nonlinear optics.[19] Inclusion of proper boundary conditions in the theory was immediately added by Bloembergen and Pershan[20] in a following paper.[21] Then, with their formalism, nonlinear wave transmission and reflection at an interface could be properly described and experimentally verified.[22] Experiments of second harmonic generation (SHG) in reflection from semiconductor surfaces were immediately performed to prove the theoretical predictions.[23] Around the same time, measurements of SHG from metal surfaces were also reported.[24] Early experiments by Bloembergen and coworkers on reflected SHG from Si and Ge[25] and by Brown and his collaborators on reflected SHG from silver[26] started the field of surface nonlinear optics. An historical perspective of this field has been given by Y R Shen.[27]

The extension to absorbing optical media where the nonlinear susceptibilities are complex quantities soon followed.[28]

In the ex-Soviet Union, the theory of nonlinear processes was also actively studied.[29] In his first paper,[19] Bloembergen derives a quantum mechanical expression for the nonlinear, induced electric-dipole moments up to terms quadratic and cubic in the field strength illustrating them with the example of the anharmonic oscillator. This paper was rather difficult to be understood at the time. Later, in his book[30] (the first one to be written on the subject) on nonlinear optics, Bloembergen gave a very intuitive explanation of the origin of nonlinearity in matter considering the classical harmonic oscillator approximation of the electron response in materials. If the applied external field has a strength not too far from the value of the Coulomb electric field suffered by the electron in the atom, this approximation breaks down and the harmonic oscillator may be replaced by a nonlinear one (an anharmonic oscillator). In this way, one may see that a nonlinear polarization arises.

Following the classical treatment of Bloembergen, at the lower level of the perturbation, we can write the anharmonic oscillator as

$$\ddot{x}(t) + \sigma\dot{x}(t) + \omega_o x(t) + Dx^2(t) = -(e/m)E(t). \qquad (8.14)$$

The solution of Equation 8.14 can be expressed as the sum of two terms

$$x(t) = x^{(1)}(t) + x^{(2)}(t), \tag{8.15}$$

in which $x^{(1)}(t)$ is obtained solving Equation 8.14 without the anharmonic term, whereas $x^{(2)}(t)$ is considered a small correction of the solution at the first order $x^{(1)}(t)$ and is obtained utilizing $x^{(1)}(t)$ in the anharmonic term

$$\ddot{x}^2(t) + \sigma\dot{x}^{(2)}(t) + \omega_o x^{(2)}(t) = -(e/m)E(t) - D[x^{(1)}(t)]^2. \tag{8.16}$$

In this way, considering the case in which the forcing electric field is formed by the sum of two fields at different frequencies

$$E(t) = E_1 \cos\omega_1 t + E_2 \cos\omega_2 t = 1/2[E_1\exp(-j\omega_1 t) + E_2\exp(-j\omega_2 t) + cc], \tag{8.17}$$

we have the solution at the first order

$$x^{(1)}(t) = 1/2[x^{(1)}(\omega_1)\exp(-j\omega_1 t) + x^{(1)}(\omega_2)\exp(-j\omega_2 t) + cc], \tag{8.18}$$

and subsequently, the solution at the second order, solving Equation 8.16 with the use of Equation 8.18 is

$$x^{(2)}(t) = 1/2\{x^{(2)}(\omega_1 + \omega_2)\exp[-j(\omega_1 + \omega_2)t] + x^{(2)}(\omega_1 - \omega_2)\exp[-j(\omega_1 - \omega_2)t] +$$
$$+ x^{(2)}(2\omega_1)\exp(-j2\omega_1 t) + x^{(2)}(2\omega_2)\exp(-j\omega_2 t) + cc\}, \tag{8.19}$$

in which

$$x^{(2)}(\omega_1 \pm \omega_2) = (-1/2)D(e/m)^2 E_1 E_2 /$$
$$\left\{(\omega_o^2 - \omega_1^2 + j\sigma\omega_1)(\omega_o^2 - \omega_2^2 + j\sigma\omega_2)\left[\omega_o^2 - (\omega_1 \pm \omega_2)^2 + j\sigma(\omega_1 \pm \omega_2)\right]\right\}$$
$$x^{(2)}(2\omega_k) = (-1/2)D(e/m)^2 E_k^2 /\left[(\omega_o^2 - \omega_k^2 + j\sigma\omega_k)^2(\omega_o^2 - 4\omega_k^2 + j\sigma\omega_k)\right], \quad k = 1,2. \tag{8.20}$$

Therefore, the solution of the second order brings to the generation of oscillations at a frequency different from the ones of the forcing field. In particular, it is possible to have frequencies equal to the sum or to the difference of the field frequencies or to the double (second harmonic). Moreover, the previous formulas remain valid also if just a single forcing field ω is present. In this case, $x^{(2)}(t)$ will be the sum of a second harmonic term (2ω) and a null pulsation term (term of optical rectification).

Remembering the expression for the polarization, we can write

$$P(t) = -Ne[x^{(1)}(t) + x^{(2)}(t)], \tag{8.21}$$

where N is the number of dipoles for volume unit. Equation 8.21 may be put in the form

$$P(t) = P_L(t) + P_{NL}(t), \tag{8.22}$$

which compared with Equation 8.20 permits to write

$$P_L = \varepsilon_o\chi^{(1)}E,$$

$$P_{NL} = \chi^{(2)}E \cdot E. \tag{8.23}$$

Therefore, the second-order polarization can be written as

$$P_i^{(2)} = \sum_{j,k} d_{ijk} E_j E_k, \qquad (8.24)$$

where i, j, and k represent the coordinates x, y, and z. The main part of the coefficients d_{ijk}, however, is usually zero and so only a few of them must be considered.

Only the non-centrosymmetric crystals can have a non-null tensor d_{ijk}. In fact, let us consider an isotropic crystal. In this case, d_{ijk} is independent from the direction and therefore it is constant. If now we invert the direction of the electric field, also the polarization must change sign, that is

$$-P_i^{(2)} = \sum d_{ijk}(-E_j)(-E_k) = \sum d_{ijk} E_j E_k = + P_i^{(2)}.$$

It is clear that, not being able to be $-P_i^{(2)} = + P_i^{(2)}$, d_{ijk} must be null. This point was already appreciated by Franken and collaborators in their first experiment.

Moreover, in materials for which $d \neq 0$, since no physical meaning can be assigned to an exchange of E_j with E_k, it must be $d_{ijk} = d_{ikj}$.

Now if we consider the Maxwell equations, writing

$$D = \varepsilon_o E + P, \qquad (8.25)$$

we have

$$\nabla \times B = \mu j + \mu \partial D/\partial t = \mu j + \mu\varepsilon_o \partial E/\partial t + \mu \partial P/\partial t$$

$$\nabla \times E = -\partial B/\partial t. \qquad (8.26)$$

Using Equation 8.22, the polarization can be written as the sum of a linear term plus a nonlinear one which for materials with second-order nonlinearity can be written as in Equation 8.24. So Equation 8.26 can be written assuming $j = 0$, as

$$\nabla \times B = \mu \partial \varepsilon E/\partial t + \mu \partial P_{NL}/\partial t, \qquad (8.27)$$

from which

$$\nabla^2 E = \mu\varepsilon \partial^2 E/\partial t^2 + \mu \partial^2 P_{NL}/\partial t^2. \qquad (8.28)$$

If we consider the unidimensional case of propagation along the direction z, we have

$$\partial^2 E_i/\partial z^2 = \mu\varepsilon \partial^2 E_i/\partial t^2 + \mu \partial^2 (P_{NL})_i/\partial t^2. \qquad (8.29)$$

Let us consider now three monochromatic fields with frequencies ω_1, ω_2, and ω_3 using the complex notation

$$E_i(\omega_1)(z,t) = 1/2[E_{1i}(z)\exp j(\omega_1 t - k_{1z}z) + cc], \qquad (8.30a)$$

$$E_k(\omega_2)(z,t) = 1/2[E_{2k}(z)\exp j(\omega_2 t - k_{2z}z) + cc], \qquad (8.30b)$$

$$E_j(\omega_3)(z,t) = 1/2[E_{3j}(z)\exp j(\omega_3 t - k_{3z}z) + cc], \qquad (8.30c)$$

where the indices i, j, and k represent the components x or y.

The polarization at frequency $\omega_1 = \omega_3 - \omega_2$, for example, from Equations 8.24 and 8.30c results in

$$P_i(\omega_1) = 1/2 \sum_{j,k} d_{ijk} E_{3j}(z) E_{2k}^*(z) \exp j\left[(\omega_3 - \omega_2)t - (k_3 - k_2)z\right] + cc. \qquad \textbf{(8.31)}$$

Substituting Equations 8.24 and 8.30 into Equation 8.29 for the component E_{1i}, it is necessary to calculate

$$\partial^2 E(\omega_1)/\partial z^2 = 1/2\partial^2[E_{1i}(z)\exp(\omega_1 t - k_1 z) + cc]/\partial z^2. \qquad \textbf{(8.32)}$$

If we assume

$$(dE_{1i}/dz)k_1 > d^2 E_{1i}/dz^2, \qquad \textbf{(8.33)}$$

we have

$$\partial^2 E_i(\omega_1)/\partial z^2 = -1/2\left[k_1^2 E_{1i}(z) + 2jk_1 dE_{1i}(z)/dz\right]\exp j(\omega_1 t - k_1 z) + cc \qquad \textbf{(8.34)}$$

with similar expressions for $\partial^2 E_j(\omega_2)/\partial z^2$ and $\partial^2 E_k(\omega_3)/\partial z^2$.

Finally, substituting Equations 8.34 and 8.31 into Equation 8.29, we have

$$dE_{1i}(z)/dz = -j(\omega_1/2)\sqrt{(\mu_o/\varepsilon_1)}\sum d_{ijk} E_{3j} E_{2k}^* \exp\left[-j(k_3 - k_2\_k_1)z\right] + cc \qquad \textbf{(8.35)}$$

and in an analogous way

$$dE_{2k}^*/dz = j(\omega_2/2)\sqrt{(\mu_o/\varepsilon_2)}\sum d_{ijk} E_{1i} E_{3j}^* \exp\left[-j(k_1 - k_3 + k_2)z\right] + cc$$
$$dE_{3j}/dz = -j(\omega_3/2)\sqrt{(\mu_o/\varepsilon_3)}\sum d_{ijk} E_{1i} E_{2k} \exp\left[-j(k_1 + k_2 - k_3)z\right] + cc. \qquad \textbf{(8.36)}$$

The **second harmonic generation** is obtained immediately from Equations 8.35 and 8.36 for the case of $\omega_1 = \omega_2$ and $\omega_3 = 2\omega_1$.

To further simplify the analysis, we can assume that the power lost by the frequency ω_1 (fundamental) is negligible, and therefore

$$dE_{1i}/dz \approx 0. \qquad \textbf{(8.37)}$$

So it is sufficient to consider just the second of Equation 8.36:

$$dE_{3j}/dz = -j\,\omega\sqrt{(\mu_o\varepsilon)}\sum d_{jik} E_{1i} E_{1k} \exp j\,\Delta k \cdot z, \qquad \textbf{(8.38)}$$

where $\omega = \omega_1 = \omega_3/2$ and

$$\Delta k = k_3^{(j)} - k_1^{(i)} - k_1^{(k)}. \qquad \textbf{(8.39)}$$

In Equation 8.39, $k_1^{(i)}$ is the constant of propagation of the beam at ω_1 polarized in the direction i. The solution of Equation 8.38 for $E_{3j}(0) = 0$ for a crystal of length L is

$$E_{3j}(L) = -j\omega\sqrt{(\mu_o/\varepsilon)}\sum d_{jik} E_{1i} E_{1k}[(\exp j\,\Delta k \cdot L - 1)/j\Delta k], \qquad \textbf{(8.40)}$$

or

$$I(L) = |E_{3j}(L)|^2 = (\mu_o/\varepsilon)\omega^2|\Sigma d_{jik}E_{1i}E_{1k}|^2 \, L^2 \, [\sin^2(\Delta k \cdot L/2)/(\Delta k \cdot L/2)^2]. \quad \textbf{(8.41)}$$

According to Equation 8.41, a requirement for an efficient second harmonic generation is that $\Delta k = 0$, that is, from Equation 8.39 with $\omega_3 = 2\omega$ and $\omega_1 = \omega_2 = \omega$:

$$k(2\omega) = 2k(\omega). \quad \textbf{(8.42)}$$

If $\Delta k \neq 0$, the second harmonic wave generated at a generic plane z_1 which propagates until another plane z_2 is not in phase with that generated in z_2. This produces an interference described by the factor

$$\sin^2(\Delta k \cdot L/2)/(\Delta k \cdot L/2)^2$$

in Equation 8.41.

Condition 8.42 is never practically satisfied because, due to dispersion, the refractive index depends on ω.

We have

$$\Delta k = k(2\omega) - 2k(\omega) = (2\omega/c)[n(2\omega) - n(\omega)], \quad \textbf{(8.43)}$$

being

$$k(\omega) = \omega n(\omega)/c, \quad \textbf{(8.44)}$$

and therefore $\Delta k \neq 0$ because $n(2\omega)$ cannot be equal to $n(\omega)$. We have no more second harmonic production when

$$\sin(\Delta k \cdot L/2)/\Delta k \cdot L/2 = 0. \quad \textbf{(8.45)}$$

This is achieved when $\Delta k \cdot L/2 = \pi$ which means $L = 2\pi/\Delta k$, which is named *coherence length*.

What happens is very easy to understand. The polarization generated by the fundamental field travels with a velocity determined by $n(\omega)$ while the second harmonic wave travels with a velocity determined by $n(2\omega)$. In general, $n(2\omega) > n(\omega)$ because of normal dispersion in the material. Since the sign of power flow from one wave to the other is determined by the relative phase between the waves, the continuous phase slip between these waves caused by their differing phase velocities leads to an alternation of the direction of the flow of power.

In Franken's experiment, efficiency was very poor (around 1 blue photon every 10^8 red photons) because the fundamental and harmonic waves were not matched and traveled at different phase velocities in the nonlinear crystal, so quickly got out of step. The coherence length in their case may be calculated around $L_{coh} \sim 10$–$20 \, \mu m$ so that only a small fraction of the quartz crystal in the experiment was participating usefully in the second harmonic generation process.

The effect of the term with Δk in Equation 8.41 was shown since the very beginning by Maker et al.[31] A ruby laser was passed through an optically plane-parallel quartz platelet, the red light filtered using a $CuSO_4$ solution filter and a grating monochrometer, and the blue light intensity measured. The sample was inclined to the beam by rotation about its z-axis, thus increasing the optical thickness and the curve shown in **Figure 8.6** was obtained.

The periodic variation of intensity of the SH with respect to thickness obtained by rotating the sample was used to obtain a measure of the nonlinearity (Maker's fringe method[32]).

A solution to the problem of phase matching using uniaxial crystals was given independently by Maker, Therune et al.,[33] and Giordmaine.[34] They showed that the phase mismatch of the phase velocity may be compensated by using the extraordinary and ordinary index values in uniaxial crystals.

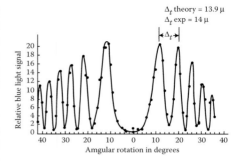

FIGURE 8.6 Blue light generation verus inclination of a quartz platelet to laser beam. Rotation axis normal to beam, parallel to the crystal z-axis. The laser beam was unfocused and polarized parallel to the z-axis. (From P D Maker et al., *Phys. Rev. Lett.* **8**, 21 (1962). With permission.)

Maker discussed the possibility to make $\Delta k = 0$ (*phase-matching condition*) taking advantage from the natural birefringence of an anisotropic crystals using potassium dihydrogen phosphate (KDP). From Equation 8.43, we see that $\Delta k = 0$ implies

$$n(2\omega) = n(\omega),$$

so that the refractive indices of second harmonic and of fundamental frequency have to be equal.

In KDP, the components of the nonlinear polarization can be written as

$$P_x = aE_yE_z, \quad P_y = aE_zE_x, \quad P_z = aE_xE_y.$$

Thus ordinary exciting rays generate extraordinary harmonic rays. As the birefringence for KDP is greater than its dispersion, in certain orientations $k(2\omega)^{(e)}$ exactly matches $2k(\omega)^{(o)}$. This matching resulted in a 300-fold increase in the second harmonic intensity.

Figure 8.7 shows in the inset a portion of the refractive index ellipse. The two curves with apex in D and B are the refractive index for extraordinary (blue) and ordinary (red) refractive index, respectively. The two indices are equal on the circle which passes from O. The main figure shows the blue light (2nd harmonic) intensity as a function of crystal orientation. Maximum output occurs at $\theta_o = 52° \pm 2°$, $\phi_o = 45°$.

The method can also be used to match more than two waves and was used in experiments on third-harmonic generation[35] and optical mixing.[36]

Another way to obtain phase matching, predating the development of birefringent phase-matching, was devised independently by Bloembergen in his first paper and by Franken and Ward who in 1963 wrote the first review paper

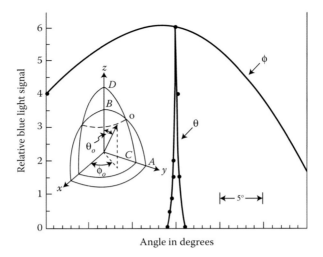

FIGURE 8.7 Blue light intensity as a function of crystal orientation for KDP. Maximum output occurs at $\theta = 52° \pm 2°$, $\varphi = 45°$. In the inset, AOB is an arc of the index of refraction surface for red ordinary rays, COD for blue extraordinary rays. (From P D Maker et al., *Phys. Rev. Lett.* **8**, 21 (1962). With permission.)

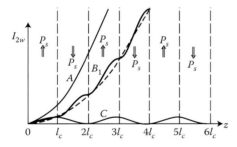

FIGURE 8.8 Calculated second harmonic intensity in the various cases of perfect phase matching (curve A), no matching (curve C), and quasi-phase matching (curve B1). (From M M Fejer et al., *IEEE J. Quantum Electron.* **QE-28**, 2631 (1992). With permission.)

on nonlinear optics[37] describing it. This method is called *quasi-phase matching* and it involves the repeated inversion of the relative phase between the fundamental and the second harmonic fields after an odd number of coherence lengths. The phase is thus "reset" periodically so that on average the proper phase relationship is maintained for growth of the second harmonic.

Figure 8.8 illustrates the dramatic improvement in second harmonic generation efficiency that phase matching delivers. It shows the intensity of second harmonic as a function of the traveled distance in the nonlinear crystal for the three cases of perfect phase matching (curve A), non-matched interaction (curve C), and quasi-phase matching obtained by flipping the sign of the spontaneous polarization every coherence length of the material (curve B1).

A particular type of quasi-phase matching is one in which the sign or magnitude of the nonlinear coefficient is modified throughout the material.

Franken, Hill, and Peters[38] obtained matching stacking thin plates of x-cut crystalline quartz with their *z*-axes alternating in direction by 180. This effectively reverses the direction of the second harmonic polarization in alternate plates. In the experiment, plates of quartz about 0.025-cm thick were used and it was found that approximately four times as much second-harmonic radiation could be produced from two plates as one, a 16-fold increase from four stacked plates and so on.

Bloembergen suggested to change periodically the sign of the nonlinear tensor. The resulting periodic structure is a grating in the nonlinear coefficient. The wave vector of this grating adds to the wave vector of the fundamental wave to equate the wave vector of the second harmonic.[39]

Ferroelectric crystals like $LiNbO_3$ form regions of periodically reversed spontaneous polarization (domains) and are suitable for the requested geometry.

It was recognized early that multidomain ferroelectric crystals could show an enhancement of second harmonic generation.[40] Rotationally twinned crystals of ZnSe, ZnS, and other materials were considered for the enhancement of second harmonic generation by several researchers in the early 1970s.[41] B F Levine et al.[42] applied a periodic electric field to liquid nitrobenzene to modulate the nonlinear susceptibility for phase matching. The geometry was realized alternating stacks of thin plates of CdTe,[43] GaAs,[44] quartz,[45] $LiNbO_3$,[46] $LiTaO_3$[47] crystals and using single-crystal fibers.[48] Quasi-phase matching was realized using $LiNbO_3$ in integrated-optic technology[49] and poled nonlinear optical polymer materials.[50] A detailed theory of quasi-phase matching was presented later by Fejer in 1992.[51]

Phase matching may also be achieved by modulating the linear susceptibility.[52]

A number of experiments were done on second harmonic generation.[53]

Observation of continuous optical harmonic generation using the 1.1526 μm wavelength of a He–Ne laser was obtained for the first time by Ashkin et al.[54] The group gave another impressive evidence of the effect of phase matching (**Figure 8.9**), contrasting the experimentally observed angular dependence of the second harmonic power (dots in **Figure 8.9**) with Equation 8.41 (the continuous curve is just $\sin^2 \psi / \psi^2$, where $\psi = \Delta k \cdot L/2$).

8.5 Physical Origin of Optical Nonlinearities

Bloembergen in his paper[55] used the semiclassical theory of the interaction between radiation and matter, due to Klein[56] and Kramers,[57] to explain the nonlinear dielectric properties of a medium. Later, in Bloembergen's book, the results were recast first in a completely classical frame making recourse to the mechanical anharmonic oscillator representation, as was discussed in the previous paragraph, and then in the full quantum mechanical treatment using the density matrix formalism.[58] It resulted in a generalization of the Kramers–Heisenberg dispersion formula in which the nonlinear susceptibilities are functions of several frequencies and have more than one resonant

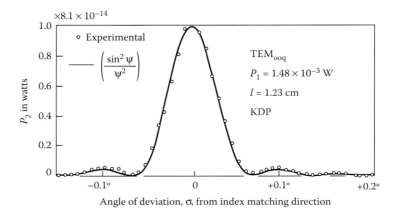

FIGURE 8.9 Variation of second harmonic power with the angle between the fundamental beam and the phase matching direction. (From A Ashkin et al., *Phys. Rev. Lett.* **11**, 14 (1963). With permission.)

denominator. They are tensors of higher order and each element has a real and an imaginary part. Akhmanov and Khokhlov[59] extended the formalism of parametric nonlinearities from the radiofrequency to the optical domain.

The physical origin of nonlinear behavior in matter was since then discussed in an innumerable number of papers.[60] Different materials exhibit different mechanism of nonlinearity which may originate from electrons, reorientation of molecules, thermal effects, and so on. Nonlinear response in metals, plasma, semiconductors, liquid crystals, and so on was derived.

The approach for calculating the nonlinear optical susceptibility entails the calculation of the atomic response using time-dependent quantum mechanical perturbation. The density matrix formalism of quantum mechanics results to be the better approach in general. The theory is well described in standard textbooks on nonlinear optics.[61]

A rigorous quantum mechanical treatment shows that the nonlinear coefficients are dependent on all the frequencies participating in a given nonlinear interaction. Hence, the second harmonic generation coefficient is, for example, not identical to the coefficient governing optical rectification.

Using the quantum mechanical photon picture, the nonlinear effects may be viewed as transformations among photons mediated by the material. So, for example, second harmonic generation is the process in which two photons at the fundamental frequency are absorbed by the nonlinear material which re-emits a single photon at twice the frequency. In the process, energy and momentum must be conserved, that is

$$h\nu_F + h\nu_F = h\nu_{SH} = h2\nu_F \tag{8.46}$$

$$\boldsymbol{k}_F + \boldsymbol{k}_F = \boldsymbol{k}_{SH} \tag{8.47}$$

Relation 8.46 was tested in the early days of second-harmonic generation by Boyne and Martin[62] using a diffraction grating to verify that the second harmonic generation in KDP and other crystals was indeed at twice the frequency of the fundamental ruby laser frequency with a precision of three parts in a million. Equation 8.47 is just the phase-matching condition.

The number of photons in the nonlinear process is obviously not conserved. A full discussion of conservation laws in nonlinear optics was given by Bloembergen.[63] In a periodic structure with period d, the structure may help to balance momentum.[64] The momentum imbalance can be taken up by the periodic structure as a whole for a quantity $2\pi n/d$ with $n = \pm 1, \pm 2, \ldots$.

8.6 Further Experiments in Nonlinear Optics

General symmetry properties of the nonlinear tensors were immediately understood. For real nonlinear susceptibilities, one may interchange the tensor indices, provided the corresponding frequencies are also interchanged in materials which are transparent to the fundamental and second harmonic frequencies. These relations were first formulated by Kleinman,[65] whereas the general permutation symmetry relations were given by Bloembergen et al.[66]

$$\chi_{ijk}(\omega_3 = \omega_1 + \omega_2) = x_{jik}(\omega_1 = \omega_3 - \omega_2) = x_{kij}(\omega_2 = \omega_3 - \omega_1).$$

Considering second harmonic production, Bloembergen and coworkers also predicted the existence of a second harmonic beam locked in phase to the fundamental,[67] propagating with the same velocity of the fundamental (inhomogeneous solution). This solution of the wave equation was later discussed and observed.[68] This inhomogeneous solution allows for propagation of the second and third harmonics in absorbing materials.[69]

As it was appreciated immediately, since the first experiment of Franken, second harmonic is only possible in media which lack of symmetry. Calcite, which is a highly symmetric crystal, should not allow second harmonic generation (SHG). However, SHG was observed in calcite as a function of a dc electric field which removes the symmetry.[70]

With a judicious use of phase matching, several researchers were able to demonstrate second harmonic generation with relatively low powers using cw gas lasers.[71]

The problem of measuring the value of the nonlinear tensor was attacked by several people.[72]

The 1960s witnessed an explosive activity in nonlinear optics. Most of the effects we know today were studied in the period 1960–1970. We mention briefly some of them now.

8.6.1 Optical Rectification

Optical rectification is the development of a steady polarization that accompanies the passage of an intense beam of light through some nonlinear crystal. It was one of the first experiments in nonlinear optics and was observed first by

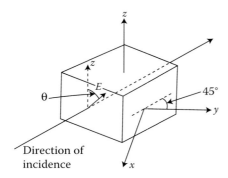

FIGURE 8.10 The experiment of rectification. The z-axis is placed normal to the capacitor plates. (From Bass et al., *Phys. Rev.* **138**, A534 (1965). With permission.)

the group of Franken (Bass et al.[73]), passing radiation from a ruby laser through potassium dihydrogen phosphate (KDP) or potassium dideuterium phosphate (KDdP) crystals. A high-power ruby laser beam passing through the crystal generated a quasi-DC polarization in the nonlinear medium that could be detected as a DC potential difference across the plate of a capacitor within which the nonlinear crystal was placed (**Figure 8.10**). The group verified the prediction that the constant which describes the rectification of light is the same as the Pockels constant, as both effects are derivable from a free energy contribution proportional to the light intensity and the dc electric field.

The results and the effect were discussed at length in a subsequent paper.[74] Later, Ward made the absolute measurement of the optical rectification coefficient.[75]

The presence of magnetic effects was also found. Van der Ziel, Pershan, and Malmstrom[76] performed an experiment in which radiation from a Q-switched ruby laser was circularly polarized by a quartz $\lambda/4$ plate and sent through a $Eu^{+2}:CaF_2$ crystal, a number of diamagnetic glasses, and several organic and inorganic liquids. A magnetization was detected by a pickup coil wound on a cylinder containing the material. This was the first observation of an optically induced magnetization in a nonabsorbing material through an effect which could be called the inverse Faraday effect. In the ordinary Faraday effect, a rotation of the plane of polarization of linear polarized light in the presence of a magnetic field is produced.

8.6.2 Optical Mixing

Optical mixing of the emission of two ruby lasers with different frequencies was first observed by Franken et al.[77] The experiment was initially devised considering mixing a neodymium–YAG and a ruby laser but was realized using two ruby lasers both rented from Trion which were taken at different temperatures (room and liquid nitrogen, respectively). The mixing was detected with a 1894 Hilgher spectrograph and a photographic plate, because Franken had no photomultiplier at that time. In the paper, they also predicted optical rectification which demonstrated soon as discussed previously.

The same year, Smith and Braslau[78] demonstrated the mixing of ruby laser radiation with both the green and yellow lines from a mercury arc lamp in a KDP crystal.

8.6.3 Parametric Oscillation and Amplification

These processes, well known in the microwave range, were extended to optical frequencies by several researchers.[79] Difference frequency generation in KH_2PO_4 with significant parametric amplification was obtained by Wang and Racette.[80] Other results were obtained from Akhmanov et al.[81] The first demonstration of optical parametric oscillation is by Giordmaine and Miller.[82] Early theoretical analyses of optical parametric oscillations were made by R H Kingston[83] and N M Kroll.[84] A review of the matter was written in 1969 by Harris.[85]

The second-order nonlinear medium can be used to mix two optical waves of different frequencies and generate (among other things) a third wave at the difference frequency or at the sum frequency. The former process is called *frequency down-conversion*, whereas the latter is known as *frequency up-conversion* or *sum-frequency generation*. Second harmonic generation is a degenerate special case of a sum-frequency generation.

How a frequency ω_3 split into two different frequencies ω_1 and ω_2 according to the relation $\omega_3 = \omega_1 + \omega_2$ or $\omega_3 = \omega_1 - \omega_2$ is governed by the conditions of phase matching that select the process which better matches the phases of the waves.

These parametric processes are known also as *three-wave mixing*. A number of effects can be obtained.

In an *optical parametric amplifier (OPA)*, waves at frequencies ω_1 and ω_3 interact so that wave ω_1 grows, and in the process an auxiliary wave at a new frequency ω_2 is created. The device operates as a coherent amplifier at frequency ω_1. Wave 3, called the *pump*, provides the required energy, whereas wave 2 is known as the *idler* wave. The amplified wave is called the *signal*. Clearly, the gain of the amplifier depends on the power of the pump.

With proper feedback, the parametric amplifier can operate as a *parametric oscillator*, in which only a pump wave is supplied.

In the phenomenon of *spontaneous parametric down-conversion*, the only input to the nonlinear crystal is the pump wave 3, and down-conversion to the lower-frequency waves 2 and 1 is spontaneous. The frequency and phase-matching conditions lead to multiple solutions, each forming a pair of waves 1 and 2 with specific frequencies and directions. The down-converted light takes the form of a cone of multispectral light.

If the nonlinear crystal is located between high reflectivity mirrors at the signal frequency one obtains an optical parametric oscillator (OPO). The first OPO was demonstrated in 1965 by Giordmaine and Miller[86] who used a crystal of $LiNbO_3$ in the wavelength range 0.97–1.15 µm with optical pumping at 0.529 µm obtained by harmonic generation from a $CaWO_4$:Nd^{3+} Q-switched laser. The Q-switching was made with a rotating prism and gave an output of several multiple pulses of 15–40 ns; the efficiency of conversion of the second harmonic was 5%.

The experimental disposition is shown in **Figure 8.11**. The nonlinear crystal 2 had entrance and exit surfaces made plane parallel and coated with dielectric layers for peak reflectivity at 1.058 µm. In Figure 8.11, the beam direction and

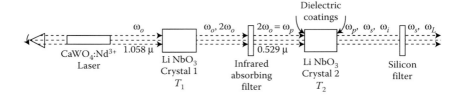

FIGURE 8.11 Layout of the OPO. (From J A Giordmaine and R C Miller, *Phys. Rev. Lett.* **14**, 973 (1965). With permission.)

the outgoing normal to the diagram are taken as x- and z-directions, respectively. The extraordinary pump field E_{pz} was coupled to ordinary signal and idler fields E_{Sy} and E_{iy} through the nonlinear polarization components. The frequencies of the signal and idler are determined by the phase-matching condition

$$k_S + k_i = k_p,$$

which was obtained by varying the temperature of crystal 2. **Figure 8.12** shows the obtained signal and idler wavelength as a function of the temperature.

The output power at 9840 Å was about 15 W.

The OPOs entered the market as efficient and reliable sources only in the 1980s when nonlinear crystals of the exceptional optical quality required became available.

The *Journal of the Optical Society of America* dedicated two issues (September and November) to the development of parametric oscillation and amplification in 1993.[87]

Parametric generation can be discussed starting from Equations 8.35 and 8.36, calling E_3 the pump and E_1 and E_2 the signal and idler, respectively, and performing the following changes:

$$A_i(z) = \sqrt{(n_i/\omega_i)}E_i,$$

$$\Delta k = k_3 - k_1 - k_2,$$

$$g = \sqrt{(\mu_o/\varepsilon_o)}(\omega_1\omega_2/n_1n_2)\, d_{jik}\, E_3(z),$$

$$\alpha_i = \sigma_i\sqrt{(\mu_o/\varepsilon_o)}.$$

We may write

$$dA_1/dz = (-1/2)\alpha_1 A_1 - i(g/2)A_2^* \exp(-i\Delta kz),$$
$$dA_2^*/dz = -(1/2)\alpha_2 A_2^* + i(g/2)A_1\exp(i\Delta kz), \qquad \textbf{(8.48)}$$
$$dA_3/dz = -(1/2)\alpha_3 A_3 - i(g/2)A_1 A_2\exp(i\Delta kz).$$

In the nondepleted pump approximation, we take $A_3(z) = A_3(0)$ and write the first two equations (8.48) as $(g' = g(z = 0))$

$$dA_1/dz = (-1/2)\alpha_1 A_1 - i(g'/2)A_2^*\exp(-i\Delta kz) \qquad \textbf{(8.49)}$$
$$dA_2^*/dz = -(1/2)\alpha_2 A_2^* + i(g'/2)A_1\exp(i\Delta kz).$$

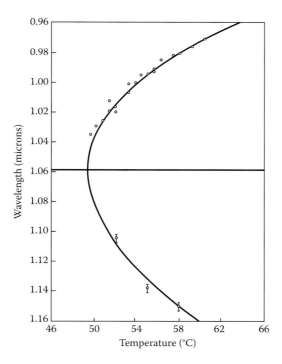

FIGURE 8.12 Observed emission frequencies tuning the temperature. The solid line represents the frequency range calculated from phase matching. (From J A Giordmaine and R C Miller, *Phys. Rev. Lett.* **14**, 973 (1965). With permission.)

In the case in which both signal and idler are present at the input with amplitudes $A_1(0)$ and $A_2(0)$ with no losses $\alpha_i = 0$, the solutions are

$$A_1(z)\exp(i\Delta kz/2) = A_1(0)\left[\cosh(bz) - i(\Delta k/2b)\sinh(bz)\right]$$
$$- i(g'/2b)A_2^*(0)\sinh(bz),$$
$$A_2^*(z)\exp(-i\Delta kz/2) = A_2^*(0)[\cosh(bz) - i(\Delta k/2b)\sinh(bz)$$
$$+ i(g'/2b)A_1(0)\sinh(bz),$$

(8.50)

where

$$b = (1/2)\sqrt{[(g')^2 - (\Delta k)^2]},$$
$$\Delta k = k_3 - k_1 - k_2.$$

In the simple case of a parametric amplifier, we have a single input, for example, $A_1(0)$. Putting $A_2(0) = 0$ and considering the phase-matched case $\Delta k = 0$, we have

$$A_1(z) = A_1(0)\cosh(g'z/2),$$
$$A_2^*(z) = iA_1(0)\sinh(g'z/2).$$

(8.51)

The increase in power of the signal and idler wave is at the expense of the pump wave. It is possible to show using Equation 8.48 that

$$-d(A_3 A_3^*)/dz = d(A_1 A_1^*)dz = d(A_2 A_2^*)/dz. \qquad (8.52)$$

Since $A_i A_i^*$ is proportional to the photon flux at ω_i, Equation 8.52 is a statement of the fact that for each photon added to the signal wave (ω_1) one photon is added to the idler wave (ω_2) and one photon is removed from the pump wave (ω_3). Since $\omega_3 = \omega_1 + \omega_2$, energy is conserved. Relation 8.52 can be extended by integration to the whole interaction volume in which case the changes in total power between the input and output planes are related by

$$-\Delta(P_3/\omega_3) = \Delta(P_1/\omega_1) = (P_2/\omega_2), \qquad (8.53)$$

where P represents the beam power. Equation 8.53 is known as the *Manley–Rowe relation*.[88]

The first report of optical parametric oscillation involving a cavity resonant at both the signal and idler wavelengths was presented by Bjorkholm.[89] In the previous experiments,[90] the maximum conversion efficiency was about 1%. In Bjorkholm's experiment, with both the signal wave at 1.04 µm and the idler wave at 2.08 µm resonant, about 22% of the maximum power was converted into signal power.

8.6.4 Third-Order Effects

When in Franken's paper the second harmonic spot appeared canceled, he remembers[91] that as a joke he wanted to prepare a quick letter to send *Physical Review Letters* showing the plate with an arrow pointing at 220 nm as evidence of the first unambiguous failure to observe third harmonics. Later, third harmonic was obtained first by Terhune, Maker, and Savage[92] in calcite and then by Maker and Terhune.[93] This last paper presented the results of a series of experiments in which a giant ruby laser was used to study third-order effects. Spatial symmetry and frequency dependence of the nonlinear tensor were also discussed. It was an excellent presentation of the effects studied up to 1964.

In media possessing centrosymmetry, the second-order nonlinear term is absent and the dominant nonlinearity is then of third order. Such media are usually called *Kerr media*. They respond to optical fields by generating third harmonics and sum and differences of triplets of frequencies. The third-order susceptibility describes, in general, a coupling between four electromagnetic waves.

The real part of $\chi^{(3)}$ describes, among others, the optical Kerr effect, self-focusing and self-defocusing. The imaginary part is responsible for two-photon absorption and stimulated Raman scattering.

The presence of a component of polarization at the frequency 3ω indicates that third-harmonic light is generated. However, in most cases, the energy conversion efficiency is low. Indeed, third-harmonic generation is often achieved via second-harmonic generation followed by sum-frequency generation of the fundamental and second-harmonic waves.

Third harmonic in noble gases was obtained in 1967 by New and Ward.[94] Reintjes et al.,[95] nearly 10 years later, generated the fifth and seventh harmonics of a laser pulse at 266 nm obtained by two consecutive frequency doublings from a Nd glass laser at 1.06 μm focused into helium gas. It was in the late 1980s that the availability of intense, short-pulse laser fields made it possible to observe high-order harmonics, as we will discuss later.

The optical Kerr effect in which the refractive index is a function of light intensity according to

$$n = n_o + n_2 I \tag{8.54}$$

was first observed by Mayer and Gires.[96] The coefficient n_2 may be positive or negative.

Measurements of changes in the real part of the index of refraction were reported independently by P D Maker et al.[97] They showed that for elliptically polarized light in isotropic centrosymmetric materials in addition to a polarization-independent change in velocity, a rotation of the axes of the vibrational ellipse as a function of distance occurred.

The refractive index change induced by the electromagnetic field produces a back-action on the field and influences its propagation characteristics producing a number of effects like self-focusing ($n_2 > 0$) or self-defocusing ($n_2 < 0$), self-phase modulation, self-trapping, spatial solitons, temporal self-phase modulation and self-chirping, self-compression or decompression, self-dispersion, self-steepening, temporal solitons.

In 1962, self-focusing in plasmas was proposed.[98] The process is described by a nonlinear parabolic equation, the solution of which describes a self-supporting waveguide[99] (so-called spatial soliton). Ostrovsky[100] in 1963 predicted shock waves and later soliton-envelope waves in an unbounded medium.[101] A bit earlier, almost simultaneously, Talanov,[102,103] Askarjan,[104] and Townes et al.[105] discussed in various ways self-focusing. Townes et al. proposed self-trapping giving a numerical solution for a two-transverse dimension beam. They calculated the critical power, the mechanisms for the nonlinearity, and used the effect to explain anomalous Raman gain and optical damage. In these papers, the one-transverse-dimension hyperbolic secant soliton solution was given. The theory was further developed the following year by Talanov.[106]

In the self-focusing case, there is the possibility that at a certain power level of the cw beam, diffraction and nonlinearity would balance such that the intensity profile remains unchanged on propagation. Such a situation is referred as *spatial soliton*. The term soliton was coined in 1965 by Zabusky and Kruskal.[107]

Spatial solitons are produced when an intense Gaussian optical beam travels through a substantial thickness of nonlinear homogeneous medium, instead of a thin sheet. Due to the light intensity distribution of the Gaussian beam, the refractive index is altered non-uniformly so that the medium can act as a graded-index waveguide. Thus, the beam can create its own waveguide. If the intensity of the beam has the same spatial distribution in the transverse plane as one of the modes of the waveguide that the beam itself creates, the beam propagates self-consistently without changing its spatial distribution. Under these conditions, diffraction is compensated by self-phase modulation, and the beam is confined to its self-created waveguide.

The possibility of a balance between diffraction and self-focusing was found in 1964 even before the word "soliton" was coined. Chiao and Townes[108] referred it as self-trapping. However, the self-trapped solution is unstable. A slight increase in the beam power leads to catastrophic self-focusing and filamentation.[109] The problem is that there are two transverse dimensions in which diffraction occurs. Indeed, if diffraction along one transverse dimension is ignored, stable solutions are allowed. There are two situations in which the diffraction effects can be assumed to remain confined in one transverse dimension. In one case, the spot size is so large along y compared with that along x that the beam diffracts little along the y-axis. In the other case, a waveguide is formed along the y-direction so that beam diffraction occurs only along the unguided x-direction.

The pertinent equation to describe the phenomena is the nonlinear Schroedinger equation, which is an equation of the kind

$$2ik_j(\partial E_j/\partial z) + (\partial^2 E_j/\partial x^2) + 2(k_j^2 n_2/n_{oj})(|E_j|^2 + k|E_{3-j}|^2) \cdot E_j = 0, \quad j = 1,2,$$

where $k_j = \omega n_{oj}/c$ is the wave vector, n_{oj} is the linear refractive index at frequency ω_j, n_2 is the nonlinear coefficient of the refractive index, and k is a coupling constant. In the nonlinear contributions to this equation, the first term in parentheses accounts for self-phase modulation, while the second term is responsible for cross-phase modulation.

Stable solutions with a hyperbolic secant amplitude profile

$$u(\xi, X) = N \operatorname{sech}(X) \exp(i\xi/2), \tag{8.55}$$

for $n_2 > 0$, have been seen experimentally in both cases using CS_2 liquid.[110]

In 1965, the first observation of self-focusing was made in liquids[111] and it was experimentally related to anomalous stimulated Raman gain.[112] A brief review of the state of work was presented by Zel'dovich.[113] In 1966, the theory of nonlinear instability was given[114] and the effect observed experimentally.[115] In this same year, multifilament structure in multimode beams was observed and the influence of spectral broadening was considered.[116] This was followed several years later by experiments in which regular filament structure was seen on carefully prepared beams.[117] A great number of papers review the work on self-focusing.[118] More can be found in Kelley.[119]

Solitons in resonant media were studied by S L McCall and E L Hahn.[120] An important step accelerating further research was made by V E Zakharov and A B Shabat.[121] They showed that the equation which describes the evolution of the envelope of the electric field amplitude in a medium with cubic nonlinearity (the nonlinear Schroedinger equation) belongs to the class of fully integrable nonlinear evolution equation which can be solved by application of the inverse scattering transform. This work, together with that of Gardner et al.[122] for the Korteweg–de Vries equation, is an important milestone in the development of soliton theory. After the publication of these ideas, it became evident that there exists a wide class of nonlinear evolution equations which are completely integrable.

The so-called bright solitons propagate as a stationary wave packet of finite extent in a (1 + 1)D nonlinear medium with $n_2 > 0$. In media with $n_2 < 0$, there exist dark solitons.[123] A dark soliton is characterized as a stationary hole on an otherwise uniform plane wave; the solution of the nonlinear Schroedinger equation is

$$u(\xi, X) = N \tanh(X) \exp(i\xi/2).$$

Temporal dark solitons, that is, intensity minimums propagating along a nonlinear fiber on a quasi-cw bright background, were observed experimentally already in 1987.[124] The first observation of a spatial dark soliton was made in 1991.[125]

The simultaneous propagation of a dark and bright spatial solitons of different wavelength (1064 and 532 nm) in a homogeneous Kerr medium of the focusing type, thanks to cross-phase modulation, was demonstrated by M Shalaby and A J Barthelemy.[126]

Many materials with nonlinearities of different origin can be considered for the production of solitons.

Spatial solitons in photorefractive crystals were considered by M Segev et al.[127] and reported experimentally by Iturbe Castillo et al.[128] They have been the object of interest, and a number of different types of solitons were investigated.[129] It was predicted that an intense pump beam can induce focusing of a weak probe beam at a different wavelength that is simultaneously propagating in a Kerr homogeneous material through cross-phase modulation. De La Fuente, Barthelemy, and Froehly demonstrated the possibility to use a soliton beam to induce a stable guiding of a probe beam[130] using a cell filled with carbon disulfide. Soliton waveguiding in lithium niobate was studied by Fazio.[131]

A possible catastrophic consequence of self-focusing is that the optical beam shrinks along the axis as already predicted by Chiao and Townes (*filamentation*)[132] causing irreversible damage to the optical components. On the other hand, the effect on an optical pulse is to cause the peak central region to travel slower than the leading and trailing wings. This affects the carrier wave structure, causing what is known as *self-phase modulation* where the local frequency is lowered ahead of the peak and raised behind it. The result is that the carrier frequency rises through the pulse, a condition known as an "up-chirp." This self-steepening effect was analyzed in detail in 1967,[133] although the effect had already been considered 2 years early by Rosen.[134]

Self-phase modulation has numerous important uses. Its most significant characteristic is the consequent increase in the spectral bandwidth[135] which is exploited for short pulse generation. These broadening effects were first reported by Stoicheff[136] and further studied by Garmire.[137]

The idea of imposing strong self-phase modulation and use negative group velocity dispersion for optical pulse compression originates from a paper by

Fisher et al.[138] A variant is the time *optical solitons*. In the time domain, group-velocity dispersion can be compensated by self-phase modulation. This leads to the formation of *temporal solitons*, which travel without changing shape creating the analogous of the spatial soliton.

In 1965, Zabusky and Krustkal[139] performed numerical simulations of soliton pulse propagation in one dimension.

These early works initiated an intense development of the theory of optical solitons which continues to the present days. It was not until 1973 that Hasegawa and Tappert[140] showed that optical fibers were ideal media for temporal soliton propagation, described by a one-dimensional nonlinear Schroedinger equation, but only in the 1980s, the envelope solitons were experimentally discovered by Mollenauer, Stolen, and Gordon.[141] The most important type of temporal solitons, from the standpoint of technological applications, is fiber solitons.

The phase shift incurred by an optical beam of power P and cross-sectional area A, traveling a distance L in the medium, is

$$\varphi = -n(I)k_o L = 2\pi n(I)L/\lambda_o = -2\pi(n + n_2 P/A)L/\lambda_o, \qquad (8.56)$$

so that with respect to what happens in a linear medium it is altered by

$$\Delta\varphi = -2\pi n_2 LP/\lambda_o A, \qquad (8.57)$$

which is proportional to the optical power P. Self-phase modulation became useful in applications in which light controls light.

Phase modulation may be converted into intensity modulation by employing one of the schemes used in conjunction with electro-optical modulators using, for example, an interferometer (a Mach-Zender, for example).

Other third-order phenomena are *stimulated Brillouin effect*[142] and *stimulated Raman effect* (for an intuitive description of these effects, see page 315).

Stimulated Raman scattering was a serendipity discovery made by E Woodbury and W K Ng. They were using two different types of detectors to measure light from a Q-switched laser. When they operated the laser without Q-switching, both detectors gave the same result, but when they activated the Q-switching, the detector readings were inconsistent. They looked at the nitrobenzene-filled Kerr cell used to obtain the Q-switching and found it was the seat of several new radiations at lower frequency that exhibited the properties of laser light. The new lines were narrowly collimated, had sharp widths, and only came into being when the ruby laser light reached a certain threshold output power.[143] Gisela Eckhardt and Robert Hellwarth eventually succeeded in identifying the effect as a new species of laser emission: "stimulated Raman scattering"[144] from the nitrobenzene.

A deliberate effort to produce stimulated Raman scattering had been initiated by H J Zeiger and P E Tannenwald at MIT but Hughes scientists were the first ones to discover it by accident.

When different liquids were inserted in the laser resonator, different frequencies were emitted. They were displaced from the ruby frequency by an amount corresponding to the vibration frequency of the liquids. In addition, the displaced radiation had the following characteristics: (1) only the frequencies belonging to the sharpest and the most intense spontaneous Raman lines showed up, (2) it had a definite threshold of excitation, (3) higher-order Stokes radiations occurred at exact harmonics of the first vibration transition. The stimulated Raman spectra of many liquids,[145] solids,[146] and gases[147] were reported. Terhune[148] put the cell outside the laser resonator and was able to detect the anti-Stokes radiation emitted forward in a characteristic ring pattern.

Also stimulated Brillouin scattering was detected by Chiao et al.[149]

G Bret and G Mayer,[150] discussing some of the first measurements on stimulated Raman scattering, made recourse to a paper by Ehrenfest and Einstein.[151] In 1923, W Pauli[152] had used the recently discovered Compton effect to explain the long unresolved problem that classical electromagnetic theory is not capable of providing an interaction mechanism leading to the establishment of thermal equilibrium between radiation and free electrons. Ehrenfest and Einstein completed the discussion showing that since the Compton scattering process involves both the disappearance (absorption) and the appearance (emission) of a light quantum, both of whose directions are fixed by conservation laws, the interaction probability should be of the form

$$(b\rho)(a' + b'\rho'),$$

where ρ and ρ' are the black-body spectral distributions corresponding to the incident and scattered frequencies ν and ν' and in which the first factor involves an induced term, while the second factor involves both a spontaneous and an induced term.

Now Bret and Mayer, representing the Raman effect as a process where one photon of frequency ν_o is transformed into one photon of different frequency ν_R by a system which takes or gives the corresponding energy difference, using Eherenfest and Einstein formula wrote the number of transformations $\nu_o \rightarrow \nu_R$ per time unit as

$$\mathrm{d}n/\mathrm{d}t = N\rho(\nu_o)[A + B\rho(\nu_R)],$$

where N is the number of systems in the state of energy which allows the transformation $\nu_o \rightarrow \nu_R$; $\rho(\nu_o)$ and $\rho(\nu_R)$ are the energy densities at frequencies ν_o and ν_R, respectively. The ratio B/A, they assumed, does not depend on ν_R. Applying this formula to the Raman effect, the term A may be interpreted as giving rise to the normal Raman effect, while the term B describes the induced Raman effect.

The theory of the effect was then discussed by R W Hellwarth[153] and E Garmire, F Pandarese, and C H Townes.[154] A much more refined theory of stimulated Brillouin and Raman scattering was done by Bloembergen[155] in 1965.

Cross-phase modulation. Two waves traveling in a third-order nonlinear medium produces a change in refractive index which depends on the sum of their intensities and therefore the phase shift encountered by wave 1 is modulated by the intensity of wave 2. This phenomenon is known as *cross-phase modulation.* It was studied in optical fibers.[156]

Four wave mixing in a third-order medium. The response of a medium to a superposition of three plane waves of angular frequencies ω_1, ω_2, and ω_3, and wavevectors k_1, k_2, and k_3 with a field

$$\mathrm{E} = \sum(1/2)E(\omega_q) \exp[j(\omega_q t + k_q \cdot r)], \qquad (8.58)$$

where $\omega_{-q} = -\omega_q$ and $E(-\omega_q) = E^*(\omega_q)$ can be described using a nonlinear polarization

$$P_{NL} = (1/8)\chi^{(3)}\sum E(\omega_q)E(\omega_r)E(\omega_l)\exp[j(\omega_q + \omega_r + \omega_l)t + (k_q + k_r + k_l) \cdot r]. \quad (8.59)$$

Thus, P_{NL} originates radiations which are the sum of harmonic components of frequencies $\omega_1, \ldots, 3\omega_1, \ldots, 2\omega_1 \pm \omega_2, \ldots \pm \omega_1 \pm \omega_2 \pm \omega_3$. The amplitude $P_{NL}(\omega_q + \omega_r + \omega_l)$ of the component of frequency $\omega_q + \omega_r + \omega_l$ can be determined by adding appropriate permutations of q, r, and l in Equation 8.59. For example, $P_{NL}(\omega_1 + \omega_2 - \omega_3)$ involves six permutations,

$$P_{NL}(\omega_1 + \omega_2 - \omega_3) = 6\chi^{(3)}E(\omega_1)E(\omega_2)E^*(\omega_3) \exp[j(\omega_1 + \omega_2 - \omega_3)$$
$$t + (k_1 + k_2 - k_3) \cdot r]. \quad (8.60)$$

Equation 8.60 indicates that four waves of frequencies ω_1, ω_2, ω_3, and ω_4 are mixed by the medium if $\omega_4 = \omega_1 + \omega_2 - \omega_3$, and $k_4 = k_1 + k_2 - k_3$ or

$$\omega_1 + \omega_2 = \omega_3 + \omega_4 \quad (8.61)$$

and

$$\mathbf{k}_1 + \mathbf{k}_2 = \mathbf{k}_3 + \mathbf{k}_4. \quad (8.62)$$

These equations constitute the frequency and phase-matching conditions for *four wave mixing* (FWM). Several processes occur simultaneously, all satisfying the frequency and phase-matching conditions. Three-wave mixing can be considered a partially degenerate case for which two of the four waves have the same frequency.

Four-wave mixing was discussed by Maker e Terhune.[157] It describes a mixing process in which three propagating light waves interact nonlinearly in a medium and generate a fourth wave. The third-order nonlinear susceptibility that governs the process should naturally exhibit resonances which are characteristics of the medium and can be probed through the resonant enhancement of the four-wave mixing output. The process was used for spectroscopic studies.[158]

In 1965, H Kogelnik[159] pointed out that conventional holographic techniques could be used for imaging through static inhomogeneous media. Experiments were performed by several researchers.[160] This may be considered the first step in an important new technique: *phase-conjugated optics*.

The first milestone in phase-conjugated optics was the demonstration, in the Soviet Union, by Zeldovich et al.[161] and Nosach et al.[162] of the cancellation of propagation distortion by stimulated backward scattering (Brillouin, Raman, or Rayleigh). In the experiment, a ruby laser beam, which was distorted by passing through a roughened plate of glass, was used as a pump for stimulated Brillouin scattering in pressurized methane gas (**Figure 8.13**). The backward propagating stimulated beam passed in reverse through the distorting plate and emerged from it with an undistorted wavefront similar to that of the pumping beam before impinging on the distorting plate.

The production of phase-conjugated waves in these processes was considered by several researchers.[163] Yariv proposed and analyzed three-wave mixing in crystals for overcoming image "loss" by modal phase dispersion in multimode fibers,[164] Hellwarth[165] suggested four-wave mixing for phase

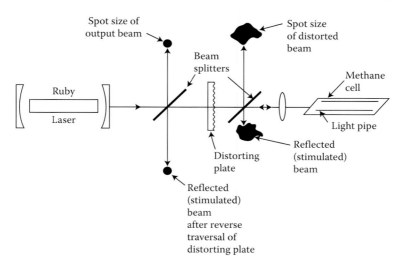

FIGURE 8.13 The experimental setup used by Zeldovich et al. using stimulated Brillouin scattering. (From B Y Zeldovich et al., *Sov. Phys. JETP* **15**, 109 (1972).)

conjugation, and Yariv and Pepper[166] showed that the four-wave process for phase conjugation was also capable of amplifying an incoming wave as well as rendering its complex conjugate version.

The first observations of phase conjugation by four-wave mixing were reported by Jensen and Hellwarth[167] and by Bloom and Bjorklund[168] using CS_2 as the nonlinear medium. Parametric amplification and oscillation in four-wave mixing was reported by Bloom et al.[169] and by Pepper et al.[170] Phase conjugation by three-wave mixing in crystals was demonstrated by Avizoms et al.[171] Wang and Giuliano[172] demonstrated the ability of the stimulated Brillouin scattering to restore high spatial frequencies. Optical phase con-

jugation and image restoration by stimulated Raman scattering was demonstrated by Zeldovich et al.[173]

Conjugated mirrors in laser resonators were demonstrated by J AuYeung et al.[174]

Phase conjugation may be made clear contrasting the process of reflection from a normal mirror and a conjugated mirror. While the normal mirror reflects the incoming wave with an angle equal to the incidence angle, the conjugated mirror reflects back the wave which retraces its initial path backward. This is shown in **Figure 8.14** in the case that

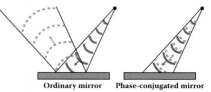

Ordinary mirror Phase-conjugated mirror

FIGURE 8.14 The different behavior of an ordinary and a phase-conjugated mirror as explained looking at a divergent wave striking the two mirrors. Full line wavefronts are the impinging wave, while the reflected wave is with dotted wavefronts. The arrows indicate the motion direction. On the left is an ordinary mirror and on the right is the conjugated one.

the spherical wave emitted by a point (diverging) is striking an ordinary mirror at an angle θ. In this case, the wave leaves the mirror at an angle $-\theta$ continuing to diverge. In contrast, the same wave striking the conjugate mirror is converted into a convergent wave that retraces the path of the incident wave and focalizes in the starting point.

To understand in the simplest way what happens we may consider that the frequency-matching condition 8.62 is automatically satisfied when all four waves are of the same frequency ω. The process is called degenerate four-wave mixing.

Assuming now that two of the waves (e.g., waves 3 and 4) are uniform plane waves traveling in opposite directions ($k_4 = -k_3$)

$$E_3 = A_3 \exp(jk_3 \cdot r), \quad E_4 = A_4 \exp(-j\, k_3 \cdot r),$$

for the field E_2, Equation 8.60 becomes

$$P_{NL}(\omega = \omega + \omega - \omega) = 6\chi^{(3)} A_3 A_4 A_1^*(\omega) \exp\left[j(\omega t + k_1 \cdot r) \right].$$

This term corresponds to a source emitting an optical wave (wave 2) of complex amplitude

$$A_2(r) \sim A_3 A_4 A_1^*(r).$$

Since A_3 and A_4 are constants, wave 2 is proportional to the conjugated version of wave 1. The conjugate wave is identical to the probe wave except that it travels in the opposite direction. The phase conjugator is therefore a special mirror that reflects the wave back onto itself without altering its wavefront.

Figure 8.15 shows an application of the complex conjugation to the compensation of phase distortions.

Phase conjugation is analogous to time reversal. Consider an optical wave of frequency ω moving in the $+z$-direction

$$E = \psi(x,y,z)e^{i\omega t},$$

where

$$\psi(x,y,z) = A(x,y)e^{i(-kz + \delta)}.$$

The phase-conjugated wave is

$$E_c = \psi(x,y,z)^* e^{i\omega t} = A(x,y)e^{i(kz - \delta)}e^{i\omega t}.$$

That is the phase-conjugated wave contains the complex conjugate of only the spatial part, leaving the temporal part unchanged. The conjugate wave corresponds to a wave moving in the $-z$-direction, with the phase δ reversed relative to the incident wave. This is equivalent to leaving the spatial part of E unchanged and reversing the sign of t, which is a *time reversal* process.

Moreover, the conjugate wave may carry more power than the probe wave because the intensity of the conjugate wave 2 is proportional to the product of the intensities of waves 3 and 4.

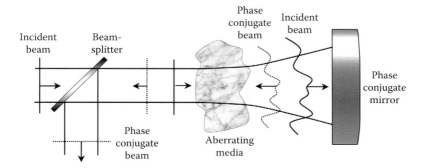

FIGURE 8.15 An example of compensation of the aberrations due to a phase distorting medium by a wavefront reflection on a phase-conjugated mirror. (From J P Huignard and A Brignon, *Phase Conjugate Laser Optics*, edited by A Brignon and J P Huignard, John Wiley & Sons, Inc, 2004.)

First good general papers were written by Yariv[175] and Hellwarth.[176] Also a good collection of papers referring to the first years of phase conjugation is the book by R A Fisher, *Optical Phase Conjugation*.[177] Applications of phase conjugation were considered in many papers. Here we remember Giuliano[178] and Huignard and Brignon.[179]

8.6.5 Down-Conversion and Entanglement

A particular case of parametric interaction is the process of down-conversion in which a pump photon is split into two photons obeying energy and momentum conservation laws (see **Figure 8.16**). In what is described as type II down-conversion,[180] the signal and idler photons have orthogonal polarizations. In the process, energy and momentum must be conserved, and birefringence effects cause the photons to be emitted along two intersecting cones, one of the ordinary (o) ray and the other of the extraordinary (e) ray (see Figure 8.16).

If one of the pair is detected at any time, then its partner is known to be present.

If we measure the polarization of a photon in a point of the extraordinary cone of **Figure 8.17a**, we always measure a vertical polarization and we know that in the corresponding point for which phase matching is fulfilled, the polarization of the incoming photon is horizontal, and if we repeat the experiment the two polarizations will always be vertical on the extraordinary cone and horizontal on the ordinary one.

There are, however, two points clearly visible in **Figure 8.17b**, in which the two polarizations are contemporarily present. A measure in one of these points will yield vertical or horizontal polarization, indifferently. In these points if we measure a vertical polarization in one point we know that in the other point polarization will be horizontal; but we could also have measured a horizontal polarization instead (with a 50:50 probability) and in the other

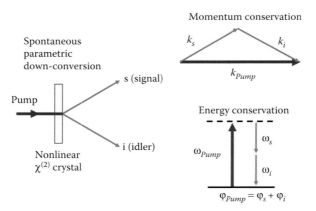

FIGURE 8.16 Scheme of spontaneous parametric down-conversion and momentum and energy conservation.

point polarization would have been vertical. The wave function of the total system of the two produced photons will collapse in one of the two situations just in virtue of the measurement and before it any of the two polarizations will equally be probable.

The structure of Figure 8.17b repeats itself at different colors, as shown in **Figure 8.18**.

In the case of the signal-idler pair of photons emitted in spontaneous parametric down-conversion, the two photons may be entangled in wavelength, momentum, angular momentum, and frequency, as well as polarization. Entanglement received precocious attention in a famous Einstein paper[181] in which the possibility that a measurement on a particle in some place at some time made another particle far away at the same time to collapse in some state was taken as an indication that quantum theory was not complete. This nonlocal property is now recognized as the most characteristics aspect of quantum mechanics.

Entanglement is further discussed in Strekalov et al.[182]

In parametric interactions, each time a pump photon creates an idler photon, it must also creates a signal photon. Thus, if a photon is detected at the idler mode, one can expect the presence of a corresponding photon at the signal mode. This was verified experimentally in parametric fluorescence by Hong and Mandel.[183] In this way, single photons are created and antibunching should occur. Such scheme of generating antibunched light was studied theoretical by Stoler and Yurke.[184]

For these sources, the creation of photon pairs is probabilistic, rather than deterministic. However, because the photons are created in pairs, one photon (the heralding photon) can be used to herald the creation of the other photon. The second detector must be activated only whenever the first one has detected a photon and not whenever a pump pulse has been emitted, therefore circumventing the problem of empty pulses. Phase matching allows one to choose

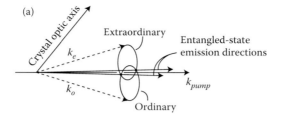

(a)

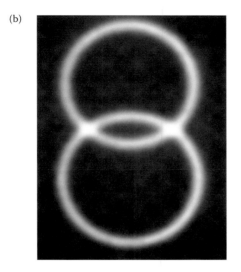

(b)

FIGURE 8.17 (a) The spontaneous down-conversion cones present with type-II phase matching. (b) Photograph of the down-conversion photons, through an interference filter at 702 nm. (From P G Kwiat et al., *Phys. Rev. Lett.* **75**, 4337 (1995). With permission.)

the wavelength and determine the bandwidth of the down-converted photons. The latter is in general rather large and varies from a few nanometers up to some tens of nanometers. For the nondegenerate case, one typically gets a bandwidth of 5–10 nm, whereas in the degenerate case (where the central frequency of both photons is equal) the bandwidth can be as large as 70 nm.

This photon pair creation is very inefficient; typically it takes some 10^{10} pump photons to create one pair in a given mode. The

FIGURE 8.18 The intersecting beams of down-conversion at different colors. (Courtesy of Paul Kwiat and Michael Reck from the University of Vienna.)

number of photon pairs per mode is thermally distributed within the coherence time of the photons and follows a Poissonian distribution for larger time windows.[185]

With a pump power of 1 mW about 10^6 pairs per second can be collected in single-mode fibers. Accordingly, in a time window of roughly 1 ns, the conditional probability of finding a second pair, having already detected one, is $10^6 \times 10^{-9} = 0.1\%$. In the case of continuous pumping, this time window is given by the detector resolution. Accepting, for example, 1% of these multipair events, one can generate 10^7 pairs per second using a realistic 10 mW pump. One must observe however that also in down-conversion there is a probability that two pairs be emitted contemporarily.

First sources of high-intensity polarization, entangled photons were produced by Kwiat et al.[186]

8.7 Nonlinearities in Optical Fibers

During the 1960s, optical fibers exhibited very high losses. In addition, the silica glass, which was the main component, had poor nonlinearity. Therefore, the fibers did not appear a useful medium for nonlinear experiments. The situation changed completely in 1970 when losses were drastically reduced[187] thanks to previous work by Charles Kao[188] (b. 1933) who was awarded the 2009 Physics Nobel Prize for *groundbreaking achievements concerning the transmission of light in fibers for optical communication.*

Conventional optical fibers consist of a glass core with a diameter ranging from 5 to 120 µm, surrounded by a cladding made of a glass with a slightly different refractive index of the core, or an organic polymer. The refractive index of the core is greater than that of the cladding, and light is guided in the core by internal reflection.

In low-loss optical fibers, nonlinear effects may be of importance because of the strong confinement of the field within the core which means that high field strengths may be obtained already with modest input powers. Moreover, the very low transmission loss of the fibers in certain infrared wavelength ranges allows nonlinear effects to become significant during propagation over long path lengths.

From the early days in the development of optical fiber technology, it was recognized that nonlinear optical process could present ultimate practical limitations on the range and on the transmission capacity of communications systems.[189]

The most interesting effects are the Raman and Brillouin stimulated effects which may limit the power that can be transmitted. Self-phase modulation gives a broadening of the transmitted pulses. Stolen and coworkers at Bell Laboratories demonstrated all these effects.[190]

The nonlinear process with the largest gain and hence the lowest threshold in fibers is stimulated Brillouin scattering. Classically it is described as a three-wave interaction involving the incident light wave (pump), a generated acoustic wave and the scattered light wave. The pump is scattered by the acoustic wave. The scattered wave has a Doppler shifted frequency because the acoustic wave acts as a moving diffraction grating. The scattered light and the pump together form a standing wave that move slowly because the two waves have a slightly different frequency. This optical standing wave leads, via the process of electrostriction, to an increase in density in the maxima and decrease in the nodes, and this mobile density structure coincides with the acoustic wave from which the initial scattering started, so that a positive feedback situation exists. The process may be initiated by the acoustic waves already existing in the medium. The pump light produces the scattered wave which, via the standing wave that is formed, drives the acoustic wave and hence stronger scattering occurs, and so on giving a stimulated process.

Due to conservation of energy and momentum, the process is maximized when the scattered wave travels in a direction opposite to the pump. Thus, a significant part of the optical power traveling in the fiber may be converted into the scattered wave which travels backward toward the transmitter. The scattered wave (known as the *Stokes wave*) is shifted to a lower frequency with respect to the pump, by an amount equal to the acoustic frequency

$$\nu_A = 2n\nu_A/\lambda,$$

where ν_A is the scattered wave frequency, ν_A is the acoustic wave velocity, n is the refractive index, and λ is the optical wavelength.

Stimulated Brillouin scattering in optical fibers was first observed by Ippen and Stolen in 1972.[191] They used 5 and 20 m lengths of fiber excited by a pulsed narrow band xenon laser at 535.5 nm. The threshold for the stimulated Brillouin effect was found to be in the region of 1 W. Later experiments in the infrared region (1.3 and 1.5 μm) using multi-km lengths of fiber showed the effect can show up at a few mW power.[192]

In a typical fiber, the threshold power is around 1–5 mW.

The stimulated Brillouin scattering is detrimental because it produces a strong attenuation, in some cases it originates multiple frequency shifts, and creates a strong coupling with the backward traveling wave. This wave if reaches sufficient power, produces fracture of the fiber input face. The gain can be quite different depending whether the pump bandwidth $\Delta\nu_P$ is much less or greater than the Brillouin linewidth $\Delta\nu_B$. In the limit when $\Delta\nu_P \gg \Delta\nu_B$ the stimulated Brillouin scattering is reduced by the ratio $\Delta\nu_B/\Delta\nu_P$. The gain was the object of an extensive research.[193]

The peak gain for stimulated Raman scattering is two to three orders of magnitude lower than for stimulated Brillouin scattering; nevertheless, it may be the dominant nonlinear process when a broadband or multimode optical signal is used.[194]

Similar to the stimulated Brillouin scattering, the simulated Raman scattering can be viewed as a coupled three-wave process in which the pump wave creates a frequency down shifted Stokes wave and a highly damped material excitation wave, which in this case corresponds to the vibrational mode of SiO_2. In the case of the stimulated Raman scattering, the gain is in the region between 0 and 1000 cm^{-1} of shift. The maximum gain is for a shift around 440 cm^{-1} at 1 μm for a linearly polarized wave.[195]

Because of the very wide gain bandwidth for stimulated Raman scattering, the threshold power would apply equally to a narrow linewidth cw laser source or one emitting pulses with duration as short as femtoseconds.

The Raman response function of silica core fibers was analyzed by Stolen et al.[196]

Conversely stimulated Raman scattering in fibers provides a very useful source of widely tunable infrared radiation. For example, using a high power Q-switched mode-locked Nd:YAG laser (1.06 µm) as the pump, the light emerging from the far end of a few tens of meters of monomode fiber comprises a spectral continuum extending from about 1.1 µm to beyond 1.6 µm. This emission is generated by a cascade of stimulated Raman scattering orders in which the first Stokes wave pumps a stimulated Raman scattering process to generate the frequency shifted second Stokes wave, which can pump a third Stokes wave, and so on. A review of Raman fiber lasers with reference also to Russian work was made by Dianov and Prokhorov in 2000.[197]

The Raman gain in fiber can be used to amplify optical signals as we already said in Chapter 7.[198]

In the case that several optical frequencies are launched into a fiber simultaneously (e.g., in a wavelength-division-multiplexed system), three waves can interact via the third-order nonlinear susceptibility to generate a fourth wave at a new frequency that could give rise to crosstalk. The effect was studied by Stolen and Bjorkholm.[199]

Surprisingly because of inversion symmetry of silica, efficient second harmonic generation was discovered in optical fibers.[200]

Solitons in optical fibers were first made by Mollenauer.[201] Because of dispersion an optical pulse propagating in an optical fiber spread over the time as the pulse propagates through the fiber. Properly balancing dispersion and self-phase modulation through the Kerr effect the pulse conserves its initial shape as if the medium were linear and dispersionless. Pulses propagating in this regime are called *solitons (temporal solitons)*. A major impediment to soliton propagation is, however, fiber loss. As the soliton propagates along the fiber, its power decreases and breaks the balance between dispersion and nonlinearity. A way to overcome this is to amplify the pulse. Mollenauer regenerated the pulse using Raman amplification, and succeeded to propagate solitons over more than 4000 km.[202] The use of an Erbium doped fiber amplifier (EDFA) to regenerate solitons was first shown by Nakazawa et al. at NTT.[203] The golden age of optical fiber amplifiers was described by Desurvire in Physics Today 1994.[204]

Optical fibers are also a good medium for the production of supercontinuum. An overview of nonlinear fiber optics can be found in several publications.[205]

A new class of optical fiber emerged in 1996 under the name of *holey fibers* or *microstructured fibers*. In the cladding of a microstructured fiber, there is a periodic alignment of holes that are approximately 1 µm in diameter. These are grouped around either a solid glass core or a large, hollow space (**Figure 8.19**). The two structures represent two types of microstructured fiber, which differ in principle and in application. In the first case, the core of these fibers is formed by a solid silica region, and the cladding is composed of air holes which run along the fiber.[206] Light guiding was initially assumed to occur because light was guided using photonic band effects.[207] It was soon realized that total internal reflection was the leading mechanism because the index of the core was higher than the effective index of the cladding. In the second type of fiber,[208]

FIGURE 8.19 Two different geometries of the microstructured fibers. On the left the geometry with a central solid core and on the right the central hollow type.

the hollow core one, light is guided via photonic band gap effects.[209]

Holey fibers, in which air holes were introduced in the cladding region, were considered since the early days of silica waveguide research.[210] They were successively not developed and the idea was resumed nearly twenty years later.

The microstructured fibers have very interesting nonlinear optical properties because of their ability to confine light in a very small area through the use of high-index contrast between holes and glass.

In addition to modifying the microstructure pattern, both the magnitude and the sign of dispersion can be tailored to suit particular application. The fibers can exhibit anomalous dispersion down to a wavelength of 550 nm, which has made soliton generation in the near IR and visible possible. Supercontinuum generation has been the most intensively investigated phenomenon.[211]

A full description with the many variants of these fibers was made by Russell.[212]

8.8 High Harmonics Generation

High harmonic generation is a highly nonlinear process in which harmonics are produced by an intense laser field when it interacts with an atomic or molecular gas.[213]

At moderate and low intensity values of the external electric field, one has the ordinary nonlinear phenomena that can be treated as a perturbation. In 1967, New demonstrated the first third harmonic generation in a gas.[214] Experiments on third- and fifth-harmonic generation continued; for many years, only low-order harmonics (second up to the fifth order) were studied, until the late 1980s. Reintjes et al.[215] at the Naval Research Laboratory, Washington, DC, generated fifth and seventh harmonics at 53.2 and 38 nm, respectively, using a Nd:YAG laser in He and Ne gas. By the late 1980s, however, much higher laser intensities were available, and a new strong field regime was entered.[216] In 1987, by interacting krypton fluoride excimer laser pulses (at 248 nm, about 20 mJ pulse energy, and 350 fs pulse duration) with neon gas,

McPherson et al. at the University of Illinois in Chicago successfully generated harmonic emission up to the 17th order.[217] The most interesting point was that the intensity of the harmonics did not, more or less continuously decrease with increasing harmonic number. Rather, after an initial, rapid decrease the harmonic intensities establish a "plateau" region of fairly constant intensities, and then drop down quickly.

By the early 1990s, harmonic orders well into the 100s were generated in neon.[218] From then onward, many other researchers have studied harmonic generation to obtain the highest possible order and conversion efficiency. As an example, we may mention that L'Huillier[219] observed the 29th harmonic in Xe, the 57th harmonic in Ar, and the 135th harmonic in Ne by using a Nd glass laser with a wavelength of 1053 nm and a pulse duration of 1 ps, and showed clearly the presence of the plateau (**Figure 8.20**). In experiments with

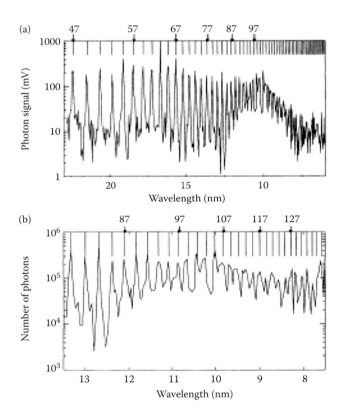

FIGURE 8.20 Experimental spectrum of the high harmonics obtained in Ne at an intensity of about 10^{15} W/cm². The position of the harmonics is indicated by the straight lines at the top of the figure. (a) Raw data, (b) data corrected from the spectrometer's response over 12–8 nm. The region from 12 to 8 nm shows the plateau. The harmonics could be resolved until about the 135th harmonic. (From A L'Huillier and Ph Balcou, *Phys. Rev. Lett.* **70**, 774 (1993).)

fs laser pulses, harmonics up to 297 have been observed with wavelengths reaching into the x-ray region.[220]

High harmonic generation is an extreme nonlinear optical process which occurs in the strong field regime. The required focused intensity of the fundamental beam for the process is at least 10^{13} W/cm^2 and can be obtained by focusing a high-power femtosecond laser beam. At this laser intensity, the contribution of very high-order nonlinearities becomes significant. However, it is worth noting that there is an upper limit of laser intensity related to optical breakdown when the plasma electron density reaches a critical density and becomes opaque to the pumping radiation. In addition, at an intensity about 10^{16} W/cm^2, the increased magnetic field prevents recombination of the generated electrons with the parent atom, which suppresses the high harmonic generation process, as described below. Self-focusing of the laser beam and the creation of plasma are also limiting factors at high intensities. At present, high harmonic generation allows us to obtain very short pulses in the region of attosecond, as described in the following section.

Since the early discovery of high harmonic generation, there were many theoretical efforts to explain the mechanism of the process. For an accurate treatment, the fully numerical solution of the time-dependent Schroedinger equation needs to be calculated. However, a quasi-classical three-step model developed by Krause[221] and Corkum[222] is able to predict the mean features.

Corkum[223] in 1993 suggested a three-step model—today known as *recollision model*—according to which high harmonic generation could be accomplished by the interaction of a very powerful and short laser pulse with an atom or molecule in a gas. The model was also discussed by Kulander et al.[224]

In this semiclassical recollision model, the possibility that an ionized electron could return to the vicinity of its parent core is considered. In a first step, the electric field of a strong laser pulse lowers the potential barrier seen by the active electrons in the atom and allows them to escape by tunneling[225] through the decreased barrier (**Figure 8.21**). The atom is ionized when the absolute electric field of the laser is close to its crest during an optical cycle and the electron is pulled away from the parent ion. The electron then interacts with the laser field and is accelerated (second step), and when the

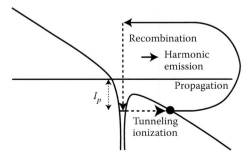

FIGURE 8.21 The three-step recollision model.

laser field changes sign about a quarter period later, the accelerated electron is pulled back and there is a significant probability that it recombines with the parent ion (third step). If the electron does not return to the core, single ionization will occur. If the electron returns to the core, it may produce single or multiple ionization or radiative recombination leading to harmonic generation.

The fraction of ionization in the tunneling regime was first studied by Ammosov et al.[226]

During the second step, the electron gains mean kinetic energy known as the *ponderomotive* energy U_p which is the cycle-averaged quivering energy of an electron in the external field

$$U_p = e^2 E^2 / 4m\omega^2. \tag{8.63}$$

When the electron returns to its parent ion, the recollision energy is the sum of the kinetic energy U_p plus the ionization potential of the ion (I_p). According to this model, the maximum energy (cutoff energy) is given by[227]

$$E_{cutoff} = \hbar\omega_{max} = I_p + 3.17 U_p. \tag{8.64}$$

The term $3.17 U_p$ is the maximum kinetic energy upon the electron returning to the nucleus and this happens for the phase of the electric field at the instant of ionization,[228] $\varphi = 17°$, which is close to the maximum of the electric field. When it returns to the ion, the electron can possess a significant amount of kinetic energy, much larger than the photon energy. This energy plus the ionization potential is transferred into emitted photon energy as soon as the electron recombines with its parent ion, which gives rise to very high harmonic orders.[229]

In the model at one half-cycle after ionization, the electron turns back toward the parent ion to recombine when the laser electric field changes sign. Hence, using multicycle IR pulses, the harmonics are produced twice each cycle and each half cycle of the driving laser produces a short (subfemtosecond) burst of XUV radiation. Thus, the harmonics are characterized by a series of bursts in the time domain separated by half the laser period.

By taking a Fourier transform, this separation of half the laser period in the time domain results in peaks separated by $2\omega_o$ in the frequency domain. In addition, since the consecutive bursts are a consequence of collisions from opposite directions, the corresponding spectral components have the same amplitude but opposite in phase. This leads to destructive spectral interference for even order harmonics ($\omega = 2q\omega_o$), and constructive interference for odd harmonics ($\omega = (2q + 1)\omega_o$). This is why, normally, only odd harmonics are observed in the high harmonic generation spectrum. For shorter pulses of only a cycle duration or less, a continuous spectrum is obtained.

The sum of the set of emission from many different electrons will have a measure of coherence due to the coherence of the original driving laser. The output flux is a combination of the single atom response and the macroscopic propagation through the medium.[230] A coherent radiation source is produced in the extreme ultraviolet and soft x-ray regions of the spectrum.[231]

A typical high harmonic spectrum can be divided into three parts: the perturbative region where the intensity of the first few order harmonics decreases quickly as the order increases, the plateau region at intermediate orders where the intensity remains almost unchanged over many harmonic orders, forming a plateau, and the cutoff region where the signal cuts off abruptly at the highest orders.[232] At moderate and low intensity values of the external electric field, one has the ordinary nonlinear phenomena that can be treated as a perturbation. Perturbation theory can be used to describe the appearance of lower order harmonics (harmonic number $q < 9$) which are produced at low intensities during the leading edge or trailing edge of the laser pulse. Basically, the harmonic yields in this region decrease as a power law.

When the electric field strength of the incident radiation is comparable to the atomic field strength (5.10^{11} V/m), then the potential barrier of the atom is strongly modified. High harmonic generation is obtained using linearly polarized ultrashort laser pulses of intensity 10^{13} W/cm^2 and time duration from ps to few fs.

The three-step model provides the basic picture of the high harmonics generation (HHG) process and has been successful in explaining the experimental observations, like the prediction of the cutoff region. However, it fails to explain the spectral characteristics of the plateau region. To deal with this problem, and to understand the high harmonic generation process more precisely, a fully quantum mechanical theory was developed within the strong field approximation by Lewenstein et al.[233] In this model, the time-dependent Schrödinger equation in a strong field was solved numerically.

The existence of the plateau was discussed before the Corkum model using various representations.[234]

From Equations 8.63 and 8.64, one sees that one approach to increase the cut-off photon energy is to use long-wavelength driving lasers. This was first demonstrated in the XUV wavelength range in 2001.[235]

The generation of high-order harmonics in laser-produced plasma obtained by shining the laser light over a solid surface was also considered since the beginning of the 1990s.[236] The research received a new impulse in 2005[237] when efficiency comparable with that obtained in gases was demonstrated. Ganeev has written in a book this side of the research in high harmonic production.[238]

8.9 Multiphoton Ionization

The three-step recollision model may be used to explain also multiphoton ionization[239] in which an atom may be ionized receiving the necessary energy through the absorption of a number of low-energy photons. First experimental demonstration of this process was produced by Damon and Tomlinson[240] in 1963 who used a ruby laser to ionize helium, argon, and a neutral air mixture. Later, Voronov and Delone[241] used a ruby laser to induce seven-photon ionization of xenon, and Hall, Robinson, and Branscomb[242] recorded two-photon electron detachment from the negative ion I$^-$. Multiphoton ionization of atoms and molecules was studied also by Prokhorov and coworkers.[243]

A crucial breakthrough was made in 1979 by Agostini et al.[244] detecting the energy-resolved photoelectrons. At high enough laser intensities ($>10^{11}$ W/cm^2), atomic systems can absorb more photons than required for ionization and therefore emit electrons with high kinetic energy.[245] The phenomenon is called *above threshold ionization*.[246]

8.10 Ultrashort Laser Pulses

The history of ultrashort pulse generation would need a book by itself. A good short review is, for example, Brabec and Krausz.[247] We may focus just on a few points.

Following the first experimental realizations of mode-locked systems, reports of many different mode-locked systems began to appear including the generation of 2 ns pulses from the ruby laser,[248] 20 ns from a passively modelocked CO_2 laser,[249] and less than 0.5 ns pulses from a Nd:glass.[250] De Maria used passive mode-locking with a saturable dye in Nd and with a simultaneous mode-locked and Q-switched regime obtained pulses ranging from 30 to 0.5 ns.[251]

In 1967, Giordmaine et al.[252] developed a two-pulse absorption fluorescence technique for the observation of picosecond pulses from a bleachable dye Q-switched, pulsed Nd:glass laser. The technique was used to show that there are picosecond pulses in the output of pulsed Nd:glass and ruby lasers,[253] and also from a Q-switched pulsed Nd:YAG laser.[254] In this last case, self-locking effects occurred and the output consisted of many picosecond pulses the smallest of which was about 19 ps. Similar results were obtained in the United Kingdom by the group in Belfast lead by D J Bradley. They obtained pulses of durations between 5 and 20 ps from neodimium,[255] ruby,[256] and organic dye[257] lasers.

Burst of short pulses were obtained also with flashlamp-pumped dye lasers.[258]

Similar results were obtained in the Soviet Union.[259]

The real starting time of the production of short pulses may be considered the year 1966, when Schaefer et al.[260] and Sorokin and Lankard[261] obtained continuous-wave operation of an organic dye laser.

The continuously pumped mode-locked dye lasers, developed in early 1970s, made possible for the first time the production of subpicosecond optical pulses in 1974.

The product of the pulse duration Δt and the spectral bandwidth $\Delta \omega$ is

$$\Delta t \cdot \Delta \omega \geq 1/2. \tag{8.63}$$

In principle, this means that in order to generate a short pulse with a specific duration (Δt) a broad spectral bandwidth ($\Delta \omega$) is required. When the product 8.63 is equal to 1/2, the pulse is called a *Fourier transform-limited* pulse.

Dye lasers were the best active material due to their large bandwidth. If the absolute bandwidth of the gain of a laser is large, the laser can amplify broad band radiation and if the radiation is in the form of a pulse, the pulse can be very short. The history of laser mode-locking is a progression of new and better ways to generate shorter and shorter pulses, and of improvements in the understanding of the mode-locking process.

The analytic theory of active mode-locking, discussed in Chapter 7, was firmly established in a classic paper by Siegman and Kuizenga.[262]

Ulrashort pulse production started in 1972 when Ippen, Shank, and Dienes[263] showed that passive mode-locking of cw dye lasers could generate subpicosecond optical pulses. The work at Bell Telephone was pivotal in the generation of ultrashort pulses.

The used configuration is illustrated in **Figure 8.22**. It was an extension of a previously realized cavity with astigmatic compensation.[264] The green output (5145 Å) of a cw Ar laser was coupled into the cavity and focused into a cell containing Rhodamine 6G. At the other end, there is a similar cell containing DODCI (diethyloxadicarbocyanine iodide). This second dye acts as a saturable absorber for the Rhodamine 6G emission. Pulses as short as 1.5 ps were obtained. The production of so short pulses remained a mystery, being that the fluorescence lifetime of DODCI is of the order of several hundred picoseconds. In the paper, a description of the method used to measure the pulses was also given.

New showed by computer modeling that the mystery could be solved considering the combined action of relaxation of the saturable gain and absorber: the saturable absorber opens a window of net gain and the gain medium shuts it off. Stable short pulse operation requires that the absorber cross-section be greater than the gain.[265]

Kryukov et al.[266] also investigated the possibility of forming an ultrashort pulse with the aid of an amplifying medium and a nonlinear absorber with fast relaxation time of the saturated state. They clearly explained that when a pulse with complicated waveform as was the case of mode-locked pulses[267] passes through a saturable absorber, strong discrimination of the amplitudes

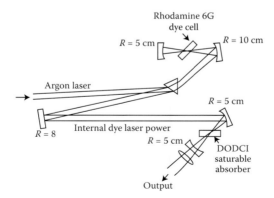

FIGURE 8.22 The experimental arrangement used by Ippen et al. The Rhodamine 6G dye laser is pumped by an argon laser. The second cell contains a saturable absorber (DODCI) that acts as a saturable absorber for the rhodamine emission. (From E P Ippen et al., *Appl. Phys. Lett.* **21**, 348 (1972).)

takes place, since the medium absorbs the weak signals and amplifies the strong ones.[268] If the absorber relaxation time is sufficiently short, then one can hope to separate a single pulse and shorten it further. The authors made experiments and calculated, observing the spectral broadening, that the duration of the pulse was around 1 ps.

The same year, the Russian group of Akhmanov[269] obtained 0.5 ps pulses in the ultraviolet (0.26–0.28 μm) by twofold doubling the emission of a Nd:glass. Parametric generation of picosecond pulses was considered for the first time in 1972.[270] The parametric generation of picosecond and femtosecond pulses is described by Laenen et al.[271]

Passive mode-locking of a cw dye laser was reported also by O'Neil[272] who obtained 4 ps pulses.

With a clever disposition, in 1974, Shank and Ippen[273] obtained pulses <1 ps. The laser configuration differed from that used in their previous work in that both the active medium (Rhodamine 6G) and the saturable absorber (DODCI) were mixed together in a single solution of ethylene glycol.

The stability criteria for pulse generation and the theory of the operation of the passively mode-locked dye laser was first described by New 1974[274] who theorized that the shortest optical pulses are obtained with a single pulse in the resonator. Nearly 10 years later, New updated the description of the field with an extended historical perspective.[275]

The work on dye lasers allowed producing shorter and shorter pulses.[276] In 1978, Diels et al.[277] obtained 200 fs.

Fork et al.[278] reported in 1981 the first production of optical pulses shorter than 100 fs with a newly developed colliding-pulse ring dye laser. The passive mode locking was obtained utilizing the interaction of two oppositely directed pulses in a thin, saturable absorber. The technique was by its authors dubbed *colliding pulse modelocking* (CPM). In contrast to the previous notion that the shortest optical pulses are obtained with a single pulse in the resonator (New), CPM requires two or more pulses in the resonator. Interaction of the counter-propagating pulses creates a transient grating in the population of absorber molecules, which synchronizes, stabilizes, and shortens the pulses in both counter-propagating trains in a surprisingly effective way. The disposition is shown in **Figure 8.23**. A special nozzle was used to generate a dye

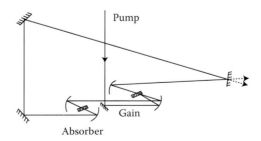

FIGURE 8.23 Schematic diagram of ring laser used for CPM by Fork et al. (From R L Fork et al., *Appl. Phys. Lett.* **38**, 671 (1981). With permission.)

stream approximately 10 μm thick. The dye laser was pumped with a cw Ar laser using 3–7 W at 5145 Å. A simple description of the colliding pulse philosophy was exposed by Shank et al.[279]

Fork's novel passive mode-locking technique in which two synchronized counter-propagating pulses interact in a thin, saturable absorber to produce a short pulse (colliding pulse mode-locking) was extended to amplify pulses as short as 70 fs to gigawatt power levels while retaining the short duration of the incident pulse.[280]

The authors observed that the short pulse traveling in the dye solvent and amplifier optics, due to group velocity dispersion, was significantly time broadened. The intracavity peak power was in fact sufficient to give rise to substantial intensity-dependent phase modulation effects. In particular, the self-phase modulation arising in the absorber dye solvent and resulting from off-resonant absorber and gain saturation effects impresses significant amounts of frequency chirp on the circulating ultrashort pulses. From an incident pulse of 75 fs in duration, the transmitted pulse was broadened to 410 fs. So the idea was to allow the relative large temporal broadening by group velocity dispersion during amplification, but restoring the short pulse duration by recompressing the pulse by a dispersion delay line after amplification. The restoring was made using a diffraction grating pair[281] which introduces a delay which increases nearly linearly with wavelength and which can thus be adjusted to compensate for the approximately linear chirp caused by group velocity dispersion.

Transient four-wave mixing by parametric scattering was studied by Fujimoto and Ippen[282] as a way to measure the parametric pulse duration and apply this process to the compression of femtosecond optical pulses, obtaining pulses as short as 40 fs.

The group of Prokhorov[283] obtained 20-ps pulses from a Nd:glass laser with a rapidly switched plasma mirror.

A new technique for the compression of picosecond light pulses in a single mode optical fiber was discussed in 1981 by Nakalnika et al.[284] With nonlinear optical fiber techniques, it is also possible to compress the picosecond pulses emitted by a variety of lasers.[285] The propagation in the fiber broadens and chirps the pulse due to the combined action of group velocity dispersion and self-phase modulation. The pulse can then be compressed to the Fourier limit by passage through a suitable dispersive delay line. Thirty femtoseconds were achieved by fiber compression of amplified pulses from a colliding-pulse method based system.[286]

Shank et al.[287] succeeded in bringing the CPM and the nonlinear fiber together and compressing the 90 fs output of a colliding pulse laser by a factor of 3, **that is,** 30 fs at a wavelength of 620 nm. Shank produced the pulses using self-phase modulation in a short 15-cm optical fiber followed by a grating compressor.

Fujimoto et al.[288] improved these results compressing a 70 fs pulse to 16 fs. The source was a CPM ring dye oscillator which produced pulses as short as 55 fs.[289] Single pulses were selected and amplified at a repetition rate of 10 Hz by the first two stages of a high-power, femtosecond dye amplifier chain.[290]

The pulses were spatially filtered and their spread was compensated by a grating pair compressor. Then the pulses were coupled into a short length of polarization preserving optical fiber.[291] The spectral broadening produced by the passage through the fiber was recompressed with a pair of gratings.

Shortly after that report, Halbout and Grischkovsky[292] compressed a single pulse embedded in a comb of amplified pulses at a repetition rate of 500 Hz from 110 to 12 fs.

The importance of self-phase modulation and group velocity dispersion pushed Martinez, Fork, and Gordon[293] to extend Haus' theory[294] of a passively mode-locked laser to include the effects of self-phase modulation and group-velocity dispersion. Short pulses using soliton-like shaping in a laser containing an optical fiber was discussed also by Mollenauer and Stolen.[295]

The use of pairs of prisms to correct for group-velocity dispersion was further analyzed by Fork, Martinez, and Gordon[296] and the following year Fork and coworkers obtained 27 fs[297] with a laser that combined and balanced, within a single resonator, the four-pulse shaping mechanisms traditionally associated with passive mode-locking and soliton-like shaping, namely self-phase modulation, group velocity dispersion, saturable absorption, and saturable gain.

In 1985, the shortest pulses generated up to that moment had duration of 8 fs.[298]

The introduction of chirped pulse amplification (CPA) in 1985 determined a true revolution in the field. To avoid the onset of self-focusing of intense light pulses which limits the amplification of ultra-short pulses, Donna Strickland and Gerard Mourou[299] introduced the idea, derived from radar operation, of deliberately producing a long pulse by stretching a short, low-energy pulse, chirping it. The stretched pulse is then amplified and then recompressed. Amplifying the stretched pulse rather than the compressed pulse allows for higher energies to be achieved before self-focusing occurs. The two authors obtained pulses at 1.06 μm with pulse width of 2 ps and energies at the millijoule level using a single mode optical fiber which linearly chirped by the combination of group velocity dispersion and self-phase modulation.[300]

A schematic diagram of the system they used is shown in **Figure 8.24**.

To minimize nonlinear effects, the pulse is first stretched several thousand times lowering the intensity accordingly without changing the input fluence (I/cm^2). The pulse is next amplified by a factor of 10^6–10^{12} and then recompresses by a factor of several thousand times close to its initial value.[301]

Two years later, in 1987, Fork et al.[302] obtained 6 fs using pulse compression external to the cavity. The problem of generating ultrashort pulses is, in general, one of the minimizing phase distortion generated during pulse compression. The pulses propagating through the medium in the system experience nonlinear phase shifts that lead to spectral broadening. The spectral broadening results in a chirp, which is a spread of frequencies. This spread can be suppressed if the different frequency components are superimposed by propagation in a dispersive medium, or by reflection from a grating pair and in this way the phase distortion can be compensated.

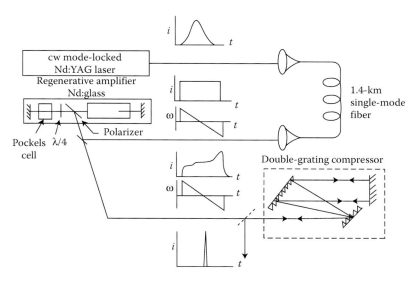

FIGURE 8.24 The 150 ps pulses from the cw mode-locked Nd:YAG laser is coupled into a single mode fiber. The average power at the output of the fiber is 2.3 W with a pulse width of around 300 ps. The stretched pulse is amplified by the regenerative Nd glass amplifier. The energy of the amplified pulse is about 2 mJ. Two gratings compressed the pulse to 1.5 ps. (From D Strickland and G Mourou, *Opt. Comm.* **56**, 219 (1985).)

Fork and collaborators in 1987 showed that a combination of prisms and diffraction gratings can be adjusted to compensate for quadratic and cubic phase distortion of an ultrashort pulse so allowing to obtain a 6 fs pulse. Writing the phase using the usual Taylor series expansion, one has

$$\varphi(\omega) = \varphi(\omega_o) + (d\varphi/d\omega)_{\omega o}(\omega - \omega_o) + 1/2(d^2\varphi/d^2\omega)_{\omega o}(\omega - \omega_o)^2$$
$$+ (1/6)(d^3\varphi/d^3\omega)_{\omega o}(\omega - \omega_o)^3.$$

A pair of diffraction gratings was shown to be useful to compensate for the quadratic term by Treacy[303] and Christov and Tomov.[304] Fork in his paper demonstrated that a combination of prisms and diffraction grating can be adjusted to compensate for the quadratic and also the cubic term. The arrangement is shown in **Figure 8.25**.

Chirped pulse amplification allowed obtaining shorter pulses and higher powers.

In the United Kingdom, 19 fs were obtained with a different scheme[305] at 630 nm.

Meantime, new broadly tunable lasers were discovered, such as the color center laser.[306] Pulse generation in the color center lasers has relied basically on two different techniques: mode locking with coupled nonlinear external cavities[307] and saturable absorber mode locking in single cavity configurations.[308]

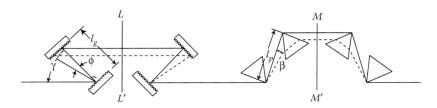

FIGURE 8.25 The combination of diffraction gratings and prisms used by Fork to compensate for the quadratic and cubic terms in phase distortion. (From Fork et al., *Opt. Lett.* **12**, 483 (1987). With permission.)

The technique of group velocity dispersion compensated colliding pulse mode-locking was also exploited in LiF:F$^+_2$ color-center lasers[309] obtaining pulses as short as 180 fs. Using a bulk InGaAsP saturable absorber for passively mode-lock, a KCl F$_A$(Tl) color center laser in a resonator configuration where the gain material was placed in a main cavity and the saturable absorber was placed in a weakly coupled external cavity, pulses as short as 320 fs were obtained at 1.50–1.55 μm.[310]

A big advance occurred when in 1986 Peter Moulton[311] built the Ti-sapphire laser which, with its very large band-width, could support femtosecond pulses, and it was generally assumed that everything was known about mode-locked Ti:sapphire lasers by the end of the 1980s.[312] However, 1990 saw two important discoveries. Ishida et al.[313] produced stable 190 fs pulses using a passively mode-locked Ti:Sapphire laser with a saturable absorber. Then Sibbet et al.[314] showed it was possible to produce 60 fs pulses from a Ti:Sapphire laser that appeared to have no saturable absorber at all (it was a self-mode locking operation). Sibbet's mode-locking approach was initially termed 'magic mode-locking' and a huge research effort was launched to understand these results. The phenomenon was soon understood and is now called Kerr lens mode-locking (KLM).[315] It also explained Ishida's result.

In Kerr lens mode-locking,[316] the rod of laser material acts as a lens for high-intensity light due to the optical Kerr effect which act as a switch. The focusing from the Kerr lens restrict the beam modifying its profile so that the losses caused by a small diaphragm put on the axis of the laser decrease as the power increases and all the light can pass.

The development of the Ti:sapphire laser with very broad gain profile in combination with intensity-dependent self-focusing through a limiting aperture in the optical cavity led to the commercial availability of femtosecond laser pulse generators during the decade of the 1990s and a very rapid progress in the generation of ultrashort pulses directly from mode-locked solid-state lasers occurred[317] and allowed the production of pulses as short as 11 fs.[318] Stable pulses of 5 nJ energy and duration of <10 fs were generated routinely from a simple laser oscillator.[319]

Pulses of two optical cycles or fewer have been generated by external pulse compression[320] or optical parametric amplification[321] or laser oscillators.[322]

Pulses shorter than 5 fs were obtained by Baltuska et al.[323] in 1997 using a self-mode-locked cavity-dumped Ti:sapphire oscillator compressed chirping in a single mode fused-silica fiber. With the same system, the group[324] obtained a complete characterization of 4.5 fs, 15 nJ pulses using the second harmonic generation version of FROG. Eventually, Schenkel et al.[325] obtained 3.6 fs pulses with energies of up to 15 µJ from a supercontinuum produced in two cascaded hollow fibers.

The development of Ti:sapphire laser systems using chirped pulse amplification made possible the generation of high-intensity few-cycle pulses in the near-infrared with a central wavelength around 800 nm with less than 6 fs.[326] The pulse spectrum covers wavelengths from above 950 nm to below 630 nm. At that wavelength, an optical cycle lasts about 2.7 fs. The shortest pulse that was obtained was 3.8 fs[327] near 800 nm.

The first demonstration of a high-repetition-rate femtosecond optical parametric oscillator which was synchronously pumped by a colliding-pulse mode-locked dye laser was given by Edelstein et al. The system was based on a crystal of $KTiOPO_4$ and made continuous tuning between 0.72 and 4.5 µm possible. Pulses of 105 fs at 840 nm and 108 Hz repetition were obtained.[328]

The self-mode-locked Ti:sapphire laser has led to a new level of performance in the high-repetition-rate femtosecond OPO.[329]

The presented discussion is not exhaustive at all. Many other methods have been omitted, as the use of semiconductor saturable absorber mirrors (SESAM) for pulse generation in solid-state lasers. This technique has been reviewed by U Keller et al.[330]

Also other solid-state materials were exploited for the production of femtosecond pulses as Yb:YAG[331] or Cr^{3+}:LiCAF which allowed generation of sub-10-fs pulses.[332]

The first short pulses were measured with various techniques such as the two-photon fluorescence, or stroboscopic camera, or nonlinear autocorrelation. The second-order autocorrelation of temporally overlapping pulses by the nonlinear mixing process of second harmonic generation was first demonstrated in 1966 by Maier et al.[333] and Armstrong.[334] The technique became the standard for measuring the duration of picosecond and subpicosecond pulses.[335] It is usually implemented using a Michelson interferometer to provide two pulse trains with an adjustable time delay between them. Subsequent mixing of these pulse trains to generate second harmonic yields a second harmonic output intensity containing the second-order autocorrelation function of the pulse intensity as a function of the time delay between the pulse trains. The technique has the disadvantage that to recover the exact width of the pulse one needs to know its shape, which is usually assumed to be a Gaussian, even if it is not. When the pulses became very short, these techniques were no more suitable and other methods were developed. A vast literature exists on the methods for measurement of such short pulses. We refer here to frequency-resolved optical grating (FROG)[336] and spectral phase interferometry for direct electric field reconstruction (SPIDER).[337] FROG yields the full intensity and phase of ultrashort pulses and guarantees the validity of the data through internal consistency checks. The second harmonic generation version of FROG[338] was soon developed. Other systems used second harmonics generation, third harmonic generation, transient gratings, polarization gratings, self-diffraction with various acronyms, SPIDER,[339] SPIRIT,[340] MIIPS,[341] GRENOUILLE,[342] and variants.[343]

By the 1990s, the bandwidth of the shortest light pulses became of the order of their central frequency (10^{15} s^{-1}) and the only way to have shorter pulses was to create a shorter wavelength source. This was done making recourse to very high harmonic generation which allowed to enter the x-ray region and shorten pulses to attosecond times (10^{-15} s).

It was so that high harmonic generation in gases was exploited and allowed to reach the attosecond region. The simple classical model developed by Corkum in 1993 and described in Section 8.9 was sufficient to explain what happens.

The attosecond burst emitted from a single atom is enormously weak. However, the process involves millions of atoms in the enlightened region and happens in perfect synchronism. As a result, the faint atomic XUV emissions add to build up an intense XUV pulse that is delivered in a highly collimated laser-like beam.

By the late 1990s, several research teams could show theoretically that they had produced attosecond pulses, but none could prove it experimentally. Fourier synthesis was proposed by several authors.[344] The basic idea is to generate a "comb" of equidistant frequencies of nearly equal amplitude with controlled relative phases over a frequency range within the plateau region. The principle is analogous to that of the mode-locked laser. However, calculations[345] showed that the harmonics in the plateau region are in general not in phase.

Also methods to obtain isolated attosecond pulses were proposed by several researchers[346] using the high sensitivity of harmonic efficiency to the laser field polarization as shown by the three-step model.[347] The harmonics are essentially produced when the polarization of the laser field is linear. By creating a laser pulse whose polarization is linear only during a short time, the harmonic emission may be limited to this interval so that single attosecond pulses are produced as experimentally demonstrated.[348] Other methods have also been proposed.[349]

The calculations made by Antoine et al.[345] showed that under certain conditions the macroscopic temporal profile for laser focusing before the gas jet yields a single peak of about 300 as duration per half-period of the laser field.

Then in June 2001 a team of researchers led by Pierre Agostini at the Saclay Research Centre in France, Philippe Balcou at the Laboratory of Applied Optics in France (LOA), and Harm Muller of AMOLF announced that they had obtained the creation of a train of 250 attosecond pulses,[350] using an experimental setup first suggested by Muller. The key to their observation was being able to confirm that the different harmonics were in phase with one another. They did this by directing the emitted ultraviolet beam and half of the original beam into a second target of rare-gas atoms and then altering the relative phase of the two beams. Observing changes in the energy spectrum of any electrons ejected from the gas provided the information they needed to work out the relative phase of the different harmonics.

In November 2001, Ferenc Krausz[351] of the Vienna University of Technology and colleagues announced that they had used a similar technique to observe pulses lasting 650 attoseconds. But unlike Agostini, Balcou, Muller and coworkers, the Vienna team was able to single out individual pulses, and it remained the only research group able to do so for some time.[352]

The technique was subsequently optimized still obtaining pulse trains.[353]

Krausz's group was able to isolate individual pulses because it used a laser that generated very short optical pulses, lasting just 5 fs, whereas other groups were using pulses lasting 50 or 60 fs. The 5 fs pulse contains only about two cycles of the optical laser light and therefore only about four attosecond pulses. It is then possible to filter out the few extraneous pulses. Krausz in collaboration with Theodor Hansch of the Max Planck Institute of Quantum Optics near Munich also controlled the temporal structure of attosecond pulses.

The isolated attosecond pulses were produced by selecting the high-energy harmonics (90 eV) generated in neon by few cycles (<7 fs) linearly polarized fundamental pulses with stabilized carrier envelope.[354] The group obtained 250 attosecond pulses one per 1 ms at 13 nm.[355]

The team had developed amplitude grating which suppressed the redundant attosecond pulses by blocking low-energy photons, and in 2008 used it to set the previous short-pulse record of 80 attosecond using phase-stabilized 3.8 fs pulses.[356]

In addition to high peak intensity and high repetition rate, the laser systems delivering few-cycle pulses must also provide reliable control over the carrier envelope phase, namely the phase of the oscillations of the electric field at the maximum of the laser pulse.

In 2006, Sansone and colleagues[357] used polarization grating to generate 130 as pulses. For these experiments, few-cycles, carrier-envelope phase-stabilized, high power driving lasers are needed. Later, in 2010, S Gilbertson et al.[358] obtained single isolated attosecond pulses with 2 mJ, 25 fs driving lasers obtaining pulses of 163 attosecond.[359]

The record up to now is 67 as.[360]

An update to 2007 discussion on the methods to obtain attosecond pulses not restricted to high harmonics generation in gases has been made by Winterfeldt, Spielmann, and Gerber.[361] A more focused review of attosecond pulse generation is Chang and Corkum.[362]

The physics at attosecond regime is a new fascinating field, but this is another story.

The impressive progress that has been realized in decreasing the pulse duration after the introduction of laser is shown in **Figure 8.26.**

8.11 Supercontinuum Generation

A supercontinuum is broadband light generated from the nonlinear frequency conversion of laser light that is focused to high intensity in a dielectric medium. The most useful medium is an optical fiber and the result typically

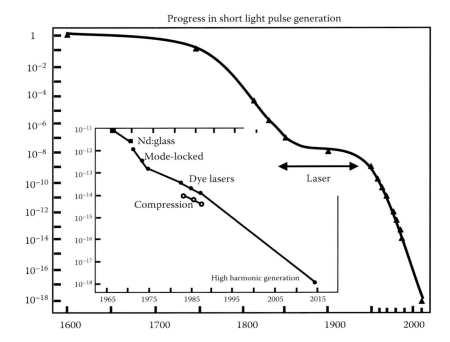

FIGURE 8.26 The progress in short pulse generation with time. On the ordinates is pulse duration in second.

spans the visible and near IR regions of the spectrum with up to several watts of average power. Because it is produced in the guided mode of an optical waveguide, the supercontinuum preserves the spatial coherence properties of the incident light and thus essentially combines the brightness and focusing properties of a laser with the bandwidth of a white light bulb. The result is an essentially arbitrary-wavelength source of spatially coherent laser light.[363]

The origin of the supercontinuum is due to a collection of nonlinear processes which act together upon the beam in order to cause severe spectral broadening of the original pump beam.

The first observation of a broadened emission was made by Boris Peter Stoicheff (1924–2010)[364] using a maser. He observed that when the maser emission was in a single sharp line, interacting with a material, the Raman emission lines were sharp; whenever the maser emission contained additional components, all of the Raman emission lines, with the exception of the first Stokes line, were considerably broadened. These weak continua allowed the first Raman absorption spectroscopy.[365]

Spectral broadening had also been reported in CS_2 by Brewer[366] and correctly interpreted in terms of the nonlinear processes of self-phase modulation.[367]

Roberto Alfano and Stanley Shapiro generated the first supercontinuum in the late 1960s by focusing picosecond pulses of green laser light into bulk crystals and glasses.[368] A series of papers treated the phenomenon.[369]

The term supercontinuum was introduced later by Manassah et al.[370]

The broadening was initially explained as due to self-phase modulation and four-wave mixing but the complete mechanisms for the broadening were long being debated.

In 1976, C Lin and A H Stolen[371] reported a nanosecond source producing continua with bandwidth of 100–180 nm at 530 nm and kW power, and 2 years later Lin, Nguyen, and French[372] obtained supercontinuum using a GeO_2-doped silica fiber. A further advance was reported by K Washio et al.[373] in 1980 pumping 150 m of single mode fiber with a 1.34 µm Q-switched Nd:YAG laser. The wavelength was inside the anomalous dispersion regime of the fiber and a continuum spanning from 1.15 to 1.6 µm was obtained with no discrete Stokes lines.

Supercontinuum generation in highly nonlinear fibers has been extensively studied, and the nonlinear dynamics leading to these ultrabroadband spectra is now well understood.[374]

Liquid, gases, optical fibers, and other media were used by many researchers.[375]

Photonic-crystal fibers[376] showed to be excellent media for obtaining supercontinua. J K Ranka et al.[377] were among the first to use photonic-crystal fibers. They used a 75-cm length fiber with zero dispersion at 767 nm pumping with 100 fs, 800 pJ at 790 nm and produced a flat continuum between 400 and 1450 nm.

Excellent reviews of the phenomenon exist.[378]

Notes

1. R A Lorentz, *Theory of Electrons* (Teubner: Leipzig, 1909).
2. J Kerr, *Phil. Mag.* **8**, 85 and 229 (1875).
3. F Pockels, *Abhandl Gesell. Wiss. Gottingen* **39**, 1 (1893); See also F Pockels, *Lehrbuch der Kristalloptik*, New York, 1968 p. 492.
4. A Einstein, *Ann. Phys.* **17,** 132 (1905).
5. M Goeppert-Mayer, "Uber Elementarakte mit zwei Quantenspringen", *Ann. Phys.* **9**(3), 273 (1931).
6. E Schroedinger, 1942, Nonlinear Optics, *Proc. Royal Irish Academy* **47A**, 77–117 (1942); A new exact solution in non-linear optics, *Proc. Royal Irish Academy* **49A**, 59 (1943). The term "nonlinear optics" appears also in the book by S J Vavilov, *Microstructures of Light* (in Russian, Academia NAUKA, Moscow 1950).
7. M Born, *Proc. Roy. Soc.* **A143**, 410 (1934).
8. M Born and L Infeld, *Proc. Roy. Soc.* **A144**, 425 (1934).
9. G Breit and E Teller, *Astrophys. J.* **91**, 215 (1940).
10. J Wheeler, *J. Opt. Soc. Am.* **37**, 813 (1947).
11. A T Forrester, W E Parkins and E G Gerjuoy, *Phys. Rev.* **72**, 728 (1947) .
12. A T Forrester, R A Gudmundsen and P O Johnson, *Phys. Rev.* **99**, 1691 (1955).
13. P A Franken, A E Hill, C W Peters and G Weinreich, *Phys. Rev. Lett.* **7**, 118 (1961).
14. P Franken, Oral History Transcription by Joan Bromberg March 8, 1985. Interview of Peter Frank by Joan Bromberg on March 8, 1985, Niels Bohr Library & Archives, American Institute of Physics, College Park, MD, USA.

15. W Kaiser and C G B Garrett, *Phys. Rev. Lett.*, **7**, 229 (1961).
16. S J R Sheppard and R Kompfner, *Appl. Opt.* **17**, 2879 (1978).
17. W Deck, J H Strickler and W W Webb, *Science* **248**, 73 (1990).
18. Th Neugebauer, *Acta Phys. Acad. Sci. Hung.* **10**, 221 (1959); D A Kleinman, *Phys. Rev.* **125**, 87 (1962); R Braunstein, *Phys. Rev.* **125**, 475 (1962); W C Hennenberger, *Bull. Am. Phys. Soc.* **7**, 14 (1962).
19. J A Armstrong, N Bloembergen, J Ducuing and P S Persan, *Phys. Rev.* **127**, 1918 (1962).
20. N Bloembergen and P S Pershan, *Phys. Rev.* **128**, 606 (1962).
21. A review was given by N Bloembergen, *Opt. Acta* **13**, 311 (1966).
22. N Bloembergen, H J Simon and C H Lee, *Phys. Rev.* **181**, 1261 (1969).
23. J Ducuing and N Bloembergen, *Phys. Rev. Lett.* **10**, 474 (1963).
24. F Brown, R E Parks and A M Sleeper, *Phys. Rev. Lett.* **14**, 1029 (1965).
25. R K Chang, C H Lee and N Bloembergen, *Phys. Rev. Lett.* **16**, 986 (1966).
26. F Brown et al., *Phys. Rev. Lett.* **14**, 1029 (1965).
27. Y R Shen, *IEEE J. Sel. Topics Quant. Electron.* **6**, 1375 (2000).
28. N Bloembergen, *Proc. IEEE* **51**, 124 (1963); N Bloembergen and Y R Shen, *Phys. Rev.* **A37**, 133 (1964).
29. R V Khokhlov, *Radiotekhnika i élektronika* **6**, 1116 (1961); S A Akhmanov and R V Khokhlov, *JETP* **43**, 351 (1962), *Sov. Phys. JETP* **16**, 252 (1963); S A Akhmanov, et al., *Sov. Phys. JETP* **18**, 919 (1964); S A Akhmanov, A I Kovrigin and N K Kulakova, *Sov. Phys. JETP* **21**, 1034 (1965). Also in the ex Soviet Union there were earlier experiments, see S I Vavilov and W L Levshin, *Z.f. Phys.* **31**, 920 (1926).
30. N Bloembergen, *Nonlinear Optics* (W A Benjamin: New York, 1965).
31. P D Maker, R W Therune, M Nisenhoff, C M Savage, *Phys. Rev. Lett.* **8**, 21 (1962).
32. P D Maker, R W Therune, M Nisenhoff, C M Savage, *Phys. Rev. Lett.* **8**, 21 (1962); J Jerphagnon and S K Kurtz, 1970, *J. Appl. Phys.* **41**, 1667.
33. P D Maker, R W Therune, M Nisenhoff, C M Savage, *Phys. Rev. Lett.* **8**, 21 (1962).
34. J A Giordmaine, *Phys. Rev. Lett.* **8**, 19 (1962).
35. R W Terhune, P D Maker and C M Savage, *Phys. Rev. Lett.* **8**, 404 (1962).
36. A W Smith and N Braslau, *IBM J. Res. Dev.* **6**, 361 (1962).
37. P A Franken and J F Ward, *Rev. Mod. Phys.* **35**, 23 (1963).
38. Reported in P A Franken and J F Ward, *Rev. Mod. Phys.* **35**, 23 (1963).
39. N Bloembergen and A J Sievers, *Appl. Phys. Lett.* **17**, 483 (1970).
40. R C Miller, *Phys. Rev.* **134**, A1313 (1964).
41. J Mazart, F Bellon, C A Arquello and R C C Leite, *Opt. Comm.* **6**, 329 (1972); C F Dewey Jr. and L O Hocker, *Appl. Phys. Lett.* **26**, 442 (1975); L O Hocker and C F Dewey Jr. *Appl. Phys. Lett.* **28**, 267 (1976).
42. B F Levine, C G Bether and R A Logan, *Appl. Phys. Lett.* **26**, 375 (1975).
43. M S Piltch, C D Cantrell and R C Sze, *J. Appl. Phys.* **47**, 3514 (1976).
44. AA Szilagyi, A Hordvin and H Schlossberg, *J. Appl. Phys.* **47**, 2025 (1976); D E Thompson, J D McMulten and D B Anderson, *Appl. Phys. Lett.* **29**, 113 (1976).
45. M Okada, K Takizawa and S Ieiri, *Opt. Comm.* **18**, 331 (1976).
46. A Feist and P Koidl, *Appl. Phys. Lett.* **47**, 1125 (1985).
47. W C Wang, Q Zhou, Z H Geng and D Feng, *J. Cryst. Growth* **79**, 706 (1986).
48. G A Magil, M M Fejer and R L Ryer, *Appl. Phys. Lett.* **56**, 108 (1990).
49. J Webjoern, F Laurell and G Arvidsson, *J. Lightwave Technol.* **7**, 1597 (1981).
50. G Khanarian, R A Norwood, D Haas, B Ferrer and D Karim, *Appl. Phys. Lett.* **57**, 977 (1990).
51. M M Fejer, G A Magel, D H Jundt and R L Byer, *IEEE J. Quantum Electron.* **QE-28**, 2631 (1992).

52. S Sonekh and A Yariv, *Appl. Phys. Lett.* **21**, 140 (1972); C L Tang and P B Bey, *IEEE J. Quantum Electron.* **QE-9**, 9 (1973).

53. R C Miller, *Phys. Rev.* **134**, A1313 (1964); J Muzart et al., *Opt. Commun.* **6**, 329 (1972).

54. A Ashkin, G D Boyd and J M Dziedzic, *Phys. Rev. Lett.* **11**, 14 (1963).

55. J A Armstrong, N Bloembergen, J Ducuing and P S Pershan, *Phys. Rev.* **127**, 1918 (1962); J A Armstrong, N Bloembergen et al., 1963, *Appl. Phys. Lett.* **3**, 68.

56. O Klein, *Z. Phys.* **40**, 407 (1927).

57. H A Kramers, *Quantum Mechanics* (North-Holland: Amsterdam, 1957), pp. 482–489.

58. See also N Bloembergen and Y R Shen, *Phys. Rev.* **133**, A37 (1963).

59. C A Akhmanov and R V Khokhlov, *Problems in Nonlinear optics* (NAUKA, Moscow 1964 in Russian), translated in English Gordon and Breach Science Publishers, New York 1972.

60. R Braunstein, *Phys. Rev.* **125**, 475 (1962); W C Henneberger, *Bull. Am. Phys. Soc.* **7**, 14(1962); P J Price and E Adler, *Bull. Am. Phys. Soc.* **7**, 329 (1962); R Loudon, *Proc. Phys. Soc.* (London) **80**, 952 (1962); P S Pershan, *Phys. Rev.* **130**, 919 (1963); D A Kleinman, *Phys. Rev.* **125**, 87 (1962); **128**, 1761 (1962); **126**, 1977 (1962); P N Butcher and T P McLean, *Proc. Phys. Soc. (London)* **81**, 219 (1963); **83**, 579 (1964); P L Kelley, *J. Phys. Chem. Solids* **24**, 607 (1963); see also P S Pershan, in *Progress in Optics*, E Wolf ed. (North-Holland: Amsterdam, 1966), vol. **5**, p. 85.

61. See, for example, R W Boyd, *Nonlinear Optics* (Academic Press: Boston: 1992) or Y R Shen, *The principles of nonlinear optics* (John Wiley and Sons, New York: 1984).

62. H S Boyne and W C Martin, *J. Opt. Soc. Am.* **52**, 880 (1962).

63. N Bloembergen, *J. Opt. Soc. Am.* **70**, 1429 (1980).

64. N Bloembergen and A J Slevers, *Appl. Phys. Lett.* **17**, 483 (1970).

65. D A Kleinman, *Phys. Rev.* **126**, 1977 (1962).

66. J A Armstrong, N Bloembergen, J Ducuing and P S Persan, *Phys. Rev.* **127**, 1918 (1962).

67. N Bloembergen and P S Persan, *Phys. Rev.* **128**, 600 (1962).

68. S L Shapiro, *Appl. Phys. Lett.* **13**, 19 (1968); W H Glenn, *IEEE J. Quantum Electron.* **5**, 284 (1969); J T Manassah and O R Cockings, *Opt. Lett.* **12**, 1005 (1987); L D Noordam et al., *Opt. Lett.* **15**, 1464 (1990); R M Rassoul et al., *Opt. Lett.* **22**, 268 (1997); W Su et al., *J. Opt. Soc. Am.* **B23**, 51 (2006).

69. M Centini et al., *Phys. Rev. Lett.* **101**, 113905 (2008); E Fazio et al., *Opt. Express* **17**, 3141 (209); E Fazio et al., *Opt. Express* **18**, 7972 (2010).

70. R W Therune et al., *Phys. Rev. Lett.* **8**, 404 (1962).

71. A Ashkin, G D Boyd and J M Dziedzic, *Phys. Rev. Lett.* **11**, 14 (1963); S L McCall and L W Davis, *J. Appl. Phys.* **34**, 2921 (1963); N I Adam and P B Schoefer, *Appl. Phys. Lett.* **3**, 19 (1963).

72. N Bloembergen and Y R Shen, *Phys. Rev.* **133**, A37 (1964); *Phys. Rev. Lett.* **12**, 504 (1964); A Ashkin, G D Boyd and J M Dziedzic, *Phys. Rev. Lett.* **11**, 14 (1963); R C Miller and A Savage, *Phys. Rev.* **128**, 2175 (1962); A Savage and R C Miller, *Appl. Opt.* **1**,661 (1962); R A Soref and H W Moos, J. *Appl. Phys.* **35**, 2152 (1964).

73. M Bass, P A Franken, J F Ward and G Weinrich, *Phys. Rev. Lett.* **9**, 446 (1962).

74. M Bass, P A Franken and J F Ward, *Phys. Rev.* **138**, A534 (1965).

75. J F Ward, *Phys. Rev.* **143**, 569 (1966).

76. J P van der Ziel, P S Pershan and L D Malmstrom, *Phys. Rev. Lett.* **15**, 190 (1965).

77. M Bass, P A Franken, A E Hill, C W Peters and G Weinrich, *Phys. Rev. Lett.* **8**, 18 (1962).

78. A W Smith and N Braslau, *IBM J. Res. Dev.* **6**, 361 (1962).

79. R H Kingston, *Proc. IRE* **50**, 472 (1962); S A Akhmanov and R V Khokhlov, *Soviet Phys. JETP* **16**, 252 (1963); J A Armstrong, N Bloembergen, J Ducuing and P S Pershan, *Phys. Rev.* **127**, 1918 (1962); N M Kroll, *Phys. Rev.* **127**, 1207 (1962).

80. C C Wang and C W Racette, *Appl. Phys. Lett.* **8**, 169 (1965).

81. S A Akhmanov et al., *Zh. Eksp. Teor. Fiz. Pis'ma* **2**, 300, 1918 (1965).

82. J A Giordmaine and R C Miller, *Phys. Rev. Lett.* **14**, 973 (1965).

83. R H Kingston, *Proc. IRE* **50**, 472 (1962).

84. N M Kroll, *Phys. Rev.* **127**, 1207 (1962).

85. S E Harris, *Proc. IEEE* **57**, 2096 (1969).

86. J A Giordmaine and R C Miller, *Phys. Rev. Lett.* **14**, 973 (1965).

87. *J. Opt. Soc. Am.* Issue September and November 1993, pp. 1656 and 2148.

88. J M Manley and H E Rowe, *Proc. IRE* **47**, 2115 (1959).

89. J E Bjorkholm, *Appl. Phys. Lett.* **13**, 53 (1968).

90. J A Giordmaine and R C Miller, *Phys. Rev. Lett.* **14**, 973 (1965); S A Akhmanov et al., *Sov. Phys. JETP* **3**, 372 (1966); J A Giordmaine and R C Miller, *Appl. Phys. Lett.* **9**, 298 (1966); R C Miller and W A Nordland, *Appl. Phys. Lett.* **10**, 53 (1987); L B Kreuzer, *Appl. Phys. Lett.* **10**, 336 (1967).

91. Franken's interview note 14.

92. R W Terhune, P D Maker and C M Savage, *Phys. Rev. Lett.* **8**, 404 (1962).

93. P D Maker and R Terhune, *Phys. Rev.* **137**, A801 (1965).

94. G H C New and J F Ward, *Phys. Rev. Lett.* **19**, 556 (1967).

95. J Reintjes et al., *Phys. Rev. Lett.* **37**, 1540 (1976); *Appl. Phys. Lett.* **30**, 480 (1977).

96. G Mayer and F Gires, *C.R.* **258**, 2039 (1964).

97. P D Maker, R W Therune and C M Savage, *Phys. Rev. Lett.* **12**, 507 (1964).

98. G A Askar'yan, *JETP* **15**, 1088 (1962); see also P L Kelley, *Phys. Rev. Lett.* **15**, 1005 (1962).

99. S A Akhmanov, A P Sukhorukov and R V Khokhlov, *Sov. Phys. Usp.* **93**, 609 (1968).

100. L A Ostrovsky, *Zh. Tekh. Fiz.* **33**, 905 (1963) (in Russian).

101. L A Ostrovsky, *Zh. Eksp. Teor. Fiz.* **51**, 1189 (1966) (in Russian).

102. V I Talanov, *Radiophys.* **8**, 254 (1964).

103. V I Talanov *Radiofizika* **7**, 564 (1964) (in Russian).

104. G A Askarjan, *Sov. Phys. JETP* **15**, 1161 (1962).

105. R Y Chiao, E M Garmire and C H Townes, *Phys. Rev. Lett.* **13**, 479 (1964); erratum ibid. **14**, 1056 (1965).

106. V I Talanov, *Radiophysics.* **9**, 138 (1965); P L Kelley, *Phys. Rev. Lett.* **15**, 1005 (1965).

107. N Zabusky and M D Kruskal, *Phys. Rev. Lett.* **15**, 240 (1965).

108. R Y Chiao, E Garmire and C H Townes, *Phys. Rev. Lett.* **13**, 479 (1964).

109. Y R Shen, *The principles of nonlinear optics* (John Wiley and Sons, New York: 1984).

110. A Barthelemy, S Maneuf and C Froehly, *Opt. Comm.* **55**, 201 (1985); S Maneuf, R Desailly and C Froehly, *Opt. Comm.* **65**, 193 (1988).

111. N F Pilipetskii and A R Rustamov, *JETP Lett.* **2**, 55 (1965).

112. Y R Shen and Y J Shaham, *Phys. Rev. Lett.* **15**, 1008 (1965); P Lallemand and N Bloembergen, *Phys. Rev. Lett.* **15**, 1010 (1965).

113. B Zel'dovich, *Sov. Phys. Uspekhi* **8**, 729 (1966).

114. V I Bespalov and V I Talanov, *JETP Lett.* **3**, 307 (1966); R Y Chiao, P L Kelley and E M Garmire, *Phys. Rev. Lett.* **17**, 1158 (1966).

115. R L Carman, R Y Chiao and P L Kelley, *Phys. Rev. Lett.* **17**, 1281 (1966).

116. R Y Chiao, M A Johnson, S Krinsky, H A Smith, C H Townes and E M Garmire, *IEEE J. Quantum Electron.* **QE-2**, 467 (1966); see also S A Akhmanov, A P Sukhorukov and R V Khkhlov, *Sov. Phys. JETP* **23**, 1025 (1966).

117. A J Campillo, S L Shapiro and B R Suydam, *Appl. Phys. Lett.* **23**, 628 (1973).
118. S A Akhmanov, R V Khokhlov and P Sukhorukov, in *Laser Handbook,* edited by F T Arecchi and E O Shultz-Dubois (Elsevier: New York, 1972), Vol. 2, p. 1151; O Svelto in *Progress in Optics*, Vol. XII, edited by E Wolf (North-Holland: Amsterdam, 1974) p. 35; Y R Shen, *Prog. Quant. Electron.* **4**, 1 (1975); J H Marburger, *Prog. Quant. Electron.* **4**, 35 (1975); Y S Kivshar and G P Agrawal, *Optical Solitons* (Academic Press: Boston, 2003); R W Boyd, S G Lukishova and Y R Shen, eds. *Self-focusing: Past and Present* (Springer: New York, 2009).
119. P L Kelley, *IEEE Sel. Top. Quant. Electron.* **6**, 1259 (2000).
120. S L McCall and E L Hahn, *Phys. Rev. Lett.* **18**, 908 (1967).
121. V E Zakharov and A B Shabat, *Zh. Eksp. Teor. Fiz.* **61**, 26 (1971) (in Russian).
122. C S Gardner et al., *Phys. Rev. Lett.* **19**, 1095 (1967).
123. V E Zakharov and A B Shabat, *Sov. Phys. JETP* **37**, 823 (1973).
124. P Empli et al., *Opt. Commun.* **62**, 374 (1987); D Kroekel et al., *Phys. Rev. Lett.* **60**, 29 (1988); A M Weiner et al., *Phys. Rev. Lett.* **61**, 2445 (1988); W J Tomlinson et al., *J. Opt. Soc. Am.* **B6**, 329 (1989).
125. G A Swartzlander et al., *Phys. Rev. Lett.* **66**, 1583 (1991); G R Allan et al., *Opt. Lett.* **16**, 156 (1991); see also D R Andersen et al., *Opt. Lett.* **15**, 783 (1990).
126. M Shalaby and A J Barthelemy, *IEEE J. Quantum Electron.* **QE-28**, 2736 (1992).
127. M Segev, B Crosignani, A Yariv and B Fisher, *Phys. Rev. Lett.* **68**, 923 (1992).
128. M D Iturbe Castillo et al., *Appl. Phys. Lett.* **64**, 408 (1994).
129. B Crosignani et al., *J. Opt. Soc. Am.* **B10**, 446 (1993); G Duree et al., *Phys. Rev. Lett.* **71**, 533 (1993); M Segev et al., *Opt. Phot. News* **4**, 8 (1993); G Duree et al., *Opt. Lett.* **19**, 1195 (1994); M Segev et al., *Opt. Lett.* **19**, 1296 (1994); M Segev et al., *Phys. Rev. Lett.* **73**, 3211 (1994); G C Valley et al., *Phys. Rev.* **A50**, 4457 (1994); G Duree et al., *Phys. Rev. Lett.* **74**, 1978 (1995).
130. R De La Fuente, A Barthelemy and C Froehly, *Opt. Lett.* **16**, 793 (1991).
131. E Fazio et al., *J. Optics A: Pure Appl. Opt.* **3**, 466 (2001); E Fazio et al., *Appl. Phys. Lett.* **85**, 2193 (2004); *Phys. Rev.* **E66**, 016605-12 (2002).
132. R Y Chiao, E Garmire and C H Townes, *Phys. Rev. Lett.* **13**, 479 (1964).
133. F DeMartini, C H Townes, T K Gustafson and P L Kelley, *Phys. Rev.* **164**, 312 (1967).
134. G Rosen, *Phys. Rev.* **139**, A539 (1965).
135. N Bloembergen and P Lallemand, *Phys. Rev. Lett.* **16**, 81 (1966).
136. B P Stoicheff, *Phys. Letters* **7**, 186 (1963); W J Jones and B P Stoicheff, *Phys. Rev. Lett.* **13**, 657 (1964).
137. E Garmire, in *Proceedings of the International Conf on the Physics of Quantum Electronics*, Puerto Rico 1965 edited by P L Kelley, B Lax and P E Tannenwald (McGraw-Hill: New York, 1965).
138. R A Fisher, P L Kelley and T K Gustafson, *Appl. Phys. Lett.* **14**, 140 (1969).
139. N J Zabusky and M D Kruskal, *Phys. Rev. Lett.* **15**, 240 (1965).
140. A Hasegawa and F Tappert, *Appl. Phys. Lett.* **23**, 142 (1973).
141. L F Mollenauer, R H Stolen and J P Gordon, *Phys. Rev. Lett.* 45, 1095 (1980).
142. Chiao, R Y, C H Townes, B P Stoicheff, *Phys. Rev. Lett.* **12**, 592 (1964).
143. E J Woodbury and W K Ng, *Proc. IEEE* **50**, 2367 (1962).
144. G Eckhart, R W Hellwarth, F J McClung, S E Schwartz, D Neiner, E J Woodbury, *Phys. Rev. Lett.* **9**, 455 (1962); R W Hellwarth, *Phys. Rev.* **130**, 1850 (1963).
145. M Geller, D P Bortfeld and W R Sooy, *Appl. Phys. Lett.* **3**, 36 (1963).
146. G Eckhardt, D P Bortfeld and M Geller, *Appl. Phys. Lett.* **3**, 137 (1963).
147. R W Minck, R W Terhune and W G Rado, *Appl. Phys. Lett.* **3**, 181 (1963).
148. R W Terhune, *Bull. Am. Phys. Soc.* **8**, 359 (1963); *Solid State Design* **4**, 38 (1963).

149. R Y Chiao, C H Townes and B P Stoicheff, *Phys. Rev. Lett.* **12**, 592 (1964).
150. G Bret and G Mayer, *C R Acad. Paris* **258**, 3265 (1964).
151. P Ehrenfest and A Einstein, *Z. Phys.* **19**, 301 (1923).
152. W Pauli, *Z. Phys.* **17**, 272 (1923).
153. R W Hellwarth, *Phys. Rev.* **130**, 1850 (1963) and *J. Appl. Opt.* **2**, 847 (1963).
154. E Garmire, F Pandarese and C H Townes, *Phys. Rev. Lett.* **11**, 160 (1963).
155. Y R Shen and N Bloembergen, *Phys. Rev.* **137**, A1787 (1965).
156. See, for example, A R Chraplyvy and J Stone, *Electron. Lett.* **20**, 996 (1984); Chraplyvy et al., *IEEE J. Lightwave Technol.* **LT-2**, 6 (1984).
157. P D Maker and R W Terhune, *Phys. Rev.* **A137**, 801 (1965).
158. See for example J-L Oudar and Y R Shen, *Phys. Rev.* **A22**, 1141 (1981).
159. H Kogelnik, *Bell Syst. Tech. J.* **44**, 2451 (1965).
160. J J Gerritsen, *Appl. Phys. Lett.* **10**, 237 (1967); J P Woerdman, *Opt. Comm.* **2**, 212 (1971); J M Amodei, *RCA Rev.* **32**, 185 (1971); B I Stepanov, E V Ivakin and A S Rubanov, *Sov. Phys. Dokl. Tech. Phys.* **16**, 46 (1971); D L Staebler and A J Amodei, *J. Appl. Phys.* **43**, 1042 (1972).
161. B Y Zeldovich et al., *Sov. Phys. JETP* **15**, 109 (1972).
162. O Y Nosach et al., *Sov. Phys. JETP* **16**, 435 (1972).
163. B Ya Zel'dovich et al., *JETP Lett.* **25**, 36 (1977); I M Bel'dyugin et al., *Sov. J. Quantum Electron.* **6**, 1349 (1976); G G Kochemasov and V D Nikolaev, *Sov. J. Quantum Electron.* **7**, 60 (1977).
164. A Yariv, *Appl. Phys. Lett.* **28**, 88 (1976); *J. Opt. Soc. Am.* **66**, 301 (1976).
165. R W Hellwarth, *J. Opt. Soc. Am.* **67**, 1 (1977).
166. A Yariv and D M Pepper, *Opt. Lett.* **1**, 16 (1977).
167. S L Jensen and R W Hellwarth, *Appl. Phys. Lett.* **32**, 166 (1978).
168. D M Bloom and G E Bjorklund, *Appl. Phys. Lett.* **31**, 592 (1977).
169. D M Bloom, P F Liao and N P Economu, *Opt. Lett.* **2**, 158 (1978).
170. D M Pepper, D Fekete and A Yariv, *Appl. Phys. Lett.* **33**, 41 (1978).
171. P V Avizoms et al., *Appl. Phys. Lett.* **31**, 435 (1977).
172. V Wang and C R Giuliano, *Opt. Lett.* **2**, 4 (1978).
173. B Ya Zel'dovich and V V Shkunov, *Sov. J. Quantum Electron.* **7**, 610 (1977).
174. J AuYeung et al., *IEEE J. Quantum Electron.* **QE-15**, 1180 (1979).
175. A Yariv, *IEEE J. Quantum Electron.* **QE-14**, 650 (1978).
176. R W Hellwarth, *Progr. Quantum Electron.* **5**, 1 (1977).
177. *Optical Phase Conjugation*, R A Fisher ed., Academic Press, New York 1983.
178. C R Giuliano, Physics Today, April 1981, p. 27.
179. J P Huignard and A Brignon, *in Phase Conjugate Laser Optics*, edited by A Brignon and J P Huignard (John Wiley & Sons, 2004).
180. The down-conversion process is called type-I if the signal and idler photons have identical polarizations. With type-II phase matching, the signal and idler photons have orthogonal polarizations. If the angle between the optical axis of the crystal and the pump beam direction is different from zero, the two cones tilt toward the pump and intersect along two rays.
181. A Einstein, B Podolky, and N Rosen, *Phys. Rev.* **47**, 777 (1935).
182. D V Strekalov et al., *Phys. Rev.* **A60**, 2685 (1999).
183. C K Hong and L Mandel, *Phys. Rev. Lett.* **56**, 58 (1986).
184. D Stoler and B Yurke, *Phys. Rev.* **A34**, 3143 (1986).
185. Walls and Milburn 1995 eds. *Quantum Optics*, Springer, Berlin.
186. Kwiat P G et al., *Phys. Rev. Lett.* **75**, 4337 (1995).
187. F P Kapron, D B Keck and R D Maurer, *Appl. Phys. Lett.* **17**, 423 (1970).
188. K C Kao and G A Hockham, *Proc. IEEE* **113**, 1151 (1966).

189. R G Smith, *Appl. Opt.* **11**, 2489 (1972); R H Stolen, *Proc. IEEE* **68**, 1232 (1980).

190. R H Stolen, E P Ippen and A R Tynes, *Appl. Phys. Lett.* **20**, 62 (1972); E P Ippen and R H Stolen, *Appl. Phys. Lett.* **21**, 539 (1972); R H Stolen and A Ashkin, *Appl. Phys. Lett.* **22**, 294 (1973); R H Stolen, J E Bjorkholm and A Ashkin, *Appl. Phys. Lett.* **24**, 308 (1974); R H Stolen, *IEEE J. Quantum Electron.* **11**, 100 (1975); R H Stolen and C Lin, *Phys. Rev.* **A17**, 1448 (1978).

191. E P Ippen and R H Stolen, *Appl. Phys. Lett.* **21**, 539 (1972).

192. D Cotter, *J. Opt. Commun.* **4**, 10 (1983); *Electron. Lett.* **18**, 495 (1982).

193. Y Aoki et al., *IEEE J. Lightwave Technol.* **6**, 710 (1988); D Cotter, *J. Opt. Comm.* **4**, 10 (1983); A Cosentino and E Iannone, *Electron. Lett.* **25**, 1459 (1989); A Bolle et al., *Electron. Lett.* **25**, 2 (1989); R G Waarts and R P Braun, *Electron. Lett.* **21**, 1114 (1985).

194. A R Chraplyvy and P S Henry, *Electron. Lett.* **19**, 641 (1983).

195. E P Ippen and R H Stolen, *Appl. Phys. Lett.* **21**, 539 (1972).

196. R H Stolen, J P Gordon, W J Tomlinson and H A Haus, *J. Opt. Soc. Am.* **B6**, 1159 (1989).

197. E M Dianov and A M Prokhorov, *IEEE J. Sel. Topics Quantum Electron.* **6**, 1022 (2000).

198. G A Koepf, D M Kalen and K H Greene, *Electron. Lett.* **18**, 942 (1982); E Desurvire, M Papuchon and J P Pocholle, *Electron. Lett.* **19**, 751 (1983); Aoki et al., *Electron. Lett.* **19**, 620 (1982); **21**, 191 (1985); *Appl. Opt.* **25**, 1056 (1986); K Nakamura et al., *IEEE J. Lightwave Technol.* **2**, 379 (1984); M Nakasaka, *Appl. Phys. Lett.* **46**, 628 (1985).

199. R H Stolen and J E Bjorkholm, *IEEE J. Quantum Electron.* **QE-18**, 1062 (1982).

200. Y Fujii, B S Kawasaki, K O Hill and D C Johnson, *Opt. Lett.* **5**, 48 (1980); Y Ohmori and Y Sasaki, *Appl. Phys. Lett.* **39**, 466 (1981); U Osterberg and W Margulis, *Opt. Lett.* **12**, 57 (1987); R H Stolen and H W K Tom, *Opt. Lett.* **12**, 585 (1987); R W Terhune and D Weinberger, *J. Opt. Soc. Am.* **B4**, 661 (1987); U Osterberg et al., *Opt. Lett.* **16**, 132 (1991).

201. L F Mollenauer, R H Stolen and J P Gordon, *Phys. Rev. Lett.* **45**, 1095 (1980); L F Mollenauer and R H Stolen, *Opt. Lett.* **9**, 13 (1984); R H Stolen, *J. Lightwave Techn.* **26**, 1021 (2008).

202. L F Mollenauer, K Smith, *Opt. Lett.* **12**, 888 (1993).

203. M Nakazawa, Y Kimura and K Suzuki, *Electron. Lett.* **25**, 199 (1989).

204. W Desurvire, *Physics Today*, January 1994, p. 20.

205. G P Agrawal, *J. Opt. Soc. Am.* **B28**, A1 (2011); G P Agrawal, *Nonlinear Fiber Optics* (Academic Press, 2012).

206. T A Birks et al., *Electron. Lett.* **31**, 1941 (1995); T A Birks, J C Knoght and P St J Russell, *Opt. Lett.* **22**, 961 (1997).

207. T A Birks et al., *Electron. Lett.* **31**, 1941 (1995).

208. J C Knight et al., Science **282**, 1475 (1998); R F Cregan et al., *Science* **285**, 1537 (1999).

209. T A Birks et al., *Electron. Lett.* **31**, 1941 (1997); R F Cregan et al., *Science* **285**, 1537 (1999).

210. P V Kaiser and H W Astle, *Bell Syst. Tech. J.* **53**, 1021 (1974).

211. J K Ranka et al., *Opt. Lett.* **25**, 25 (2000); W J Wadsworth et al., *J. Opt. Soc. Am.* **B19**, 2148 (2002); S Coen et al., *J. Opt. Soc. Am.* **B19**, 753 (2002); J Herrmann et al., *Phys. Rev. Lett.* **88**, 173901 (2002).

212. P St J Russell, *J. Lightwave Technol.* **24**, 4729 (2006).

213. W Becker, S Long and J K McIver, *Phys. Rev.* **A41**, 4112 (1990); J L Krause, K J Schafer and K C Kulander, *Phys. Rev. Lett.* **68**, 3535 (1992); *Phys. Rev.* **A45**, 4998 (1992); J J Macklin, J D Kmetec and C L Gordon III, *Phys. Rev. Lett.* **70**, 766

(1993). See also C J Joachain, N J Kystra and R M Polvliege, *Atoms in Intense Laser Fields* (Cambridge University Press: Cambridge, 2012).

214. G H C New and J F Ward, *Phys. Rev. Lett.* **19**, 556 (1967).
215. J Reintjes et al., Phys. *Rev. Lett.* **37**, 1540 (1976); *Appl. Phys. Lett.* **30**, 480 (1977).
216. A McPherson et al., *J. Opt. Soc. Am.* **B4**, 595 (1987); M Ferray et al., *J. Phys.* **B21**, L31 (1988).
217. A McPherson, G Gibson, H Jara, U Johann, T S Luk, I A McIntyre, K Boyer and C K Rhodes, *J. Opt. Soc. Am.* **B4**, 595 (1987).
218. J J Mcklin et al., *Phys. Rev. Lett.* **70**, 766 (1993); A L'Huillier and Ph Balcou, *Phys. Rev. Lett.* **70**, 774 (1993).
219. A L'Huillier and P Balcou, *Phys. Rev. Lett.* **70**, 774 (1993).
220. Z Chang et al., *Phys. Rev. Lett.* **79**, 2967 (1997).
221. J L Krause, K J Schafer and KK C Kulander, *Phys. Rev. Lett.* **68**, 3535 (1992).
222. P B Corkum, *Phys. Rev. Lett.* **71**, 1994 (1993); P B Corkum et al., *Opt. Lett.* **19**, 1870 (1994).
223. P B Corkum *Phys. Rev. Lett.* **72**, 1994 (1993); K C Kulander, K J Schafer and J L Krause, *Super-Intense Laser-Atom Physics,* edited by B Piraux, A L'Huillier and K Rzazewski (Plenum Press: New York, 1993), p. 95.
224. K C Kulander, K J Schafer and J L Krause, *Super-Intense Laser-Atom Physics*, edited by B Piraux, A L'Huillier and K Rzazewski (Plenum Press: New York, 1993), p. 95.
225. This optical tunneling had been documented by experiments by P B Corkum, N H Burnett and F Brunel, *Phys. Rev. Lett.* **62**, 1259 (1989); S Augst et al., *Phys. Rev. Lett.* **63**, 2212 (1989).
226. M V Ammosov, N B Delone and V P Krainov, *JETP,* **64**, 1191 (1986).
227. J L Krause, K J Schafer and KK C Kulander, *Phys. Rev. Lett.* **68**, 3535 (1992); P B Corkum, *Phys. Rev. Lett.* **71**, 1994 (1993).
228. C Winterfeldt, C Spielman and G Gerber, *Rev. Mod. Phys.* **80**, 117 (2008).
229. T Pfeifer, C Spielmann and G Gerber, *Rep. Prog. Phys.* **69**, 443 (2006).
230. A L'Huillier, K J Schaefer and K C Kulander, *Phys. Rev. Lett.* **66**, 2200 (1991); A L'Huillier, P Balcou, S Candel, K J Schafer and K C Kulander, *Phys. Rev.* **A46**, 2778 (1992).
231. J Zhou et al., *Phys. Rev. Let.* **76**, 752 (1996); Z Chang et al., *Phys. Rev. Lett.* **79**, 2967 (1997).
232. M Ferray, A L'Huillier, X F Li, L a Lomprè, G Mainfray and C Manus, *J. Phys.* **B21**, L31 (1988).
233. M Lewenstein, Ph Balcou, M Yu IIvanov, A L'Huillier and P B Corkum, *Phys. Rev.* **A49**, 2117 (1994).
234. See, for example, W Becker, S Long and J K McIver, *Phys. Rev.* **A41**, 4112 (1990).
235. B Shan and Z Chang, *Phys. Rev.* **A65**, 011804 (2001).
236. Y Akiyama et al., *Phys. Rev. Lett.* **69**, 2176 (1992); S Kubodera et al., *Phys. Rev.* **A48**, 4576 (1993); C G Wahlstrom et al., *Phys. Rev.* **A51**, 585 (1995).
237. R Ganeev et al., *Opt. Lett.* **30**, 768 (2005); *Phys. Rev.* **A74**, 063824 (2006); *J. Appl. Phys.* **99**, 103303 (2006).
238. R A Ganeev, *High-Order Harmonic Generation in Laser Plasma Plumes* (Imperial College Press: London, 2013).
239. L V Keldysh, *JETP* **20**, 1307 (1965); F H M Faisal, *J Phys.* **B6**, L89 (1973) and H R Reiss, *Phys. Rev.* **A22**, 1786 (1980).
240. E K Damon and R G Tomlinson, *Appl. Opt.* **2**, 546 (1963).
241. G S Voronov and N B Delone, *Sov. Phys. JETP Lett.* **1**, 66 (1965).
242. J L Hall, E J Robinson and L M Branscomb, *Phys. Rev. Lett.* **14**, 1013 (1965).

243. F V Bunkin and A M Prokhorov, *Zh. Eksp. Teor. Fiz.* **46**, 1090 (1964); F V Bunkin, R V Karapatyan and A M Prokhorov, *Zh. Eksp. Teor. Fiz.* **47**, 216 (1964).

244. P Agostini et al., *Phys. Rev. Lett.* **42**, 1127 (1979).

245. R Rosman et al., *J. Opt. Soc. Am.* **B5**, 1237 (1988); M Ferray et al., *J Phys.* **B21**, L31 (1988); X F Li et al., *Phys. Rev.* **A39**, 5751 (1989).

246. P Agostini and H G Mueller, *J. Phys.* **B21**, 4097 (1988).

247. T Brabec and F Krausz, *Rev. Mod. Phys.* **72**, 545 (2000).

248. T Deutsch, *Appl. Phys. Lett.* **7,** 80 (1965).

249. O R Wood and S E Schwarz, *Appl. Phys. Lett.* **12**, 263 (1968); G W Flynn, L O Hocker, A Javan, M A Kovacs and C K Rhodes, *IEEE J. Quantum Electron.* **QE-2**, 378 (1966).

250. A J De Maria, C Ferrar and G E Danielson, *Appl. Phys. Lett.* **8**, 22 (1966).

251. A J DeMaria, D A Stetser and H Heynau, *Appl. Phys. Lett.* **8**, 174 (1966).

252. J A Giordmaine, P M Rentzepis, S L Shapiro and K W Wecht, *Appl. Phys. Lett.* **11**, 216 (1967).

253. S L Shapiro, M A Duguay and L B Kreuzer, *Appl. Phys. Lett.* **12**, 36 (1968); M A Duguay, S L Shapiro and P M Rentzepis, *Phys. Rev. Lett.* **19**, 1014 (1967).

254. M Bass and D Woodward, *Appl. Phys. Lett.* **12**, 275 (1968).

255. D J Bradley, S J Caughey and G H C New, *Phys. Lett.* **30A**, 78 (1969); *Opt. Comm.* **2**, 41 (1970).

256. D J Bradley, T Morrow and M S Petty, *Opt. Comm.* **2**, 1 (1970).

257. D J Bradley, A J F Durrant, F O'Neill and B Sutherland, *Phys. Lett.* **30A**, 535 (1969).

258. E G Arthurs, D J Bradley and A G Roddie, *Appl. Phys. Lett.* **20,** 125 (1972).

259. P G Kryukov, Yu A Matveets, S A Churilova and O B Shatberashvili, *JETP* **35**, (1972) manca la pagina; V I Malyshev, A A Sychev and V A Badenko, *JETP Lett.* **13**, 422 (1970); P G Kryukov, Yu A Matveets, S V Chekalin and O B Shatverashvili, *JETP Lett.* **16**, 81 (1972); S D Fanchenko and B A Frolov, *JETP Lett.* **16**, 101 (1972); A A Malyutin and M Ya Shchelev, *JETP Lett.* **9**, 266 (1969).

260. F P Schaefer et al., *App. Phys. Lett.* **9**, 306 (1966).

261. P P Sorokin and J R Lankard, *IBM J. Res. Dev.* **10**, 162 (1966).

262. D I Kuizenga and A E Siegman, *IEEE J. Quantum Electron.* **QE-6**, 803 (1970).

263. E P Ippen, C V Shank and A Dienes, *Appl. Phys. Lett.* **21**, 348 (1972).

264. A Dienes, E P Ippen and C V Shank, *IEEE J. Quantum Electron.* **QE-8**, 388 (1972); H W Kogelnik, E P Ippen, A Dienes and C V Shank, *IEEE J. Quantum Electron.* **QE-8**, 373 (1972).

265. G H C New, *IEEE J. Quantum Electron.* **QE-10**, 115 (1974); H A Haus, *IEEE J. Quantum Electron.* **QE-11**, 736 (1975).

266. P G Kryukov et al., *JETP Lett.* **16**, 81 (1972).

267. Reference was given to the papers by M A Duguay, J W Hansen and S L Shapiro, *IEEE J. Quantum Electron.* **QE-6**, 725 (1970); P G Kryukov et al., *Sov. Phys. JETP* **35**, (1972); V I Malyshev et al., *JETP Lett.* **13**, 419 (1971).

268. V S Letokhov, *JETP Lett.* **7**, 35 (1968).

269. S A Akhmanov, R Yu Orlov, I B Skidan and L S Telegin, *JETP Lett.* **16**, 335 (1972).

270. T A Rabson et al., *Appl. Phys. Lett.* **21**, 129 (1972).

271. R Laenen et al., *J. Opt. Soc. Am.* **B10**, 2151 (1993).

272. F O'Neil, *Opt. Comm.* **6**, 360 (1972).

273. C V Shank and E P Ippen, *Appl. Phys. Lett.* **24**, 373 (1974).

274. New, G H C, *IEEE J. Quantum Electron.* **QE-10**, 115 (1974).

275. G H C New, *Rep. Prog. Phys.* **46**, 877 (1983).

276. E G Arthurs, D J Bradley and T J Glynn, *Opt. Comm.* **12**, 136 (1974); E G Arthurs, D J Bradley, P N Puntambek and I S Ruddock, *Opt. Comm.* **12**, 360 (1974).

277. J C Diels, E van Stryland and G Benedict, *Opt. Com.* **25**, 93 (1978).

278. R L Fork, B I Greene and C V Shank, *Appl. Phys. Lett.* **38**, 671 (1981).

279. C V Shank, R L Fork and F Beisser, *Laser Focus*, June 1983, p. 59.

280. R L Fork, C V Shank and R T Yen, *Appl. Phys. Lett.* **41**, 223 (1982).

281. E B Treacy, *IEEE J. Quantum Electron.* **QE-5**, 454 (1969).

282. J G Fujimoto and E P Ippen, *Ot. Lett.* **8**, 448 (1983).

283. A N Malkov, A M Prokhorov, V B Fedorov and I V Fomenkov, *JETP Lett.* **33**, 617 (1981).

284. Nakalnika et al., 1981, *Phys. Rev. Lett.* **47**, 910.

285. H Nakatsuka and D Grischkowsky, *Opt. Lett.* **6**, 13 (1981); H Nakatsuka, D Grischkowsky and A C Balant, *Phys. Rev. Lett.* **47**, 910 (1981); D Grischkowsky and A C Balant, *Appl. Phys. Lett.* **41**, 1 (1982); L F Mollenauer, R H Stolen, J P Gordon and W J Tomlinso, *Opt. Lett.* **8**, 289 (1983); B Nikolaus and D Grischkowsky, *Appl. Phys. Lett.* **42**, 1 (1983); **43**, 228 (1983).

286. C V Shank, R L Fork, R Yen, R H Stolen and W J Tomlinson, *Appl. Phys. Lett.* **40**, 761 (1982).

287. C V Shank et al., *Appl. Phys. Lett.* **40**, 761 (1982).

288. J G Fujimoto, A M Weiner and E P Ippen, *Appl. Phys. Lett.* **44**, 832 (1984).

289. J M Halbout and C L Tang, *IEEE J. Quantum Electron.* **QE-19**, 487 (1983); W Dietel, J J Fontaine, and J C Diels, *Opt. Lett.* 4 (1983).

290. R L Fork, C V Shank and R T Yen, *Appl. Phys. Lett.* **41**, 223 (1982).

291. R H Stolen et al., *Appl. Phys. Lett.* **33**, 699 (1978).

292. J M Halbout and D Grischkowsky, *Appl. Phys. Lett.* **45**, 1281 (1984).

293. O E Martinez, R L Fork and J P Gordon, *Opt. Lett.* **9**, 156 (1984).

294. H A Hauss, *IEEE J. Quantum Electron.* **QE-11**, 736 (1975).

295. L F Mollenauer and R H Stolen, *Opt. Lett.* **9**, 13 (1984).

296. R L Fork, O E Martinez and J P Gordon, *Opt. Lett.* **9**, 150 (1984).

297. J A Valdmanis, R L Fork and J P Gordon, *Opt. Lett.* **10**, 131 (1985).

298. W H Knox, R L Fork, M C Downer, R H Stolen, J A Valdmanis and C V Shank, *Appl. Phys. Lett.* **46**, 1120 (1985).

299. D Strickland and G Mourou, *Opt. Comm.* **56**, 219 (1985).

300. B Nikolaus and D Grischkowsky, *Appl. Phys. Lett.* **42**, 1 (1983).

301. G Mourou, T Tajima, S V Bulanov, *Rev. Mod. Phys.* **78**, 309 (2006).

302. R L Fork, C H B Cruz, P C Becker and C V Shank, *Opt. Lett.* **12**, 483 (1987).

303. E B Treacy, *IEEE J. Quantum Electron.* **QE-5**, 454 (1969).

304. II P Christov and I V Tomov, *Opt. Comm.* **58**, 338 (1986).

305. A Finch, G Chen, W E Sleat and W Sibbett, *J. Mod. Opt.* **35**, 345 (1988).

306. L F Mollenauer and D M Bloom, *Opt. Lett.* **4**, 247 (1979).

307. L F Mollenauer and R H Stolen, *Opt. Lett.* **9**, 13 (1984); K J Blow and B P Nelson, *Opt. Lett.* **13**, 1026 (1988); P N Kean et al., *Opt. Lett.* **14**, 39 (1989); J Mark et al., *Opt. Lett.* **14**, 48 (1989); C P Yakymyshyn et al., *Opt. Lett.* **14**, 621 (1989); R S Grant et al., *Opt. Lett.* **16**, 384 (1991).

308. M N Islam et al., *Appl. Phys. Lett.* **56**, 2177 (1990); C E Soccolich et al., *Appl. Phys. Lett.* **61**, 886 (1992).

309. N Langford, K Smith and W Sibbett, *Opt. Lett.* **12**, 903 (1987); N Langford et al., *Opt. Lett.* **14**, 45 (1898).

310. K Mollmann et al., *Opt. Lett.* **18**, 42 (1993).

311. P Moulton, *J. Opt. Soc. Am.* **B3**, 125 (1986).

312. U Keller, *Nature* **424**, 834 (2003).

313. Y Ishida, N Sarukura and H Nakano, *Opt. Lett.* **16**, 153 (1991).
314. D E Spencer, F N Kean and W Sibbett, *Opt. Lett.* **16**, 42 (1991).
315. U Keller *Opt. Lett.* **16**, 1024 (1991).
316. G Cerullo, S De Silvestri and V Magni, *Opt. Lett.* **19**, 1040 (1994); S Gatz, J Herrmann and M Muller, *Opt. Lett.* **21**, 1573 (1996); B E Bouma and J G Fujimoto, *Opt. Lett.* **21**, 134 (1996); I D Jung et al., *Opt. Lett.* **22**, 1009 (1997); Z Zhang et al., *Opt. Lett.* **22**, 1006 (1997); I P Bilinsky, R P Prasankumar and J G Fujimoto, *J. Opt. Soc. Am.* **B16**, 546 (1999); J Jasapara et al., *J. Opt. Soc. Am.* **B17**, 319 (2000).
317. A non-exhaustive list is N Sarukura, Y Ishida and H Nakano, *Opt. Lett.* **16**, 153 (1991); C P Huang et al., *Opt. Lett.* **17**, 139 (1992); C P Huang et al., *Opt. Lett.* **17**, 1289 (1992); J M Jacobson et al., *Opt. Lett.* **17**, 1608 (192); B Lemoff and C Barty, *Opt. Lett.* **17**, 1367 (1992); P Curley et al., *Opt. Lett.* **18**, 54 (1993); R J Ellinngson and C L Tang, *Opt. Lett.* **18**, 438 (1993); K Yamakawa et al., *Opt. Lett.* **23**, 1468 (1998); J P Chambaret et al., *Opt. Lett.* **21**, 1921 (1996); A Sullivan et al., *Opt. Lett.* **21**, 603 (1996).
318. M T Asaki et al., *Opt. Lett.* **18**, 977 (1993).
319. J Zhou et al., *Opt.Lett.* **19**, 1149 (1994).
320. N Nisoli et al., *Opt. Lett.* **22**, 522 (1997).
321. A Shirakawa et al., *Appl. Phys. Lett.* **74**, 2268 (1999).
322. U Morgner et al., *Opt. Lett.* **24**, 411, 920 (1999); D H Sutter et al., *Opt. Lett.* **24**, 631 (1999).
323. A Baltuska et al., *Appl. Phys.* **B65**, 175 (1997).
324. A Baltuska et al., *Opt. Lett.* **23**, 1474 (1998).
325. B Schenkel et al., *Opt. Lett.* **28**, 1987 (2003).
326. D H Sutter et al., *Opt. Lett.* **24**, 631 (1999); R Ell, *Opt. Lett.* **26**, 373 (2001).
327. E Goulielmakis et al., *Science* **320**, 1614 (2008).
328. D C Edelstein et al., *Opt. Lett.* **15**, 136 (1990); D C Edelstein, E S Wachman and C L Tang, *J. Appl. Phys.* **54**, 1728 (1989); E S Wachman, W S Pelouch and C L Tang, *J. Appl. Phys.* **70,** 1893 (1991).
329. Q Fu et al., *Opt. Lett.* 17, 1006 (1992); W S Pelouch et al., *Opt. Lett.* **17**, 1070 (1992); G Mak et al., *Appl. Phys. Lett.* **60**, 542 (1992); R J Ellingson and C L Tang, *Opt. Lett.* **18**, 438 (1993).
330. U Keller et al., *IEEE J. Sel. Top. Quantum Electron.* **2**, 435 (1996).
331. Hoenninger et al., *Opt. Lett.* **20**, 2402 (1995).
332. P C Wagenblast et al., *Opt. Lett.* **27**, 1726 (2002).
333. M Maier, W Kaiser and J A Giordmaine, *Phys. Rev. Lett.* **17**, 1275 (1966).
334. J A Armstrong, *Appl. Phys. Lett.* **10**, 16 (1967).
335. J Kuhl, H Klingenberg and D von der Linde, *Appl. Phys.* **18**, 297 (1979); P G May, W Sibbett and J R Taylor, *Appl. Phys.* **B26**, 179 (1981); R L Fork et al., *Appl. Phys. Lett.* **38**, 671 (1981); D B McDonald, D Waldeck and G R Fleming, *Opt. Comm.* **34**, 127 (1980); D Welford and B C Johnson, *Opt. Comm.* **45**, 101 (1983); J C Diels et al., *Appl. Opt.* **24**, 1270 (1985).
336. D J Kane and R Trebino, *IEEE J. Quantum Electron.* **29**, 571 (1993); R Trebino and D J Kane, *J. Opt. Soc. Am.* **A10**, 1101 (1993); J Paye, M Ramaswamy, J G Fujimoto, E P Ippen, *Opt. Lett.* **18**, 1946 (1993); Baltuska et al., *J. Quantum Electron.* **QE-35**, 459 (1999).
337. C Iaconis and I A Walmsley, *Opt. Lett.* **23**, 79 (1998).
338. G Taft et al., *Opt. Lett.* **20**, 743 (1995); K W DeLong et al., *J. Opt. Soc. Am.* **B11**, 2206 (1994).
339. C Iaconis and I Walmsley, *Opt. Lett.* **23**, 792 (1998); C Dorrer, *Opt. Lett.* **24**, 1532 (1999); L Gallmann et al., *Opt. Lett.* **24**, 1314 (1999).

340. M Lelek, F Louradour, A Barthelemy and C Froehly, *Opt. Comm.* **261**, 124 (2006).

341. V V Lozovoy, I Pastirk and M Dantus, *Opt. Lett.* **29**, 775 (2004).

342. S Akturk, M Kimmel, P O'Shea and R Trebino, *Opt. Lett.* **29**, 1025 (2004).

343. J Y Zhang et al., *Opt. Express* **11**, 601 (2003); D N Fittinghoff et al., *Opt. Lett.* **21**, 884 (1996); S Linden, J Kuhl and H Giessen, *Opt. Lett.* **24**, 569 (1999).

344. T W Hansch, *Opt. Comm.* **80**, 70 (1990); G Farkas and C Toth, *Phys. Lett.* **A168**, 447 (1992); S E Harris, J J Macklin and T W Haensch, *Opt. Comm.* **100**, 487 (1993).

345. P Antoine, A L'Huillier and M Lewenstein, *Phys. Rev. Lett.* **77**, 1234 (1996).

346. P B Corkum, N H Burnett and M Y Ivanov, *Opt. Lett.* **19**, 1870 (1994); M Ivanov et al., *Phys. Rev. Lett.* **74**, 2933 (1995); V T Platonenko and V W Strelkov, *J. Opt. Soc. Am.* **B16**, 435 (1999).

347. K S Budil et al., *Phys. Rev.* **A48**, R3437 (1993).

348. C Altucci et al., *Phys. Rev.* **A58**, 3934 (1998); M Kovacev et al., *Eur. J. Phys.* **D26**, 79 (2003); O Tcherbakoff et al., *Phys. Rev.* **A68**, 043804 (2003).

349. E Gouielmakis et al., *Science* **317**, 769 (2007); A L CCavalieri et al., *Nature* **449**, 1029 (2007); A Baltuska et al., *Nature* **421**, 611 (2003); P B Corkum et al., *Opt. Lett.* **19**, 1870 (1994); V T Platonenko and V V Strelkov, *J. Opt. Soc. Am.* **B16**, 435 (1999); O Tcherbakoff et al., *Phys. Rev.* **A68**, 043804 (2003); H Mashiko et al., *Phys. Rev. Lett.* **100**, 103906 (2008); S Gilbertson et al., *Appl. Phys. Lett.* **92**, 071109 (2008); Y Oishi et al., *Opt. Express* **14**, 7230 (2006); M Hentschel et al., *Nature* **414**, 509 (2001).

350. P M Paul, E S Toma, P Breger, G Mullot, F Augè, P Balcou, H G Muller and P Agostini, *Science* **292**, 1689 (2001).

351. M Hentschel, R Kienberger, C Spielmann, G A Reider, N Milosevic, T Brabec, P Corkum, U Heinzmann, M Dresher and F Krausz, *Nature* **414**, 509 (2001).

352. Generation of single attosecond pulses was obtained by G Sansone et al., *Science* **314**, 443 (2006); K Zhan et al., *Opt. Lett.* **37,** 3891 (2012); E Goulielmakis et al., *Science* **320**, 1614 (2008).

353. R Kienberger et al., *Nature* **427**, 817 (2004); Y Mairesse et al., *Science* **302**, 1540 (2003); Y Mairesse et al., *Phys. Rev. Lett.* **93**, 163901 (2004).

354. M Hentschel et al., *Nature* **414**, 509 (2001); Kienberger, R et al., 2002, *Science* **297**, 144.

355. A Baltuska et al., *Nature* **421**, 611 (2003).

356. E Goulielmakis et al., *Science* **320**, 1614 (2008).

357. G Sansone et al., *Science* **314**, 443 (2006).

358. S Gilbertson et al., *Phys. Rev.* **A81**, 043810 (2010).

359. X Feng et al., *Phys. Rev. Lett.* **103**, 183901 (2009).

360. K Zhao et al., *Opt. Lett.* **37**, 3891 (2012).

361. C Winterfeldt, C Spielmann and G Gerber, *Rev. Mod. Phys.* **80**, 117 (2008).

362. Z Chang and P Corkum, *J. Opt. Soc. Am.* **B27**, B9 (2010).

363. R R Alfano ed. *The Superconyinuum Laser Source: Fundamentals with Updated References*, Springer, New York 2006; J M Dudley, G Genty, S Coen, *Rev. Mod. Phys.* **78**, 1135 (2006); J BM Dudley, J R Taylor eds., *Superconytinuum Generation in Optical Fibers* (Cambridge Univ. Press: New York, 2010).

364. B P Stoicheff, *Phys. Lett.* **7**, 186 (1963).

365. W J Jones and B P Stoicheff, *Phys. Rev. Lett.* **13**, 657 (1964).

366. R G Brewer, *Phys. Rev. Lett.* **19**, 8 (1967).

367. F Shimizu, *Phys. Rev. Lett.* **19**, 1097 (1967).

368. R R Alfano and S L Shapiro, *Phys. Rev. Lett.* **24**, 584, 592 (1970).

369. R Alfano, *Phys. Rev. Lett.* **24**, 592, 584 and 1217 (1970).

370. J T Manassah, P P Ho, A Katz and R R Alfano, *Photonic Spectra* **18**, 53 (1984); J T Manassah, R R Alfano and M Mustafa, *Phys Lett.* **107A**, 305 (1985).

371. C Lin and R H Stolen, *Appl. Phys. Lett.* **28**, 216 (1976).

372. C Lin, V Nguyen and W French, *Electron. Lett.* **14**, 822 (1978).

373. K Washio, K Inoue and T Tanigawa, *Electron. Lett.* **16**, 331 (1980).

374. J M Dudley, G Genty and S Coen, *Rev. Mod. Phys.* **78**, 1135 (2006).

375. C Lin and R H Stolen, *Appl. Phys. Lett.* **28**, 216 (1976).

376. J C Knight, T Birks, P Russell and D Atkin, *Opt. Lett.* **21**, 1547 (1996).

377. J K Ranka, R S Windeler and A J Stenz, *Opt. Lett.* **25**, 25 (2000).

378. J M Dudley, G Genty and S Coen, *Rev. Mod. Phys.* **78**, 1135 (2006) and J M Dudley and G Genty, *Physics Today,* July 2013, p. 29.

9

More Exotic Lasers

9.1 Introduction

Since the very beginning of the maser and laser realization, there was an explosion of research on innovative methods to produce lasing. It would be very difficult to give an account of all the ideas that sprang up and were developed and pursued. Here we give a few examples.

9.2 Lasing Without Inversion

The traditional approach to maser and laser has been to obtain a population inversion necessary to make stimulated emission overcoming absorption. The threshold of the device is obtained when the stimulated emission equals the absorption. One way to lower the threshold would be to lower and asymptotically to eliminate absorption.

As early as 1957, Javan[1] showed, in the context of the theory of a three-level maser, the possibility of inversionless amplification involving discrete energy levels in the so-called V-type level system shown in **Figure 9.1**.

In Javan's scheme, an intense maser field saturates the transition of one arm, while another microwave coherent field of smaller frequency probes the adjacent transition in the other arm. A net-induced emission at some portion of the resonant line and a net absorption at another frequency within the line width could be obtained due to the contribution of an interference term arising from the overlap between the doublet built up by the driving field.

Lasing without population inversion was later suggested in two-level systems making use of the asymmetry between the recoil-induced shift for stimulated emission and absorption processes.[2] A simple energy and momentum balance shows that due to atomic recoil the difference between the resonance frequency for stimulated emission and absorption is given by $\hbar\omega^2/Mc^2$, $\hbar\omega$ being the photon energy and Mc^2 the rest mass of the atom under

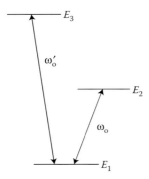

FIGURE 9.1 The V-level scheme considered by Javan. (From A Javan, *Phys. Rev.* **107**, 1579 (1957). With permission.)

consideration. Due to this shift, there is a frequency range in which stimulated emission overcomes absorption even if the excited population is smaller than the corresponding one of the ground level. In the optical domain, this splitting is very small but can be larger than Doppler broadening in the X-ray range.

Preparing an atomic system in a coherent superposition of states may, under certain conditions, allow us to cancel absorption. Resonant coherence effects in three-level media have been investigated beginning from the 1950s[3] and the exploration continued in the following years,[4] until in the 1980s it was seriously pursued by O Kocharovskaya and Khamin,[5] Harris,[6] Scully,[7] and others[8] using different atom schemes.

In lasing without inversion, the essential idea is the absorption cancellation by atomic coherence and interference.[9] This phenomenon is also the essence of electromagnetic-induced transparency.[10] Usually, this is accomplished in three-level atomic systems in which there are two coherent routes for absorption that can destructively interfere, thus leading to the cancellation of absorption. A small population in the excited state can then lead to net gain.

In Kochakovska model,[5] a three-level medium in the so-called Λ configuration in which two lower levels $|b\rangle$ and $|c\rangle$ are coupled to a single upper level $|a\rangle$ interact with two monochromatic fields, each field interacting with its own resonant transition only (**Figure 9.2**). Using straightforward quantum mechanics, it can be shown that there exists a linear superposition of the two low-lying states such that no transitions can be induced between the upper state and the superposition state due to destructive interference.[11] The population is trapped in the lower states and there is no absorption even in the presence of the field.

The system was further investigated[12] but did not work.

Light amplification without inversion is possible if an external source is used to produce the coherent superposition state.

In Harris' study, the amplifying medium consists of four-level atoms with

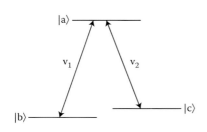

two upper levels which are homogeneously broadened and decay to an identical continuum. In the work of Scully et al., three-level atoms pass through a resonant cavity and interact with the same monochromatic field at both optical transitions.

Scully examined several possibilities and concentrated on the scheme shown in Figure 9.2 (Λ scheme). He considered a microwave cavity containing

FIGURE 9.2 The Λ scheme.

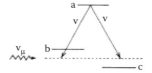

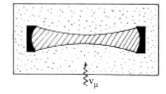

FIGURE 9.3 The Λ-type laser. The upper diagram shows the energy-level scheme. Laser action takes place between the two mirrors, and laser radiation exists only in the cross-hatched region. Atoms pass through this region with an average transit time τ. (From M A Scully et al., *Phys. Rev. Lett.* **62**, 2813 (1989). With permission.)

such three-level atoms, the lower two levels of which are driven by an intense microwave field. There is also a weak, incoherent, pumping mechanism which populates level a. The lasing field exists only between the mirrors, as indicated in **Figure 9.3**. The medium was considered to be a dilute gas so that one is in the collisionless regime, and the levels are long lived compared to the average transit time t for an atom to cross the lasing region.

The main result was the linear gain equation for the average photon number, n, of the laser field

$$dn/dt = a\rho^\circ_{aa}(n+1) - a\rho^\circ_{cc}\, n(1 - \cos \varphi), \tag{9.1}$$

where a is the linear gain, ρ°_{ll} denotes the initial population in the $|1\rangle$ state, and φ is the phase of the microwave field. For $\varphi = 0$, the absorption of light by the lower levels vanishes; however, the linear gain (the ρ°_{aa} term) is not affected. Hence, we can have lasing even if only a small fraction of the atoms are in the excited state $|a\rangle$.

Physically, the lack of absorption in the Λ quantum bet laser is a manifestation of quantum interference phenomena. When an atom makes a transition from the upper level to the two lower levels, the total transition probability is the sum of the a → b and a → c probabilities; however, transition probabilities from the two lower levels to the single upper level are obtained by squaring the sum of the two probability amplitudes. When there is coherence between the two lower levels, this can lead to interference terms yielding a null in the transition probability corresponding to photon absorption.

A different scheme has been considered by Imamoglu et al.[13] using the system shown in **Figure 9.4**: a strong electromagnetic field of frequency ω_c couples a metastable state $|2\rangle$ to a state $|3\rangle$ which decays radiatively to states $|2\rangle$ and $|1\rangle$. This coupling field creates a pair of dressed states which a priori

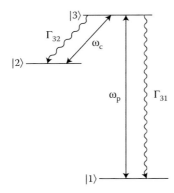

FIGURE 9.4 The scheme considered by Imamoglu et al. (From A Imamoglu et al., *Phys. Rev. Lett.* **66**, 1154 (1991). With permission.)

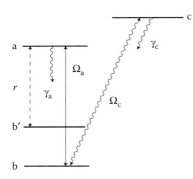

FIGURE 9.5 Simplified four-level model for lasing without inversion. (From A S Zibro et al., *Phys. Rev. Lett.* **75**, 1499 (1995). With permission.)

decay to the same final state. The absorption of state $|1\rangle$ atoms as measured by a probe of frequency ω_p will exhibit a destructive interference with a minimum absorption at the frequency $\omega_2 + \omega_c - \omega_1$.

Eventually, absorption cancellation due to atomic coherence using two strong beams to prepare the atomic coherence in the Na D_1 line and gain were demonstrated experimentally by Fry in 1993.[14] Laser amplification without population inversion was also reported.[15] Laser oscillation without population inversion was first reported by Zibrov et al.[16] using a V-type system in Rb.[87]

Two sublevels of the ground state (**Figure 9.5**) are coupled to a pair of excited states via three fields. A strong driving field with Rabi frequency Ω_c and a weak probe field with Rabi frequency Ω_a are assumed to be quasi-monochromatic. These fields have linewidths $\Delta\upsilon_c$ and $\Delta\upsilon_a$, both of which are much less than atomic radiative decay rates. The third (pump) field is taken to be incoherent, i.e. it has a very broad linewidth ($\Delta\upsilon_{pump} \gg \gamma_a, \gamma_c$) and is represented by an incoherent pumping rate r.

In the absence of the incoherent pump field, almost all the population is optically pumped into state b′ by the strong driving field. The incoherent pump destroys this optical pumping by populating the upper state a, and thus state b, via spontaneous decay. Hence, the population difference $\rho_{aa} - \rho_{bb}$ is determined by the rates with which atoms leave and decay into states a and b. Note that in the limit of weak probe field, atoms can leave state b only via state c; thus, the population difference above is determined by the rate of decay rates from level a to level b and from level c to level b′. If level c decays more slowly than level a, an inversion on the transition coupled by the weak probe field cannot be created.

Amplification of a weak probe field on the a → b transition is, however, possible, due to the presence of atomic coherence between upper levels a and c. The physical origin of this mechanism is to be interpreted as a quantum interference cancellation of absorption.

Several experiments have shown transient and continuous wave amplification without inversion.[16,17]

Continuous wave amplification and laser oscillation without population inversion was observed in a Λ scheme within the sodium D line by Padmabandu et al.[18] in 1996. An extension of that experiment conducted on the D_1 line of the Cs atom by use of the Λ scheme was performed[19] with a single-pass inversion gain of ~11%.

Although the principle at the basis of lasers without inversion may be exploited also for other phenomena, for example, negative refractive index, the research on this field has been rather meagre.

It could be worth noting that Raman lasers also operate without inversion. However, the scheme is very different. The Raman laser is a device which parametrically transfers energy from one pump frequency to the Stokes or anti-Stokes line, and not from the atoms to the field.

9.3 Random Lasers

Scattering is considered a detrimental effect in ordinary lasers; however, scattering may provide suitable conditions for a new kind of lasers, dubbed *random lasers*.

If a gain medium contains a large number of scattering centers, light is scattered many times before it escapes. The multiple scattering increases the path length of light inside the gain medium, enhancing light amplification. In this case, mirrors are no more needed to trap the light inside the medium. Since strong light scattering usually occurs in highly disordered media, the word "random" has been used to describe lasers operating in this way.

Basov's group at Lebedev Institute, Moscow, realized the ancestor of this kind of lasers with non-resonant feedback, replacing one mirror of the Fabry–Perot cavity of a He–Ne laser with a scattering surface.[20] In this way, light is backscattered in the gain medium with a random direction avoiding the creation of a round trip and therefore resonances of the electromagnetic field are absent. Feedback provides simply to return energy to the system and so all the gain line is used, and the mean frequency of emission does not depend on the dimensions of the laser cavity but only on the center frequency of the amplification line. After reaching a threshold, the oscillation spectrum narrows continuously toward the center of the amplification line.

In a system of this kind, instead of individual high-Q resonances, a large number of low-Q resonances may be considered to appear which spectrally overlap and form a continuous spectrum.

The Russian group found that the process of spectral narrowing was much slower than in an ordinary laser with resonant feedback and that this limits the width of the spectrum in the pulsed mode operation. In cw working, the spectrum width is determined by fluctuations and the statistical properties of the emitted radiation are the ones of the radiation of an extremely bright black body in a narrow range of the spectrum.[21] The radiation has no spatial coherence and is not stable in phase.

In 1967, Letokhov[22] (see Chapter 4) proposed self-generation of light in an active medium filled with scatterers and even suggested it as a mechanism to

explain astrophysical masers.[23] In this system, one must consider the generation length L_{gen}, that is, the average distance a photon travels before generating a second photon by stimulated emission, and the mean path length L_{pat},
that is, the distance a photon travels in the gain medium before escaping. The
stronger the scattering, the longer the L_{pat}. When

$$L_{pat} > L_{gen},$$

every photon generates another photon before escaping the medium, trigger-
ing a chain reaction in which one photon generates two photons, which generate four photons and so on. Since this process is analogous to the process
of neutron generation in an atomic bomb, the device was named "photonic
bomb."

In the 1980s, intense stimulated radiation was observed in neodymium-doped glass and titanium-doped sapphire powders pumped by laser
pulses.[24] Similar results were obtained by Gouedard et al.[25] and Noginov
et al.[26]

In 1994, Lawandy et al.[27] reported laser-like emission from a methanol
solution of Rodhamine 640 perchlorate dye and titanium dioxide microparticles. Unlike the case of the powder samples, the gain medium (the
dye) and the scattering element (microparticles) were separated in the liquid sample. Optical amplification took place outside the scatterer. It was
found experimentally that the threshold was reduced by more than two
orders of magnitude when the density of scattering particles was increased
from 5×10^9 to 2.5×10^{12} cm^{-3} at a fixed dye concentration. The strong
dependence of the threshold on the density of scatterers revealed that the
feedback was related to scattering. The system realized what may be considered a "laser paint."

Since this work, many experimental and theoretical studies of random
lasers have been performed.[28]

Later, Cao[29] showed that when the effective size of the cavity was reduced
by decreasing the scattering mean free path, a limited number of modes could
be identified that mediated coherent laser action over fixed, closed trajectories. The phenomenon received considerable attention and is described as random laser with coherent (or resonant) feedback (also called *coherent random
lasers*).[30]

Cao[31] considered the case of ZnO nanorods grown on a sapphire substrate
of uniform height but randomly located on the substrate. When the system
is pumped with a tripled Nd:YAG frequency (355 nm), discrete narrow peaks
emerge on top of a broad spontaneous emission and additional peaks appear
increasing the pumping power, whose frequency depended on which part of
the sample was pumped (**Figure 9.6**).

The explanation of this behavior is that after multiple scattering, light can
return to a nanorod from which it was scattered before, forming a closed
loop and when the optical amplification along the loop exceeds the loss,
lasing oscillation occurs in the loop that serves as the cavity. Lasing frequencies are determined by the condition that the phase delay along the

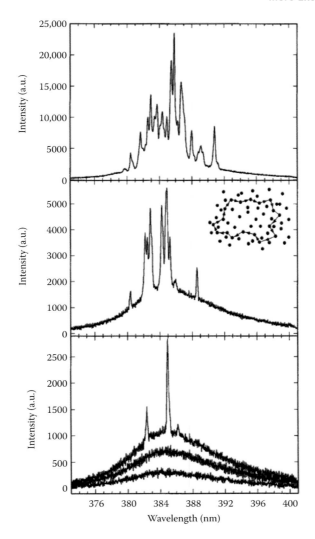

FIGURE 9.6 Spectra of emission from ZnO powder when the excitation intensity is (from bottom to top) 400, 562, 763, 875, and 1387 kW/cm². The inset is a schematic diagram showing the formation of a closed-loop path for light through recurrent scattering in the powder. (From Cao et al., *Phys. Rev. Lett.* **82**, 2278 (1999). With permission.)

closed loop is equal to an integer multiple of 2π. Because light can return to its original position through many different paths which interfere among themselves and only at certain frequencies the interference is constructive, lasing occurs at these frequencies, producing discrete peaks in the emission spectrum.[32]

The behavior of a random laser with coherent feedback can thus be very similar to that of a conventional laser; however, a random laser does not have directional output: because the disorder-induced scattering is random in direction, the output is in all directions. The statistical properties of random lasers with coherent feedback were found similar to those of a traditional laser but very different from the ones of a random laser with incoherent feedback.[33]

In experiments on ZnO powders, Thareja and Mitra[34] obtained emission from closed loops in the ultraviolet. Also electrically pumped random lasers have been considered[35] and random lasers electrically pumped have been extended to the mid-infrared region ($\lambda \approx 10~\mu m$).[36] There exist also random lasers in weakly disordered media such as π-conjugated polymer films.[37]

Microrandom lasers in which light is confined in a volume of the dimension of a wavelength have been realized by Cao et al.[38]

The activity in this kind of lasers both from the theoretical and the applicative side is rather large.[39] The working of random lasers has been the subject of debate for several years and their description has given rise to a large literature.[40] Increasing pumping rate nonlinear effects also come into play.[41]

These lasers are not expensive and may have some interesting application due to the lack of a need for a perfectly reflecting mirror or to the possibility of making coating with a lasing paint, just to give two examples. They may have high temporal coherence due to coherent scattering and interference and low spatial coherence owing to the existence of independent lasing modes with uncorrelated wavefronts and these two properties make random lasers an ideal light source for imaging with high photon radiance while free of coherent artefacts such as speckles.[42] From a more fundamental point of view, they are interesting for studying the interplay between localization and coherent amplification.

9.4 Nanolasers

Nanolaser is a miniature laser that has nanoscale dimensions. Somehow paradoxically while at the beginning one of the problems encountered in designing lasers was that the cavity could not be made on a wavelength scale, as in the case of microwaves; nowadays, with the improvement of the technique, people have not been satisfied until it was able to build cavities smaller than the wavelength (one says to *bet the diffraction limit*). Eventually, the way to obtain this has been to couple light to *plasmons*, because a plasmon has a wavelength much shorter than light.

With the advances in technology, in the 1990s, construction of objects of dimensions of a few nanometers has become possible. Accordingly, what previously appeared impossible, the construction of a resonant cavity of the dimension of the visible wavelength, became a routine affair. Microscopic lasers that can reach the diffraction limit have been built based on photonic crystals,[43] metal-clad cavities,[44] and nanowires.[45]

The vertical cavity surface emitting laser[46] (VCSEL) in which light is confined between two epitaxially grown distributed Bragg reflectors was one of the first semiconductor cavities with dimensions in the order of the wavelength of light. Another microcavity laser, the microdisk laser,[47] uses total internal reflection at the edge of a high refractive index disk to form low-loss whispering gallery-type modes.

However, these lasers have dimensions larger than half the wavelength of the optical field.

The first nanolasers have been developed using photonic crystal cavity lasers. In the microlaser realized by Painter et al.,[48] the light is confined strongly in two dimensions by what are effectively two-dimensional Bragg mirrors formed by making a pattern of holes in a high-refractive-index semiconductor. In all cavities of this type, the refractive index differences of dielectric materials are used to confine the light.

A smaller laser requires less power and can potentially be switched on and off faster, Altug et al.[49] realized a photonic crystal laser with response times as short as a few picoseconds.

The smallest dimensions to which the optical mode of such cavities can be confined is related to the diffraction limit and is in the order of one-half of the wavelength in size in order to be effective, implying that the total laser size is still large compared to the wavelength of light. Therefore, such lasers, albeit of extremely small dimensions, are restricted, both in optical mode size and in physical device dimension, to being larger than half the wavelength of the optical field. However, with the increased miniaturization of all electronic devices, lasers with the dimension of wavelength became too large.

A way to obtain shorter confinement is to make use of surface plasmons[50] which are able to localize light.

In theory, it is possible to confine light with the use of metals to form the laser resonator. The metal may form either strong compact mirrors, able to confine light to about the size of the diffraction limit, or the light may interact strongly with the free electrons in the metal, being guided in the form of surface plasmon polaritons (SPPs) at the interface between a metal and a dielectric material such as air.

Plasmon nanolasers are new lasers smaller than the 3D diffraction limit. Although this division may be somehow arbitrary, we may consider two types of such lasers both working on the use of plasmons: *plasmon lasers* and *spasers*.

9.5 Plasmon Lasers

On the nanoscale, optical fields are almost purely electric oscillations at optical frequencies where the magnetic field component is small and does not significantly participate in the nano-optical physics. The collective oscillations of the sea of conduction electrons in a metal at optical frequencies are what is called a *surface plasmon*.[51] A surface plasmon is excited at the boundary between a dielectric and a metal in the range of frequencies where the metal has negative susceptibility. The most important characteristic is that the wavelength of a plasmon with respect to the wavelength of a light wave of the same frequency is orders of magnitude smaller.

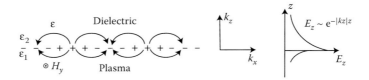

FIGURE 9.7 The charges and the electromagnetic field of surface plasmons propagating on a surface in the x-direction are shown schematically. The exponential dependence of the field E_z is seen on the right. H_y shows the magnetic field in the y-direction of this p-polarized wave.

The electron charges on a metal boundary perform coherent fluctuations which are called *surface plasma oscillations* with a frequency ω that is tied to their wave vector k_x by a dispersion relation $\omega(k_x)$.

A surface plasmon may be described as follows. The charge fluctuations of the metal can be localized in the z-direction within the Thomas–Fermi screening length of about 1 Å and are accompanied by a mixed transversal and longitudinal electromagnetic field which disappears at $z \to \infty$ (**Figure 9.7**) and has its maximum in the surface $z = 0$, typical for surface waves.

The field is described by

$$E = E_0^\pm \exp[i(k_x x \pm k_z z - \omega t)] \tag{9.2}$$

with the sign + for $z \geq 0$ and − for $z \leq 0$, and with imaginary k_z, which causes the exponential decay of the field E_z.

The wave vector k_x lies parallel to the x-direction; $k_x = 2\pi/\lambda_p$, where λ_p is the wavelength of the plasma oscillation. The wave propagates in the x-direction and the problem does not depend on y. Applying Maxwell equations with the continuity conditions at the interface, one obtains the dispersion relation

$$k_x = (\omega/c)[\varepsilon_1\varepsilon_2/(\varepsilon_1 + \varepsilon_2)]^{1/2},$$

where ε_1 and ε_2 are the dielectric constants of the metal and the dielectric, respectively. The wave vector k_x is continuous through the interface. If we assume, besides a real ω and ε_2, a complex ε_1 with $\varepsilon_1'' < |\varepsilon_1'|$ we obtain a complex $k_x = k_x' + ik_x''$ with

$$k_x' = (\omega/c)[\varepsilon_1'\varepsilon_2/(\varepsilon_1' + \varepsilon_2)]^{1/2}$$

and

$$k_x'' = (\omega/c)[\varepsilon_1'\varepsilon_2/(\varepsilon_1' + \varepsilon_2)]^{3/2}[\varepsilon_1''/2(\varepsilon_1')^2].$$

For real k_x', one needs $\varepsilon_1' < 0$ and $|\varepsilon_1'| > \varepsilon_2$ which can be fulfilled in a metal and also in a doped semiconductor near the eigenfrequency; the imaginary part k_z'' determines the internal absorption.

The minimum size scale of structures employing this form of light confinement is related to the penetration depth of light into the metal, which is typically in the order of tens of nanometers. Initially, many researchers doubted if it would be possible to overcome the losses in plasmonic or metallic waveguides and cavities with the currently available optical gain materials. The metal's conduction electrons, which oscillate in synchronism with the optical field, dissipate energy through collisions with the metal's atomic lattice. This energy dissipation leads to high optical losses, so surface plasmons can travel only short distances, and metal mirrors have higher losses than dielectric ones. However, a number of researchers did examine the possibility of using gain in metallic waveguides[52] and microscopic lasers that can reach the diffraction limit, based on metal-clad cavities, have been successfully reported notwithstanding the high losses.

The first report of lasing in metallic nanostructures occurred in 2007[53] with what was dubbed the "gold-finger" laser. It was the first experimental attempt using metals to confine the optical energy to lasing using a round semiconductor pillar encapsulated in a dielectric and then gold, pumped electrically. Successively, the structure was repeated as a cylindrical semiconductor pillar with a height of 1.6 μm and a diameter of about 250 nm, consisting of an InGaAs active medium sandwitched by InP pillar sections, surrounded by a thin insulating Si_3N_4 layer and followed by a silver cladding. The metallic cavity acted as the n-contact itself, and a p-contact was laterally inserted (**Figure 9.8**).[54]

The pillar had an overall diameter of approximately 260 nm and an InGaAs active region height of 300 nm. The InGaAs heterostructure in the pillar and the metal formed a resonator with an optical mode trapped on the InGaAs

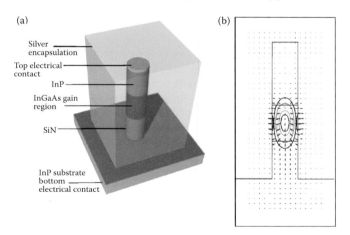

FIGURE 9.8 (a) A semiconductor heterostructure pillar is encapsulated in an insulator, then a low-optical-loss metal. Light escapes through the bottom of the pillar and the substrate. (b) The optical mode trapped in the InGaAs as shown by the electric field squared contour lines. (From M T Hill et al., *J. Opt. Soc. Am.* **B27**, B36 (2010). With permission.)

gain medium (**Figure 9.8b**) in a region with dimensions approximately one-half the wavelength of light in the medium, though it was diffraction limited because of its nonplasmonic nature. The large resistive losses associated with the metal required cryogenic temperatures to obtain sufficient gain in InGaAs to overcome losses in the resonator and achieve lasing at a wavelength of approximately 1400 nm. This small laser was electrically pumped with a threshold current of 6 µA at 77 K. Later, a nanolaser showing plasmonic character with one-dimensional confinement was demonstrated[55] and we will discuss it in the *spaser* section.

A number of other groups also observed in 2009 nanolasers made from metal semiconductor structures lasing in the near-infrared. A metal–dielectric encapsulated semiconductor heterostructure showed room-temperature operation for an optically pumped device that was less than a free space wavelength of light in all three dimensions.[56] At the California Institute of Technology, evidence of lasing in a micro-disk cavity having one side coated with metal was obtained.[57] Another realization was at cryogenic temperatures in a small patch of metal–semiconductor–metal sandwich by Yu et al.[58]

With a different philosophy, Oulton et al.[59] showed lasing in a section of a hybrid dielectric-plasmonic waveguide to overcome the limitations due to losses. The waveguide was constructed from a flat metal substrate (silver) coated with a thin dielectric layer on which a thin high-index semiconductor cadmium sulfide nanowire was placed. A hybrid plasmon mode was concentrated in the insulator gap of 5–10 nm with its tail overlapping with the semiconductor gain.[60] In this approach, the electromagnetic field is lifted from the metal into the dielectric gap, resulting in low loss operation, yet maintaining the plasmonic nature of high confinement. The optical mode propagating in the silver-dielectric–semiconductor waveguide has a significant amount of its energy squeezed into the thin dielectric gap between the nanowire and the metal substrate, leading to a highly localized mode which has an area significantly below the diffraction limit. The nanowire forms a Fabry–Perot cavity with the plasmonic modes resonating between its two ends which are a few micrometers apart. Using short optical pulses to pump the semiconductor nanowire, lasing was shown in the Fabry–Perot cavity at low temperature in the blue part of the visible spectrum.

Plasmon lasers of extremely small mode area at cryogenic and room temperatures have been demonstrated.[61] Other nanolasers were built on deep subwavelength metal nanocavities.[62]

Review papers on metallic and plasmonic nanolasers,[63] photonic crystal lasers,[64] and plasmon lasers[65] have been published.

9.6 Spaser

Making use of surface plasmons (SP), which are capable of tightly localizing light on dimensions much smaller than the wavelength in vacuum, has the drawback that ohmic losses at optical frequencies inhibits the realization. A source of energy must be provided. A new device, *spaser*, was devised for this purpose by Stockman.[66]

Spaser is the acronym for surface plasmon amplification by stimulated emission of radiation. It was first described by Bergman and Stockman in 2003[67] and is a nanoplasmonic counterpart of the laser in which photons are replaced by surface plasmons.[68] A spaser consists of a metal nanoparticle, which plays the role of the laser cavity (resonator), and the gain medium which is an active medium with an inverted population and supplies energy to the lasing.[69]

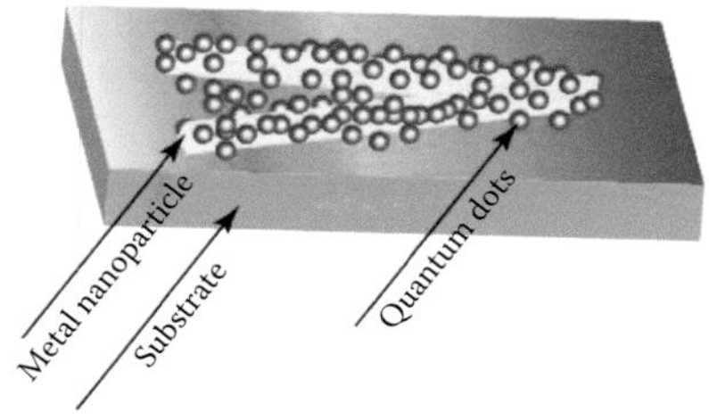

The physical principle is the amplification of surface plasmons at the expenses of an excited material, typically a semiconductor or a dye. **Figure 9.9** shows the original geometry proposed by Stockman which considered a V-shaped metal nanoparticle surrounded by a layer of semiconductor nanocrystal quantum dots.

FIGURE 9.9 Schematic of the spaser as originally proposed by Bergman and Stockman. The resonator of the spaser is a metal nanoparticle shown as a gold V-shape structure. It is covered by the gain medium depicted as nanocrystal quantum dots. This active medium is supported by a neutral substrate. (From M I Sockman, *Opt. Express* **19**, 22029 (2011). With permission.)

The laser has two principal elements: a resonator (or cavity) that supports photonic mode(s) and the gain (or active) medium that is population-inverted and supplies energy to the lasing mode(s). An inherent limitation of a traditional laser is that the size of the laser cavity in the propagation direction is at least half wavelength and practically more than that even for the smallest developed lasers, which we have described previously. In a true spaser, this limitation is overcome. The spasing modes are surface plasmons (SPs) whose localization length is on the nanoscale.[70]

One of the simplest and potentially most promising types of nanoparticles to function as a spaser resonator is a metal–dielectric nanoshell. Such nanoshells have been introduced by Naomi Halas and collaborators and have since found a very wide range of applications.[71]

To better understand its working, we follow the Stockman papers and show in **Figure 9.10** the field around a metallic nanoparticle (**Figure 9.10a** and **b**) together with two possible spaser realizations (**Figure 9.10c** and **d**). In Figure 9.10c, the gain medium is surrounding the metallic nanosphere and the plasmonic mode develops at the interface between the nanosphere and the gain medium. In **Figure 9.10d**, the gain medium is inside the metallic nanosphere. Let us consider the geometry in which a metallic (gold) sphere is covered with a suitable material such as a semiconductor or a dye. Upon excitation of the dye with a suitable pumping laser, a transfer of an optical excitation to a plasmon should occur. A plasmon mode in the shell is therefore feed at the expenses of the optically excited dye.[72]

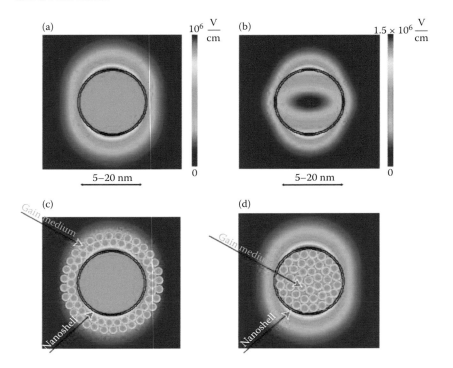

FIGURE 9.10 Schematic of spaser geometry, local fields, and fundamental processes leading to spasing. (a) Nanoshell geometry and the local optical field distribution for one SP in an axially symmetric dipole mode. The nanoshell has an aspect ratio $h = 0.95$. The local field magnitude is color-coded by the scale bar in the right-hand side of the panel. (b) The same as (a) but for a quadrupole mode. (c) Schematic of a nanoshell spaser where the gain medium is outside of the shell, on the background of the dipole mode field. (d) The same as (c) but for the gain medium inside the shell. (From M Stockman, *Opt. Express*, **19**, 22029 (2011). With permission.)

Figure 9.11 shows the physical operation. The gain medium, for example, a semiconductor, is excited and then decays on an intermediate level which has energy very near to the plasmon energy.

The gain medium chromophores may be semiconductor nanocrystal quantum dots,[73] dye molecules,[74] rare-earth ions,[75] or electron–hole excitations of an unstructured semiconductor.[76] Let us consider quantum dots.

The pump excites electron–hole pairs in the chromophores (Figure 9.11), as indicated by the vertical black arrow, which relax to form excitons. The excitons constitute the two-level systems that are the donors of energy for the SP emission into the spasing mode. In vacuum, the excitons would recombine emitting photons. However, in the spaser geometry, the photoemission is strongly quenched due to the resonance energy transfer to the surface plasmon modes, as indicated by the red arrows in the panel.

The plasmons already in the spaser mode create the high local fields that stimulate more emission to this mode by the gain medium, which is the feedback mechanism. If this feedback is strong enough, and the life time of the spaser SP mode is long enough, then an instability develops leading to the avalanche of the SP emission in the spasing mode and spontaneous symmetry breaking, establishing the phase coherence of the spasing state.

The theory of the device was first presented by Bergman and Stockman and has been further developed theoretically.[77]

When the device works, a narrowing of the emission linewidth occurs and a collimated beam may be produced.

As Stockman pointed out, one of the advantages of the spaser compared with existing sources of local fields, which also sets it apart from the laser, is that it can generate dark modes that do not couple to the far-zone optical fields. In other words, a spaser generates coherent, strong local fields, but does not necessarily emit photons.

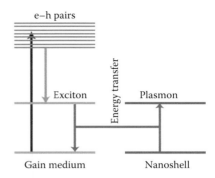

FIGURE 9.11 Schematic of the spasing process. The gain medium is excited and population-inverted by an external source, as depicted by the black arrow, which produces electron–hole pairs in it. These pairs relax, as shown by the green arrow, to form the excitons. The excitons undergo decay to the ground-state emitting SPs into the nanoshell. The plasmonic oscillations of the nanoshell stimulates this emission, supplying the feedback for the spaser action. (From M I Stockman, *Nature Photon.* **2**, 327 (2008).)

This is a potentially great technological advantage because it offers a source of nanolocalized optical fields that does not emit any background radiation. The source can still act on molecules in its near field and excite their radiation (such as fluorescence and Raman effects) as conventional nano-optical sources do.

Spaser effect was observed experimentally[78] in 2009, using 44-nm-diameter nanoparticles with a gold core and a dye-doped silica shell containing the organic dye Oregon Green 488 (OG-488), providing for gain (see **Figure 9.12**).

A modified synthesis technique for producing high-brightness luminescent core-shell silica nanoparticles, known as Cornell dots, was used. The number of dye molecules per nanoparticle was estimated to be 2.7×10^3 and the nanoparticles were in a water suspension with a concentration of 3×10^{11} cm^{-3}. Emission was expected at 525 nm with a quality factor $Q = 14.8$. The measured extinction spectrum, shown in **Figure 9.13**, is dominated by the surface plasmon resonance band at ~520 nm. The samples were pumped at 488 nm with 5-ns pulses. Once the pumping energy exceeded a critical threshold value, a narrow peak appeared at 531 nm.

The emission wavelength corresponds to the lower-order localized plasmon mode of the gold sphere.

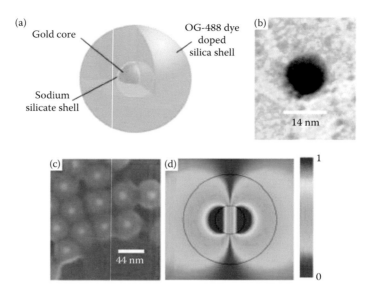

FIGURE 9.12 Spaser design. (a) Diagram of the hybrid nanoparticle architecture (not to scale), indicating dye molecules throughout the silica shell. (b) Transmission electron microscope image of Au core. (c) Scanning electron microscope image of Au/silica/dye core–shell nanoparticles. In (d) spaser mode (in false color), with $\lambda = 525$ nm and $Q = 14.8$; the inner and the outer circles represent the 14-nm core and the 44-nm shell, respectively. The field strength color scheme is shown on the right. (From Noginov et al., *Nature* **460**, 1110 (2009). With permission.)

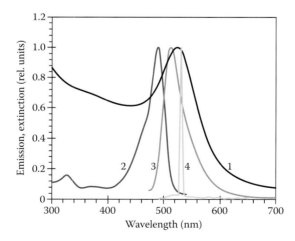

FIGURE 9.13 Spectroscopic results. Normalized extinction (1), excitation (2), spontaneous emission (3), and stimulated emission (4) spectra of Au/silica/dye nanoparticles showing the peak of lasing emission at 530 nm. (From Noginov et al., *Nature* **460**, 1110 (2009). With permission.)

Although this nanoparticle approach provides the ultimate scaling down in all three dimensions, its optical mode extends appreciably outside the structure, and electrical connections are difficult to implement.

A number of surface plasmon polariton (SPP) spasers with different geometries have been experimentally observed[79] and we already discussed some of them. Some more examples are described in the following. A room-temperature semiconductor spaser at 1.46 μm wavelength has been demonstrated by Flynn et al.[80] by sandwiching a gold-film plasmonic waveguide between optically pumped InGaAs quantum-well gain media (**Figure 9.14**). The spaser exhibits gain narrowing, the expected transverse magnetic polarization, and mirror feedback provided by cleaved facets in a 1-mm long cavity fabricated with a flip-chip approach. The 1.06-μm pump threshold of ~60 kW/cm² is in good agreement with calculations.

Another laser emitting in the green[81] has been built with an MOS structure consisting of a bundle of green-emitting InGaN/GaN semiconductor

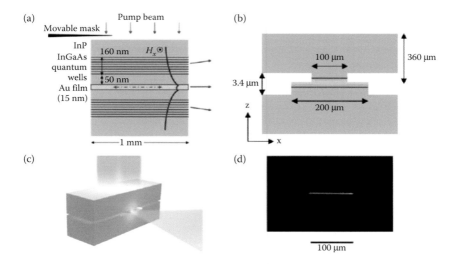

FIGURE 9.14 A room-temperature spaser at 1.5 μm. (a) Side view (not to scale), showing the Au-film plasmonic waveguide sandwiched between two dies containing the quantum well stacks. The stacks are immersed in the evanescent field of the long-range SPP mode represented by the transverse optical magnetic field H_x. Arrows to the right indicate the diagnostic radiation emitted when SPPs reflecting from the end facet. (b) End-on view (not to scale) showing the ridge architecture after dies are flip-chip bonded. (c) Rendering of the bonded dies pumped from above and emitting diagnostic radiation to the right. (d) End-on micrograph of the output facet, showing above-threshold diagnostic radiation at 1.46 μm emitted from the end of the SPP waveguide over the width of 100-μm-wide confining ridge. Output power was obtained by integrating the calibrated micrograph intensity. (From R A Flynn et al., *Opt. Express* **19**, 8954 (2011). With permission.)

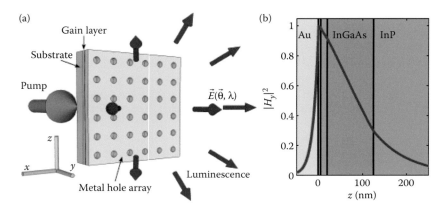

FIGURE 9.15 (a) A semiconductor layer in close proximity of a metal hole array is pumped optically. (b) The simulation of the absolute value squared of the magnetic field of the surface plasmon mode. (From F van Beijnum et al., *Phys. Rev. Lett.* **110**, 206802 (2013). With permission.)

nanorods which is coupled to an underlying colloidal gold triangular plate through a dielectric layer.

A beautiful planar spaser[82] is shown in **Figure 9.15**.

Figure 9.16a shows the output spectrum as a function of pump power. At 40 mW pump power, the peak at 1480 nm increases dramatically and the presence of a threshold is clearly shown in the plot of **Figure 9.16b**.

A spaser emitting continuously in the green has been made by Yu-Jung Lu et al.[83] The laser plasmonic nanocavity is formed between a smooth silver film and a single optically pumped nanorod consisting of epitaxial gallium nitride shell and an indium gallium nitride core (gain medium). The atomically smooth silver layer decreases the modal volume, which reduces surface plasmonic losses.

An electrical injection subwavelength metallic-cavity semiconductor laser operating cw at room temperature was made by Ding et al.[84]

Other proposals to realize spasers have been presented.[85] Stockman has suggested graphene as a suitable material and has discussed an electric spaser in the extreme quantum limit.[86]

9.7 One- and Two-Photon Laser

We discussed already in Chapter 4 the micromaser, a one-atom maser. A microlaser would allow the detection of photons, as well as atoms, whereas only atoms can be probed in the microwave region, making the study of photon statistics in the microlaser quite promising. One major obstacle in the development of a microlaser was, however, the technical difficulty of fabricating a very high-Q cavity in the optical regime. In 1994, An et al.,[87] at MIT, took advantage of the supercavity technology and built a microlaser using

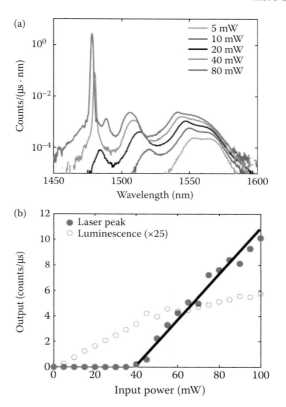

FIGURE 9.16 (a) Upper diagram, luminescence spectra as a function of pump power. Increasing pump power the bandwidth of the luminescence increases until the device starts lasing. (b) Lower diagram shows that the power in the lasing peak exhibits a clear threshold at 40 mW. (From F van Beijnum et al., *Phys. Rev. Lett.* **110**, 206802 (2013). With permission.)

the $^3P_1(m = 0) \rightarrow {}^1S_0$ transition of ^{138}Ba atoms at a wavelength of 791 nm. In the experiment, a beam of ^{138}Ba atoms traversed a single-mode cavity made with two mirrors with a finesse of 8×10^5. The atoms were excited by a π pulse from the $^1S_0$ ground state to the $^3P_1(m = 0)$ excited state before they enter the cavity by means of a cw Ti:sapphire pump laser. Laser oscillation at 791 nm was observed, with the mean number of atoms inside the mode ranging from 0.1 to 1.0, resulting in the mean number of photon inside the cavity changing from 0.14 to 11. The results were in good agreement with theory. The excited atom in the upper energy level entered the cavity with no photon inside. Once inside the cavity, the atom underwent a downward transition due to vacuum Rabi oscillation emitting a photon in the cavity mode. The next excited atom, interacting with the photon in the mode, experienced a larger Rabi frequency, contributing a second photon to the field and so on. This led to an equilibrium situation in which, despite the dynamical atom–field interaction, the number of photons remained constant, and the small energy loss due to the cavity decay and atomic spontaneous decay was exactly balanced by the mean energy transfer from the excited atom to the field mode.

Two-photon lasers in which the stimulated emission involves two photons at the time is another story. If we consider a two-level atom with an upper and a lower level which possess the same parity so that one-photon transition between the two states is electric-dipole forbidden, spontaneous emission of the upper state can occur with the emission of two photons of frequency ω' and ω'', such that $\omega' + \omega'' = \omega_{ab}$, where ω_{ab} is the frequency corresponding to the jump from the two states, through a real or virtual intermediate state with the opposite parity. In the stimulated emission process, one incident photon of frequency $\omega' \# \omega_{ba}$ induces the emission of two photons at ω' and a single photon at the complementary frequency $\omega'' = \omega_{ba} - \omega'$ or two incident photons force the atom to the lower state so that four photons result.

Breit and Teller[88] already considered in 1940 the decay of the hydrogen $2S_{1/2}$ state via two-photon spontaneous emission to the $1S_{1/2}$ ground state, and research on effects involving spontaneous emission or absorption of two photons were subsequently considered.[89]

The first public discussion of a two-photon laser was made by A M Prokhorov[90] who mentioned in his speech for the Nobel Prize, previous work done at Lebedev Institute of Moscow. He discussed the possibility of building a two-photon laser that should produce simultaneously two different frequencies ω' and ω'' with $\omega' + \omega'' = \omega_{ab}$ which could be selected by choosing properly the resonator so that a broadly tunable source could be obtained. He quoted a Russian patent[91] filed in December 1963 in the former USSR by researchers of the Lebedev Institute in Moscow. The following year, a short paper by Selivanenko[92] appeared followed by a longer discussion by Kirsanov and Selivanenko.[93]

At IBM, Sorokin and Braslau in 1964, independently, treated the operating characteristics of a degenerate two-photon laser.[94] A generalization for non-degenerate operation was soon provided by Garwin.[95] The subject was summarized in the Smith and Sorokin book on lasers.[96]

Laser beam amplification due to the stimulated process was observed in some experiments.[97] The first demonstration of a two-photon maser was by Brune, Raimond et al.[98] who used a rubidium atomic transition as we have described in Chapter 4.

The realization of a two-photon laser met with greater difficulty because of the extremely low two-photon gain in usual transitions and to the existence of very strong competing nonlinear processes. Lasers operating on two-photon atomic transitions were widely studied in theoretical papers.[99]

Success was obtained when Gauthier et al.[100] in 1992 realized a continuous-wave two-photon laser employing a driven two-level-atom system made by a Barium atom beam and gave the theoretical interpretation using the dressed-atom picture.[101]

The dressed-atom energy eigenstates are shown in **Figure 9.17a**, where the dressed-state doublets are separated in energy by $\hbar\omega_d$, ω_d being the driving-field frequency, and are split by $\hbar\Omega_d'$, where $\Omega_d'd$ is the generalized Rabi frequency of the pump field

$$(\Omega_d')^2 = (\Omega_d)^2 + \Delta_d^2,$$

being $\Omega_d = 2\mu Ed/h$ the resonant Rabi frequency (μ is the transition matrix element) and $\Delta_d = \omega_d - \omega_o$ is the detuning of the driving field from the atomic transition frequency ω_o.

In **Figure 9.17b**, a schematic diagram of the experiment is shown. A collimated atomic beam of barium was made to pass through the center of a spherical-mirror confocal cavity orthogonally to the cavity axis and was driven by a pump beam of frequency ω_d. The pump laser had a wavelength of 553.5 nm corresponding to the transition $^1S_o \rightarrow {}^1P_1$ of barium. The power was adjusted so that typically $\Omega_d'/2\pi \sim 410$ MHz. A second dye laser was employed to produce a probe beam that propagated along the cavity axis. The cavity mirror spacing (and hence the resonant frequency) could be piezoelectrically scanned.

In the experiment, the total output power emitted out one end of the cavity was measured as a function of the cavity-pump-laser detuning $\Delta_c = \omega_c - \omega_d$, where ω_c is the empty cavity resonance frequency. A second actively stabilized laser was employed to produce a probe beam that propagated along the cavity axis. In the experiment, the output power observed in **Figure 9.18a** arises from the one-photon-gain based laser for $\Delta_c = -\Omega_d'$, corresponding to the peak of driven-atom one-photon gain. To make two-photon gain as well as one-photon gain visible, a cw probe laser (approximately matched to the fundamental mode of the cavity) is injected into one end of the cavity and the power emitted from the other end is probed. The vertical scale of **Figure 9.18b** is calibrated so that unit transmission represents the empty cavity probe

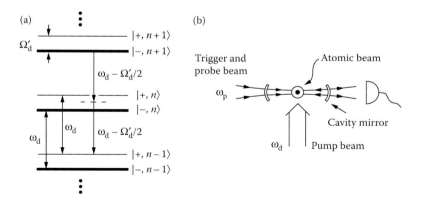

FIGURE 9.17 (a) Dressed-atom doublets represented by lines whose thicknesses indicate the relative populations. Degenerate two-photon gain occurs at $\omega_d - \Omega_d'/2$. (b) Schematic of the two-photon laser. (From Gauthier et al., *Phys. Rev. Lett.* **68**, 464 (1992). With permission.)

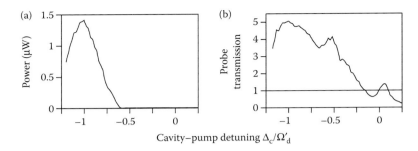

FIGURE 9.18 (a) Power emitted out from one end of the cavity as a function of the cavity-pump laser detuning Δ_c/Ω'_d. The observed signal is due to one-photon-based signal. There is no power emitted for $\Delta = -0.5\,\Omega'_d$. (b) Maximum value of the power emitted from the cavity as a saturating probe beam (Rabi frequency ~240 MHz) is scanned through the cavity resonance for various values of the cavity frequency. The pronounced peak at $\Delta = -0.5\,\Omega'_d$ is due to two-photon gain.

transmission. The output power has a pronounced peak at $\Delta_c = -\Omega'_d/2$ that was attributed to two-photon gain.

The threshold condition for a two-photon laser is quite different from the one of a one-photon laser. The two-photon gain increases not only with the number of active-gain atoms, N, but also with the intracavity light intensity (until saturation occurs). A threshold number of atoms N_{th} was defined as the number of atoms needed to satisfy the gain–equals–loss condition with an intracavity intensity just sufficient to saturate the two-photon gain. However, even if one starts with $N > N_{th}$, gain will not exceed losses until some critical intracavity light intensity is reached. If the laser is initially off, it cannot turn on unless a fluctuation brings the intracavity light intensity above the necessary critical value.[102] Alternatively, a field may be injected into the cavity to start the two-photon laser.[103]

It was predicted[104] and later demonstrated[105] that the two-photon absorption rate is highest for a thermal field (because of bunching) than for a coherent laser field. The properties of the emission of a two-photon laser formed the object of a lively series of studies resumed by Gauthier.[106] In general, the statistical properties of a two-photon laser are very different from those of a one-photon laser and the possible generation of squeezed states was predicted[107] giving rise to a strong controversy.

Raman two-photon lasers were also made.[108]

9.8 High-Power Lasers

Laser power has increased by more than 19 orders of magnitude since Maiman's first demonstration of 10^3 W in 1960. Shortening the pulses and increasing the total energy has brought to very powerful and very large sized lasers. The high power has been obtained thanks to chains of ever more sophisticated amplifiers obtained with new improved technologies. Here we briefly discuss, without pretending to be exhaustive, some of the high-power facilities that were built around the world. These facilities were assembled to study the interaction of high-energy laser light with matter and principally, even if not exclusively, for inertial fusion research. Their number has enormously increased with time and practically every country has today at least one facility.

Beginning 1961, Basov and Prokhorov, disjoint in separated groups, started to be interested in high-power lasers and their application to plasma fusion.[109] Laser-driven fusion was discussed by Basov and Krokhin in March 1961 at a session of the Presidium of the Soviet Academy of Sciences.[110] Later, Basov and Krokhin discussed laser fusion at the 3rd International Quantum Electronics Conference in Paris in 1963.[111] Prokhorov promoted a project, started in 1967, called "Omega", in the future General Physics Institute and studied the interaction of high-energy laser radiation on metal targets with a 10 kJ millisecond Nd:glass laser. CO_2 lasers were also used and a gas-dynamic CO_2 laser was realized. In 1968, Basov group[112] reported the first thermonuclear neutrons and in 1972 new experiments, using spherical targets, with the nine-beamline laser KALMAR, were started at the Lebedev Institute.

In 1971, three laser media—atomic iodine, CO_2, and Nd:glass—were available for scaling to sizes capable of producing 10 TW of energy in pulses shorter than 10^{-7} s. The atomic iodine laser system was initially developed at Lebedev Institute, Moscow, and in the United States CO_2, Nd:glass, and excimer lasers were preferred.

Late in the 1960s, a number of programs started to make high-energy lasers to be used for laser fusion. In 1972, John Nuckolls,[113] in the United States, predicted that ignition could be achieved with laser energies of about 1–10 kJ. This fuelled the construction of a number of high-energy lasers to be used for laser fusion, even if the Nuckolls' prediction resulted immediately wrong, and starting from the 1980s an explosion of big facilities sprang. We briefly mention some of them.

In the United States, the most renowned laboratory for laser fusion studies was the Lawrence Livermore National Laboratory (LLNL), a research laboratory of the United States Department of Energy (DOE), managed by the University of California, established in 1952. Here a number of lasers of increasing power were built and dismantled.[114] The attention was focused on Nd:glass lasers of which were assembled a number of systems of ever-larger powers. In the early 1970s, pulsed lasers with 0.1-ns pulse duration were only capable of delivering about 1.0 J of laser light to a target surface (10^{10} W). They had poor beam amplitude and focusing qualities. However, the Nd ion is capable of holding energy in a metastable atomic state for many hundreds of microseconds, thereby allowing efficient stimulated emission in very short pulses. Disks of Nd in optical glass plates could be fabricated into large diameter which could be excited uniformly by arrays of flash lamps to significant gain levels. A disk laser amplifier could be made[115] so that a low-energy pulse could be amplified to kilojoules passing through 1–2 m of disk amplifiers. All this had to be designed and implemented and constituted a formidable challenge that was gradually solved step by step. After some lasers built starting mid-1960s, the first powerful lasers were the two-beam JANUS and CYCLOPS, both built in 1975, the two-beam Nd:glass ARGUS built the following year and the 20-beam SHIVA built in 1977 with 30-TW power. All these lasers were made to explore the problem of inertial fusion with laser and to give information for the design of a definitive successful big laser. Earlier lasers operated at the fundamental wavelength of 1.06 μm, whereas later lasers were operated at one or more of the harmonic wavelengths because it was established that short-wavelengths reduce deleterious plasma instabilities.

Early in the late 1970s, the design of a new system with the hope to succeed in obtaining ignition conditions in a nuclear fusion project was initiated and in 1984, using Nd:glass, the Nova system was built a high-power laser with 10 beamlines, designed to be a versatile experimental system capable of irradiating targets with pulse duration from 0.05 ns to over 10 ns at wavelengths of 1.05, 0.53, and 0.35 μm.[116] The construction started with a two-beam Nd:glass (phosphate glass) called Novette, assembled as a test proof in about 15 months throughout 1981 and 1982 and completed in January 1983. It was made by recycling parts from Shiva and Argus. With it, the first experiments were done.[117] The Nova complete system entered operation in 1984 and was dismissed in 1999. During its construction, it was proved that Nuckolls' calculations were wrong and there was no hope Nova would reach ignition. The system was then modified into a smaller design and the possibility to make frequency conversion was added.

At the end, Nova as a whole was capable of delivering approximately 100 kJ of infrared light at 1064 nm, or 40–45 kJ of frequency-tripled light at 351 nm in a pulse duration of about 2–4 ns and thus was able to produce a UV pulse of about 16 TW. **Figure 9.19** shows a vision of the 10 beams of the system and gives an idea of the large size of these high-power lasers.

Nova has also played a fundamental role in the study conducted at the Laboratory of X-ray lasers, as we will see later.

Starting in 1992, using chirped pulse amplification (CPA), one of Nova's existing arms was modified to build an experimental CPA laser that produced up to 1.25 PW. Known simply as Petawatt,[118] it operated until 1999 when Nova was dismantled to make way for NIF (National Ignition Facility) which was completed in March 31, 2009, with an energy of 2 MJ distributed in 192 beams with a pulse duration of about 3–10 ns.[119] NIF in 2013 produced a meagre 1% of the laser 1.8 MJ input remaining far from ignition.[120]

Another large laboratory is the Los Alamos National Laboratory (LANL) in New Mexico which was established in 1943 for the atomic bomb realization in the Manhattan project. At Livermore, a number of different lasers were built and dismantled. The laboratory choose to investigate CO_2 lasers because of their high efficiency (2% which they hoped to increase) and because the gaseous medium allows to operate at high repetition rates, the gas cannot be damaged irreversibly by the laser beam, and the required cooling was easily attained. The first CO_2 laser system was the 1 kJ *single-beam system*, designed in 1970 that began operation in 1973 and was in service until 1977. Between 1974 and 1984, a series of short-pulse (about 1 ns) CO_2 laser systems with

FIGURE 9.19 The 10-beam Nova laser (in the photo are visible 5 beams on each side structure). The man in the floor gives an idea of dimensions.

increasing energy and power were designed and constructed. To the *single-beam system* followed GEMINI, a two-beam, 2.5 kJ laser, in service from 1976 to 1982, HELIOS a 8-beam, 10 kJ laser which became operative in April 1978, and ANTARES with its 40 kJ of energy was the largest CO_2 laser ever built. Fusion experiments demonstrated that the long CO_2 wavelength produced plasma instabilities too severe to overcome, and the construction of this type of lasers stopped.

The Laboratory turned its attention to short-wavelength lasers. In August 1989 became operational a KrF laser, AURORA, which was a 96-beam, 1.3 kJ laser system. It also was dismantled.[121] In the early 1990s, a solid-state facility was moved to Los Alamos, the TRIDENT laser[122] with 3 beams, two of them with energy of 400 J with a pulse duration between 100 ps and 1 µs. The other beam was about 100 J with a pulse duration varying between 600 fs up to 2 ns.

The Naval Research Laboratory in Washington, DC, also developed a number of lasers. The Pharos laser is a 2 beam, 1 kJ/pulse infrared Nd:glass laser. A krypton fluoride (KrF) excimer electron beam pumped gas laser, named NIKE, had 56 beams, 4–5 kJ per pulse in the ultraviolet at 248 nm with pulse width of a few nanoseconds. It was completed in the late 1980s. It was built to study the physics of direct drive inertial confinement fusion. The two final amplifiers of the NIKE laser were both electron beam pumped systems. The smaller amplifier produces an output laser beam energy over 100 joules. The larger amplifier amplifies the laser beam up to 5 kJ. The laser will deliver up to 2.2 kJ of 248 nm light onto a planar target with intensities of 2×10^{14} W/cm^2 in a 4 ns pulse.[123] Full operation and target experiments were begun in May 1995.

At the Lawrence Berkeley National Laboratory, University of California, there exists a laser dubbed BELLA. In July 2012, the laser produced 1 PW (10^{15} W) pulse long 40 fs (40×10^{-15} s) at a frequency of 1 Hz achieving the world record power at one pulse per second. The BELLA system was born from years of studies on laser plasma accelerators and should be used to accelerate electrons to 10 GeV in 1 m length.

At the Laboratory for Laser Energetics, an interdepartmental center for advanced study of laser fusion and laser matter interaction, founded by M J Lubin at the University of Rochester, a 24-beam laser OMEGA was completed in 1980 and then dismantled to build the system OMEGA 60 that has been operated since 1995. This system has 60 beams with 40 kJ and pulse width of around 1 ns. In 2008, OMEGA EP has been constructed with 4 beams, 1 kJ. Two beams may operate in ps pulses regime. OMEGA 60 and EP may have joint operation.

At the University of Michigan, the HERCULES laser was upgraded to become a 300 TW Ti:sapphire system, amplified via CPA and capable of producing a repetition rate of 0.1 Hz with pulses of about 30 fs, reaching 2×10^{22} W/cm^2 thanks to the group of Mourou.[124]

The Texas Petawatt Laser (TPL) at the University of Texas, Austin (Texas, USA), has exceeded the Petawatt threshold thanks to the OPCPA technique by compressing an energy of 186 J in a pulse lasting 167 fs.

The Rutherford Appleton Laboratory, located in Chilton, Oxfordshire, UK, was formed in 1957. It hosts the Central Laser Facility which operates high-power glass and Ti:sapphire laser installations (including the Vulcan and Astra lasers) and a number of small-scale, tunable lasers. Previous laser plasma facilities at Culham were also transferred here.

Vulcan formerly was a Nd:glass 6-beam system producing 70 ps to 1 ns pulses with power up to 2 TW or 1 kJ energy at 1053 nm.[125] It was upgraded adding a single beam with chirped pulse amplification (CPA) generating up to 40 TW in ps to sub-ps pulses.[126] It was further upgraded in 1996 to an 8-beam, 2.5 kJ with 2 ns pulse system. It may operate in the frequency doubled mode giving 1 kJ at 532 nm in 2 ns pulse. One of Vulcan's beams is also available as an ultra-short pulse (700 J in 700 fs) using chirped pulse amplification and is employed for inertial confinement fusion. In 2005, Vulcan was the highest intensity focused laser in the world capable of producing a beam with a focused intensity of 10^{21} W/cm^2. It was soon overcome by Hercules. In a near future, Vulcan will be upgraded to 10 PW using OPCPA. In this project, two additional kilojoule-level beamlines would be added.

A KrF laser named SPRITE operated in CPA to produce 1 TW pulses of 300 fs duration[127] was also in service for some time. SPRITE was shut down on April 1, 1995, to upgrade with the new name TITANIA, which reached its first state of operation on April 1, 1996. In CPA mode, the system delivers 10 TW in 300 fs.

Another system, ASTRA Gemini laser, consists of two independent Ti:sapphire laser beams of 0.5 PW each (energy 15 J pulse duration 30 fs).

In France, the Commissariat à l'Energie Atomique (CEA) operated the laser PHEBUS in service between 1985 and 1999 in the Center of Limeil-Brévannes.

The laser had two beams giving energy in the order of 10–20 kJ. The initial wavelength of 1.053 μm was tripled to 351 nm with KDP crystals. The laser was utilized for fusion experiments. A new laser, the Laser Megajoule[128] (LMJ) using Nd:glass as the lasing material was expected to be completed in 2014 (LMJ) as the most energetic laser with 1.8 MJ of energy thanks to 240 converging beams (but the number of beams could be decreased) and pulses of 20 ns. It will be located at the Centre d'études scientifiques et techniques d'Aquitaine (Cesta) in Gironde, 25 km from Bordeaux. A second laser dubbed PETAL (Petawatt Aquitaine Laser) with 3 kJ and 1 ps will be utilized to start the fusion reaction in the plasma created by the Laser Megajoule.

At the Laboratoire d'Utilisation des Lasers Intenses (LULI), Palaiseau, France, there was a 450 J, 600 ps Nd laser at 1.06 μm that was used also for X-ray laser experiments.

In the Soviet Union at Lebedev Institute, under the direction of Basov and Skilizkov, a nine-beamline laser KALMAR was built in 1971 with a power 10^{14}–10^{15} W/cm^2 (energy 100–200 J). In 1981, experiments started at Lebedev with a new 108-beam Nd:glass laser named DELFIN.[129]

Iodine photodissociation lasers were developed at the Lebedev[130] Institute under Basov and jointly produced by the Lebedev Institute and the All-Soviet-Union Scientific Institute for Experimental Physics in Arzamas-16, a Russian city on the Tesa river, 100 km south of Gorkij. In 1968, a iodine photodissociation laser capable of generating up to 1 MJ pulses was assembled. A XeF laser with energy up to 1 kJ in the UV and in the visible region was also built. Other lasers, ISKRA-4 and 5 at the Russian Federal Nuclear Center VNIIEF in Arzamas-16, were also built. ISKRA-4 was completed in 1979. ISKRA-5 was completed in 1989. The main uses of these lasers were in inertial confinement fusion, high-energy density physics, and nuclear weapons research. The ISKRA-4 laser was a spatially filtered 8-beam photolytically pumped iodine gas laser capable of producing laser pulse energies of around 2 kJ with pulse width of about 1 ns at 1.315 μm. It could operate also in double frequency at 658 nm with pulse energy around 500 J. It produced the first thermonuclear neutrons from imploding DT fuel capsules in 1981.

ISKRA-5 is a spatially filtered 12-beam photolytically pumped iodine gas laser capable of producing laser pulse energies around 30 kJ and pulse powers of around 100 TW (pulsewidth about 0.25 ns) at 1.315 μm. It may be frequency doubled.

A new system, ISKRA-6 is under investigation. It would be a 128-beam Nd:glass laser capable of irradiating targets with about 300 kJ at the third harmonics (351 nm) with pulses of around 1 to 3 ns.

Iodine lasers were built also at Max Planck Institute at Garching, Germany.

In Germany, there is a Petawatt High-Energy Laser for heavy Ion eXperiments (PHELIX) which is a Nd;YAG laser at the Gesellschaft fur Schwerionenforschung in Darmstadt, capable of delivering an energy of 120 J in about 500 fs.

In Jena is located the Petawatt Optical Laser Amplifier for Radiation Intensive Experiments (POLARIS)[131] with a power of 100 TW, energy 10 J, and pulses of 100 fs.

The Gekko XII system at the Institute of Laser Engineering, Osaka University, in Osaka, Japan,[132] is a high-power 12-beam Nd:glass laser. It was completed in 1983 and is used for high-energy density physics and inertial confinement fusion research. GEKKO was only frequency doubled at 532 nm and delivered about 10 kJ per 1–2 ns pulse (10–20 TW). From 1996 to 1997, the system was upgraded with a 0.4 kJ, 0.5 PW ultrashort pulse beam used to investigate fast ignition. GEKKO is currently being upgraded with the addition of a second side-by-side laser, the LFEX (Laser for Fast Ignition Experiment) part of the FIREX-1 program, in order to deliver a 10 kJ pulses in 10 ps.[133]

Asterix IV now PALS at the Academy of Sciences of the Czech Republic is a iodine laser at 1.315 μm. The name is the acronym for Prague Asterix Laser System. In Prague, there was a small iodine laser reconstructing the laser Perun 1 created in Lebedev's Physical Institute in Moscow into a more efficient iodine laser Perun 2 in 1992. At that time in Garching, Germany, was working a high-power iodine photodissociation laser system ASTERIX, which was upgraded in the Max Planck Institute for Quantum Optics to the terawatt laser ASTERIX IV. Around 1997, the Max-Planck Institute committed ASTERIX IV to the Academy of Sciences of the Czech Republic and the laser was dismantled and transferred to Prague in 2000 where it was reassembled in an improved version named PALS. In 2003, PALS became a part of the Laserlab-Europe Consortium.

In Europe, with the help of the European Union, some projects are progressing. HiPER is one of the experimental system under development.

We end up mentioning the ELI (Europe's Extreme Light Infrastructure) project. This project started as a bottom-up initiative of the European scientific laser community and the network of large national facilities, Laserlab-Europe. From 2007 to 2010, ELI entered into an European Commission funded preparatory phase, comprising 40 laboratories from 13 countries. Gerard Mourou was the initiator of the project and the coordinator of the preparatory phase. ELI is a large-scale laser facility consisting of four pillars: one devoted to nuclear physics, one to attosecond physics, one to secondary beams (photon beams, ultra-relativistic electron and ion beams; a High Repetition-Rate Advanced Petawatt Laser System (HAPLS) is under construction) and one to high-intensity physics. The construction has started in Czech Republic in Dolni Brezany, near Prague (beamlines facility; April 20, 2011), Hungary, in Szeged (attosecond facility; 2014) and Romania, in Magurele (nuclear physics facility; September 18, 2012). All three facilities are expected to start operation in early 2018. The fourth pillar, the highest intensity laser which should exceed that of the current pillars by about one order of magnitude, is still to be decided.

The list of facilities is not complete and many systems have not been mentioned; however, the ones that were mentioned may testify of the large effort that was made to develop very high-energy and -power lasers. Most of them for laser fusion, and they constituted a very large money effort without the expected return.

We have frequently mentioned upgrading of some of the previous structures with chirped pulse amplification (CPA) or optical parametric chirped pulse amplification (OPCPA). It was only after the experimental implementation of these techniques that it was possible to reach very high intensities. We already explained CPA in Chapter 8; the other amplification technique called optical parametric chirped pulse amplification (OPCPA) is based on the nonlinear interaction among laser beams in crystals and was suggested almost at the same time as CPA.[134] The properties of optical parametric amplifiers can be used to provide tuneable pulses that can form the basic component in a chirped pulse amplifier to deliver powers exceeding 10 PW.[135]

Chirped pulse amplification (CPA) had a dramatic impact in short-pulse amplification allowing to fully use the high-energy storage media[136] such as neodymium glass, alexandrite, titanium sapphire, and chromium $LiSrAlF_6$.

In these systems, CPA stretches an impulse by 10^3–10^5, amplifies it by 10^{11} (from nanojoules to tens of joules) and recompress by 10^3–10^5. The ability to compress a short pulse over so many orders of magnitude maintaining its characteristics came with the discovery of the matched stretcher and compressor. The first CPA system used the positive group velocity dispersion (in which the red wavelength goes faster than the blue) of a single mode fiber to temporally spread the frequency components of the ultrashort pulse.[137] After passing through the fiber, the pulse is stretched with the red component first, followed by the blue. It is then amplified to the desired level and recompressed by a pair of parallel diffraction gratings. This pair exhibits a negative group velocity dispersion (blue goes faster than red). However, the fiber stretcher and diffraction grating compressor did not have their dispersive characteristics exactly matched, so the recompression was not perfect, leading to temporal wings in the pulse, and hence limiting the stretching compression ratio to about 100.

In 1987, Mourou became aware[138] of a compressor designed by Oscar Martinez at the University of Buenos Aires for optical communications, to compress pulses at a 1500-nm wavelength.[139] In this spectral domain, light in optical fibers experiences a negative group velocity dispersion. The proposed dispersive system was a telescope of magnification 1 placed between two anti-parallel gratings. Mourou noted that the device had the exact same dispersive function as the Treacy compressor,[140] but with the opposite sign and demonstrated a terawatt system using a Nd glass laser system.[141] Successively in a collaboration of Mourou with CEA Limeil and Palaiseau, generation of 50 TW with femtosecond pulses was obtained using a femtosecond Ti:sapphire source and a regenerative amplifier connected to an existing Nd glass power chain at 1064 nm by using CPA.[142]

An alternative was to use very short-length amplifying media to minimize the stretching-compression ratio.[143]

Ultrashort pulses have enormous spectra and all the spectral components must be amplified equally over many orders of magnitude. For ultrashort pulses, the ultrabroad band of Ti:sappire was pivotal.[144] This laser has a gain bandwidth that can theoretically support the amplification of pulses of less than 5 fs in duration. It was first used in high-peak-power systems by J Kmetec, J J Macklin, and J F Young[145] at Stanford and by the R Falcone group[146] at Berkeley.

CPA or OPCPA techniques have been applied in many of the big lasers we have discussed previously.

9.9 High-Power Fiber Lasers

The first operation of a fiber laser was demonstrated by Snitzer et al.[147] in the 1960s and revisited in the 1970s by Stone and Burrus.[148] The subject received major interest 10 years later.[149] The work on neodymium-doped fibers demonstrated a gain of several decibels per milliwatt of absorbed pump power and made these systems promising for optical telecommunications in the wavelength region around 1.06 μm. Interest was later focused on the erbium-doped fiber amplifier at 1.55 μm.[150] After these first works, cw fiber lasers, specifically silica-host Yb-doped fiber-based lasers (YDFLs), have been developed with high power.[151] Since 1999, YDFLs have been developed for pulsed high-power operation.[152] The subject has been reviewed by Richardson, Nilsson, and Clarkson.[153]

9.10 X- and Gamma-Ray Laser

X-ray sources would enable to study macromolecules in living tissues, increase the power of X-ray crystallography, and allow us to obtain very short pulses with which "freeze" molecular vibrations and electronic motion in matter, just to mention some of the advantages. For this reason, the possibility of creating such sources was pursued since the very beginning.

However, the problem of decreasing laser wavelength using traditional schemes is a formidable one. One of the main reasons is that the probability of spontaneous emission scales with the third power of frequency (see Chapter 2). As a consequence, the pump power needed for operation increases at very high levels. In the traditional scheme, the threshold pump power for laser action may be written as

$$P_{\text{th}} = h\nu\Delta n_{\text{th}}/\tau,$$

where Δn_{th} is the population difference at threshold and τ is the upper state lifetime. Under reasonable assumptions, P_{th} turns out to be proportional to ν^6. Thus, going from green wavelength (500 nm) to the soft X-ray region (10 nm), the threshold has an increase by a factor 50^6. Another obstacle is the inefficiency of cavity reflectors, because in the X-ray region the refractive index of any material is practically 1, and special optics must be used, with the exception of the longer wavelengths of soft X-ray region (tens to approximately 0.2 nm) in which multi-layer structures can be used.

Nowadays, the X-ray laser action is principally realized in highly ionized plasmas or with free-electron lasers. High harmonics production was considered in Chapter 8.

In the case of plasma produced x-rays, considering the gain (g) length (L) product (gL) of the system, one may speak of laser action when in the spectroscopic regime the lasing signal is relatively high compared with non-lasing lines ($gL < 5$). Increasing the gL product, $5 < gL < 15$ is the stage in which the lasing line dominates in the spectrum. At last, the saturated amplification regime $gL > 15$ means that a significant part of the population inversion is converted into the laser signal.

People started to speak of X lasers practically since the very beginning of the laser story and the literature on the subject is huge; we will limit here to present only what at our knowledge and judgement were some of the most significant results.

A first good proposal for making X-ray lasers was made by the Soviet researchers Gudzenko and Shelepin[154] who suggested in 1965 a scheme for the development of soft X-ray laser in a recombining plasma (*recombination scheme*).

Nearly 10 years later, in July 1972, a first announcement of the making of an X-laser was made by John G Kepros et al.[155] at the University of Utah who reported to have obtained X-rays irradiating copper sulfate by pulses of energy up to 30 J from a 20 ns Nd-glass laser. The experiment, however, was not reproducible.

The literature on the possible schemes for far-ultraviolet and X-ray laser making in the meanwhile was enormously growing.[156] Vinogradov and Sobel'man[157] considered that in the region of a few Ångstrom one may use a plasma with multiply charged ions expanding into a neutral gas. The plasma could be produced using a powerful laser pulse onto a solid target. Using a high-power laser pulse, it is possible to create a plasma from which the laser may be obtained. In the equilibrium plasma, ions having specific number of electrons such as 2 (helium-like), 10 (neon-like), 28 (nickel-like), and 46 (palladium-like) are relatively stable. These ions survive in a wide range of temperatures and densities. In such relatively stable plasmas, the electron collisional excitation may create the population inversion and a source emitting around 1–10 Å could be produced.

Zherikhin, Koshelev, and Letokhov[158] first described a mechanism for obtaining an inversion between $2p^53p$ and $2p^53s$ levels in neon-like ions. Vinagradov et al.[159] refined the theoretical description. To produce the inversion, the $n = 3$ excited levels are populated by electron impact excitation from the ground ($2p^6$) state of the neon-like ion, which is itself produced in the plasma heated by the optical laser. The population inversion between $2p^53p$ and $2p^53s$ levels develops because of the large difference between the radiative decay rates. Subsequent calculations extended the scheme to other ions.[160] **Figure 9.20** shows a simplified energy diagram of neon-like and nickel-like ion scheme.

Meanwhile, population inversion in C^{5+} in a recombining plasma was observed by Irons and Peacock.[161]

In 1975, George Chapline and Lowell Wood of the Lawrence Livermore National Lab produced a review paper[162] on the subject that until that moment had many announces but no clear results. The paper considered a number of possible schemes and resumed some of the experiments. Chapline later recalled that he and Wood first thought about a bomb-driven X-ray laser as a result of discussion with Sobel'man. Some tests were executed.[163] Peter Hagelstein made a number of proposals, but no positive result was obtained. As a consequence, in 1976, DARPA stopped supporting the research in the United States, and Livermore remained the only center of US X-ray research.

In the Soviet Union, at the beginning of the 1970s, in the Quantum Radiophysics Laboratory of the Physics Institute of the USSR Academy

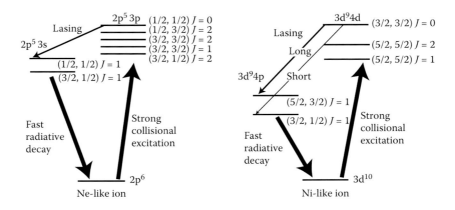

FIGURE 9.20 Schematic diagrams of the electron collisional excitation for neon- and nickel-like ion schemes. (From H Daido, *Rep. Progr. Phys.* **65**, 1513 (2002). With permission.)

of Sciences in Moscow, led by Basov, work begun on finding ways to make lasers that operate in the far ultraviolet and soft X-ray regions of the spectrum. As already said, a number of theoretical possible schemes were suggested for producing inversion on the transitions of multiply charged ions in a laser plasma.[164] Two schemes were considered as being the best: creating an inversion via electronic excitation in the transitions of Ne-like ions, and the method of selective photopumping.[165] At the same time, experimental efforts to achieve amplification and generation on the transitions of the neon-like CaXI begun. In 1977, preliminary positive results were obtained for the first time in the range around 600 Å.[166]

A review of these and other efforts was made by Bunkin et al.[167]

Starting from the suggestion of Gudzenko and Shelepin,[154] an extensive theoretical model for the recombination lasers in rapidly expanding cylindrical plasmas was developed by Geoffrey Pert,[168] at the University of Hull, UK, who shortly reviewed the situation and in 1980 reported laser gain at 18.2 nm.

Eventually, in 1984, quasi-continuous wave operation, limited only by the duration of the pumping impulse, could be achieved using outer-shell "optical" transitions in highly ionized atoms. It was so that a first laser was obtained by Matthews et al.[169] using a 24 times ionized Se. The population inversion is maintained by rapid depletion of the lower-state density and this is usually accomplished through radiative decay to the ground state.

In Matthew's experiment, pumping was achieved by the powerful second-harmonic beam (532 nm) of the Novette laser (pulse energy around 1 kJ, pulse duration about 1 ns), consisting of one arm of the Nova laser at Lawrence Livermore Laboratory. The beam was focused to a fine line ($d \sim 200$ μm, $l = 1.2$ cm) on a thin stripe (75-nm thick) of selenium, evaporated on a 150 nm thick foil of Formvar (**Figure 9.21**). The foil could be irradiated from one or both sides. Exposed to the high intensity of the pump beam ($\sim 5 \times 10^{13}$ W/cm²), the foil exploded to form a highly ionized, approximately cylindrical, selenium plasma whose diameter was $d \sim 200\ \mu m$. The model explaining the results was reported in a previous paper in the same issue of *Physical Review Letters*.[170] During the electron–ion recombination process, a particularly long lived constituent of this plasma is formed, which consists of Se^{24+}. This ion has the same ground-state electronic configuration as neutral Ne ($1s^2 2s^2 2p^6$), accordingly it is usually referred as neon-like selenium (**Figure 9.22**). Impact collisions with hot-plasma electrons ($T_e \sim 1$ keV) raise Se^{24+} from its ground state to excited states, thereby achieving population inversion between

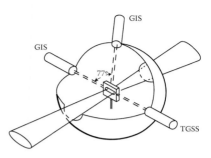

FIGURE 9.21 Sketch of the Matthews experiment. (From Matthews et al., *Phys. Rev. Lett.* **54**, 110 (1985). With permission.)

the states $2p^5 3p$ and $2p^5 3s$ because the lifetime of the 3s to ground-state transition is much shorter than the lifetime of the 3p to 3s transition. With the pump configuration in Figure 9.21, a strong longitudinal emission due to amplified spontaneous emission (ASE) was observed on two lines (20.63 and 20.96 nm) of the $2p^5 3p$ to $2p^5 3s$ transition. Due to the much higher nuclear charge of Se compared to Ne, these lines fall in the soft X-ray region.

From the length dependence of the emitted energy, one deduced that a maximum single-pass gain $G = \exp(\sigma Nl)$ of about 700 was obtained. The X-ray output energy produced was only 10^{-10} of the pump energy.

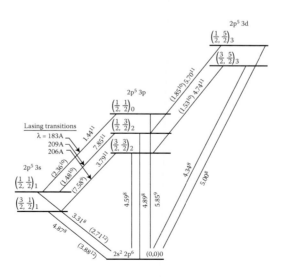

FIGURE 9.22 Simplified energy-level diagram for Ne-like Se. (From Rosen et al., *Phys. Rev.* **54**, 106 (1985). With permission.)

Matthews et al. achieved gain on 3p–3s transitions in neon-like selenium. The dominant line expected[171] to be amplified was the $J = 0$ to 1 transition at 182.4 Å. Instead, the group did not observe large gain on this line but saw gains of order 5 cm^{-1} on $J = 2$ to 1 transitions. An alternative picture of the operating mechanism was discussed by Apruzzese et al.,[172] to explain this anomaly, and received support by later calculations and experimental results by Nilsen and Moreno.[173]

Contemporarily, Szymon Suckewer et al. of Princeton University measured a gain-length product of 6.5 on the 18.2 nm carbon line and an amplification of about 100 using a 300 J CO_2 laser confining the plasma magnetically.[174] This was the first clear demonstration of recombination type soft-X-ray lasing. The recombination pumped scheme was originally based on the rapid cooling of the medium by free expansion which also caused rapid decrease of the electron density.

The Livermore group was very active. They obtained gain with Y^{29+} at 155.0 and 157.1 Å and with Mo^{32+} at 131.0 and 132.7 Å.[175] They also demonstrated soft X-ray amplification in nickel-like ions Eu^{35+} at 71.0 Å.[176] The scheme was repeated in Yb[177] decreasing the emission wavelength to 50.26 Å. This wavelength was further shortened to 43.18 Å.[178]

The theory of a Ni-like, analogue to the Ne-like scheme, had previously been discussed by Maxon et al.[179] Later, the Livermore group extended the Ni-like scheme to Ta at 44.83 Å and W at 43.18 Å.[180]

Collisionally pumped Ni-like lasers are 4d–4p transitions in high-Z Ni-like ions. The 4d levels are populated through a combination of direct collisional excitation from the ground state and cascading from upper levels. The 4d–4p population inversion is maintained by fast radiative decay from the 4p levels to the 3d Ni-like ground state, while the 4d levels are metastable to radiative decay to the ground state. The largest gain in expected on the $J = 0 \rightarrow 1$ transition $(\underline{3d}_{3/2}4d_{1/2})_0 \rightarrow (\underline{3d}_{3/2}4p_{1/2})_1$ at 44.83Å in Ta, where $\underline{3d}_{3/2}$ denotes a vacancy in an otherwise full $\underline{3d}$ shell. The upper state of this transition is populated mainly by collisional excitation from the ground state. In the experiment Ta^{45+} and W^{46+} were used. The W transition $(\underline{3d}_{3/2}4d_{1/2})_0 \rightarrow (\underline{3d}_{3/2}4p_{1/2})_1$ produced the shortest wavelength at 43.18 Å. A Ni-like amplifier was produced by irradiating a thin foil of Ta with the Nova laser at Livermore. The heated foil then became a plasma that expands to form a large, uniform gain medium. At a density of 10^{21} cm^{-3} and an electron temperature of the order of 1 keV, a significant Ni-like population is produced with a measured gain of 2.3 cm^{-1} with a duration of 250 ps at 44.83 Å. Ni-like W produced a gain of 2.6 cm^{-1} with a total $gL = 7$ at 43.18Å.

Observation of high gain in Ne-like Ag was also reported by the group[181] at 9.9 nm with a gain of 9.4 cm^{-1}. They obtained also the shortest wavelength yet observed at 8.15 nm for a Ne-like laser. The group developed an improved excitation scheme using two laser pulses. The experiment was conducted at the Nova 2-beam facility at Lawrence Livermore National Lab. Two opposite beams of the Nova laser, operating at 0.53 μm with 500 ps width pulses were focused into a line on an Ag foil. Also a transition at 8.15 nm showed gain, being the shortest wavelength Ne-like laser measured to that time.

The X-ray laser performance is significantly improved by the use of a soft X-ray mirror as a reflector for double pass amplification. The first demonstration of half cavity (double pass amplification) and full cavity operation was performed by Ceglio et al.[182]

The question of why in the Ne-like Se laser, the $J = 2 \rightarrow 1$, $3p \rightarrow 3s$ transitions at 206 and 209 Å dominated the spectra while the $J = 0 \rightarrow 1$, $3p \rightarrow 3s$ transition at 182 Å which was predicted to be the dominant laser line was always found very weak and was not even observed in the early experiments, remained a mystery for several years. In 1995, Nilsen and Moreno[183] instead of using a long pulse employed a series of short 100 ps pulses from the Nova laser to illuminate slab targets of selenium. With this technique, the 182 Å line appeared to be dominant (**Figure 9.23**). The explanation that the two authors gave was very simple. The multiple pulses created a larger, more uniform density plasma, at the densities required for lasing at the short wavelengths. The first pulse heats and expands the plasma. During the second and third pulses, a much larger and more uniform plasma is created which can directly absorb the optical pulse and which is at the right densities for gain and laser propagation. Computer simulations confirmed the model. Since then multi pulse technique was mostly used.[184]

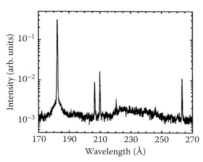

FIGURE 9.23 Spectrum of Se illuminated by a series of three 100 ps pulses, 400 ps apart. The Ne-like $J = 0 \rightarrow 1$ laser line at 182 Å dominates the spectrum. The $J = 2 \rightarrow 1$ lines at 206 and 209 Å as well as the other laser lines at 220 and 262 Å are quite visible. (From J Nilsen and J C Moreno, *Phys. Rev. Lett.* **74**, 3376 (1995). With permission.)

A number of successful experiments adopting different elements, different target design, and different pumping schemes were demonstrated in many laboratories around the world. The advances were relied in the development of high-power lasers and on a better understanding of the atomic and plasma physics.[185] Neon-like ions (from Ag^{37+} to Ag^{8+}), hydrogen-like ions (from Al^{12+} to C^{5+}), lithium-like ions (from Si^{11+} to Al^{10+}), and nickel-like ions (from Au^{51+} to E^{35+}) all were demonstrated to work.[186] For the same level of pumping laser intensity, Ni-like, $4d \rightarrow 4s$ transitions produce wavelengths shorter than Ne-like, $3p \rightarrow 3s$ systems, as obvious.

The collisional radiative equilibrium model formed the object of refined calculations[187] by the Japan group at Osaka University and Daido et al.[188] obtained emission at 6–8 nm with Ni-like lanthanide ions (Nd, Sm, Gd, Tb, and Dy) pumping with multiple infrared pulses at 1.053 μm. The laser pulse was composed of two or three 100 ps pulses separated by 400 ps. A gain coefficient of 3.1 cm⁻¹ and a $gL = 7.8$ was achieved at 7.97 nm in the Nd ions with 250 J of pumping energy on a target long 2.5 cm. The energy was an order of magnitude less than the value in the previous Ni-like lasers. The output power was estimated to be of the order of 10 kW (energy 1 µJ). The group used a curved slab target. The use of curved slab targets was introduced by Lunney in 1992[189] and was an important advance helping to compensate for the refractive losses in the plasma.

The basic idea of the prepulse, as we already said, is that it helps create a large-scale length plasma with smaller gradients. Without the prepulse, refractive losses dominate. Relaxing the density gradient in collisionally excited X-ray lasers through multiple pulse pumping proved the most significant efficiency enhancing technique since the introduction of curved slab targets.

The possibility to make compact source, indicated as table-top system was demonstrated both using collisional excitation[190] and recombination schemes.[191]

Table-top lasers, that is, compact lasers with a reasonable dimension, were developed since the 1990s.[192] Rocca et al.[193] at Colorado University reported large amplification ($gL = 7.2$) at 46.9 nm in a discharge-created plasma using Ne-like Ar. At Livermore, Dunn et al.[194] demonstrated a Ni-like laser pumped with less than 10 J of laser energy from a "table-top laser" utilizing two laser pulses: one to form the plasma with 5 J energy and 800 ps duration and a pump beam of 5 J energy in a pulse of 1.1 ps. A gain of 35 cm^{-1} and a $gL = 12.5$ were measured on the 4d to 4p $J = 0$ to 1 transition for Ni-like Pd at 147 Å. Significant progress was since then made[195] developing high repetition rate soft X-ray lasers.[196]

A dramatic improvement of the spatial coherence and beam divergence of a 13.2 nm Ni-like Cd table-top laser was obtained by the group at University of Colorado[197] by injection seeding with high harmonics pulses (the 59th harmonic) from a Ti:sapphire laser generated in a Ne gas jet. The harmonic was injected into a grazing incidence pumped Ni-like Cd plasma amplifier that was irradiated with a sequence of Ti:sapphire laser pulses. Successively,[198] the group obtained gain-saturated 10.9 nm tabletop laser operating at 1 Hz repetition rate with an average power of 1 µW and pulse energy of up to 2 µJ.

Rocca et al.[199] reported saturated laser operation with 5 Hz repetition rate of a transient collisional electron excitation of Ni-like Cd at 13.2 nm. A gain length product $gL = 17.6$ was obtained heating a pre-created plasma with 8 ps pulses of only 1 J energy from a table-top Ti:sapphire laser. The laser had an average power of 1 µW.

In 2008, the group sent high harmonics from a Ti:sapphire laser into the excited plasma. Where plasma gain matched a harmonic frequency, the plasma amplified the harmonic seed pulse, generating gain-saturated output at 18.9 nm from nickel-like molybdenum and 13.9 nm from nickel-like silver.[200] The technique of amplification of coherent high-harmonic pulses in the population inversions created in dense plasmas had been previously demonstrated.[201]

An all-diode-pumped soft X-ray laser in the 18.9 nm line of Ni-like Mo at a repetition rate of 10 Hz using a compact CPA Yb:YAG pump laser at 1.03 µm was described by the Rocca group.[202]

More recently, the group demonstrated a 100 Hz repetition rate gain saturated diode pumped table-top X-ray laser at 18.9 nm, generating an average power of 0.15 mW, in a Mo plasma.[203]

The study of X-ray lasers was and continues to be the object of a series of dedicated conferences. In Aussois, France, in 1986, was held an International Colloquium on X-ray lasers which was the first of a series of conferences devoted to X-ray lasers. The series continued in 1990 with the 2nd Symposium held in York, the 3rd in Scliersee in 1992, and then in many other localities with a cadence of 2 years.

One of the first books on X-ray lasers was written by Elton in 1990.[204] A partial list of the facilities for X-ray lasers was given by Kato in 1996.[205] An updated review on soft X-ray laser has been made in 2002 by H Daido.[206]

Other approaches to obtain X-ray lasers have been through high harmonic generation and frequency mixing. High harmonic generation was discussed in Chapter 8. X-ray lasing may be obtained also by channelling high-energy electrically charged particles in crystals, as discussed, for example, in Elton's book.

At very short wavelengths, gamma ray lasers (*Grasers*) were also considered. A gamma-ray laser operating on a nuclear transition in a solid was discussed as early as 1961.[207] Much of the basic understanding was formulated in the Soviet Union in the 1970s.[208] Several review papers have discussed the various proposals and possibilities without any experimental device working.[209] Also optical pumping of a gamma-ray laser was proposed.[210]

The other way to obtain soft X-ray lasers is through the use of free-electron lasers. The first proposal for a soft X-ray free-electron laser producing radiation at about 1–20 nm was made by Brian Newnam from Los Alamos National Laboratory in 1988.[211]

Since the electrons in a free-electron laser are not bound to atoms, the wavelength of the laser is not limited to specific atomic transitions. It can therefore be tuned by adjusting the energy of the accelerator or the magnetic field strength and modulation frequency of the undulator. The X-ray laser will also be several orders of magnitude more intense than any current X-ray source. In addition, pulses will be short in the region of femtoseconds.[212]

The amplification medium of a free-electron laser is made up of a beam of free electrons that are forced to move in a strong magnetic field. Bunches of free electrons are first generated in an electron gun and travel in the undulator emitting radiation. The photons are initially incoherent and concentrated over a narrow range of wavelengths, with the peak wavelength depending on the energy of the electrons and the properties of the magnetic field. After leaving the undulator, the electron beam is finally separated from the light beam by a bending magnet. In general, the incoherent light can be made coherent by using an optical resonator with mirrors at both ends of the undulator. The photon pulses then pass repeatedly through the undulator and meet up on each round trip with an electron bunch. The problem with X-rays is that appropriate mirrors for such a resonator do not exist in this spectral region. A way of amplifying the spontaneously emitted light in one single pass is needed. One solution is self-amplified spontaneous emission (SASE).[213]

SASE is based on the fact that while the synchrotron radiation emitted by the electrons moves through the undulator at the speed of light, the electrons themselves actually travel slightly slower. The electrons therefore lag a little behind their emitted radiation, which can catch up with—and interact with—earlier electrons. The interaction will either accelerate or decelerate the electrons depending on their exact position and the phase of the light wave with which they interact.

The net result is that the light wave pushes the electrons into smaller so-called micro-bunches, which are separated by a distance corresponding to the wavelength of the undulator magnetic field. Several electrons now start to emit light in tandem, producing light of a higher intensity. This light then groups the electrons into tighter and tougher bunches, and causes them to radiate in phase. As a result, the radiation power rises exponentially with distance along the undulator, until it finally saturates. As the SASE scheme amplifies and saturates the radiation in a single pass, it does not require an optical cavity. For this process to work, however, electron beams of extremely high quality are needed.

Therefore, with measured gains of a few percent per meter, the realization of a SASE FEL awaited the development of sufficient bright electron sources. In 1992, Claudio Pellegrini proposed using a recently developed RF photocathode electron gun coupled to the two-mile Stanford linear accelerator to produce lasing in the hard-X-ray region.[214] The facility was dubbed the Linac Coherent Light Source (LCLS). Many groups followed developing similar sources.

After many years of development, efforts started to come to success. In September 2000, at the Deutsches Elektronen-Synchrotron (DESY) Laboratory in Hamburg, Germany, the first demonstration was obtained of saturation in a visible SASE FEL[215] with an output wavelength down to 80 nm. The group subsequently raised the device to 150 nm.[216]

The first FEL in the soft X-ray range was FLASH which began operation in 2005 at DESY in Hamburg, using SASE[217] at wavelengths down to 65Å. It was since then upgraded to run from 6.7 to 47 nm.

Subsequently, the VISA collaboration between the Brookhaven National Lab, UCLA, and the Stanford Linear Accelerator Center demonstrated lasing to saturation with SASE at 800 nm with a very short wiggler, only 4 m in length. Then Linac Coherent Light Source (LCSL)[218] which uses the last kilometer of SLAC linear accelerator to pump electrons to 14.3 GeV and is provided with an undulator 112-m long and generates pulses at 0.15–1.5 nm lasting 1–230 fs was the world's first hard X-ray FEL since October 1, 2009. The X-ray wavelength is easily tunable from 22 to 1.2 Å by varying the electron energy in the range 3.5–15 GeV.[219] It completed its first year of operation in 2010. Construction was initiated in 2005 as a collaborative effort by SLAC, Argonne National Laboratory, and Lawrence Livermore National Lab.[220]

In 1999, a Brookhaven National Lab/Argonne National Lab/University of California Los Angeles collaboration showed lasing to saturation using a sub-harmonic seed laser.[221] This technique called *high gain harmonic generation* relies on the fact that FELs emit harmonics of the lasing wavelength. The harmonics are preserved in the electron density distribution and can be used to seed a harmonic amplifier in the form of a second wiggler tuned to one of the harmonics. The seed laser for an amplifier can thus be at a subharmonic of the eventual laser frequency.

In 2013, the only other FEL in the hard X-ray spectrum using SASE was Japan's Spring-8 the Angstrom Compact Free Electron Laser (SACLA).[222] SACLA was built at the RIKEN Laboratory in Japan[223] and on December 1, 2011, gave the first demonstration of intense full-coherent soft-X-ray laser.

However, the temporal coherence of SASE is intrinsically poor because of the start-up process from random noise and the lack of a spectral purifier.

A seeded FEL using a high-order harmonic-generation source was successfully demonstrated at vacuum ultraviolet wavelengths.[224] However, the reduced intensity of the seed pulse at shorter wavelengths limits the implementation of this scheme. Instead, a self-seeding scheme in which monochromatized SASE was used as a coherent seed drastically shortened the wavelength, enabling hard x-ray to be produced at LCLS.[225] In this scheme, a crystal is installed in the middle of the undulators to monochromatize the SASE emitted from the first half of the undulators. The monochromatized light is then reused to seed the amplification process in the second half of the undulators.

Another approach is to use frequency upconversion.[226] Generation of fully coherent soft-X-ray pulses by a two-stage high-gain harmonic generation FEL scheme was reported in Italy by the Free Electron Laser for Multidisciplinary Investigations (FERMI) in Trieste.[227] FERMI obtained lasing at wavelengths in the range 20–65 nm in 2012, pushed down to 4.3 nm in 2013. To prevent degradation of the electron beam quality, the electron bunch is divided into two portions, one is used in the first stage and the other in the second stage. The seed laser pulse overlaps the first portion of the electron bunch in the first stage and the higher harmonic radiation obtained in the first stage is then suitably adjusted to be used in the second stage.[228] The FERMI apparatus is also able to produce variably polarized light using helical undulators.

Other hard X-ray FELs are under construction in Germany, South Korea, and Switzerland.

An overview of X-ray sources and a preview of X-ray quantum optics have been presented by B W Adams et al.[229]

Notes

1. A Javan, *Phys. Rev.* **107**, 1579 (1957).
2. D Marcuse, *Proc. IEEE* **51**, 849 (1963); H K Holt, *Phys. Rev.* **A16**, 1136 (1977).
3. A M Clogston, *J. Phys. Chem. Solids* **4**, 271 (1958); V M Kontorovich and A M Prokhorov, *JETP* **6**, 1100 (1958).
4. T W Haensh and P E Toschek *Z. Phys.* **236**, 213 (1970); T Popov, A Popov and S Rautian, *JETP Lett.* **30**, 466 (1970); V Arkhipkin and Yu Heller, *Phys. Lett.* **40A**, 12 (1983).
5. O Kocharovskaya and Ya I Khamin, *JETP Lett.* **48**, 630 (1988).
6. S E Harris, *Phys. Rev. Lett.* **62**, 1033 (1989); S E Harris and J H Macklin, *Phys. Rev.* **A40**, 4135 (1985).
7. M O Scully, S Y Zhu and A Gavrielides, *Phys. Rev. Lett.* **62**, 2813 (1989).
8. A Imamoglu, *Phys. Rev.* **A40**, 2835 (1989); A Lyras, X Tang, P Lambropoulos and J Zhang, *Phys. Rev.* **A40**, 4131 (1989).

9. H Lee and M O Scully, *Z. Naturforch.* **A34**, 33 (1999); M O Scully and M S Zubairy, *Quantum Optics* (Cambridge University Press: Cambridge, 1997); M O Scully and M Fleischhauer, *Science* **263**, 337 (1994); J Mompart and R Corbalan, *J. Opt. B Quantum Semiclassical Opt.* **2**, R7 (2000).

10. For a treatment of self-induced transparency, see S McCall and E Hahn, *Phys. Rev. Lett.* **18**, 908 (1967); *Phys. Rev.* **183**, 457 (1969); R Brewer and E Hahan, *Phys. Rev.* **A8**, 464 (1973).

11. E Arimondo and G Orriols, *Nuovo Cimento Lett.* **17**, 333 (1976); G Alzetta, A Gozzini, L Moi and G Orriols, *Nuovo Cimento* **B36**, 5 (1976); G Alzetta, L Moi and G Orriols, *Nuov. Cim.* **B52**, 209 (1979); G Orriols, *Nuov. Cim.* **53**, 1 (1979); H M Gray, R M Whitly and C R Stroud Jr., *Opt. Lett.* **3**, 218 (1978); H J Yoo and J H Eberly, *Phys. Rep.* **118**, 239 (1985).

12. O Kocharovskaya and P Mandel, *Phys. Rev.* **A42**, 523 (1990).

13. A Imamoglu, J E Field and S E Harris, *Phys. Rev. Lett.* **66**, 1154 (1991).

14. E S Fry et al., *Phys. Rev. Lett.* **70**, 3235 (1993).

15. X Li et al., *Proceedings of the International Conference on Lasers* 92, Houston, 1992 (STS Press: McLean, VA, 1992), p. 446; A Nottelmann, C Peters and W Lange, *Phys. Rev. Lett.* **70**, 1783 (1993); E S Fry et al., ibidem **70**, 3235 (1993); W E van der Veer et al., ibid. **70**, 3243 (1993).

16. A S Zibrov, M D Lukin, D E Nikonov, L Hollberg, M O Scully, V L Velichansky and H G Robinson, *Phys. Rev. Lett.* **75**, 1499 (1995).

17. G G Padmabandu, G H Welch, I N Shubin, E S Fry, D E Nikonov, M D Lukin and M O Scully, *Phys. Rev. Lett.* **76**, 2053 (1996); F B de Jong, A Mavromanolakis, R J C Spreeuw and van Linden, *Phys. Rev.* **A57**, 4869 (1998).

18. G G Padmabandu et al., *Phys. Rev. Lett.* **76**, 2053 (1996).

19. P S Bhatia, G R Welch and M O Scully, *J. Opt. Soc. Am.* **B18**, 1587 (2001).

20. R V Ambartsumyan et al., *Progress in Quantum Electronics*, J H Sanders and K W H Stevens, eds. Vol. 1 (Pergamon Press: London, 1970); R V Ambartsumyan, N G Basov, P G Kryukov and V S Letokhov, *JETP Lett.* **3**, 167 (1966); *Sov. Phys. JETP* **24**, 481 (1966); *IEEE J. Quantum Electron.* **QE-2**, 442 (1966).

21. R V Ambartsumyan, P G Kryukov, V S Letokhov and Yu A M Natveets, *JETP Lett.* **5**, 312 (1967); *Sov. Phys. JETP* **26**, 1109 (1967).

22. V S Letokhov, *Sov. Phys. JETP* **26**, 1246 (1968); V S Letokhov, *Sov. Astronomy— A.J.* **16**, 604 (1973); V S Letokhov, in *Amazing Light, A volume dedicated to Charles Hard Townes on his 80th birthday*, edited by R Y Chiao (Springer-Verlag: New York, 1996), p. 409.

23. V S Letokhov, *Sov. Phys. JETP* **26**, 835 (1968).

24. V M Markushev, V F Zolin and Ch Briskina, *Sov. J. Quantum Electron.* **16**, 281 (1986); V M Markushev, N E Ter-Gabrielyan, Ch M Briskina, V R Belan and V F Zolin, *Sov. J. Quantum Electron.* **20**, 773 (1990); N E Ter-Gabrielyan et al., *Sov. J. Quantum Electron.* **21**, 840 (1991).

25. C Gouedard et al., *J. Opt. Soc. Am.* **B10**, 2358 (1993).

26. M A Noginov et al., *J. Opt. Soc. Am.* **B13**, 2024 (1996).

27. N M Lawandy, R M Balachandran, A S I Gomes and E Sauvain, *Nature* **368**, 436 (1994).

28. See, for example, H Cao in *Optics & Photonic News*, January 2003, p. 24.

29. H Cao et al., *Phys. Rev. Lett.* **82**, 2278 (1999).

30. H Cao in *Progress in Optics*, edited by E Wolf (Elsevier: Amsterdam, 2003), Vol. 45, p. 317.

31. H Cao, *Phys. Rev. Lett.* **82**, 2278 (1999).

32. A Yu Zyuzin, *Europhys. Lett.* **26**, 517 (1994); *Europhys. Lett.* **46**, 160 (1999).

33. G Zacharakis et al., *Opt. Lett.* **25**, 923 (2000).
34. R K Thareja and A Mitra, *Appl Phys.* **B71**, 181 (2000).
35. E S P Leong and S F Yu, *Adv. Mater.* **18**, 1685 (2006).
36. H K Liang et al., *Adv. Mater.* **25**, 6859 (2013).
37. S V Frolov et al., *Phys. Rev.* **B59**, R5284 (1999); R C Polson et al., *Synth. Met.* **119**, 7 (2001); K Yoshino et al., *Appl. Phys. Lett.* **74**, 2590 (1999).
38. H Cao, J Y Xu, E W Seeling and R P H Chang, *Appl. Phys. Lett.* **76**, 2997 (2000).
39. A good review of the initial work may be found in H Cao, in *Progress in Optics*, edited by E Wolf (Elsevier: Amsterdam, 2003), Vol. 45, p. 317.
40. S John and G Pang, *Phys. Rev.* **A54**, 3642 (1996); D S Wiersm and A Lagendijk, *Phys. Rev.* **E54**, 4256 (1996); H Cao et al., *Phys. Rev.* **E61**, 1985 (2000); X Jiang and C M Soukoulis, *Phys. Rev. Lett.* **85**, 70 (2000); C Vanneste and P Sebbah, *Phys. Rev. Lett.* **87**, 183903 (2001); R C Polson, M E Raikh and Z V Vardeny, *C.R. Phys.* **3**, 509 (2002); V M Apalkov, M E Raikh and B Shapiro, *Phys. Rev. Lett.* **89**, 016802 (2002); A Tulek et al., *Nature Phys.* **6**, 303 (2010); S Mujumdar et al., *Phys. Rev. Lett.* **93**, 053903 (2004).
41. J Andreasen, P Sebbah and C Vanneste, *J. Opt. Soc. Am.* **B28**, 2947 (2011).
42. B Redding, M A Choma and H Cao, *Nature Photon.* **6**, 355 (2012).
43. H Altug, D Englund and J Vuckovic, *Nature Phys.* **2**, 484 (2006).
44. M T Hill et al., *Nature Photon.* **1**, 589 (2007).
45. J C Johnson et al., *Nature Mater.* **1**, 106 (2002); X Duan, Y Huang, R Agarwal and C M Lieber, *Nature*, **421**, 241 (2003); M A Zimmler, J Bao, F Capasso, S Muller and C Ronning, *Appl. Phys. Lett.* **93**, 051101 (2008); M H Huang et al., *Science* **292**, 1897 (2001); J C Johnson et al., *J. Phys. Chem.* **B105**, 11387 (2001); A H Chin et al., *Appl. Phys. Lett.* **88**, 163115 (2006); F Xia et al., *Nature Nanotechnol.* **3**, 609 (2008).
46. I I Jewel, I P Harbison, A Scherer, Y H Lee and L T Florez, *IEEE J. Quantum Electron.* **27**, 1332 (1991).
47. S I McCall et al., *Appl. Phys. Lett.* **60**, 289 (1992); see also A C Tamboli et al., *Nature Photon.* **1**, 61 (2007).
48. O Painter, R K Lee, A Scherer, A Yariv, J D O'Brien, P D Dapkusand and I Kim, *Science* **284**, 1819 (1999).
49. H Altug, D Englund and J Vuckovic, *Nature Phys.* **2**, 484 (2006).
50. R F Oulton et al., *Nature* **461**, 629 (2009).
51. R H Ritchie, *Phys. Rev.* **106**, 874 (1957) studied this kind of excitations that were demonstrated by C J Powell and J B Swan, *Phys. Rev.* **118**, 640 (1960).
52. M P Nezhad, K Tetz and Y Fainman, *Opt. Express* **12**, 4072 (2004); S A Maier, *Opt. Comm.* **258**, 295 (2006); A Maslov and C Z Ning, *Proc. SPIE* **6468**, 646801 (2007).
53. M T Hill et al., *Nature Photon.* **1**, 589 (2007).
54. T M Hill, *J. Opt. Soc. Am.* **B27**, B36 (2010).
55. M T Hill et al., *Opt. Express* **17**, 11107 (2009).
56. M P Nezhad et al., *Nature Photon.* **4**, 395 (2010).
57. R Perahia et al., *Appl. Phys. Lett.* **95**, 201114 (2009).
58. K Yu, A Lakhani and M C Wu, *Opt. Express* **18**, 8790 (2010).
59. R F Oulton et al., *Nature* **461**, 629 (2009).
60. P F Oulton et al., *Nature* **461**, 629 (2009).
61. P F Oulton et al., *Nature Photon.* **2**, 496 (2008); V J Sorger et al., *Nature Commun.* **2**, 331 (2011); R M Ma et al., *Nat. Mater.* **10**, 110 (2010).
62. K Ding et al., *Phys. Rev.* **B 85**, 041301-1-5 (2012); S H Kwon, J H Kang, S K Kim, and H G Park, *IEEE J. Quantum Elect.* **47**, 1346–1353 (2011); M J H Marell et al., *Opt. Express* **19**, 15109 (2011).
63. M T Hill, *J. Opt. Soc. Am.* **B27**, B36 (2010).

64. S Noda, *J. Opt. Soc. Am.* **B27**, B1 (2010).

65. R M Ma et al., *Laser Photonics Rev.* **7**, no. 1, 1 (2013).

66. M I Stockman, *Phys. Today* 64, 39 (2011); M I Stockman, *Opt. Express* 19, 22029 (2011); M I Stockman and D J Bergman, Surface plasmon amplification by stimulated emission of radiation (spaser), US Patent 7, 569, 188 (2009); K Li, X Li, M I Stockman, and D J Bergman, *Phys. Rev.* **B 71**, 115409 (2005).

67. D J Bergman and M I Stockman, *Phys. Rev. Lett.* **90**, 027402 (2003).

68. M. I. Stockman, *Nat. Photon.* **2**, 327–329 (2008).

69. M Stockman has publicized his proposal in a number of papers. Here we quote some of them. D J Bergman and M I Stockman, *Laser Phys.* **14**, 409 (2004); M I Stockman, *Physics Today*, **64** February 2011, p. 39; M I Stockman, *Phil. Trans. R. Soc.* **A369**, 3510 (2011); M I Stockman, *Opt. Express* **19**, 22029 (2011); M I Stockman, *J. Opt.* **12**, 024004 (2010).

70. M I Stockman, S V Faleev and D J Bergman, *Phys. Rev. Lett.* **87**, 167401 (2001).

71. R D Averitt, D Sarkar and N J Halas, *Phys. Rev. Lett.* **78**, 4217 (1997).

72. For a discussion of the field in metal nanoparticles, see for ex. S D Campbell and R W Ziolkowski, *Adv. Optoelectron.* 2012 article ID368786 doi: 10.1155/2012/368786.

73. S Kim et al., *Nature* **453**, 757 (2008); E Plum, V A Fedotov, P Kuo, D P Tsai and N I Zheludev, *Opt. Express* **17**, 8548 (2009).

74. J Seidel, S Grafstroem and L Eng, *Phys. Rev. Lett.* **94**, 177401 (2005); M A Noginov et al., *Phys. Rev. Lett.* **101**, 226806 (2008).

75. J A Gordon and R W Ziolkowski, *Opt. Express* **15**, 2622 (2007).

76. M T Hill et al., *Nat. Photon.* **1**, 589 (2007); M T Hill et al., *Opt. Express* **17**, 11107 (2009).

77. M I Stockman, *Nature Photon.* **2**, 327 (2008); *J. Opt.* **12**, 024004 (2010); M I Stockman, *Phys. Rev. Lett.* **106**, 156802 (2011); *Phil. Trans. Roy. Soc.* **A369**, 3510 (2011); M I Stockman, *Phys. Rev. Lett.* **106**, 156802 (2011); E S Andrianov et al., *Opt. Lett.* **36**, 4302 (2011).

78. M A Noginov et al., *Nature* **460**, 1110 (2009).

79. M T Hill et al., *Opt. Express* **17**, 11107 (2009); R F Oulton et al., *Nature* **461**, 629 (2009); R M Ma et al., *Nat. Mater.* **10**, 110 (2010); M J H Marell et al., *Opt. Express* **19**, 15109 (2011); V J Sorger and X Zhang, *Science* **333**, 709 (2011); R M Ma et al., *Rev. Sci. Instrum.* **82**, 033107 (2011); M Ma et al., *Laser Photon. Rev.* **7**, 1 (2013); M T Holl, *J. Opt. Soc. Am.* **B27**, B36 (2010); M T Hill et al., *Nature Photon.* **1**, 589 (2007); J Y Suh et al., *Nano Lett.* **12**, 5769 (2012); S Y Liu et al., *Opt. Lett.* **36**, 1296 (2011).

80. R A Flynn et al., *Opt. Express* **19**, 8954 (2011).

81. C Y Wu, C T Kuo, C Y Wang, C L He, M H Lin, H Ahn and S Gwo, *Nano Lett.* **11**, 4256 (2011).

82. F van Beijnum et al., *Phys. Rev. Lett.* **110**, 206802 (2013).

83. Y J Lu et al., *Science* **337**, 450 (2012).

84. K Ding et al., *Opt. Express* **21**, 4728 (2013).

85. N I Zheludev, S L Prosvirnin, N Papasimakis and V A Fedotov, *Nature Photon.* **2**, 351 (2008).

86. V Apalkov and M I Stockman, Graphene spaser, arXiv:1303.0220v1(2013); D Li and M I Stockman, *Phys. Rev. Lett.* **110**, 106803 (2013).

87. K An, J J Childs, R R Dasari and M S Feld, *Phys. Rev. Lett.* **73**, 3375 (1994).

88. G Breit and E Teller, *Astrophys. J.* **91**, 215 (1940).

89. See, for example, D J Gauthier in *Progress in Optics*, edited by E Wolf (Elsevier: Amsterdam, 2003), Vol. 45, p. 205.

90. A M Prokhorov, *Science* **10**, 828 (1965).

91. A M Prokhorov and A C Selivanenko, USSR Patent Application No. 872, 303 priority December 24, 1963.
92. A S Selivanenko, *Opt. Spectrosc.* **21**, 54 (1966).
93. B P Kirsanov and A S Selivanenko, *Opt. Spectrosc.* **23**, 242 (1967).
94. P P Sorokin and N Braslau, *IBM J. Res. Dev.* **8**, 177 (1964).
95. R L Garwin, *IBM Res. J. Res. Dev.* **8**, 338 (1964).
96. W V Smith and P P Sorokin, *The Laser* (McGraw-Hill: New York, 1966).
97. M T Loy, *Phys. Rev. Lett.* **41**, 473 (1978); H Schlemmer, D Froelich and H Welling, *Opt. Comm.* **32**, 141 (1980); B Nikolaus, D Z Zhang and P E Toschek, *Phys. Rev. Lett.* **47**, 171 (1981).
98. M Brune et al., *Phys. Rev. Lett.* **59**, 1899 (1987): *Phys. Rev.* **A35**, 154 (1987).
99. V S Letokhov, *JETP Lett.* **7**, 221 (1968); L R Estes, L M Narducci and B Shammas, *Lett. Nuovo Cim.* **1**, 775 (1971); L M Narducci, W W Eidson, P Furcinitti and D C Eteson, *Phys. Rev.* **A16**, 1665 (1977); R L Carman, *Phys. Rev.* **A12**, 2048 (1975); H P Yuen, *Appl. Phys. Lett.* **26**, 505 (1975); *Phys. Lett.* **51A**, 1 (1975); *Phys. Rev.* **A13**, 226 (1976); K J McNeil and D F Walls, *J. Phys.* **A8**, 104, 111 (1975); T Hoshimiya et al., *Jpn. J. Appl. Phys.* **17**, 2177 (1978); N Nayak and B K Mohanty, *Phys. Rev.* **A19**, 1204 (1979); Z C Wang and H Haken, *Z. Phys.* **B55**, 361 (1984); **B56**, 77 (1984); **B56**, 83 (1984).
100. D J Gauthier et al., *Phys. Rev. Lett.* **68**, 464 (1992).
101. C Cohen-Tannouji and S Reynaud, *J. Phys.* **B10**, 365 (1977).
102. M Brune et al., *Phys. Rev.* **A35**, 154 (1987).
103. Z C Wang and H Haken, *Z. Phys.* **B55**, 361 (1984).
104. P Lambropoulos, C Kikuchi and R K Osborn, *Phys. Rev.* **144**, 1081 (1966).
105. F Shiga and S Imamura, *Phys. Rev. Lett.* **25**, 706 (1967).
106. D J Gauthier in *Progress in Optics*, edited by E Wolf (Elsevier: Amsterdam, 2003) Vol. 45, p. 205.
107. H P Yuen, *Appl. Phys. Lett.* **26**, 505 (1975); *Phys. Rev.* **A13**, 2226 (1976).
108. D J Gauthier in *Progress in Optics*, edited by E Wolf (Elsevier: Amsterdam, 2003), Vol. 45, p. 205.
109. See, for example, N V Karlov, O N Krokhin and S G Lukishova, *Appl. Opt.* **49**, F32 (2010).
110. S Yu Gus'kov and V B Rozanov, *P N Lebedev Institute activity on inertial confinement physics* in *Inertial Confinement Nuclear Fusion: A Historical Approach by Its Pioneers*, edited by G Velarde and N Carpintero-Santamaria (Foxwell & Davies Scientific, 2007), p. 313 and N G Basov and O N Krokhin, *Sov. Phys. JETP* **19**, 123 (1964).
111. N G Basov and O N Krokhin, in *Electronique Quantique-Quantum Electronics*, Comptes Rendus de la 3eme Conference Internationale (Dunod, 1964), Vol. 2, p. 1373.
112. N G Basov et al., *JETP Lett.* **8**, 14 (1968).
113. J Nuckolls et al., *Nature,* **239**, 129 (1972).
114. For a history of the initial efforts toward laser fusion in America, see J L Bromberg, *The Laser in America: 1950–1970* (The MIT Press: Cambridge, 1991), p. 214.
115. J M McMahon et al., *IEEE J. Quant. Electron.* **QE-9**, 992 (1973).
116. J F Holzrichter, *Nature* **316**, 309 (1985).
117. K R Maries and W W Simmons, *J. Opt. Soc. Am.* **A2**, 528 (1985).
118. Perry et al., *Opt. Lett.* **24**, 160 (1999).
119. J F Holzrichter has written an excellent recollection of the Livermore work for a chapter in *Inertial Confinement Nuclear Fusion: A Historical Approach by Its Pioneers,* edited by G Velarde and N Carpintero-Santamaria (Foxwell & Davies Scientific, 2007).

120. O A Hurricane et al., *Nature* **506**, 343 (2014).
121. D B Harris et al., *Laser and Particle Beams*, **11**, 323 (1993).
122. N K Moneur et al., *Appl. Opt.* **34**, 4274 (1995); S H Bata et al., *Rev. Sci. Instrum.* **79**, 10F305 (2008).
123. T Lehecka et al., *Opt. Comm.* **117**, 485 (1995).
124. V Yanovsky et al., *Opt. Express* **16**, 2109 (2008).
125. C L S Lewis et al., *Opt. Comm.* **91**, 71 (1992).
126. J Zhang et al., *Phys. Rev. Lett.* **74**, 1335 (1995).
127. I N Ross et al., *Opt. Comm.* **109**, 288 ((1994).
128. C Sauteret et al., *Opt. Lett.* **18**, 214 (1993).
129. I V Aleksandrova et al., *Sov. J. Quantum Electron.* **13**, 1103 (1983).
130. See N V Karlov, O N Krokhin and S G Lukishova, *Appl. Opt.* **49**, F32 (2010).
131. J Hein et al., *AIP Conf. Proc.* **1228**, 159 (2010).
132. Y Kato *Appl. Phys. Lett.* **38**, 72 (1981); K Yamakawa, H Shiraga and Y Kato, *Opt. Lett.* **16**, 1593 (1991).
133. R Kodama & the Fast-Ignitor Consortium, Nature, **418**, 933 (2012).
134. A Piskarkas et al., *Sov. Phys. Uspeki*, **29**, 869 (1986); A Dubietis, G Jonusauskas and A Piskarkas, *Opt. Comm.* **88**, 437 (1992).
135. I N Ross et al., *Opt. Comm.* **144**, 125 (1997).
136. P Maine et al., *IEEE J. Quantum Electron.* **24**, 398 (1988); M D Perry et al., *J. Opt. Soc. Am.* **B8**, 2384 (1991); C Sauteret et al., *Opt. Lett.* **18**, 214 (1993); C N Danson, L N Barzanti and Z Chang, *Opt. Commun.* **103**, 392 (1993); K Yamakawa, H Shiraga and Y Kato, *Opt. Lett.* **16**, 1593 (1991).
137. D Strickland and G Mourou, *Opt. Comm.* **56**, 219 (1985).
138. G A Mourou, C P J Barty and M D Perry, *Physics Today*, January 1998, p. 22.
139. O E Martinez, *IEEE J. Quantum Electron.* **QE-23**, 1385 (1987).
140. E B Treacy, *IEEE J. Quantum Electron.* **QE-5**, 454 (1969).
141. P Maine and G Mourou, *Opt. Lett.* **13**, 467 (1988).
142. C Rouyer et al., *Opt. Lett.* **18**, 214 (1993).
143. B E Lem off and P J Barty, *Opt. Lett.* **18**, 1651 (1993); J Zhou et al., *Opt. Lett.* **20**, 64 (1995).
144. D E Spence, P N Kean and W Sibbett, *Opt. Lett.* **16**, 42 (1991); J P Chamberet et al., *Opt. Lett.* **21**, 1921 (1996); J Zhou et al., *Opt. Lett.* **20**, 64 (1995); C P Barty et al., *Opt. Lett.* **21**, 668 (1996); A Stingl et al., *Opt. Lett.* **19**, 204 (1996); B C Stuart et al., *Opt. Lett.*, **22**, 242 (1997).
145. J D Kmetec, J J Maclin and J F Young, *Opt. Lett.* **16**, 1001 (1991).
146. A Sullivan et al., *Opt. Lett.* **16**, 1406 (1991).
147. E Snitzer, *J. Appl. Phys.* **32**, 36 (1961); *Phys. Rev. Lett.* **7**, 444 (1961); C J Koester and E Snitzer, *Appl. Opt.* **3**, 1182 (1964).
148. J Stone and C a Burrus, *Appl. Phys. Lett.* **23**, 388 (1973).
149. S B Poole, D N Payne and M E Fermann, *Electron. Lett.* **21**, 737 (1985); R J Mears, L Reekie, I M Jauney and D N Payne, *Electron. Lett.* **21**, 738 (1985).
150. R J Mears, L Reekie, I M Jauncey and D N Payne, *Electron. Lett.* **23**, 1026 (1987).
151. H M Pask et al., *IEEE J. Sel. Top. Quantum Electron.* **1**, 2 (1995); R Paschotta et al., *IEEE J. Quantum Electron.* **33**, 1049 (1997); D C Hanna et al., *J Mod. Opt.* **37**, 517 (1990).
152. Y Jeong et al., *Opt. Express* **12**, 6088 (2004).
153. D J Richardson, J Nilsson and W A Clarkson, *J. Opt. Soc. Am.* **B27**, B63 (2010).
154. L I Gudzenko and L A Shelepin, *Sov. Phys. JETP* **18**, 998 (1964); *Sov. Phys. Dokl.* **10**, 147 (1965).
155. J G Kepros et al., *Proc. Natl. Acad. Sci. USA* **69**, 1744 (1972).

156. L Gold, *Quantum Electronics*, Proc. 3rd Int. Congress, Paris, 1963, Vol. 2, Paris, 1964, p. 1155; M A Duguay and P M Pentzepis, *Appl. Phys. Lett.* **10**, 350 (1967); A G Molchanov, *Sov. Phys. Uspekhi* **15**,124 (1972); P Jaegle et al., *Phys. Lett.* **36A**, 167 (1971).

157. A V Vinogradov and I I Sobel'man, *Sov. Phys. JETP* **38**, 1115 (1973).

158. A Zherikhin, K Koshelev and V Letokhov, *Sov. J. Quant. Electron.* **6**, 82 (1976).

159. A V Vinagradov and Shiwaptsev, *Sov. J. Quantum Electron.* **13**, 1511 (1983).

160. U Feldman, A Bhatia and S Suckewer, *J. Appl. Phys.* **54**, 2188 (1983); U Feldman, J F Seely and A K Bhatia, *J. Appl. Phys*, **56**, 2475 (1984); J P Apruzese and J Davis, *Phys. Rev.* **A28**, 3686 (1983).

161. F E Irons and N J Peacock, *J. Phys.* **B7**, 1109 (1974).

162. G Chapline and L Wood, *Phys. Today*, June 1975, p. 40.

163. See J Hecht, *OPN*, 19, May 2008, p. 26.

164. A V Vinogradov, I I Sobelman and E A Yukov, *Sov. Phys. JETP* **36**, 1115 (1972); *Sov. J. Quantum Electron.* **5**, 59 (1975).

165. A V Vinogradov, I I Sobelman and E A Yukov, *Sov. J. Quantum Electron.* **5**, 59 (1975); Photopumping consisted in using the laser radiation to pump inner shell atomic electrons to create population inversion on inner levels. It was first proposed by M A Duguay and P M Rentzepis, *Appl. Phys. Lett.* **10**, 350 (1967); It was later considered by several authors among whom are B A Norton and N J Peacock, *J. Phys.* **B8**, 989 (1975); V A Bhagavatula, *J. Appl. Phys.* **47**, 4535 (1976); W E Alley et al., *J. Quantum Spectrosc. Radiat. Transfer* **27**, 257 (1982); N Qi and M Krishnan, *Phys. Rev. Lett.* **59**, 2051 (1987); J. Nilsen et al., *Appl. Opt.* **31**, 4950 (1992); W T Silfvast et al., *Opt. Lett.* **8**, 551 (1983); H C Kapteyn, *Appl. Opt.* **31**, 4931 (1992); J Nilsen, *Appl. Opt.* 31, 4957 (1992) and references therein.

166. A A Ilyukhin, G V Peregudov, E N Ragozin, I I Sobel'man and V A Chirkov, *JETP Lett.* **25**, 535 (1977).

167. F V Bunkin, V I Derzhiev and S I Yakovlenko, *Sov. J. Quantum Electron.* **11**, 981 (1981).

168. G J Pert, *J. Phys. B: Atom. Mol. Phys.* **9**, 3301 (1976).

169. D L Matthews et al., *Phys. Rev. Lett.* **54**, 110 (1985).

170. M D Rosen et al., *Phys. Rev. Lett.* **54**, 106 (1985).

171. A N Zherikhin et al., *Sov. J. Quantum Electron.* **6**, 82 (1976); A V Vinogradov and V N Shylaptsev *Sov. J. Quantum Electron.* **13**, 1511 (1983); U Feldman et al., *J Appl. Phys.* **54**, 2188 (1983) and **56**, 2475 (1984); P L Hagelstein, Ph.D. thesis unpublished.

172. J P Apruzzese et al., *Phys. Rev. Lett.* **55**, 1877 (1985).

173. J Nilsen and J Moreno, *Phys. Rev. Lett.* **74**, 3376 (1995).

174. S Suckewer et al., *Phys. Rev. Lett.* **55**, 1753 (1985).

175. B J MacGowan et al., *J. Appl. Phys.* **61**, 5243 (1987).

176. B J MacGowan et al., *Phys. Rev. Lett.* **59**, 2157 (1987).

177. B J MacGowan et al., *J. Opt. Soc. Am.* **B5**, 1858 (1988).

178. B J MacGowan et al., Phys. Rev. Lett. **65**, 420 (1990).

179. S Maxon, P Hagelstein, K Reed and J Scofield, *J. Appl. Phys.* **57**, 971 (1985); **59**, 239 (1986); see also L Hagelstein, *Phys. Rev.* **A34**, 874 (1986).

180. B J McGowan et al., *Phys. Rev. Lett.* **65**, 420 (1990).

181. D J Fields et al., *Phys. Rev.* **A46**, 1606 (1992).

182. N M Ceglio et al., *Opt. Lett.* **27**, 108 (1988); *Appl. Opt.* **27**, 5022 (1988).

183. J Nilsen and J Moreno, *Phys. Rev. Lett.* **74**, 3376 (1995).

184. The prepulse technique had already introduced, see T Boehly et al., *Phys. Rev.* **A42**, 6962 (1990); J Nilsen, B J MacGowan, L B Da Silva and J C Moreno, *Phys.*

Rev. **A48**, 4682 (1993); J Nilsen et al., *Appl. Phys.* **B57**, 309 (1993) and was immediately used Y Li, G Pretzler and E E Fill, *Phys. Rev.* **A52**, R3433 (1995).

185. *X-Ray Lasers 1996*, edited by S Svanberg and C G Wahlstrom (Institute of Physics Conference Series N 151, Bristol, 1996).
186. See H Daido, *Rep. Prog. Phys.* **65**, 1513 (2002).
187. H Takabe and T Nishikawa, *J. Quant. Spectrosc. Radiat. Transfer* **51**, 379 (1994).
188. H Daido et al., *Phys. Rev. Lett.* **75**, 1074 (1995).
189. J G Lunney, *Appl. Phys. Lett.* **48**, 891 (1986); R Kodama et al., *Phys. Rev. Lett.* **73**, 3215 (1994).
190. S Basu, P L Hagelstein and J G Goodberlet, *Appl. Phys.* **B57**, 303 (1993).
191. T Hara et al., *Jpn. J. Appl. Phys.* **28**, L1010 (1989); Y Nagata et al., *Phys. Rev. Lett.* **71**, 3774 (1993).
192. See S Basu, P L Hagelstein and J G Goodberlet, *Appl. Phys.* **B57**, 303 (1993); T Hara et al., *Jpn. J. Appl. Phys.* **28**, L1010 (1989); Y Nagata et al., *Phys. Rev. Lett.* **71**, 3774 (1993).
193. J J Rocca et al., *Phys. Rev. Lett.* **73**, 2192 (1994).
194. J Dunn et al., *Phys. Rev. Lett.* **80**, 2825 (1998).
195. R Smith et al., *Phys. Rev.* **A59**, R47 (1999).
196. B R Benware et al., *Phys. Rev. Lett.* **81**, 5804 (1998); C D Macchietto, B R Benware and J J Rocca, *Opt. Lett.* **24**, 1115 (1999); M Frati, M Seminario and J J Rocca, *Opt. Lett.* **25**, 1022 (2000); S Sebban et al., *Phys. Rev. Lett.* **86**, 3004 (2001); *Phys. Rev. Lett.* **89**, 253901 (2002); D V Korobkin et al., *Phys. Rev. Lett.* **77**, 5206 (1996); K A Janulewicz et al., *Phys. Rev.* **A68**, 051802 (2003); Y Wang et al., *Phys. Rev.* **A72**, 053807 (2005); J J Rocca et al., *Opt. Lett.* **30**, 2581 (2005).
197. F Pedaci et al., *Opt. Lett.* **33**, 491 (2008).
198. D Alessi et al., *Opt. Lett.* **35**, 414 (2010).
199. J J Rocca et al., *Opt. Lett.* **30**, 2581 (2005); see also Y Wang et al., *Phys. Rev.* **A72**, 053807 (2005).
200. Y Wang et al., *Nature Photon.* **2**, 94 (2008).
201. H C Kapteyn et al., *Phys. Today* **58**, 39 (2005); X Zhang et al., *Opt. Lett.* **29**, 1357 (2004); Y Nagata et al., *Opt. Lett.* **32**, 722 (2007); T Ditmire et al., *Phys. Rev.* **A51**, R4337 (1995).
202. F J Furch et al., *Opt. Lett.* **34**, 3352 (2009).
203. B A Reagan et al., *Opt. Lett.* **37**, 3624 (2012).
204. R C Elton, *X-Ray Lasers* (Academic Press Inc.: Boston, 1990).
205. Y Kato: *International Laser Facilities* paper presented at *X-ray Lasers 1996 Conf., Lund, Sweden*, Inst. Phys. Conf. Ser. No 151 (IoP, Bristol:1996) p 274.
206. H Daido, *Rep. Prog. Phys.* **65**, 1513 (2002).
207. R C Elton, *X-Ray Lasers*, (Academic Press Inc.: Boston, 1990), p. 216.
208. V I Gold'danskii and Yu Kagan, *Sov. Phys. JETP* **37**, 49 (1973); W A Bushnev and R N Kuzmin, *Sov. Phys. Uspeki* **17**, 942 (1975); B V Chirikov, *Sov. Phys. JETP* **17**, 1355 (1963); Yu A Il'inskii and R V Khokhlov, *Sov. Phys. Usp.* **16**, 565 (1974); V L Gold'danskii, S V Karyagin and V A Namiot, *Sov. Phys. Solid State* **16**, 1640 (1975); V S Letokhov, *Sov. J. Quantum Electron.* **3**, 360 (1974); see also G C Baldwin and R V Khokhlov, *Phys. Today* **28**, 32 (1975); V Vali and W Vali, *PIEEE* **51**, 182 (1963); G T Trammell and J P Hannon, *Opt. Comm.* **15**, 325 (1975).
209. G C Baldwin, J C Solem and V I Gol'danskii, *Rev. Mod. Phys.* **53**, 687 (1981); G C Baldwin and J C Solem, *Rev. Mod. Phys.* **69**, 1085 (1997).
210. D Marcuse, *Proc. IEEE* **51**, 849 (1963); V S Letokhov, *Sov. Phys. Quantum Electron.* **3**, 360 (1974); E V Baklanov and V P Chebotaev, *JETP Lett.* **21**, 131 (1975); L A Rivkin, *Sov. J. Quantum Electron.* **7**, 380 (1977); V I Vysotskii, *Sov. Phys. JETP* **50**,

250 (1979); B Arad et al., *Phys. Lett.* **A74**, 395 (1979); C B Collins et al., *Phys. Rev.* **C20**, 1942 (1979); S Olariu et al., *Phys. Rev.* **C23**, 50 (1981); 1007 (1981).

211. D A G Deacon and B E Newnam *J. Opt. Soc. Am.* **B6**, 1061 (1989).
212. A short but very informative paper is P G O'Shea and H P Freund, *Science* **292**, 1853 (2001).
213. R Bonifacio, C Pellegrini and L M Narducci, *Opt. Comm.* **50**, 359 (1984).
214. C Pellegrini, *Nucl. Inst. Meth. Phys. Res.* **A475**, 1 (2001) provides an excellent summary of SASE FELs.
215. J Andruszkow et al., *Phys. Rev. Lett.* **85**, 3825 (2000).
216. A general information on DESY and SLAC may be found in E Ploenjes, J Feldhaus and T Moeller, *Physic World,* July 2003 p. 33.
217. W Ackerman et al., *Nature Photon.* **1**, 336 (2007).
218. J N Galayda et al., *J. Opt. Soc. Am.* **B27**, B106 (2010). The paper gives also the history of the project.
219. P Emma et al., *Nature Photon.* **4**, 641 (2010); *Proc. of the 2009 Part. Acc. Conf., Vancouver,* p. 3115.
220. See S Jamison, *Nature Photon.* **4**, 589 (2010) for a short information.
221. L H Yu et al., *Science* **289**, 932 (2000); *Phys. Rev. Lett.* **65**, 420 (1990).
222. T Ishikawa et al., *Nature Photon.* **6**, 540 (2012).
223. H Tanaka, *Nature Photon.* **6**, 540 (2012).
224. G Lambert et al., *Nature Phys.* **4**, 296 (2008).
225. J Aman et al., *Nature Photon.* **6**, 693 (2012).
226. L H Yu et al., *Science* **289**, 932 (2000); G Stupakov, *Phys. Rev. Lett.* 102. 074801 (2009).
227. E Allaria et al., *Nature Photon.* **7**, 913 (2013).
228. E Allaria et al., *Nature Photon.* **6**, 699 (2912); **7**, 913 (2013); *Nature Commun.* **4**, 2476 (2013).
229. B W Adams et al., *J. Mod. Opt.* **60**, 2 (2013).

10

The Statistical Properties of Light

10.1 Introduction

The most fundamental property of laser radiation is its coherence. The very high spectral brightness, the monochromaticity, and the directionality of a laser beam are all properties connected with the coherence of its emission. Spectral narrowing and directionality are manifestations of what is nowadays called *first-order coherence*. However, the physical possibility of obtaining both a very high brightness and a much narrower linewidth than the Doppler-broadened line is due to the circumstance that the light emission in a laser takes place in conditions of thermodynamic nonequilibrium. It was consideration of the properties of radiation emitted in conditions of thermodynamic nonequilibrium or, what amounts to the same thing, in conditions in which stimulated emission dominates, that led R J Glauber (b 1925) to extend the concept of classical coherence, thereby disclosing the particular statistical properties of radiation emitted by an ideal laser. These peculiar, statistical properties are perhaps the most important general feature of lasers; therefore, they deserve special mention in this book.

10.2 Introduction of the Concept of the Photon

We will not be concerned with the old dispute among scientists about the nature of light—the wave theory which was proposed in the seventeenth century by R Hooke (1635–1703) and C Huygens (1629–1695) versus the corpuscular theory put forward by Isaac Newton (1642–1727). The steps leading to the first quantum theory introduced by M Planck and to the concept of quantization of energy can be found in the excellent book by E Whittaker.[1] The formulation of the quantum concept is also well illustrated and discussed by M Jammer.[2]

The black-body radiation law was discovered by Max Karl Ernst Ludwig Planck (1858–1947) in 1900,[3] just a few months after the publication of the Rayleigh formula.[4] Within a few weeks, Planck observed that, in order to derive his formula, it was necessary to introduce the notion of a *quantum of energy*, which represents the smallest amount of energy that an oscillator can either emit or absorb.[5]

According to Planck's theory, an oscillator of frequency υ can emit or absorb energy only in multiples of $h\upsilon$. Planck regarded the quantum property as belonging essentially to the interaction between radiation and matter: free radiation, he supposed to, consist of electromagnetic waves, in accordance with Maxwell's theory.

The next important advance was made by Einstein[6] in 1905, who clearly recognized Planck's discovery, which had until then attracted little attention. The paper entitled *On a heuristic point of view concerning the creation and conversion of light* was published in the same volume (17) of *Annalen der Physik* in which the theory of Brownian motion and the special theory of relativity were published,[7] which probably makes this volume one of the most remarkable in the whole scientific literature, as Max Born noted!

This paper is usually referred to nowadays as the Einstein paper on the photoelectric effect,[8] but in fact carries a much broader relevance. In it Einstein deduced, from statistical thermodynamics, that the entropy of radiation described by Wien's distribution law has the same form as the entropy of a gas of independent particles. Einstein used this result to argue for the heuristic viewpoint that light consists of quanta, each possessing an amount of energy $\varepsilon = h\upsilon$ and then applied this conclusion to explain the photoelectric effect (indicated as the emission of cathode rays through illumination of solid bodies). He observes:

> The wave theory of light, which operates with continuous spatial functions, has worked well in the representation of purely optical phenomena and will probably never be replaced by another theory. It should be kept in mind, however, that the optical observations refer to time averages rather than instantaneous values. In spite of the complete experimental confirmation of the theory as applied to diffraction, reflection, refraction, dispersion, etc, it is still conceivable that the theory of light which operates with continuous spatial functions may lead to contradictions with experience when it is applied to the phenomena of emission and transformation of light.
>
> It seems to me that the observations associated with black-body radiation fluorescence, the production of cathode rays by ultraviolet light, and other related phenomena connected with the emission or transformation of light are more readily understood if one assumes that the energy of light is discontinuously distributed in space. In accordance with the assumptions to be considered here, the energy of a light ray spreading out from a point source is not continuously distributed over an increasing space, but consists of a finite number of energy quanta which are localized at points in space, which move without dividing, and which can only be produced and absorbed as complete units.

Einstein's view was, in fact, in contrast with that of Planck, that radiation was quantized not only in the emission process, but also that it remained quantized also in the propagation process.[9]

Einstein uses the word *quantum* of light or of energy. The name *photon* was introduced much later in 1926 by G N Lewis[10] (1875–1946), who thought it inappropriate to speak of quanta of light:

> …
> if we are to assume that it spends only a minute fraction of its existence as a carrier of radiant energy, while the rest of the time it remains an important structural element within the atom … I therefore take the liberty of proposing for this hypothetical new atom which is not light but plays an essential part in every process of radiation, the name photon.

The explanation of the photoelectric effect in terms of photons took, however, a considerable time to be fully accepted. At that time, a resonance theory due to Lenard[11] was commonly accepted.

Einstein's theory was verified in 1912 by O W Richardson (1879–1959) and K T Compton (1881–1954)[12] and by A L Hughes[13] and with great care by R A Millikan (1868–1958)[14] who started by completely disbelieving it and instead gave the best confirmation of it, through 10 years work, being awarded the Nobel Prize in 1923 partly for this.

Surprisingly, no mention or connection with the "softening" of x-rays scattered by a substance of low atomic weight was made in the discussion. This effect, discovered by C A Sadler and P Mesham in 1912,[15] was eventually experimentally confirmed and explained by A H Compton (1892–1962) in 1922,[16] and is nowadays one of the better known examples of the corpuscular behavior of light.[17]

This phenomenon soon became known as the *Compton effect*, and for this discovery Compton received the Noel Prize in 1927 sharing it with C T R Wilson (1869–1959), the inventor of the famous cloud chamber.

To get an insight into the difficulties encountered by Einstein's theory of the photoelectric effect, it may help recalling the following two examples also reported by M J Klein.[8]

Still in 1913 in a letter which proposed Einstein for membership in the Prussian Academy and for a Research Professorship and where Einstein's work was fully appreciated, Planck wrote[18]:

> That he may sometimes have missed the target in his speculations, as for example in his hypothesis of light quanta, cannot really be held against him

Some years later, in 1916, describing his experimental confirmation of the Einstein equation for the photoelectric effect, Millikan wrote of the same hypothesis[19]

> I shall not attempt to present the basis for such an assumption, for, as a matter of fact, it had almost none at the time.

Eventually, Albert Einstein was awarded the Nobel Prize in 1921 just for his theory of the photoelectric effect!

Finally, it may be observed that it is commonly believed that Einstein developed his hypothesis of light quanta as an extension of Planck's theory of black-body radiation. That belief is not supported by a careful reading of the work of both physicists, as M Klein has shown.[8] Rather, Einstein postulated the existence of quanta of light on pure thermodynamic grounds (he had previously carefully studied thermodynamics and statistical mechanics writing three works on these subjects) without using Planck's distribution law or his discrete, quantized oscillator energies in his own arguments. He also wrote the magnitude of his light quanta as $(R/N_o)\beta\upsilon$, R being the gas constant, N_o Avogadro's number, and β the exponential coefficient appearing in Wien's radiation formula, and not as $h\upsilon$.

10.3 Fluctuations of Radiant Energy

In 1909, four years after his "photoelectric paper," Einstein published a paper[20] in which he showed that Planck's radiation law itself implies that the radiation field exhibits not only wave features, but also corpuscular features. This result, which was the first clear indication of the so-called *wave-particle duality*, has notable importance in our history.

In this paper, Einstein calculated the fluctuations of radiant energy in a partial volume V of an isothermal enclosure at temperature T. The subject of radiation fluctuation played a key role in the development of quantum statistics which followed in the period up to 1930. Later, much interest was centered on the practical problem of attaining the maximum detectivity in spectrometric investigations.

Like other problems of fluctuations connected with departure from equilibrium, radiation fluctuations in a cavity can be treated with the method of statistical mechanics.

By denoting by E the instantaneous energy in the cavity within the frequency interval υ to $\upsilon + d\upsilon$, with an average value

$$\bar{E} = \int E \exp(-E/kT)\,\mathrm{d}p\,\mathrm{d}q/\int \exp(-E/kT)\,\mathrm{d}p\,\mathrm{d}q, \tag{10.1}$$

and with $\varepsilon = E - \bar{E}$ the energy fluctuation, Einstein was easily able to show that the mean-square fluctuation $\overline{\varepsilon^2}$ which is defined as

$$\overline{\varepsilon^2} = \overline{E^2} - \bar{E}^2, \tag{10.2}$$

where

$$\overline{E^2} = \int E^2 \exp(-E/kT)\mathrm{d}p\,\mathrm{d}q/\int \exp(-E/kT)\mathrm{d}p\,\mathrm{d}q \tag{10.3}$$

is given by

$$\overline{\varepsilon^2} = kT^2 \mathrm{d}\bar{E}/\mathrm{d}T. \tag{10.4}$$

The mean energy $\bar{E}$ can also be obtained by Planck's law by substituting in

$$\bar{E} = V\rho_\upsilon d\upsilon, \tag{10.5}$$

the expression for ρ_υ given by Planck, Einstein found

$$\overline{\varepsilon^2} = (8\pi h^2 \upsilon^4 d\upsilon/c^3)\{1/[\exp(h\upsilon/kT) - 1] + 1/[\exp(h\upsilon/kT) - 1]^2\}$$
$$= h\upsilon E + c^3 E^2/8\pi\upsilon^2 d\upsilon. \tag{10.6}$$

Einstein observed that if, instead of Planck's law of radiation, one had taken Wien's law, one should have obtained

$$\overline{\varepsilon^2}_W = h\upsilon E, \tag{10.7}$$

while if one had taken Rayleigh's law one should have obtained

$$\overline{\varepsilon^2}_R = c^3 E^2/8\pi\upsilon^2 d\upsilon. \tag{10.8}$$

Therefore, the mean-square value of the fluctuations according Planck's law is the sum of the mean squares of the fluctuations according to Wien's law and Rayleigh's law. The result, looked at in the light of the principle that fluctuations due to independent causes are additive, suggests that the causes operative at high frequencies (for which Wien's law holds) are independent of those operative in the case of low frequencies (for which Rayleigh's law is valid). Now, Rayleigh's law is based on the wave theory of light, and, in fact, Lorentz[21] showed that the value $c^3 E^2/8\pi\upsilon^2 d\upsilon$ is a consequence of the interference of wave trains, which, according to the classical picture, are crossing the cavity in every direction; whereas the value $h\upsilon E$ for the mean-square fluctuations is what would be obtained if one were to take the formula for the fluctuations of the number of molecules in a unit volume of an ideal gas and suppose that each molecule has energy $h\upsilon$, that is, the expression which would be obtained by a corpuscular quantum theory.

Moreover, the ratio of the particle term to the wave term in the complete expression for the fluctuations is

$$\exp(h\upsilon/kT) - 1.$$

When $h\upsilon/kT$ is small, that is, at low frequencies and high temperatures, the wave term is predominant, and when $h\upsilon/kT$ is large, that is, when the energy density is small, the particle term predominates.

The formula therefore suggests that light cannot be completely represented either by waves or by particles, although for some phenomena the wave representation is practically sufficient, and for other phenomena the particle representation is good.

Einstein was greatly attracted by the problem of light, and later, in 1915–1916, published the paper in which he introduced the concept of stimulated emission and which once again was based on the corpuscular concept of light.[22] Meanwhile, the photon hypothesis was carefully examined by several people. P Ehrenfest,[23] A Joffé,[24] L Natanson,[25] and G Krutkow[26] showed that if we assume that each of the light quanta of frequency υ has an energy $h\upsilon$ and that they are completely independent of each other, Wien's law of radiation is obtained instead of Planck's law. In order to obtain Planck's formula, it is necessary to assume that elementary photons of energy $h\upsilon$ form aggregates of energy $2h\upsilon$, $3h\upsilon$, and so on and that the total energy of radiation is distributed, on average, in a regular manner between them.

L de Broglie (1892–1987), later in 1922,[27] made a similar remark concerning the energy fluctuations. He observed that Planck's formula

$$E = (8\pi h\upsilon^3/c^3)\{d\upsilon/[\exp(h\upsilon/kT) - 1]\} \qquad (10.9)$$

may be written as

$$E = (8\pi h\upsilon^3/c^3)[\exp(-h\upsilon/kT) + \exp(-2h\upsilon/kT) + \exp(-3h\upsilon//kT) + \cdots]d\upsilon$$
$$= E_1 + E_2 + E_3 + \cdots + E_s + \cdots, \qquad (10.10)$$

where

$$E_s = (8\pi h\upsilon^3/c^3)\exp(-sh\upsilon/kT)d\upsilon. \qquad (10.11)$$

Now, Einstein formula (10.6) can be written as

$$\overline{\varepsilon^2} = (8\pi h^2\upsilon^4 d\upsilon/c^3)[\exp(-h\upsilon/kT) + 2\exp(-2h\upsilon/kT) + 3\exp(-3h\upsilon/kT) + \cdots]$$
$$= \sum_{s=1}^{\infty} sh\upsilon E_s. \qquad (10.12)$$

This resembles the first term $h\upsilon E$ in formula 10.6, but it is now summed for all values of s. So it is precisely the result we should expect if the energy E_s was made up of light quanta, each of energy $sh\upsilon$. Thus, de Broglie suggested that the term E_1 should be regarded as corresponding to energy existing in the form of quanta of amount $h\upsilon$; that the second term E_2 should be regarded as corresponding to energy existing in the form of quanta of amount $2h\upsilon$; and so on.

So, the Einstein formula for the fluctuations may be obtained on the basis of a purely corpuscular theory of light, provided the total energy of the radiation is suitably allocated among corpuscles of different energies $h\upsilon$, $2h\upsilon$, $3h\upsilon$, and so on.

The following year, W Bothe[28] gave the calculation of the number of quanta $h\upsilon$ in black-body radiation which are associated as "photo-molecules" in pairs, $2h\upsilon$, trios $3h\upsilon$, and so on, which has already been discussed in Chapter 3.[29] These associations are responsible for the bunching properties of radiation we shall discuss below.

10.4 Bose and the Statistics of Radiation

A further step forward was made by Satyendra Nath Bose[30] in 1924 at Dacca University, India, in a short paper wherein a new proof of Planck's formula was given.

Bose (**Figure 10.1**) was born on January 1, 1894, in Calcutta and died on February 4, 1974. In 1916, he and M N Saha (1893–1956) became two of the first lecturers in the new University College of Science built in 1914 by Vice-Chancellor Sir Asutosh Mookerjee in Calcutta, a college where the professorial posts had to be filled by Indians. The following year, with the creation of a department, lecturers in physics were nominated.

Bose immediately showed his interest in statistical physics. In 1919, together with Saha, he published an anthology of works by Einstein on relativity, one of the first of such collections published in English. In 1921,

FIGURE 10.1 Satyandra Nath Bose (1924–1974).

Bose left Calcutta to become Reader in Physics in the reorganized University of Dacca in East Bengal. There he lectured, read, thought, and spent many sleepless nights thinking about Planck's law. In late 1923, he submitted a paper on the subject to the prestigious English review, *Philosophical Magazine*. Six months later, the editors of that magazine informed him that the referee's report on his paper was unfavorable.

He then sent the manuscript to Einstein (on June 4, 1924), who was then Professor of Physics at Berlin. The paper was accompanied by a letter which began:[31]

> Respected Sir,
>
> I have ventured to send you the accompanying article for your perusal and opinion. I am anxious to know what you think of it. You will see that I have tried to deduce the coefficient $8\pi\upsilon^2/c^3$ in Planck's law independently of the classical electrodynamics, only assuming that the ultimate elementary region in the phase space has the content h^3. I do not know sufficient German to translate the paper. If you think the paper worth of publication I shall be grateful if you arrange for its publication in Zeitschrift fur Physik. Though a complete stranger to you, I do not hesitate in making such a request. Because we are all your pupils though profiting only from your teachings through your writings ...

Einstein translated the paper and sent it in July 1924 to the *Zeitschrift* in Bose's name, where it was published under the title *Plancks Gesetz und Lichtquanten-hypothese* (Planck's law and the hypothesis of light quanta).[30] He also added a note stating:

> In my opinion Bose's derivation of the Planck formula signifies an important advance. The method used also yields the quantum theory of the ideal gas, as I will work out in detail elsewhere.

He then sent a postcard to Bose signifying that he considered the work a most important contribution.

Bose's method is based on arguments relative to phase space. He considered the radiation as being composed of photons, which for statistical purposes can be treated as particles of a gas, with the important difference that photons are indistinguishable one from the other; so instead of considering the disposition of photons as being individually distinguishable in an ensemble of states, he paid attention to the number of states which contain a given number of photons. He assumed that the total energy E of the photons was known and that they were enclosed in a cavity of unit volume.

The frequency distribution of the radiation at an absolute temperature T was then deduced by finding the distribution in phase space that maximizes the entropy of the system.

To this end, Bose observed that a photon $h\upsilon$ can be specified by its coordinates (x,y,z) and by the three components of its momentum (p_x, p_y, p_z). Because the total momentum is $h\upsilon/c$, we have

$$p_x^2 + p_y^2 + p_z^2 = r^2, \text{ where } r = h\upsilon/c.$$

By using a phase space of six dimensions (x,y,z,p_x,p_y,p_z), for the frequency range υ to $\upsilon + d\upsilon$, the corresponding volume is

$$\int dx \, dy \, dz \, dp_x \, dp_y \, dp_z = 4\pi r^2 dr V = 4\pi V(h^3\upsilon^2/c^3)d\upsilon. \qquad \textbf{(10.13)}$$

Bose now assumed that this space is divided into cells, each of volume h^3, so that in it there are $(4\pi\upsilon^2 d\upsilon/c^3)V$ cells. To take account of polarization, one must then multiply this number by 2.

Now, let N_s be the number of quanta belonging to the frequency domain $d\upsilon_s$ and then let us consider the number of ways in which these quanta can be distributed among the cells belonging to $d\upsilon_s$. Let p_{so} be the number of vacant cells; p_{s1} the number of those which contain one quantum; p_{s2} the number of cells containing two quanta, and so on; then the number of possible ways of choosing an ensemble of p_{so}, p_{s1} cells, and so on from the total of $8\pi\upsilon^2 d\upsilon/c^3$ cells is

$$A_s!/p_{so}!p_{s1}!p_{s2}! \ldots,$$

where

$$A_s = (8\pi\upsilon^2/c^3)d\upsilon_s,$$

and we have

$$N_s = \sum_r r p_{sr}.$$

As a fundamental hypothesis of his statistics, Bose now assumed that if one considers a particular quantum state, all the values for the number of particles in that state are equivalent, so that the probability of each distribution specified by p_{sr} is measured by the number of different ways in which it can be realized. Therefore, the probability of a state specified by p_{sr} (taking into account all the intervals of frequency) is

$$W = \prod_s A_s! / p_{s0}! p_{s1}! p_{s2}! \ldots.$$

Because the p_{sr} are large, one may use Stirling's approximation $\log n! = n \log n - n$. So

$$\log W = \sum_s A_s \log A_s - \sum_s \sum_r p_{sr} \log p_{sr},$$

where

$$A_s = \sum_r p_{sr}.$$

This expression should be a maximum satisfying the auxiliary conditions

$$E = \sum_s N_s h \upsilon_s, \qquad N_s = \sum_r r p_{sr}.$$

Carrying out the variation gives the conditions

$$\sum_s \sum_r \delta p_{sr} (1 + \log p_{sr}) = 0, \qquad \sum_s \delta N_s h \upsilon_s = 0$$
$$\sum_s \delta p_{sr} = 0, \qquad \delta N_s = \sum_r r \delta p_{sr}.$$

It follows that

$$\sum_s \sum_r \delta p_{sr} [(1 + \log p_{sr} + \lambda_s) + (r h \upsilon_s / \beta)] = 0,$$

where β and λ_s are constant, so

$$p_{sr} = B_s \exp(-r h \upsilon_s / \beta),$$

where B_s is constant.
Therefore,

$$A_s = \sum_r p_{sr} = \sum_r B_s \exp(-r h \upsilon_s / \beta) = B_s [1 - \exp(-h \upsilon_s / \beta)]^{-1},$$

or

$$B_s = A_s [1 - \exp(-h \upsilon_s / \beta)],$$

while

$$N_s = \sum_r r p_{sr} = A_s \sum_r r \exp(-h \upsilon_s / \beta)[1 - \exp(-h \upsilon_s / \beta)]$$
$$= A_s \exp(-h \upsilon_s / \beta)/[1 - \exp(-h \upsilon_s / \beta)].$$

Therefore

$$E = \sum_s N_s h \upsilon_s = \sum_s [8 \pi h (\upsilon_s)^3 / c^3] \exp(-h \upsilon_s / \beta)\left[1 - \exp(h \upsilon_s / \beta)\right]^{-1} d\upsilon_s V. \quad \textbf{(10.14)}$$

Using the preceding results, one also finds that the entropy is

$$S = k \log W = k\{(E/\beta) - \sum_s A_s \log[1 - \exp(-h\upsilon_s/\beta)]\}. \qquad (10.15)$$

Because

$$\partial S/\partial E = 1/T,$$

where T is the absolute temperature, one as $\beta = kT$. Therefore

$$E = \sum_s [8\pi h(\upsilon_s)^3/c^3][\exp(h\upsilon_s/kT) - 1]^{-1}d\upsilon_s, \qquad (10.16)$$

which is equivalent to Planck's formula.

Bose's work showed that to obtain Planck's law one has to suppose that photons obey a given kind of statistics.[32]

Bose's discovery was immediately extended by Einstein to the study of a monoatomic ideal gas.[33] His analysis brought the following conclusions: the average number of particles of mass m in a unit volume with energies between ε and $\varepsilon + d\varepsilon$ is

$$\bar{n}d\varepsilon = (2\pi/h^3)(2m)^{3/2}\varepsilon^{1/2}d\varepsilon/[\exp(\varepsilon/kT + \mu) - 1], \qquad (10.17)$$

where μ is a constant, and the total number of particles per unit volume is

$$N = (2\pi/h^3)(2m)^{3/2} \int_0^\infty \varepsilon^{1/2}/[\exp(\varepsilon/kT + \mu) - 1]$$

and the total energy per unit volume is

$$E = (2\pi/h^3)(2m)^{3/2} \int_0^\infty \varepsilon^{3/2}d\varepsilon/[ep(\varepsilon/kT + \mu) - 1]. \qquad (10.18)$$

These are the fundamental formulae of what is generally called *Bose–Einstein statistics*. Since Equation 10.17 differed from Boltzmann factor $\exp(-\varepsilon/kT)$ of ordinary statistical mechanics, all the thermodynamic properties of the gas were correspondingly more complicated. Einstein was able to show, however, that when the temperature was high and the density was low, his equation returned to those for the classical gas.

In 1924, Bose received support for a study period of 2 years in Europe and arrived in Paris in September. He recounts that on the strength of the postcard Einstein had sent him, the German Consulate in Calcutta issued his visa without requiring payment of the customary fee!

On arrival in Paris, he met P Langevin (1872–1946), who suggested to him the possibility of working in Mme Curie Laboratory. Bose recalled this meeting with amusement in an interview with W A Blanpied.[31] Mme Curie spoke in English all the time and did not let him say a single word.

She spoke to him about another Indian student who worked with her and had encountered serious difficulties because he did not speak good French.

Then she suggested that Bose should concentrate on the language for 6 months and then come back to her. Bose did not even get the chance to tell her that he had studied French for 15 years!

After this discouraging contact, Bose met the de Broglie brothers and was with Maurice for some time. However, he was still very anxious to go to Einstein. On October 26, 1924, he had written Einstein a letter which began:

> Dear Master,
> My heartfelt gratitude for taking the trouble of translating the paper your-self and publishing it. I just saw it in print before I left India. I have sent you about the middle of June a second paper entitled Thermal Equilibrium in the Radiation Field in the Presence of Matter.
>
> I am rather anxious to know your opinion about it as I think it to be rather important. I don't know whether it will be possible also to have this paper pub-lished in Zeitschrift fur Physik.
>
> I have been granted study leave by my university for two years. I have arrived just a week ago in Paris. I don't know whether it will be possible for me to work under you in Germany. I shall be glad however, if you grant me the permission to work under you, for it will mean for me the realization of a long cherished hope.
>
> I shall wait for your decision as well as your opinion of my second paper here in Paris ....

Einstein had already translated this second paper and sent it to the *Zeitschift* which published it.[34] This time, however, Einstein added a remark stating that he could not agree with his authors' conclusions and went on to give his rea-sons.[35] Apparently, he communicated these objections privately because in a letter from Paris dated January 27, 1925, Bose thanks Einstein for his commu-nication of November 3 and states he is about to send off to him another paper which he hopes will satisfy his objections. This work, however, was never pub-lished not can any mention of it be found in Bose's correspondence. There is no copy of Einstein's letter available and so his plans for Bose's visit in Berlin are unknown; however, Bose had not given up the idea of coming, for in the letter dated January 27, he closes by again stating that he hopes to be able to work with Einstein.

Nowadays, it may be said that the opportunity of a collaboration between Bose and Einstein had already evaporated by January 1925.

In July 1924, about the time Bose was finally settling the study leave ques-tion with the Dacca University authorities, Einstein was reading a paper[33] before the Prussian Academy where he applied Bose's statistical method to an ideal gas with particles of nonzero rest mass and suggested some of its non-chemical thermodynamic properties.

The similarity in the statistical behavior between photons and gas particles found in the paper was further investigated by Einstein. This nonchemical gas continued to fascinate him for several months. In September 1924 (while Bose was at sea), Einstein wrote to Ehrenfest stating that at low temperatures the molecules of the gas would condense into the zero-energy state even in the absence of attractive forces between them.[36]

At the beginning of December (while Bose was pondering Einstein's objections to his second work), Einstein wrote to Ehrenfest

The matter of the quantum gas is getting very interesting.[36,37]

Finally, in January 1925, Einstein published a second paper[38] where he observed that in expression 10.6 both terms appear if one assumes that particles satisfy Bose statistics, and made clear that the particles were not treated independent in the Bose–Einstein counting procedure. He derived the analog of Equation 10.6 as

$$(\Delta s)^2 = n_s + n_s^2/z_s,$$

where Δs is the fluctuation about the average value n_s of the number of molecules in a specified energy interval, and z_s is the number of cells. He also predicted the now well-known Bose–Einstein condensation and tried to rationalize Bose's method, after which he turned his attention to other matters.[39]

When Bose arrived in Berlin, and on October 8, 1925, he wrote to Einstein asking for an appointment, Einstein had ceased to be interested in the argument and was in Leyden. He came back only after several weeks. When finally the two men met, the encounter was very disappointing. As a result, Bose obtained a letter allowing him to enjoy some privileges common to students in Berlin, including permission to take books from the university library!

Later, in the summer of 1926, Bose returned to Dacca and, one year later, was appointed Physics Professor. He held this position until 1945 when he went back to Calcutta University, again as Physics Professor.

During all those years in Dacca, he worked and dedicated himself to Indian and Western philosophy and literature. He also interested himself in politics and was an ardent supporter of the campaign for Bengali independence.

He returned to France in 1951 and was often abroad, but he never wanted go back to Germany. He occupied himself with topics connected with Einstein. Of the six papers he published between 1953 and the time of his retirement, five treat the unified theory of Einstein. The last of Einstein's letters to him is dated October 22, 1953, containing thanks for a preprint received and for the enclosed friendly letter.[31] It can be stated, however, that after the paper on statistics, Bose did not contribute any more to the forefront of physics. He had arrived in Europe in 1924, met Langevin and de Broglie in a period when the new physics was boiling: so why did he lose the opportunity of participating? Partly, perhaps, because of his placid temperament: Einstein understood everything on the quantum gas in 7 months, partly also because he was very intimidated by most Europeans. The heavy contribution that Einstein had given to the determination of Bose statistics was soon recognized, and the statistics was named *Bose–Einstein*. Dirac, who first coined the word "boson", seems to have been among the few European physicists who gave Bose full credit for his achievement, as Bose himself noted in an interview given three years before his death.

10.5 A Few More Remarks on Bose– Einstein Condensation

We have mentioned that in his study on the behavior of a quantum gas, Einstein[40] recognized that at low temperatures molecules can condensate.

When a gas of bosonic atoms (zero or integer spin particles) is cooled below a critical temperature T_c, a large fraction of the atoms condenses in the lowest quantum state. Atoms at temperature T and with mass m can be regarded as quantum-mechanical wave packets that have a spatial extent of the order of a thermal de Broglie wavelength

$$\lambda_{dB} = (2\pi\hbar^2/mk_BT)^{1/2}.$$

The value of λ_{dB} is the position uncertainty associated with the thermal momentum distribution and increases with decreasing temperature. When atoms are cooled to the point where λ_{dB} is comparable to the interatomic separation, the atomic wave packets "overlap" and the gas starts to become a "quantum soup" of indistinguishable particles. Bosonic atoms undergo a quantum-mechanical phase transition and form a Bose–Einstein condensate, a cloud of atoms all occupying the same quantum-mechanical state at a precise temperature (if the atoms are Fermions, cooling gradually brings the gas closer to being a "Fermi sea" in which exactly one atom occupies each low-energy state). More precisely to have a Bose–Einstein condensate,[41] the dimensionless phase-space density, $\rho_{ps} = n(\lambda_{dB})^3$, must be greater than 2.612, where n is the number density. It means that the gas should contain at least 2.62 particles inside a cube with sides equal to the average de Broglie wavelength of the thermal distribution.

It is worth noting that at the time Einstein elaborated this calculations, the existence of the difference in the behavior of particles with integer or zero spin (*bosons*) and semi-integer spin (*fermions*) was not known. This distinction came the following year when Pauli enunciated his exclusion principle.[42] Then Enrico Fermi[43] published the statistics of particles that obey the exclusion principle and Paul Dirac[44] connected the Bose–Einstein and Fermi statistics of particles with the symmetry properties of their wave functions and named the particles *bosons* and *fermions*.

Bose–Einstein condensation (BEC) is unique in that it is a purely quantum-statistical phase transition, that is, it occurs even in the absence of interactions. Einstein described the transition as condensation without attractive forces. At very low but finite temperature, a large fraction of the atoms would go into the lowest energy quantum state. In Einstein words,[45] "A separation is effected; one part condenses, the rest remains a 'saturated ideal gas.'

This prediction was not taken terribly seriously, even by Einstein himself, until Fritz London[46] and Laszlo Tisza[47] resurrected the idea as a possible mechanism underlying superfluidity in liquid helium 4. Although it was a source of debate for decades, it is now recognized that the remarkable properties of superconductivity and superfluidity in both helium 3 and helium 4 are related to BEC, even though these systems are very different from the ideal gas considered by Einstein because of strong interactions.[48]

Excitons in semiconductors were also considered, but unambiguous signature for BEC in this system has proved difficult to find.[49]

The advent of laser cooling in the 1980s opened up a new approach to ultralow-temperature physics. Microkelvin samples of dilute atom clouds were generated and used for precision measurements and studies of ultracold collisions. Nanokelvin temperatures allowed to explore quantum-degenerate gases, such as Bose–Einstein condensate that was first realized in 1995.

In 1976, Stwalley and Nosanow emphasized that because spin-polarized atomic hydrogen will stay a gas at temperatures down to absolute zero it offered a possibility for observing BEC in a weakly interacting system.[50] Their paper stimulated a session dedicated to spin-polarized hydrogen at the meeting of the American Physical Society in 1978.[51] This pushed a number of experimental groups[52] to work using traditional cryogenics to cool a sample of polarized hydrogen. Spin-polarized hydrogen was first stabilized by Silvera and Walraven[53] in 1980 and by the mid-1980s spin-polarized hydrogen had been brought within a factor of 50 of condensing.[54] The initial approach was to stabilize the atoms in a high magnetic field. The cooling was obtained in a cryogenic environment, while the confinement was provided by the helium covered wall of the trapping cell. The group at MIT (Tom Greytak and Dan Kleppner's laboratories) came closest to the degenerate regime at temperatures of 0.55 K and densities of 4.5×10^{18} cm^{-3}. However, when the cell was made very cold, the hydrogen stuck to the helium surface and recombined. Hess[55] then suggested magnetic trapping of atoms[56] so that atoms had no contact with a physical surface and cooling by evaporation. In this way, far lower temperatures than previously obtained could be possible. By 1988, the MIT group had demonstrated evaporation cooling.[57] The Colorado University hydrogen group also considered evaporative cooling.[58] At roughly the same time, but independently from the hydrogen work, major advances were made in the optical trapping and laser cooling, magneto-optical trapping, and Zeeman slowing of alkali metal atoms, and high phase-space densities became accessible for these atoms.[59] In the mid-1980s, Carl Wieman (b 1951) began investigating the technology of laser trapping and cooling which resulted in the creation of a useful magneto-optical trap (MOT) in a simple glass vapor cell[60] and eventually, after much more work, inspired by the hydrogen work and learning from the efforts on this material, the group at JILA (Joint Institute for Laboratory Astrophysics, University of Colorado and National Institute of Standards and Technology, Boulder, CO) decided that heavy alkali atoms would likely have more probability of success, even if the temperatures required to achieve BEC with them are far more lower than those needed for the same density of hydrogen. Alkali atoms (Li, Na, K, Rb, Cs) have one valence electron which makes them spectrally similar to hydrogen and very apt to laser cooling. Starting from 1990, the group carried out a series of experiments exploring the various magnetic traps and measuring the relevant collision cross sections.[61]

In the fall of 1992, Eric Cornell (b 1961), a postdoc who collaborated with the experiments, accepted a position at JILA/NIST and with his startup money built a new experimental apparatus with a simple quadrupole trap choosing to use rubidium for the experiment. With the help of some more students,

Mike Anderson, Jason Ensher, and Mike Matthews, Wieman, and Cornell succeeded[62] in June 1995 to have evidence of BEC.[63]

The evidence emerged from time-of-flight measurements. The atoms were left to expand by switching off the confining trap and then imaged with optical methods. A sharp peak in the velocity distribution was then observed below a certain critical temperature, providing a clear signature for BEC (**Figure 10.2**).

In the same years at MIT work on trapping and cooling was progressing. Wolfang Ketterle (b 1957) had entered as a postdoc with Dave Pritchard at the Center for Ultracold Atoms at MIT. The initial goal was to build an intense source of cold atoms to study collisions and pure long-range molecules. They studied a number of solutions to overcome some limitation of magnetic traps.[64] In 1991, at a Varenna summer school, the two men presented a new three-level cooling scheme.[65] After further work, the idea for a spontaneous force optical trap (Dark SPOT) was developed[66] and after some discussions, the group decided to study evaporative cooling of sodium. Work was progressing when the group heard about the breakthrough in Boulder in June 1995 and increasing the efforts eventually, in September 1995, observed BEC in sodium.[67] Their condensate was obtained with 500,000 atoms, 200 times more than in Boulder with a cooling cycle of only 9 s.[68] The atom trap they used offered a superior combination of light confinement and capture volume and allowed to attain unprecedented densities of cold atomic gases. The three researchers, Cornell, Wieman, and Ketterle, were awarded the 2001 Nobel Prize in Physics "for the

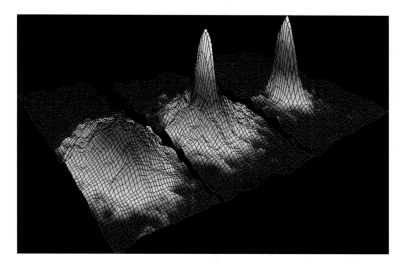

FIGURE 10.2 Velocity distribution for a gas of Rb atoms. The three distributions can be found (left) to just before the appearance of Bose–Einstein condensation, (center) just before the appearance of the condensate, and (right) after further evaporation. The false color indicates the number of atoms at each velocity, with red being the fewer and white being the most. The areas appearing white and light blue are at the lowest velocities. (From E Cornell, *J. Res. Nat. Inst. Stand. Technol.* **101**, 419 (1996).)

achievement of Bose–Einstein condensation in dilute gases of alkali atoms, and for early fundamental studies of the properties of the condensate."

Cornell and Wieman obtained the condensate by cooling a dilute vapor consisting of approximately 2000 Rb[87] atoms to below 170 nK using a combination of laser cooling and magnetic evaporative cooling.[69] Figure 10.2 shows the appearance of BEC in their experiment. A month later, Randal Hulet[70] at Rice University announced the creation of a condensate of Li atoms and about 4 months later, W Ketterle[71] created a condensate of Na[23].

Bose–Einstein condensation of hydrogen was obtained in June 1998 by Kleppner, Greytak, and collaborators at MIT.[72] A full account of the search for BEC in atomic hydrogen was written by Kleppner et al.[73] Hydrogen is of particular interest because of its simplicity and unique properties. Its low mass allows for the highest transition temperature, 50 μK, and the weak interaction between the atoms results in a condensate density more than an order of magnitude larger than the critical density, even at small condensate fractions.

The observation of BEC represented a new era in the study of properties of matter. One development involved the manipulation of coherent matter waves. Ketterle group studied the coherence of the condensation developing a tool to split a condensate into two halves that were made to interfere obtaining clear interference fringes on November 1996.[74] The group used a pulsed outcoupler; strong radiofrequency pulses were used to flip magnetically trapped atoms to untrapped states such that they could leak out of the trap.[75] Great interest arose in creating coherent atomic beams dubbed *atom lasers*.

10.6 Further Developments in the Theory of Fluctuations of Radiation Fields

In the year following 1924 up to about 1930, the subject of radiation fluctuations was extensively considered by A Smekal,[76] W Bothe,[77] and R Furth.[78] Furth performed an important generalization, showing that Equation 10.6 can be obtained directly by the application of Bose–Einstein statistics to photons, irrespective of their spectral distribution.[79] In these papers, two expressions for the statistics appear. The probability that a state with energy E is occupied as

$$W(E) = 1/[\exp(E/kT) - 1], \tag{10.19}$$

and the probability $p(n)$ of having n photons of a distribution with mean value n as

$$p(n) = 1/[(1 - \bar{n})(1 - 1/\bar{n})^n]. \tag{10.20}$$

All these treatments were related to closed systems rather than to propagating light beams.

In the 1940s, during the war, the development and application of radiation detectors focused attention on the importance of investigating the fluctuations of electromagnetic fields propagating far from their sources.

The first studies by Burgess[80] were focused on fluctuations of radio waves and noise induced in radio antennas.

Other works[81] on a large variety of radiation detectors showed that the fluctuations of the radiation field imposed an ultimate limit on the measurements in every case.

In particular, it was shown by Burgess[82] and later by others[83] that the spectral density of the radiation fluctuations obeyed the quantum form of the well-known Nyquist formula for Johnson's noise.[84]

Lewis,[85] Fellgett,[86] and Jones[87] also obtained expressions by a thermodynamic argument for the mean-squared fluctuation $\overline{(\Delta n)^2}$ of the number of photon counts n recorded by an illuminated photocell. Although some of these expressions were later shown to be inapplicable, after some discussion[88] the fact was recognized that the fluctuations of n depart from the classical counting statistics and that $(\Delta n)^2$ can be expressed as the sum of two contributions in close analogy with Equation 10.6.

In 1948, in a review article, MacDonald[89] considered it "... clear *a priori* that no information about the energy distribution of the incident photons can be derived...." from counting measurements with a photocell. This is nowadays known to be incorrect.

Investigations into coherence as a manifestation of correlations between interfering fields began somewhat earlier, in the nineteenth century[90]; the historical path these investigations took has been followed in the classical textbook by M Born and E Wolf[91] and the review paper by L Mandel and E Wolf.[92] The main outcome of this research before the 1950s was twofold: the introduction of precise measure of correlation between the field variables at two space–time points, and the formulation of dynamical laws which the correlations obey.

Emil Wolf (b 1922) (**Figure 10.3**) was born in Prague but was forced to leave when Germans invaded Czechoslovakia. After brief periods in Italy and France, he came to the United Kingdom in 1940 and received his BSc in Mathematics and Physics (1945) and PhD in Mathematics from Bristol University in 1948. Between 1951 and 1954, he worked at the University of Edinburgh with Max Born writing a famous textbook on Optics. He then moved to the United States in 1959 to take a position at the University of Rochester where he worked in optics giving many fundamental contributions. Wolf may be considered the man who more contributed to the theory of classical coherence.

FIGURE 10.3 Emil Wolf.

The *classical coherence theory* which is now well established is set up by introducing an analytical representation of the field variables (e.g., the electric field of the wave) through the complex function $V(t)$, known as the *analytic signal*.[93] Any correlation experiment due to interference could, before the 1950s, be treated by introducing the correlation function

$$\Gamma(r_1, r_2, \tau) = \langle V^*(r_1, t) V(r_2, t + \tau) \rangle = \Gamma_{12}(\tau)$$

$$= \lim_{T \to \infty} (1/2T) \int_{-T}^{+T} V^*(r_1, t) V(r_2, t + \tau) \mathrm{d}t. \tag{10.21}$$

The ensemble correlation function Γ was first introduced by Zernike (1888–1966) in 1938[94] (although not of course in terms of the analytic signal), together with its normalized quantity

$$\gamma_{12}(\tau) = [\Gamma_{12}(\tau)][\Gamma_{11}(0)]^{-1/2} \ [\Gamma_{22}(0)]^{-1/2}, \tag{10.22}$$

called by him the *complex degree of coherence*. γ_{12} is 1 for a fully coherent source and 0 for a fully incoherent source. The time τ at which $\gamma(\tau)$ reaches the value $1/e$ is called the *coherence time*.

The use of these functions was further developed by Hopkins.[95] Several important properties and extensions of correlation functions were derived by Wolf[96] and, independently, by Blanc-Lapierre and Dumontet[97] in 1955. Some of these are:

1. The value of γ_{12} for a plane illuminated by circular source of diameter a, at distance R,[95] is

$$\gamma_{12} = [2J_1(z)/z]\exp[\mathrm{i}k(a\rho/2R)], \tag{10.23}$$

where $J_1(z)$ is the Bessel function of the first kind and first order and

$$z = (2\pi/\lambda)(\rho a/2R) \tag{10.24}$$

with ρ being the distance from the two observation points r_1 and r_2.

2. The demonstration of the fact that the coherence factor is essentially the normalized integral over the source of the Fourier (frequency) transform of the intensity function of the source.[96]

The clarification of the concept of coherence which resulted from these discussions enabled Forrester to develop an old idea which he had several years before: the production of beats from light waves,[98] or *photoelectric mixing* as he called it. Photoelectric mixing is a direct analogue of the familiar mixing of AC electrical signals in nonlinear circuit elements. In the Forrester et al. experiment, a ^{198}Hg lamp was placed in a magnetic field of 3300 G, and the Zeeman-split 5461 Å light illuminated the photocathode of a microwave phototube. The two strong σ components of the light (for which the Zeeman splitting was about 10^{10} Hz) beat each other in the square-law photoelectric emission process, and the photocurrent was found to have an AC component at the Zeeman-difference frequency. As we mentioned in Chapter 8, this was the first experiment in nonlinear optics of frequency mixing.

The greatest concern in obtaining beats was to get the two waves coherent. Forrester argued that the different spectral components of a homogeneously broadened line should beat between themselves and derived a very simple relation which showed that to obtain a well-defined beat pattern, it was necessary that the coherence time of the light should be long compared with the period of the bet. Forrester's experiment was very difficult at the time, due to the low intensity and large bandwidth of the light source.

The topic found no immediate application: however, some years later, in 1959, Alkemade[99] pointed out that a single optical spectral line falling on a phototube would produce a photocurrent whose low-frequency noise spectrum would provide information on the shape, and particularly the width, of the optical line. The same conclusion was reached by Forrester in 1961[100] in a paper entitled *Photoelectric Mixing as a Spectroscopy Tool*. He considered a spread spectral line as a series of "slices," and treated beats between all possible pairs of these slices as a simple extension of the two-component case. With this approach, Forrester deduced the photocurrent spectra which would be produced by several different types of optical spectra.

He also discussed the two different approaches that can be used, namely producing beats between two different sources and between different spectral parts of the same line. He called these *superheterodyne* and *low-level detection*, respectively.[101]

These two approaches were subsequently used to study the spectral output of lasers.[102] They form the basis of the now well-known field of *light beating spectroscopy*.[103]

10.7 The Hanbury Brown and Twiss Experiment

Robert Hanbury Brown (1916–2002) and Richard Quentin Twiss (1920–2005) made the surprising discovery of quantum correlations in ordinary light and started a revolution that brought to the end the construction of a quantum theory of coherence.

Before the early 1950s, interest in radiation fluctuations was largely based on the unavoidable limitations which they imposed on the accuracy of radiation measurements. However, in the years 1952–1956, Hanbury Brown et al., working in the field of radioastronomy, developed a new interferometry technique for the determination of angular diameters of radio sources[104] in which the measurement of radiation fluctuations played a new and essential role.

Later, Hanbury Brown and Twiss applied this technique to light waves.[105] They showed, among other things, that the degree of coherence at two points in a radiation field can be inferred from correlation measurements of the fluctuating signals appearing at radio antennas placed at the two points.

Hanbury Brown[106] tells us he had the idea for the intensity interferometer on a night in 1949, while he was designing a radio interferometer to measure the angular dimensions of two important radio sources, Cygnus A and Cassiopeia A. An ordinary radio interferometer would need a base of several thousand kilometers. Would it be possible to construct such an apparatus? He wondered:

> The immediate technical difficulty in modifying a conventional system was to feed a coherent oscillator to the two distant points … and I started to wonder if this was really necessary.

Could one perhaps compare the radio waves received at two points by some other means? As an example, I imagined a simple detector which demodulated waves from the source and displayed them as the usual noise which one sees on a cathode-ray oscilloscope. If one could take simultaneous photographs of the noise at two stations, would the two pictures look the same? This question led directly to the idea of the correlation of intensity fluctuations and to the principle of intensity interferometer.[106]

After some days of meditation, Hanbury Brown convinced himself that the idea was good and asked the help of Richard Twiss to give him a mathematical basis for his idea. The principle of the interferometer was born and was published later in 1954.[107] In the meantime, Hanbury Brown and Twiss, together with R C Jennison, assembled an intensity radio interferometer at Jodrell Bank at a frequency of 125 MHz and used it to measure the diameter of the sun. The next step was, in 1952, the measurement of the radius of Cygnus and Cassiopeia.[108] These measurements showed that the radii of these stars were in fact much larger than expected and could have been measured with a normal interferometer without the large base prepared by Hanbury Brown. This is when something new was observed[106]:

At the beginning of this programme we had thought that the sole advantage of an intensity interferometer, compared with the radio version of Michelson's interferometer, was that it did not require mutually coherent local oscillators at the separated stations and was therefore peculiarly suitable for extremely long baselines … However, as we watched our interferometer at work, we noticed that when the radio sources were scintillating violently, due to ionospheric irregularities, the measurements of correlation were not significantly affected. Richard Twiss investigated the theory of this surprising effect and confirmed that it was to be expected. We had overlooked one of the principal features of an intensity interferometer: the fact that it can be made to work through a turbulent medium.

The team went on to consider the possibility of constructing a light interferometer and Twiss studied its theory. Calculations showed that, to measure a first magnitude star, two telescopes of at least 2.5 m diameter would be required, one of which would need to be mobile.

It took us six months to realise that although we should certainly need two very large telescopes, they could be extremely crude by astronomical standards

adds Hanbury Brown. In practice, in fact, the telescopes were only needed to collect the light and not to give an excellent quality image as for astronomical telescopes.[106] At this point, Hanbury Brown recalls:

With renewed enthusiasm we returned to establish the detailed theory of an optical interferometer, and immediately ran into a barrage of criticism …

Our original theory was clearly correct at radio wavelengths but when it came to light waves there were one or two lingering doubts in our own minds and several firmly entrenched doubts in the minds of others. The trouble of course was due to worrying about photons. As I have already pointed out, radio engineers in those days looked on radio waves as being simply waves and our theory of the radio intensity interferometer was accepted without question. But when we came to deal with physicists, all sort of queries were raised. One group of objections was concerned with the validity of our semiclassical model of photoelectric emission. We had assumed that the probability of emission of a photoelectron is proportional to the instantaneous square of the electric vector of the incident light wave treated classically. Further, we had assumed that there is no significant delay in the photoelectric process and that in the output current all the components of the envelope of light, at least up to 100 MHz, would be present with their correct phases and amplitudes. At that time there was no sufficient detailed quantum mechanical treatment of photoelectric emission ... another stream of objections about photons were both instructive and entertaining. Our whole argument was based on the idea that the fluctuations in the output of two photoelectric detectors must be correlated when they are exposed to a plane wave of light. We had shown that this must be so by a semiclassical analysis in which light is treated as a classical wave and in this picture there is no need to worry about photons—the quantisation is introduced by the discrete energy levels in the detector. However, if one must think of light in terms of photons then, if the two pictures are to give the same result, one must accept that the times of arrival of these photons at the two separated detectors are correlated—they tend to arrive in pairs. Now, to a surprising number of people, this idea seemed not only heretical but patently absurd and they told us so in person, by letter, in publications, and by actually doing experiments which claimed to show that we were wrong. At the most basic level they asked how, if photons are emitted at random in a thermal source, can they appear in pairs at two detectors? At a more sophisticated level the enraged physicist would brandish some sacred text, usually by Heitler, and point out that the number n of quanta in a beam of radiation and its phase Φ are represented by noncommuting operators and that our analysis was invalidated by the uncertainty relation

$$\delta n \cdot \delta \Phi \sim 1. \tag{10.25}$$

We tried as best we could to answer all these objections and to quieten people down. We were certainly interested in seeking the truth but in raising money to build an interferometer it was desirable that our proposal should be widely regarded as sound. These difficulties about photons troubled physicists who had been brought up on particles and had not fully appreciated that the concept of a photon is not a complete picture of light. Thus many people are reluctant to accept the notion that a particular photon cannot be regarded as having identity from emission to absorption. These objections can, in fact, be answered straight out of text-books and we developed some considerable skill in expounding the orthodox paradoxical nature of light, or, if you like, explaining the incomprehensible—an activity closely, and interestingly, analogous to preaching the Athanasian Creed. In answer to the more sophisticated objection that our proposal was inconsistent with the uncertainty relation in equation (10.25) we pointed out that we were proposing to measure only the relative phase $(\Phi_1 - \Phi_2)$ between two beams of radiation; the total energy of two beams $(n_1 + n_2)$ and their relative phase $(\Phi_1 - \Phi_2)$ can be represented by commuting operators and can be represented classically.

Finally, there were the objections based on laboratory experiments which claimed to show that photons are not correlated at two separate detectors. Here we were on sure ground because we had already done our own carefully laboratory test of the principle. In 1955, I had borrowed the dark-room which housed the spectro-heliograph at Jodrell Bank and set up our first optical interferometer. An artificial star was formed by focusing the brightest part of a high-pressure mercury arc onto a pinhole. The light from this pinhole was then divided into two beams, by a half-aluminized mirror, to illuminate two photomultipliers mounted so that their photocathodes could be optically superimposed or separated by a variable distance as seen from the pinhole. The whole system [see **Figure 10.4**] simulated the measurement by two detectors on the ground of a star with a surface temperature of about 8000 K. After the usual troubles with the equipment we observed the expected correlation and successfully measured it …

We were therefore able to face with confidence objections based on two independent experiments claiming to show that there is no correlation between photons in coherent light beams. The first was performed in Budapest by Adam, Janossy and Varga (1955)[109] and was published at the same time as our own test was being made. In the introduction to their paper they stated that, according to quantum theory, the pulses produced in two separate detectors illuminated by coherent light should be independent of one another. Their aim was "to investigate the validity of this prediction of quantum theory." They illuminated two photomultipliers with coherent light from a single source and also with incoherent light from separate sources, and they counted the coincidences of the pulses produced by individual photons in the two phototubes. In an observation lasting ten hours they found no significant correlation between the arrival times of photons and they claimed that this showed that "in agreement with quantum theory, the photons of two coherent light beams are independent of each other or at least that the biggest part of such photons are independent of each other." Since the results of this experiment were welcomed by our critics as evidence that an intensity interferometer was fundamentally unsound, we took closer look at these claims. It was at once obvious, from a quantitative

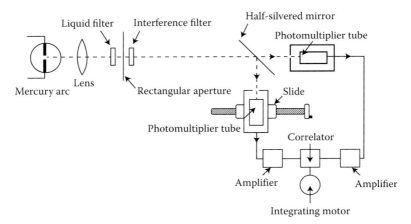

FIGURE 10.4 Simplified diagram of the apparatus used by Hanbury Brown and Twiss. (From Hanbury Brown and Twiss, *Nature* **177**, 27 (1956); Reprinted by permission from Macmillan Journals Ltd. *Nature* **177**, 27, copyright 1956.)

analysis of the parameters of their experiment—light intensity, resolving time, etc.—that there was no hope whatever of observing correlation within 10 h or of testing the predictions of quantum theory. In reply we published a brief note[110] drawing attention to the fact that in order to observe a significant correlation (three times RMS noise) Adam et al. would have had to observe for 10^{11} years—somewhat longer than the age of the Earth.

The second experimental objection was made in 1956 at the University of Western Ontario by Brannen and Ferguson[111] just after the publication of our own laboratory work. They designed their optical system to resemble as far as possible the one we had used at Jodrell Bank. Two photomultipliers were illuminated by coherent light from a high-pressure mercury arc via a half-silvered mirror and the outputs of the two phototubes were taken, not to a linear multiplier, but to a coincidence counter. They conclude that "there is no correlation (less than 0.01%) between photons in coherent light rays." They added that "if such a correlation did exist it would call for a major revision of some fundamental concepts in quantum mechanics." Again, we analysed these conclusions and found that the parameters, as before, were hopelessly inadequate to allow the detection of correlation between photons, within a reasonable time. In this case the essential point is that in order to achieve a practical signal-to-noise ratio with a coincidence counter one needs an intense source of light with an extremely narrow bandwidth, and this they did not have. We published a short note[112] showing that it would have taken Brannen and Ferguson 1000 years to observe a significant correlation.

Both these experiments were beyond reproach from an experimental point of view, but since they had been planned without an adequate theoretical foundation they were far too insensitive to be of any significant use. Nevertheless, they did provide our opponents with ammunition.

"Bloody but unbowed" Hanbury Brown and his collaborators assembled the first stellar intensity interferometer by using military relics and were able to measure throughout the winter of 1955–1956 the angular diameter of Sirius.[113]

The year before they had performed the laboratory test[114] already referred to and another was undertaken in 1957 at the Jodrell Bank Experimental Station of the University of Manchester[115] as Hanbury Brown recalls above, in response to the many objections raised by Adam et al.[109] and Brannen and Ferguson.[111]

The apparatus used in the first experiment is shown in Figure 10.4.[114] Light from a mercury lamp was filtered and split into two beams by a half-silvered mirror and fell on two photocells whose outputs were sent through band-limited amplifiers to a correlator. The theory of this experiment was developed later[116] in a rather clumsy manner: the main results can be quite easily explained as follows.

The signal currents at the two photocells are proportional to the light intensities $I_1(t)$ and $I_2(t)$ falling on them, and therefore the correlation measured by Hanbury Brown and Twiss was proportional to $\langle I_1(t + \tau)I_2(t)\rangle$ which can be shown to obey the following relation, for a classical thermal source[117]

$$\langle I_1(t + \tau)I_2(t)\rangle = I_1 I_2[1 + (1/2)|\gamma_{12}(\tau)|^2]. \tag{10.26}$$

Alternatively, defining

$$\Delta I = I - \langle I \rangle, \tag{10.27}$$

we have

$$\langle \Delta I_1(t + \tau)\Delta I_2(t) \rangle = \langle (I_1(t + \tau) - \langle I_1 \rangle)(I_2(t) - \langle I_2 \rangle) \rangle$$
$$= \langle I_1(t + \tau)I_2(t) \rangle - \langle I_1 \rangle\langle I_2 \rangle, \tag{10.28}$$

so that

$$\langle \Delta I_1(t + \tau)\Delta I_2(t) \rangle = (1/2) \langle I_1 \rangle\langle I_2 \rangle |\gamma_{12}(\tau)|^2. \tag{10.29}$$

The foregoing discussion does not describe the experimental situation fully. In fact, shot noise from the photocurrent is not taken into account. Moreover, although $I(t)$ changes slowly compared with the electric field, its fluctuations are too rapid for the electronic correlator to follow. In practice, the signals to be correlated in the Hanbury Brown and Twiss experiment are derived from band-limited electronic amplifiers, characterized by a certain frequency response $B(\upsilon)$. So, if we indicate with $S(t)$, the output of the linear filter, then we have

$$S(t) = \alpha \int_0^\infty I(t - t')b(t')dt', \tag{10.30}$$

where $b(t)$ is the response of the system to a short impulse at $t = 0$ and is the Fourier transform of $B(\upsilon)$. The measured correlation is therefore

$$\langle S_1(t)S_2(t) \rangle = \alpha_1\alpha_2 \int\int_0^\infty \langle I_1(t - t')I_2(t - t'') \rangle b(t')b(t'')dt' \, dt''$$
$$= \alpha_1\alpha_2 \int\int_0^\infty \langle I_1(t + t'' - t')I_2(t) \rangle b(t')b(t'')dt' \, dt'', \tag{10.31}$$

and

$$\langle \Delta S_1(t)\Delta S_2(t) \rangle = \tfrac{1}{2}\alpha_1\alpha_2\langle I_1 \rangle\langle I_2 \rangle |\gamma_{12}(0)|^2 \int\int |\gamma_{11}(t'' - t')|^2 \, b(t')b(t'')dt' \, dt''$$
$$= \tfrac{1}{2}\alpha_1\alpha_2\langle I_1 \rangle\langle I_2 \rangle |\gamma_{12}(0)|^2 \int_0^\infty \int_{-t'}^\infty |\gamma_{11}(t''')|^2 \, b(t')b(t' + t''')dt''' \, dt',$$
$$\tag{10.32}$$

where use has been made of the relation[118]

$$\gamma_{12}(\tau) = \gamma_{12}(0)\gamma_{11}(\tau). \tag{10.33}$$

Now $|\gamma_{11}(\tau)|$ vanishes for τ much in excess of $1/\Delta\upsilon$. On the other hand, if the highest frequency passed by the electrical filter is still small compared with the bandwidth $\Delta\upsilon$ of the light, as is normally the case, then $b(t)$ is nearly constant over intervals in the order $1/\Delta\upsilon$. Hence, for those values of t''' for which the integrand does not vanish, $b(t' + t''') \sim b(t')$. Therefore,

$$\langle \Delta S_1(t)\Delta S_2(t)\rangle = \tfrac{1}{2}\alpha_1\alpha_2\langle I_1\rangle\langle I_2\rangle\left|\gamma_{12}(0)\right|^2 \xi(\infty)\int_0^{\infty} b^2(t')dt', \qquad \textbf{(10.34)}$$

where

$$\xi(\infty) = \int_{-\infty}^{+\infty}\left|\gamma_{11}(\tau)\right|^2 d\tau. \qquad \textbf{(10.35)}$$

Provided $\langle S_1\rangle$ and $\langle S_2\rangle$ are not zero, this can be put in the form

$$\langle \Delta S_1(t)\Delta S_2(t)\rangle = \tfrac{1}{2}\langle S_1\rangle\langle S_2\rangle[\xi(\infty)/T']|\gamma_{12}(0)|^2, \qquad \textbf{(10.36)}$$

where

$$\left\{\int_0^{\infty} b(t')dt'\right\}^2 \Big/ \left\{\int_0^{\infty} b^2(t')dt'\right\} = T' \qquad \textbf{(10.37)}$$

is a rough measure of the time spread of the impulse response function $b(t)$, and therefore has the nature of a *resolving time T'*.

If noise is taken into account, the normalized correlation coefficient ρ can be written as

$$\rho = [\langle \Delta S_1(t)\Delta S_2(t)\rangle][\langle \Delta S_1\rangle^2\langle \Delta S_2\rangle^2]^{-\frac{1}{2}}$$
$$= [\tfrac{1}{2}\alpha\bar{I}\xi(\infty)][1 + \tfrac{1}{2}\alpha\bar{I}\xi(\infty)]^{-1}\left|\gamma_{12}(0)\right|^2. \qquad \textbf{(10.38)}$$

The factor $1/2\ \alpha\bar{I}\xi(\infty)$, which gives half the average number of photoelectrons emitted in a time equal to the coherence time by a coherently illuminated photocathode, is called the degeneracy parameter δ.[119]

For all classical light beams δ is very much less than unity. The result of the experiment can now be understood by looking at the meaning of $\gamma_{12}(0)$. This correlation function is defined as

$$\gamma_{12}(0) = \langle V(P_1,t)V^*(P_2,t)\rangle/[\langle V(P_1,t)V^*(P_1,t)\rangle]^{-1/2}[\langle V(P_2,t)V^*(P_2,t)\rangle]^{-1/2}, \qquad \textbf{(10.39)}$$

and gives the correlation of the disturbance at two different points at equal times. For a circular thermal source (such as a star), it had already been calculated by Hopkins[95] and is given in Equation 10.23. Its square modulus is shown in the full line of **Figure 10.5.** By varying the separation d between the two photomultipliers, the function $|\gamma_{12}(0)|^2$ is measured and therefore the parameters of the source (its angular magnitude) can also be measured. The result of the Hanbury Brown and Twiss experiment is shown in Figure 10.5.[114]

The result of this experiment confirmed the operational principle of an intensity interferometer as shown in **Figure 10.6.** When using it for observing a star, the correlation will decrease with increasing baseline in a manner similar to that shown in Figure 10.5 and then a measurement of the curvature will give the angular magnitude of the star.

The difference between such an interferometer and the conventional stellar Michelson interferometer is that the intensity interferometer measures the

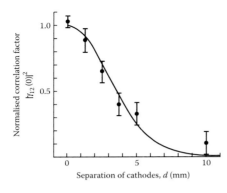

FIGURE 10.5 The experimental and theoretical values of the normalized correlation factor $|\gamma_{12}(0)|^2$ for different values of separation between the photocathodes. The full line is the theoretical curve and the experimental results are plotted as points with their associated probable errors. (R Hanbury Brown and R Q Twiss, *Proc. R. Soc.* **A243**, 291 (1957).)

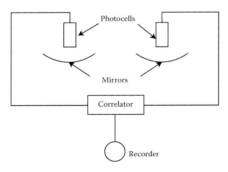

FIGURE 10.6 The Hanbury Brown and Twiss intensity interferometer at optical wavelengths. (R Hanbury Brown and R Q Twiss, *Nature* **177**, 27 (1956); Reprinted by permission from Macmillan Journals Ltd. *Nature* **177**, 27, copyright 1956.)

square of the modulus of the complex degree of coherence—while the Michelson one gives the complex degree of coherence itself—and so the phase of this complex function is lost. Roughly speaking, this means that one cannot reconstruct the angular distribution across an asymmetrical source without ambiguity; for example, when observing a double star with two unequal components, one cannot tell which star is on the left and which is on the right.

The first demonstration of time coincidence between photons was mentioned earlier (in Hanbury Brown's recollections). To obtain an adequate signal-to-noise ratio in a coincidence-counting experiment in which individual photons can be counted, an intense and narrow-band source of light is needed. Hanbury Brown and Twiss thought of using a mercury isotope lamp and found this type of lamp in Sydney, where the experiment was performed in 1957 by Twiss, Little, and Hanbury Brown.[112] They used an electrode-less radio-frequency discharge in mercury-198 vapor as a source of light, 1P21 phototubes, and a coincidence counter with a resolving time of 3.5×10^{-9} s. The arrangement was similar to the one shown in Figure 10.4. Both phototubes were mounted on movable slides, so that, as seen by the source, they could be optically superimposed or separated by a distance of 5 mm transverse to the line of sight; at this separation the pinhole source was completely resolved, so that the incident light beams were uncorrelated.

They compared the coincidences between photons arriving at these two phototubes with the coherent illumination (when the two phototubes were optically superposed) and the incoherent illumination (when the two phototubes were displaced). In a test lasting 8 h, they found that, with coherent illumination, the number of coincidences was increased by $1.93 \pm 0.17\%$, which was in satisfactory agreement with the theoretical estimate of 2.07%.

Subsequently, the correlation between photons was confirmed, using coincidence counters, by Rebka and Pound,[120] and others.[121]

The correlation observed between the photocurrent fluctuations of two photoelectric detectors immediately rules out the possibility that these currents consist of statistically independent pulses corresponding to the arrival of statistically independent photons at the photocathodes. Purcell[122] in 1956 was the first to point out that the effects observed could be explained quantitatively in terms of the quantum statistical behavior of the photons; and these considerations were later followed up and extended by several authors.[123] Purcell began by considering one beam of light falling onto one photomultiplier and examined the statistical fluctuations in the counting rate. By assuming that the probability that a photoelectron is ejected in time dt can be written as $\alpha P \, dt$, where α is a constant and P is the square of the electric field in the light, he derived the mean number of photoelectrons (n_T) which will be counted in a time interval T as being

$$\bar{n}_T = \alpha \bar{P} T, \tag{10.40}$$

and showed that the variance

$$\overline{(\Delta n)^2} = \overline{n^2} - (\bar{n})^2 \tag{10.41}$$

satisfies the relation

$$\overline{(\Delta n)^2} = \bar{n}[1 + \bar{n}(\tau_o / T)], \tag{10.42}$$

where

$$\tau_o = \int_{-\infty}^{+\infty} |\gamma_{11}(\tau)|^2 \, d\tau. \tag{10.43}$$

Assuming a quasi-monochromatic light with a Lorentzian lineshape of full width $\Delta \upsilon$ at half intensity

$$\tau_o \sim 1/\Delta \upsilon, \tag{10.44}$$

it results

$$\overline{(\Delta n)^2} = \bar{n}[1 + (\bar{n}/\Delta \upsilon T)]. \tag{10.45}$$

This result shows that the fluctuations in the output current of a photoelectric detector are greater than the expected value $(\Delta n^2 = n)$ in a simple random stream.

He adds:

If one insists on representing photons by wave packets and demands an explanation in those terms of the extra fluctuations, such an explanation can be given ...

Think, then of a stream of wave packets, each about $c/\Delta v$ long, in a random sequence. There is a certain probability that two such trains accidentally overlap. When this occurs they interfere and one may find (to speak rather loosely) four photon, or none, or something in between as a result. It is proper to speak of interference in this situation because the conditions of the experiment are just such as will ensure that these photons are in the same quantum state. To such interference one may ascribe the "abnormal" density fluctuations in any assemblage of bosons.

Turning now to the split-beam experiment of Hanbury Brown and Twiss, let n_1 be the number of counts of one photomultiplier in an interval T, and let n_2 be the number of counts in the other in the same interval. As regards the fluctuations in n_1 alone, from interval to interval, we face the situation already analyzed, except that we shall now assume both polarizations present, which changes Equation 10.42 into

$$\overline{(\Delta n_1^2)} = \overline{n_1}[1 + \tfrac{1}{2}\overline{n_1}(\tau_0/T)]. \tag{10.46}$$

A similar relation holds for n_2. Now, if we were to connect the two photomultiplier output together, we would revert to a single-channel experiment with count $n = n_1 + n_2$, or

$$\overline{(\Delta n^2)} = \overline{n}[1 + \tfrac{1}{2}\overline{n}(\tau_0/T)], \tag{10.47}$$

but

$$\overline{(\Delta n^2)} = \overline{(\Delta n_1 + \Delta n_2)^2} = \overline{n_1}[1 + \tfrac{1}{2}\overline{n_1}(\tau_0/T)] + \overline{n_2}[1 + \tfrac{1}{2}\overline{n_2}(\tau_0/T)] + 2\overline{\Delta n_1 \Delta n_2}. \tag{10.48}$$

From Equations 10.47 and 10.48, it follows that

$$\overline{\Delta n_1 \Delta n_2} = \tfrac{1}{2}\overline{n_1 n_2}(\tau_0/T), \tag{10.49}$$

which is the positive cross-correlation effect of Hanbury Brown and Twiss. It was therefore demonstrated that the fluctuations in the counts recorded by a photoelectric detector which is illuminated by light from a thermal source are those to be expected from a boson assembly.

In 1958, Mandel[124] derived classically the relation between the statistics of the photons of the field and the statistics of photoelectrons emitted by a photodetector.

Leonard Mandel (**Figure 10.7**) was born in Berlin on May 9, 1927. His family was pursued by the Gestapo and escaped to London where he attended school and later applied to University of London earning there his BSc in mathematics and physics in 1947. He worked shortly on cosmic rays

FIGURE 10.7 Leonard Mandel (1927–2001).

for the PhD (1951) and joined the Research Laboratories of Imperial Chemical Industries Ltd, returning to academia with a lectureship in the Physics Department at Imperial College where he moved in 1955 and met Emil Wolf with whom developed a lifelong friendship. In 1964, Leonard was offered a professorship in the Department of Physics at the University of Rochester where he remained until his death in 2001. His interest in optical coherence was stimulated by the Brown and Twiss experiment in 1956 which brought him to develop the classical papers in 1958–1959 that we discuss now.

By developing Purcell idea,[125] Mandel began by saying:[125]

> We shall now associate photons with the Gaussian random wave $y(t)$, by defining a probability that a photoelectron is ejected in a short time interval between t and $t + dt$. If we consider first-order transitions only significant, in which one photon gives rise to one photoelectron, then this probability will be given by $\alpha P(t)dt$, where α is the quantum sensitivity of the photoelectric detector, assumed constant over the narrow frequency range Δv_o. The observable $P(t)$ provides the only link between the wave and the particle description of the beam.
>
> The fluctuations of the number of particles n_T therefore have two causes. There are first of all fluctuations of the wave intensity $P(t)$, determined by the spectral lineshape and there is the stochastic association of particles with the wave intensity. This twofold source of the fluctuations results in the departure from classical statistics, as we shall show.[117]

He then wrote $p_n(t,T)$ for the probability that n photoelectrons are ejected in the interval between t and $t + T$. In particular, from the definition

$$p_1(t,T) = \alpha P(t)dt, \tag{10.50}$$

the expected value of n in the interval t to $t + T$ is

$$\alpha \int_t^{t+T} P(t')dt'. \tag{10.51}$$

We may therefore expect from first principles to have

$$p_n(T) = \overline{p_n(t,T)} = (1/n!)\left\{ \alpha \int_t^{t+T} P(t')dt' \right\}^n \exp\left\{ -\alpha \int_t^{t+T} P(t')dt' \right\}, \tag{10.52}$$

where the bar indicates time averaging.

The two papers[124] are concerned with the calculation, on purely classical grounds, of $\bar{n}, \overline{n^2}$, and the problem of determining the statistics $p_n(T)$ for different values of T with respect to the coherence time τ of the light ($\tau \sim 1/\Delta v_o$). In particular, when $\Delta v_o T \ll 1$, $P(t)$ does not vary much in the time T, and it follows that

$$\alpha \int_t^{t+T} P(t')dt' \sim \alpha P(t)T. \tag{10.53}$$

Therefore,

$$\overline{p_n(t,T)} = (1/n!)\overline{(\alpha PT)^n \exp(-\alpha PT)}. \tag{10.54}$$

If the probability distribution P is known, $p_n(t,T)$ can be evaluated by averaging over the ensemble. Since $y(t)$ is a narrow-band Gaussian random variable, the local average is

$$\langle y^2(t) \rangle = P(t) = 1/2 w^2(t), \tag{10.55}$$

where $w(t)$ is the envelope of $y(t)$. Now, the probability density of $w(t)$ was shown by Rice[126] to be of the form

$$(w/\overline{P}) \exp(-w^2/2\overline{P}).$$

Hence

$$P'(P)\mathrm{d}P = (1/\overline{P}) \exp[-(P/\overline{P})]\mathrm{d}P, \tag{10.56}$$

and

$$\overline{p_n(t \cdot T)} = \left\{ (1 + \alpha \overline{PT})[1 + (1/\alpha \overline{PT})]^n \right\}^{-1} \tag{10.57}$$

with

$$\overline{\Delta n^2} = \overline{n}(1 + \overline{n}). \tag{10.58}$$

This is the well-known formula for the fluctuations of the occupation numbers of a *single cell* in phase space for an assembly of bosons. Mandel says

> The photoelectrons in the interval T therefore obey "pure" Bose–Einstein statistics. The reason for this can be seen at once if we examine the size of the elementary cell in phase space. In the direction of the beam this extends over a distance $c\Delta t$, where $\Delta t \sim 1/\Delta v_o$. Thus, the photons in an interval $T \ll 1/\Delta v_o$ as above, i.e. much shorter than the so-called coherence time of the light ..., occupy the same cell in phase space. By the uncertainty principle they are therefore intrinsically indistinguishable and n obeys pure Bose–Einstein statistics.

When $\Delta v_o T \gg 1$, Mandel[124] still derived

$$\overline{\Delta n^2} = \overline{n}(1 + k\overline{n}/\Delta v_o T), \tag{10.59}$$

where k is a number depending on the spectral density; this is the relation obtained by Purcell.

In this case, we are now dealing with a volume of phase space containing roughly $\Delta v_o T = s$ cells. The mean number of photons per cell is therefore $n/s = \overline{m}$, and Equation 10.59 can be written as

$$\overline{\Delta n^2} = \overline{n}(1 + k\overline{m}) = \overline{n}(1 + ks\overline{m^2}/n). \tag{10.60}$$

This is the expression in the conventional form for the density fluctuations in a larger volume of phase space which was first found by Furth.[78,79]

These considerations could be perfected considering the degeneracy parameter δ. L Mandel in 1961[127] illustrated this point quite well. δ is the average number of photons in the light beam which are to be found in the same quantum state, or in the same cell of phase space. The correlation between the fluctuations of the counts recorded by two photomultipliers illuminated by partially coherent light from a thermal source is proportional to δ. The degeneracy of black-body radiation in an enclosure at a temperature T was first shown by Einstein[128] to be

$$\delta = [\exp(h\upsilon/kT) - 1]^{-1}, \qquad (10.61)$$

at a frequency υ. It was not immediately obvious that this relation was also valid for a light beam far removed from its equilibrium source, but Mandel showed that this was the case. These results on equilibrium photon statistics received full support some years later in 1964, when Mandel, Sudarshan, and Wolf[129] derived, on a semiclassical basis, that the probability $p(t)\Delta t$ of photoemission of an electron is proportional to the classical measure of the instantaneous light intensity

$$p(t)\Delta t = \alpha I(t)\Delta t, \qquad (10.62)$$

which is the same as Equation 10.50.

10.8 The Quantum Theory of Coherence

In the meantime, a revolution had taken place in coherence theory due to a few papers written by R J Glauber (**Figure 10.8**).

Roy J Glauber was born in New York in 1925. He did undergraduation at Harvard University and was recruited to work in the Manhattan Project where at the age of 18 he was one of the youngest scientists at Los Alamos. His work involved calculating the critical mass for the atomic bomb. After 2 years at Los Alamos, he returned to Harvard, receiving his BSc in 1946 and his PhD in 1949 with a quantum field theoretical thesis. He then spent a postdoctoral year at the Institute for Advanced Study in Princeton and worked in different places becoming involved in studying nuclear diffraction theory obtaining important results which he continued to elaborate also during a sabbatical year at CERN in Geneva. He then was interested in quantum optics with a quantum theory of coherence. In 2005, he was awarded the Nobel Prize in Physics together with J L Hall (b 1934) and T W Hansch (b 1941). One half of the prize was awarded to him "for his contribution to the quantum theory of optical coherence." The other half jointly was given to Hall and Hansch "for their contribution to the development of laser-based precision spectroscopy, including the optical frequency comb technique."

FIGURE 10.8 Roy Glauber (b 1925).

The main problem, as we have seen, was to understand the essence of the Hanbury Brown and Twiss experiment. The revolution started quietly with a letter to *Phsical Review Letters.*[130] It begins:

> In 1956 Hanbury Brown and Twiss reported that the photons of a light beam of narrow spectral width have a tendency to arrive in correlated pairs. We have developed general quantum mechanical methods for the investigation of such correlation effects and shall present here results for the distribution of the number of photons counted in an incoherent beam. The fact that photon correlations are enhanced by narrowing the spectral bandwidth had led to a prediction (by Mandel and Wolf[131]) of large-scale correlations to be observed in the beam of an optical maser. We shall indicate that the prediction is misleading and follows from an inappropriate model of the maser beam.

In making this last statement, Glauber was rather sharp, and this issue raised a vigorous controversy between Glauber and the Rochester group which lasted several years, as we shall see later.

In his paper, Glauber then sketches a quantum mechanical representation of the field, introducing the *coherent states* representation and defining the density operator appropriate to the problem. He then discusses the photon correlations and shows that coherent states of the field led to no photoionization correlations whatsoever. A correlation between photons only appears when either incoherent mixtures or superpositions of the coherent states are present. Although he has no precise idea of the density operator for an actual laser, he then supposed it is more likely to be of the kind produced by a product of coherent states.

In two subsequent, very beautiful, papers in *Physical Review* of the same year,[132] Glauber fully develops the quantum theory of optical coherence. He takes as his starting point the separation of the electric field operator $E(rt)$ into its positive and negative frequency parts:

$$E(rt) = E^{(+)}(rt) + E^{(-)}(rt). \tag{10.63}$$

The positive part is associated with photon absorption and the negative part with photon emission. In particular, the positive frequency part $E^{(+)}(rt)$ may be shown[133] to be a photon annihilation operator. Applied to an n-photon state, it produces a $(n-1)$-photon state.

The Hermitian adjoint of it, $E^{(-)}(rt)$, creates a photon. Applied to an n-photon state, it produces an $(n+1)$-photon state.

This decomposition of the electric field has a very close connection with classical treatment. Before second quantization, the positive frequency part $E^{(+)}$ can indeed be identified with the classical analytical signal discussed above.

Glauber then shows that a detection process such as photoionization, in which the field makes a transition from the initial state $|i\rangle$ to a final state $|f\rangle$ in which one photon has been absorbed, is described by the matrix element

$$\langle f|E^{(+)}(rt)|i\rangle,$$

and the probability per unit time that a photon be absorbed by an ideal detector at point r at time t is proportional to

$$\sum_f \left| \langle f | E^{(+)}(rt) | i \rangle \right|^2 = \langle i | E^{(-)}(rt) E^{(+)}(rt) | i \rangle. \tag{10.64}$$

Recording photon intensities with a single detector does not exhaust the measurements one can make on the field, though it does characterize, in principle, virtually all the classic experiments of optics. A second type of measurement consists of the use of two detectors situated at different points, r and r' to detect photon coincidences or, more generally, delayed coincidences. The field matrix element for such transitions takes the form

$$\langle f | E^{(+)}(r't') E^{(+)}(rt) | i \rangle,$$

and the total rate at which such transitions occur is proportional to

$$\sum_f \left| \langle f | E^{(+)}(r't') E^{(+)}(rt) | i \rangle \right|^2 = \langle i | E^{(-)}(rt) E^{(-)}(r't') E^{(+)}(r't') E^{(+)}(rt) | i \rangle. \tag{10.65}$$

Such a rate is the one connected with experiments of the type performed by Hanbury Brown and Twiss.

More elaborate experiments can also be considered, in which detection of n-field delayed coincidences of photons for arbitrary n is considered. The total rate per unit time for such coincidences will be proportional to

$$\langle i | E^{(-)}(r_1 t_1) \ldots E^{(-)}(r_n t_n) E^{(+)}(r_n t_n) \ldots E^{(+)}(r_1 t_1) | i \rangle.$$

At this point, Glauber observes that the electromagnetic field may be regarded as a dynamical system with an infinite number of degrees of freedom. Our knowledge of the condition of such a system is, in practice, virtually never so complete or so precise as to justify in its description the use of a particular quantum state $|\rangle$.

We have therefore to consider averages over the distribution of the unknown parameters. Such averages can be constructed quantum mechanically by making recourse to a Hermitian operator known as a density operator ρ which is constructed as an average over the uncontrollable parameters of an expression which is bilinear in the state vector.[134] If $|\rangle$ is a precisely defined state of the field corresponding to a particular set of random parameters

$$\rho = \{|\rangle\langle|\}_{av}.$$

The average of an observable θ in the quantum state $|\rangle$ over the randomly prepared states is the quantity of interest in experiments and it is given by

$$\{\langle|\theta|\rangle\}_{av} = \text{Tr}\{\rho\theta\},$$

where the symbol Tr stands for the trace, or sum, of the diagonal matrix elements.

The average counting rate of an ideal photodetector, for example, is therefore proportional to

$$G(rt, rt) = \text{Tr}\{\rho E^{(-)}(rt) E^{(+)}(rt)\}. \tag{10.66}$$

Glauber now introduced the set of functions

$$G^{(n)}(x_1,\ldots,x_n,x_{n+1},\ldots,x_{2n}) = \mathrm{Tr}\{\rho E^{(-)}(x_1)\cdots E^{(-)}(x_n)E^{(+)}(x_{n+1})\cdots E^{(+)}(x_{2n})\}, \qquad (10.67)$$

which represent the correlation functions of the field at different space–time points, $x_j = r_j t_j$ being the $G^{(n)}$ function necessary to discuss an n-photon coincidence experiment.

By introducing the normalized forms

$$g^{(n)}(x_1 \ldots x_{2n}) = G^{(n)}(x_1 \ldots x_{2n})/\prod_{j=1}^{2n}\left\{G^{(1)}(x_j x_j)\right\}^2, \qquad (10.68)$$

Glauber now extends the notion of coherence, defining different orders of coherence—first-, second-, third-order of coherence and so on—according to which the sequence

$$\left|g^{(n)}(x_1 \ldots x_{2n})\right| = 1, \qquad (10.69)$$

for every n value, that is, for $n = 1, 2, 3$, and so on.

First-order coherence, or

$$\left|g^{(1)}(x_1 x_2)\right| = 1, \qquad (10.70)$$

is what is required for classical optics coherence experiments. The introduction of ensemble averages in place of time averages as used previously extended the notion of coherence to nonstationary fields.

In the next paper,[135] Glauber retraces the full treatment of quantization of the electromagnetic field and introduces the notion of *coherent states*, and the *P representation*,[136] surmising that the laser could be in a coherent state. The starting point of the standard technique of quantization of the electromagnetic radiation *in vacuo* is Maxwell's equations written in the absence of charges and currents. The vector potential $A(rt)$ is expanded in terms of a complete set of real orthogonal mode functions $u_1(r)$ with real coefficients $q_1(t)$

$$A(rt) = \sum_l q_l(t)u_1(r), \qquad (10.71)$$

and it is shown that the total energy of the field

$$H_o = (1/8\pi)(\int_{\mathrm{Cavity}} (E^2 + H^2)\,dr \qquad (10.72)$$

can be expressed as

$$H_o = \tfrac{1}{2}\sum_l \left[p_l^2(t) + \omega_l^2 q_l^2(t)\right], \qquad (10.73)$$

where

$$p_l(t) = dq_l/dt. \qquad (10.74)$$

In this way, the electromagnetic field is described in terms of a set of independent couples of conjugated variables, q_l and p_l, relative to a set of independent harmonic oscillators. The quantization of the electromagnetic field is now achieved by regarding q_l and p_l as Hermitian operators obeying the commutation relations

$$[p_l, p_m] = [q_l, q_m] = 0, \quad [q_l, p_m] = i\hbar\delta_{lm}, \tag{10.75}$$

according to a basic postulate of quantum mechanics. The q_l and p_l will be explicitly time-independent or not, according to whether or not the Schroedinger picture is used. The standard procedure of quantization of the harmonic oscillator consists, then, in the introduction of a pair of non-Hermitian operators $a_l^\dagger$ and a_l by means of the equations

$$q_l = (\hbar/2\omega_1)^{1/2}[a_l^\dagger + a_l], \tag{10.76}$$

$$p_l = i(\hbar\omega_1/2)^{1/2}[a_l^\dagger - a_l]. \tag{10.77}$$

It can easily be seen from Equations 10.76 and 10.77 that a_l and $a_l^\dagger$ are Hermitian conjugate operators, while Equation 10.75 shows that they obey the commutation relations

$$[a_l, a_m^\dagger] = \delta_{lm}, \quad [a_l, a_m] = [a_l^\dagger, a_m^\dagger] = 0. \tag{10.78}$$

The Hamiltonian of the system immediately follows from Equations 10.73, 10.76, and 10.78 as

$$H_o = \sum_l H_1 = \sum_l \hbar\omega_1(a_l^\dagger a_l + a_l a_l^\dagger). \tag{10.79}$$

If we now choose to use the Heisenberg picture, which is the most appropriate one if a comparison between classical and quantum treatment is to be made, the time evolution of $a_l(t)$ and $a_l^\dagger(t)$ is determined by the Heisenberg equations of motion

$$i\hbar(da_l(t)/dt) = [a_l(t), H_o] = \hbar\omega_l a_l(t) \tag{10.80}$$

$$i\hbar[da_l^\dagger(t)/dt] = [a_l^\dagger(t), H_o] = -\hbar\omega_l a_l^\dagger(t), \tag{10.81}$$

having taken into account Equations 10.78 and 10.79. One has, according to Equations 10.80 and 10.81,

$$a_l(t) = a_l \exp(-i\omega_l t), \quad a_l^\dagger(t) = a_l^\dagger \exp(i\omega_l t) \tag{10.82}$$

where $a_l = a_l(0)$ and $a_l^\dagger = a_l^\dagger(0)$ are, from now on, the operators in the Schroedinger picture.

The eigenvalues n_l of the operator associated with the eigenvalue equation

$$a_l^\dagger a_l |n_l\rangle = n_l |n_l\rangle \tag{10.83}$$

are furnished by all non-negative integers, so that n_l can be interpreted as the number of energy quanta in the mode l.

This allows us to give $a_l^\dagger a_l$ the meaning of a numerical operator. Furthermore, $a_l^\dagger$ and a_l are usually termed "creation" and "annihilation" operators, since they can be shown, respectively, to increase and decrease by the number of quanta, when they are operating on the eigenstates of $a_l^\dagger a_l$, according to

$$a_l^\dagger \left| n_l \right\rangle = (n_l + 1)^{1/2} \left| n_l + 1 \right\rangle, \tag{10.84}$$

$$a_l \left| n_l \right\rangle = (n_l)^{1/2} \left| n_l - 1 \right\rangle. \tag{10.85}$$

The preceding considerations summarize the properties of the vector potential operator, which, according to Equations 10.71 and 10.76, can be written in the Heisenberg picture as

$$A(rt) = \sum_l (\hbar/2\omega_l)^{1/2} [a_l \exp(-i\omega_l t) + a_l^\dagger \exp(i\omega_l t)] u_l(r). \tag{10.86}$$

The electromagnetic radiation field is then completely specified, from a quantum mechanical point of view, once the initial state of the system is assigned in terms of the $\left| n \right\rangle$.

The coherent states $\left| \alpha \right\rangle$ are now defined,[137] for a single-mode case, in terms of the eigenstates $\left| n \right\rangle$ of the numerical operators (see Equation 10.83) as

$$\left| \alpha \right\rangle = \exp(-|\alpha|^2/2) \sum_{n=0}^{\infty} [\alpha^n/(n!)^{1/2}] \left| n \right\rangle, \tag{10.87}$$

and verify the eigenvalue equation

$$\alpha \left| \alpha \right\rangle = \alpha \left| \alpha \right\rangle, \tag{10.88}$$

for any complex number α. They are not mutually orthogonal and form an overcomplete set, in the sense that every $\left| \alpha \right\rangle$ can be expanded as a linear combination of the others.

Glauber[138] provided also an alternative approach to the coherent states introducing a unitary operator D which is a function of a complex parameter β and acts as a displacement operator upon the amplitudes $a^\dagger$ and a according to

$$D^{-1}(\beta) a D(\beta) = a + \beta,$$

$$D^{-1}(\beta) a^\dagger D(\beta) = a^\dagger + \beta^*. \tag{10.89}$$

Then if $\left| \alpha \right\rangle$ obeys Equation 10.88, it follows that $D^{-1}(\beta) \left| \alpha \right\rangle$ is an eigenstate of a corresponding to the eigenvalue $\alpha - \beta$,

$$a D^{-1}(\beta) \left| \alpha \right\rangle = (\alpha - \beta) D^{-1}(\beta) \left| \alpha \right\rangle. \tag{10.90}$$

In particular, if we choose $\beta = \alpha$, we find

$$a D^{-1}(\alpha) \left| \alpha \right\rangle = 0 \tag{10.91}$$

The coherent states are just displaced forms of the ground state of the oscillator

$$\left| \alpha \right\rangle = D(\alpha) \left| 0 \right\rangle. \tag{10.92}$$

Glauber then shows that D may be represented as

$$D(\alpha) = \exp(\alpha a^\dagger - \alpha^* a). \qquad (10.93)$$

And the coherent state $|\alpha\rangle$ may therefore be written in the form

$$|\alpha\rangle = \exp(\alpha a^\dagger - \alpha^* a)\,|0\rangle. \qquad (10.94)$$

The generalization to the multi-mode case is easily performed by the introduction of the coherent state $|\{\alpha_l\}\rangle$, defined as the product of the coherent states $|\alpha_l\rangle_l$ relative to each mode

$$\left|\{\alpha_l\}\right\rangle = \prod_l \left|\alpha_l\right\rangle_l. \qquad (10.95)$$

It is then easy to show that they are eigenkets of the positive frequency part of the electric field operator

$$E^{(+)}(rt)|\{\alpha_l\}\rangle = \varepsilon\{\alpha_l\}(rt)|\{\alpha_l\}\rangle, \qquad (10.96)$$

with complex eigenvalues $\varepsilon\{\alpha_l\}(rt)$ given by

$$\varepsilon\{\alpha_l\}(rt) = i\sum_l (\hbar\omega_l/L^3)^{1/2}\,\alpha_l\,\exp[i(k_l\cdot r - \omega_l t)], \qquad (10.97)$$

and, consequently, that

$$\langle\{\alpha_l\}|E^{(-)}(rt) = \varepsilon^*\{\alpha_l\}(rt) < \{\alpha_l\}|. \qquad (10.98)$$

According to Equations 10.96 and 10.98, the correlation function $G^{(n)}$ defined by

$$G^{(n)}(r_1 t_1,\ldots,r_n t_n,r_n t_n,\ldots,r_1 t_1) = \{\langle i|E^{(-)}(r_1 t_1)\cdots E^{(-)}(r_n t_n)E^{(+)}(r_n t_n)\cdots E^{(+)}(r_1 t_1)|i\rangle\}, \qquad (10.99)$$

averaged over $|i\rangle$, assumes, for a coherent state, the particularly simple form

$$\begin{aligned} G^{(n)}(r_1 t_1,\ldots,r_n t_n,r_n t_n,\ldots,r_1 t_1) \\ = \langle\{\alpha_l\}|E^{(-)}(r_1 t_1)\cdots E^{(-)}(r_n t_n)E^{(+)}(r_n t_n)\cdots E^{(+)}(r_1 t_1)|\{\alpha_l\}\rangle \\ = \varepsilon^*\{\alpha_l\}(r_1 t_1)\cdots\varepsilon^*\{\alpha_l\}(r_n t_n)\varepsilon\{\alpha_l\}(r_n t_n)\cdots\varepsilon\{\alpha_l\}(r_1 t_1). \end{aligned} \qquad (10.100)$$

Thus, in this case, for which no statistical uncertainty is present to our knowledge of the state of the system, the quantum mechanical expression for $G^{(n)}$ is equivalent to the classical expression given by

$$G^{(n)}(r_1 t_1,\ldots,r_n t_n,r_n t_n,\ldots,r_1 t_1) = \langle \hat{E}^*(r_1 t_1)\cdots\hat{E}^*(r_n t_n)\hat{E}(r_n t_n)\cdots\hat{E}(r_1 t_1)\rangle, \qquad (10.101)$$

for a prescribed electromagnetic field with analytic signal $\varepsilon\{\alpha_l\}(rt)$. This equivalence holds independent of the average number of photons associated with the state $|\{\alpha_l\}\rangle$, the ultimate difference between the quantum mechanical and classical cases lying in the fact that $|\{\alpha_l\}\rangle$ is an eigenstate of $E^{(+)}(rt)$ and not of the electric field operator $E(rt) = E^{(+)}(rt) + E^{(-)}(rt)$.

The preceding analogy was then extended to the situation in which one deals with a statistical mixture of coherent states, introducing the density matrix ρ given by

$$\rho = |\{\alpha_l\}\rangle\langle\{\alpha_l\}|. \tag{10.102}$$

More generally, we can consider the situation in which ρ is expressed as a linear combination of operators $|\{\alpha_l\}\rangle\langle\{\alpha_l\}|$ in the form

$$\rho = \int P(\{\alpha_l\})|\{\alpha_l\}\rangle\langle\{\alpha_l\}|d^2\{\alpha_l\}, \tag{10.103}$$

with $d^2\{\alpha_l\} = \prod_l d(\text{Re } \alpha_l)d(\text{Im } \alpha_l)$. This representation for the density operator was introduced by Glauber and Sudarshan[136] and is known as *P representation*.[139] An example of this representation was given by Glauber for thermal (chaotic) occupation of the mode for Gaussian light.

In order to preserve the Hermitian and unitary character of the operator ρ, the *P* representation must be real and satisfy the normalization condition

$$\int P(\{\alpha_l\})d^2\{\alpha_l\} = 1, \tag{10.104}$$

while it may take on negative values. The possibility of expressing the density matrix in terms of the *P* representation actually exists for a large class of physical situations and allows us to write the correlation function $G^{(n)}$ as

$$
\begin{aligned}
G^{(n)} &= \text{Tr}[\rho E^{(-)}(r_1 t_1)\cdots E^{(-)}(r_n t_n)E^{(+)}(r_n t_n)\cdots E^{(+)}(r_1 t_1)] \\
&= \int P(\{\alpha_l\})|\{\epsilon\alpha_l\}(r_1 t_1)|^2\cdots|\{\epsilon\alpha_l\}(r_n t_n)|^2 d^2\{\alpha_l\}. \tag{10.105}
\end{aligned}
$$

The expression is formally equivalent to what would be written in a classical situation for which Equation 10.101 is valid, bearing in mind the fact that one would deal with a nonnegative weight function having the meaning of probability density.

These considerations make clear the important role played by the *P* representation, which, whenever it exists, enables us to perform, in a formally identical way, classical and quantum mechanical calculations relevant to the evaluation of the hierarchy of the correlation functions $G^{(n)}$. In any event, $P(\{\alpha_l\})$ cannot be interpreted, even when it assumes only positive values, as a probability density of finding the system in a given state $|\{\alpha_l\}\rangle$, since the coherent states are not mutual orthogonal. This interpretation becomes practically valid in the classical limit of a large number of photons in which $P(\{\alpha_l\})$ tends to coincide with the corresponding probability density of finding a set of mode amplitudes $\{\alpha_l\}$, the structure of the coherent states being such that $\langle\{\alpha_l\}|\{\alpha'_l\}\rangle$ approaches zero in the classical limit for $\{\alpha_l\} \neq \{\alpha'_l\}$.[140]

Later, Glauber[141] fully discussed the detection process, showing that, if one considers a broadband photoelectric detector, the probability $p^{(1)}(t)$ of counting a photon is given by

$$p^{(1)}(t) = s\int_{t_0}^{t_1} G^{(1)}(rt',rt')\,dt', \tag{10.106}$$

where the *sensitivity* s summarizes the response of the detecting atomic system.

The expression given by Equation 10.106 is fully quantum mechanical, since both the radiation and the detecting system have been quantized. In the classical limit for the radiation field, one expects the quantum mechanical operators $E^{(+)}$ and $E^{(-)}$ to be, respectively, replaced by the analytic signal $\hat{E}$ and its complex conjugate. This was rigorously proved by Glauber,[138] and it is in perfect agreement with the semiclassical treatment in which only the atomic system is quantized. In this situation, the quantum mechanical quantity $G^{(1)}$ can be approximated by the classical ensemble average

$$G^{(1)}(rt,rt) = \langle \hat{E}^*(rt) \hat{E}(rt) \rangle. \tag{10.107}$$

We now observe that, in a real counting device, $t - t_o$ contains a great number of oscillation periods of the radiation field, so that $p^{(1)}(t)$ is, in practice, an average of $G^{(1)}(rt,rt)$ over many periods. On the other hand, it can be easily shown[91] that $\hat{E}^*(rt)\hat{E}(rt)$ is, for quasi-monochromatic fields, proportional to the time average over a few periods of the instantaneous intensity $I(rt)$; so that Equation 10.106 is, in the classical limit, equivalent to

$$p^{(1)}(t) = s' \int_{to}^{t} \langle I(rt') \rangle \, dt', \tag{10.108}$$

with $s' = (8\pi/c)s$.

In the case of an actual detector containing a great number N of independent identical atoms, which are supposed to be uniformly illuminated by the radiation field, the average number of counts $\langle C(t) \rangle$ recorded in the time interval $t - t_o$ is given by

$$\langle C(t) \rangle = Np^{(1)}(t). \tag{10.109}$$

Equations 10.108 and 10.109 furnish a justification for the classical statement that the response of a photoelectric detector is proportional to the incident intensity. The quantum generalization given by Equation 10.106 shows that the relevant operator corresponding to the classical intensity is $E^{(-)}(rt)E^{(+)}(rt)$. This is a significant result since it provides an example in which the quantization of a classical law is not *a priori* unique and requires a precise physical insight for determining the correct ordering of the operators involved.

Under the same assumptions which lead to Equation 10.106, Glauber showed that the probability $p^{(n)}(t)$ of counting n photons in a time interval $t - t_o$ with n identical one-atom detectors placed at different positions r_i ($i = 1,2,\ldots, n$) is

$$p^{(n)}(t) = s^n \int_{t_o}^{t} dt_1 \cdots \int_{t_o}^{t} dt_n G^{(n)}(r_1t_1,\ldots,r_nt_n \cdot r_nt_n,\ldots,r_1t_1). \tag{10.110}$$

This in turn implies the generalization of Equation 10.108 in the form

$$p^{(n)}(t) = s'^n \int_{t_o}^{t} dt_1 \cdots \int_{t_o}^{t} dt_n \langle I(r_1t_1)\cdots I(r_nt_n) \rangle. \tag{10.111}$$

Glauber then considered the probability $p(m,t)$ of counting m photons in a given time interval 0 to t (where m is any integer smaller than or equal to N), and showed[141] that a way of expressing $p(m,t)$ in terms of the statistical properties of the field can be made very elegantly by introducing a generating function $Q(\lambda,t)$ defined through the expectation value

$$Q(\lambda,t) = \text{Tr}[\rho(1 - \lambda)^{C(t)}] = \langle(1 - \lambda)^{C(t)}\rangle, \tag{10.112}$$

where $C(t)$ is the operator number of photons registered in the time interval $(0,t)$. Accordingly, the meaning of $p(m,t)$ allows us to write

$$Q(\lambda,t) = \sum_{m=0}^{\infty} (1 - \lambda)^m\, p(m,t), \tag{10.113}$$

from which it follows that

$$p(m,t) = [(-1)^m/m!][d^m/d\lambda^m)Q(\lambda,t)]\lambda = 1. \tag{10.114}$$

Introducing now the factorial moments M_k, defined as

$$M_k = \langle C(C - 1)\cdots(C - k + 1)\rangle = \sum_{m=0}^{\infty} m(m - 1)\cdots(m - k + 1)p(m,t), \tag{10.115}$$

he showed that

$$M_k = \beta^k \int_0^t dt_1 \cdots \int_0^t dt_k \int_{Vc} dr_1 \cdots \int_{Vc} dr_k\, G^{(k)}(r_1 t_1,\ldots,r_k t_k, r_k t_k,\ldots,r_1 t_1). \tag{10.116}$$

In the same year, F Ghielmetti[142] showed that $p(n)$ may also be written as

$$p(n) = \int P(\{\alpha_k\})(U^n/n!)\exp(-U)d^2\{\alpha k\},$$

where

$$U = \sum_k |\alpha_k|^2,$$

which may alternatively be cast in the form

$$p(n) = \int_0^{\infty} P(U')(U'/n!)\exp(-U')dU',$$

where

$$P(U') = \int P(\{\alpha_k\})\delta(U' - U)d^2\{\alpha k\},$$

which can be compared with Equation 10.54 as obtained by Mandel.[143] It is interesting to observe that somewhat similar results had already been obtained by Bothe.[77]

The probability distribution $p_m(T)$ of m photons in the quantum field during an observation time T (photon number distribution) can be connected to the probability of counting n photons at a detector $p(n,T)$ by the Bernouilli transformation[144]

$$p(n,T) = \sum[m!/n!(m - n)!]\eta^n(1 - \eta)^{m-n}\, p_m(T).$$

Further insights into the properties of correlation functions and statistics were given by Mandel,[145] who observed that the quantity

$$k_n = \int_{\delta V} \cdots \int \langle E^{(-)}(x_1)E^{(+)}(x_1) \cdots E^{(-)}(x_n)E^{(+)}(x_n) \rangle \mathrm{d}^3 x_1 \cdots \mathrm{d}^3 x_n$$

represents the nth moment of the number operator for photons in a volume δV. So, unordered products of creation and annihilation operators correspond to counting moments, while normal, ordered products of the kind of Equation 10.99 are measures of the n-fold coincidence counting rate for photoelectric detectors. Later Mandel[146] again considered the problem and showed that antinormally ordered operators of the form

$$G'^{(nn)}(x_1 \cdots x_n) = \langle E^{(+)}(x_1)E^{(-)}(x_1) \cdots E^{(+)}(x_n)E^{(-)}(x_n) \rangle$$

can be connected to photon counters functioning by stimulated emission rather than absorption of photons and discussed some of the properties of these correlations.

The theory of quantum coherence is exhaustively and authoritatively expounded in an excellent book by Mandel and Wolf.[147]

10.9 The Discussion of the Need for Quantum Optics

At first the beautiful construction of Glauber's was attacked on all sides. L Mandel and E Wolf immediately replied to the *Physical Review Letters* paper,[130] and wrote[148]:

> In an interesting recent note published in this journal, Glauber[130] refers to one of our papers[131] with a comment that a prediction made in that paper is misleading and follows from an inappropriate model of the maser beam. To the extent that we refer to the possible use of our results in connection with optical masers, our remarks were indeed misleading …

and they follow to clarify this point. They then added:

> There is also an implication in reference (1)[130] that the stochastic semiclassical description of light, initiated by Bothe[77] and Purcell[149] and developed further elsewhere,[150] ceases to be useful when the wave field cannot be represented classically by a Gaussian probability distribution. We do not believe this to be the case …

This was the start of a controversy which was to last several years. Indeed, E Wolf[151] and L Mandel,[152] at the *3rd Quantum Electronics Conference*, held in Paris in 1963, presented a classical treatment in which correlation functions in terms of the analytic signal were introduced, showing that this treatment was sufficient to explain all experimental results obtained until that moment, including the Hanbury Brown and Twiss experiment. No mention was made of Glauber's work. Glauber immediately reacted and, being concerned about the logical inconsistency of mixed treatments of classical and quantum mechanics, stated, in a comment to Wolf's paper, "I think you should treat the problem quantum mechanically".

This issue at once became the subject of the controversy. Wolf replied cautiously

> … Of course, one should try to formulate a full quantum mechanical treatment of coherence, but this may not be very easy to do. For many purposes the classical and the semiclassical treatments are quite good approximations and in fact have been extremely successful in predicting the results of experiments.
>
> One should also bear in mind that the classical theory arose from an attempt to understand certain types of phenomena with light from thermal sources. Of course, as new problems arise, the theory has to be extended and this is precisely what is now being done with the help of higher order correlation functions. But my guess is that for maser light classical theories will be even more useful than for thermal light.

In fact, however, a considerable effort was made by the Rochester group to prove that everything could be treated perfectly in a fully classical way. A point in favor of this was the propounding of the so-called "optical equivalence theorem" by E C G Sudarshan,[153] in which it was shown that every time the ρ matrix can be reduced to a diagonal representation, one is able to formulate the quantum theory of optical coherence in a language formally equivalent in all essential respects to the language of classical theory, as expressed in terms of analytic signals. However, Cahill[154] showed that, whenever there is no upper bound to the number of quanta present, the P representation is missing.

With time, the quarrel died down and, in their beautiful review paper on coherence, Mandel and Wolf[92] also used Glauber's quantum treatment. The elegant construction of the quantum theory of coherence was, from that moment on, accepted unconditionally.

Meanwhile the meaning of quantum correlation functions was fully investigated by Titulaer and Glauber.[155] The first-order coherence condition implies maximum fringe contrast in interference patterns. In this paper, they investigated the mathematical consequences of assuming the condition for maximum fringe contrast. They showed that this condition in turn implies factorization of the first-order correlation function. They were then able to show that all of the higher-order correlation functions factorize into forms similar to those required for full coherence, but differing from them through the inclusion of a sequence of constant numerical factors. These coefficients furnish a convenient description of the higher-order coherence properties of the field.

An interesting properties of fields with positive-defined P functions was then shown, that is, that the combination of first- and second-order coherence implies coherence for all orders.

For some time, only chaotic fields, coherent laser fields, and mixtures of these were considered in the discussion. It was only later that the possibility of other kinds of fields was demonstrated. The first ones to be studied were types of super-Gaussian fields[156] for which the coherent properties were different from both the Gaussian and the coherent fields.

In addition, the first practical application of the theory to some real problems was only given in 1967[157] with the demonstration that, in a scattering experiment where a fully coherent field is send as a probe (plane wave), the correlation functions of the scattered field are reminiscent of the correlation functions of the fluctuations in the medium. This result, applied to fluids, may enable the testing of hypotheses on the origin of fluctuations.[158] Moreover, there were cases where the previous considerations had led to interesting new kinds of measurements. The first of these cases was shown to exist in turbulence.[159]

Further examples were scattering from a small number of particles.[160]

The enormous amount of work performed on light statistics considering all the possible different cases is expounded in a number of textbooks.[161]

10.10 Experimental Studies of the Statistical Properties of Light

The theoretical papers on coherence and statistics which we looked at in Section 10.8 led to several predictions concerning the statistical properties of fields, which are summarized here.[141,162]

10.10.1 Polarized Thermal Light

In the case of polarized thermal light, the wave field is described by a scalar random process $V(t)$ in the form of an analytic signal which obeys a Gaussian distribution.

The instantaneous intensity $I(t)$ is given by

$$I(t) = V^*(t)V(t), \tag{10.117}$$

and the probability density of I is an exponential function

$$p(t) = (1/\langle I \rangle) \exp(-I/\langle I \rangle), \tag{10.118}$$

where

$$\langle I \rangle = \lim_{T \to \infty} (1/T) \int_0^T I(t')dt' \tag{10.119}$$

is a mean intensity.

The integrated light intensity $U = \int_o^T I(t')dt'$ has mean value and variance given by

$$\langle U \rangle = \langle I \rangle T, \tag{10.120}$$

$$\langle (\Delta U)^2 \rangle = \int\int_o^T |T(t - t')|^2 \, dt \, dt'. \tag{10.121}$$

When T is very small compared with the coherence time τ_c, one obtains

$$p(U) = (1/\langle U \rangle)\exp(-U/\langle U \rangle), \tag{10.122}$$

or

$$p(I) = (1/\langle I \rangle)\exp(-I/\langle I \rangle), \tag{10.123}$$

$$p(n) = \langle n \rangle^n/(1 + \langle n \rangle)^{1+n}, \tag{10.124}$$

and

$$\langle n \rangle = \alpha \langle I \rangle T, \tag{10.125}$$

$$\langle (\Delta n)^2 \rangle = \langle n \rangle [1 + \langle n \rangle], \tag{10.126}$$

$$\langle n!/(n - m)! \rangle = m! \langle n \rangle^m. \tag{10.127}$$

Finally, the conditional probability $p_c(t/\tau)d\tau$ that a photoelectric count be registered in a time interval $d\tau$ at $t + \tau$, given that one count has been registered at time t, and assuming that the light is stationary, is given by

$$p_c(t/\tau) = \alpha \langle I(t)I(t + \tau) \rangle/\langle I(t) \rangle. \tag{10.128}$$

This probability equation is well suited to observing the bunching effects of photons. For polarized thermal light, it becomes

$$p_c(t/\tau) = \alpha \langle I \rangle [1 + |\gamma(\tau)|^2]. \tag{10.129}$$

For very large T, compared with τ_c, one obtains

$$p(U) = \delta(U - \langle U \rangle), \tag{10.130}$$

$$p(I) = \delta(I - \langle I \rangle), \tag{10.131}$$

$$p(n,T) = [\langle n \rangle^m/n!]\exp(-\langle n \rangle), \tag{10.132}$$

which is the Poisson distribution with

$$\langle (\Delta n)^2 \rangle = \langle n \rangle, \tag{10.133}$$

and

$$\langle n!/(n - m)! \rangle = \langle n \rangle^m. \tag{10.134}$$

For arbitrary T approximations must be used. In particular, Glauber[163] examined more carefully the problem of counting distributions for $T \gg \tau_c$ for the case of Gaussian noise of Lorentzian spectral profile and found that the relevant quantity, rather than being T/τ_c, is in fact the ratio of $\langle n \rangle$ and T/τ_c. If $\langle n \rangle \ll T/\tau_c$, then the counting distribution follows a Poisson law, as expressed in Equations 10.132 through 10.134. If, however, the source is very intense, so that $\langle n \rangle \gg 1$ it is possible that, even though $\tau_c/T < 1$, the parameter $\mu = \langle n \rangle \tau_c/T > 1$.

In this case, the counting distribution observed for Gaussian light deviates from a Poisson law and is given asymptotically by

$$p(n) = \langle n \rangle (2\pi n)^{-1/2} n^{-3/2} \exp[-(1/2 \mu)(n^{1/2} - \langle n \rangle n^{-1/2})^2]. \tag{10.135}$$

10.10.2 Laser Light

An *ideal* laser emits light of well-stabilized intensity. This implies that, even if fluctuations in the phase are possible, the amplitude of the wave field remains constant. Therefore,

$$p(I) = \delta(I - \langle I \rangle), \tag{10.136}$$

$$p(n; T) = \langle n \rangle^n / n!, \tag{10.137}$$

$$p_c(t/\tau) = \alpha \langle I \rangle. \tag{10.138}$$

These formulas can be casted in a more general form writing the moments of the distribution as

$$\langle n \rangle = \langle W \rangle,$$

$$\langle n^2 \rangle = \langle W \rangle + \langle W^2 \rangle, \tag{10.139}$$

where $\langle W \rangle$ is defined as

$$\langle W \rangle = \langle a^\dagger a \rangle.$$

In this way, the variance of the number of absorbed photons is expressed by

$$\langle (\Delta \tilde{n})^2 \rangle = \langle \tilde{n} \rangle + \langle (\Delta W)^2 \rangle. \tag{10.140}$$

For a coherent field $\langle (\Delta W)^2 \rangle = 0$ and $\langle (\Delta \tilde{n})^2 \rangle = \tilde{n}^2$, that is, such a field is Poissonian. Classical fields have $\langle (\Delta W)^2 \rangle > 0$ and they are said *super-Poissonian*. For example, for the chaotic field of a natural source $\langle (\Delta W)^2 \rangle = \langle I \rangle^2 T^2$ and

$$\langle (\Delta n)^2 \rangle = \langle n \rangle (1 + \langle n \rangle), \tag{10.141}$$

corresponding to the Bose–Einstein distribution 10.20.

For quantum fields having no classical analogs, it may be $\langle (\Delta W)^2 \rangle < 0$ and $\langle (\Delta n)^2 \rangle < \langle n \rangle$, and such light is *sub-Poissonian*.

These results were all discussed during the 1960s, and several laboratories started to plan experimental verification.

Intensity fluctuations can be studied using two different techniques. In the first one, the current pulses which are produced by single photoelectrons emitted by a photocathode are detected. In this kind of pulsed technique, one is interested in finding the distribution of time intervals between the arrivals of successive pulses of photoelectrons and also in investigating the number of pulses registered as a function of observation time. From these data, spectra and moments can be derived. This kind of experiment is usually called *photon counting* or *photoelectron counting*.

The second group of experimental methods is based on analog techniques, using the continuous photocurrent produced by the superposition of many photoelectron pulses without trying either to resolve the discrete pulses or to measure the time interval between pulse arrivals. These techniques are usually referred to as *intensity correlation methods*.

Three basic kinds of photon counting experiments have been undertaken in the investigation of intensity fluctuations. In the first type, the probability distribution $p_T(n)$ that a number n of pulses is counted in a given time interval T is determined.

In the second kind of experiment, the second- and higher-order moments of the distribution are studied as a function of T.

In the third type, the joint probability that n_1 photoelectrons are counted in the time interval T at t_1 and n_2 after time T at t_2 is studied.

Intensity correlation methods are in essence similar to the Hanbury Brown and Twiss experiment.

It was the development of photomultiplier and of the fast, transistorized electronic circuitry needed for data processing which had occurred in the 1950s which made the use of photon counting techniques possible.

Initial experiments, begun in 1964, demonstrated the amplitude-stabilized nature of the well-above-threshold laser output by using single detector measurements undertaken on gas lasers.[164] The excess noise was shown to be much smaller than that expected for a Gaussian noise source of comparable bandwidth. The effect of noise in the gas discharge on the intensity fluctuations of a gas laser was also demonstrated[164,165]; such extraneous noise masked the underlying intensity fluctuations.

Prescott and van der Ziel[165] measured excess noise on the 6328 Å line of the He–Ne laser just above threshold and observed that both the average output noise power P_n, and the coherence time of the noise τ_n decreased with increasing laser output power. More extensive measurements, both above and below threshold, were made the following year by Freed and Haus.[166] Their results are shown in **Figure 10.9**, where the ratio S_e/S_s of excess noise to shot noise is shown.

As threshold is approached from below (see **Figure 10.9a**), the bandwidth of excess noise decreases from the passive cavity value of 470–0.85 kHz for the curve closest to threshold. At the same time, the peak excess noise increases. As the laser excitation increased above threshold (**Figure 10.9b**), the bandwidth increased and the peak noise decreased. The results above threshold were compared with the theory of noise in Van der Pol oscillators[167] and confirmed the theoretical picture of the gas laser as an oscillator perturbed by spontaneous emission.[168]

Armstrong and Smith considered GaAs lasers operated continuously at 10 K.[169] In the first of these papers, received by *Physical Review Letters* on November 19, 1964, they observed intensity fluctuations by varying the injection current from a value below the threshold for coherent oscillation to a value well above threshold. The spectrum of the lasers studied had a single family of axial modes whose envelope narrowed in the manner to be expected for a homogeneously broadened fluorescent line. Only the strongest mode lased; the power in the other modes became saturated at threshold. As the injection current was varied the change in noise properties of the single-mode output was studied. Below threshold, they found that the mode emitted random noise like a narrow-band black-body source; above threshold, its noise was characteristic of a damped, amplitude-stabilized oscillator.

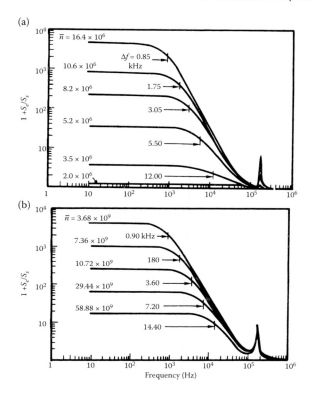

FIGURE 10.9 Photomultiplier current spectra observed for a gas laser. (a) Below threshold operation and (b) above threshold operation; n is the average photoelectron emission rate per second at the photocathode. (From C Freed and A H Haus, *Appl. Phys. Lett.* **6**, 85 (1965).)

Measurements of intensity fluctuations using the two-detector coincidence-counting technique of intensity interferometry developed by Hanbury Brown and Twiss were also made.

In a subsequent paper[170] received by *Physics Letters* on April 3, 1965, they measured the noise in the single lasing mode using the single detector method of excess photon noise and found excellent quantitative agreement between the two methods. Typical values of the relative fluctuations ρ for the three strongest modes derived from single detector measurements are shown in **Figure 10.10**. Only mode 1 was lasing, and its noise goes through a peak at threshold and then falls to very low values above threshold. The noise for the two other nonlasing modes (2 and 3 in Figure 10.10) remains high above threshold, as expected. The noise for mode 2 does not increase indefinitely but instead goes through a maximum, because of coupling between modes. This coupling also implies a correlation between noise in different modes, which is verified by the nonzero value of ρ_{12} in Figure 10.10.

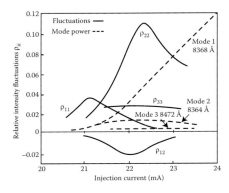

FIGURE 10.10 Variation with injection current of the relative intensity fluctuation ρ_{ii} for the lasing and two nonlasing modes, derived from single detector measurements. Also shown is the current dependence of the correlation ρ_{12} between the intensity fluctuations in the first (lasing) and the second (nonlasing) modes. (From A W Smith and J A Armstrong, in *Proc. Int. Conf.* on the Physics of Quantum Electronics, Puerto Rico, 1965, edited by P Kelly, B Lax and P Tannenwald (McGraw-Hill: New York, 1965), p. 701).)

The last paper[171] gave a full account of all these measurements.

Later on, in 1966, Geusic[172] investigated intensity fluctuations in a YAl garnet:Nd laser operating CW near threshold.

The problem of experimentally determining the form of the distribution law of photons was now exposed, and ready to be tackled. The first work, still not at the experimental stage, but clearly pointing towards an experiment, was by McLean and Pike[173] and was received by *Physics Letters* on March 19, 1965. In this paper, the authors gave an explicit expression for the counting distribution $p(n,t)$ for Gaussian light in the region which is of practical interest, when $T \gg \tau_c$, and this under conditions which can normally be obtained in practice.

At nearly the same time, in France, efforts were being made to measure some of the statistical properties of laser light. In a short note, J Marguin et al.[174] presented some experimental evidence that the ratio $\langle \Delta n^2 \rangle / \langle n \rangle$ was around unity when the counting time T was shorter than the inverse bandwidth of a He–Ne laser. This result, however, attracted little attention and was not pursued.

In June 1965, there was a Conference in San Juan, Puerto Rico, on *The Physics of Quantum Electronics*. At this conference, important papers on photon counting were presented: a theoretical paper by Glauer;[175] and three experimental papers by Johnson, McLean, and Pike,[176] by Freed and Haus,[177] and by Armstrong and Smith.[178] In these last papers, photon counting experiments are described. H A Haus[179] had already presented his results in a report to MIT the year before, and E R Pike[180] had announced the experiment under way at Malvern in a research report bulletin early in March 1965.[181]

Freed and Haus[182] considered the normalized factorial moments

$$F(k) = [\overline{N(N-1)\cdots(N-k+1)} - \overline{N}^k]/(\overline{N})^{k-1}, \qquad (10.142)$$

below and above threshold of a He–Ne gas laser as a function of counting interval, up to $F(4)$.

The assumption made by the authors was that below threshold the amplitude of the electric field components of the laser output had a Gaussian

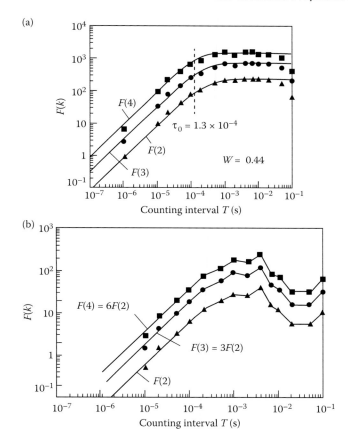

FIGURE 10.11 Normalized factorial moments: (a) below threshold (photon rate 2.81×10^9 s^{-1}; photoelectron rate 4.31×10^6 s^{-1}); (b) above threshold (photon rate 8.15×10^{11} s^{-1}; photoelectron rate 11.09×10^6 s^{-1}). (From C Freed and H A Haus in *Physics of Quantum Electronics*, edited by P L Kelley et al. (McGraw-Hill: New York, 1966), p. 715.)

probability distribution of zero mean, and that above threshold, operation of the laser occurred in one single mode. With these assumptions, they calculated the expected $F(k)$ and then made a comparison with experiment (**Figure 10.11**), which gave rather good agreement between experiment and theory.

Pike[183] reported ...

In essence, we measure the number of photons counted in a time T (using the photomultiplier for single-electro counting rather than as a current multiplier), store this number away electronically, and repeat a large number of times.

It is easily shown that to measure a shift of variance of one part in 10^3 of a Poisson distribution, with 99% confidence, one needs to sample the distribution to the order of 10^7 times.

The variances of photon counting distributions have been used in two recent investigations, but only small numbers of samples obtained has ranged from 5.10^7 to over 10^{10}, taking times from a few minutes to several hours.

In a number of circumstances one can calculate explicitly not only the excess photon noise to be expected but the actual photon-counting distribution. With the numbers of samples which we accumulate, this distribution can be found accurately, and as we shall show, the form of such distribution when compared with the expected results leads to a simple extraction of the variance or a noise measurement.

For the experimental analysis, they worked in terms of the quantity

$$F(n) = [(n + 1)p(n + 1,T)]/[p(n,T)] = \langle n \rangle (1 - \alpha\delta) + \alpha\delta n, \quad \textbf{(10.143)}$$

where α was the quantum efficiency and δ the degeneracy parameter.

In these first photon-counting distribution measurements presented at Puerto Rico, a number of new technical considerations prevented them obtaining exactly the expected results. Among these were the necessarily finite resolution time, the source and circuit stability, and multiplier afterpulsing. After mastering these problems, distributions were obtained which permitted moments as high as n^6 to be later successfully compared with theory for thermal and laser sources.[184]

In November 1965, two papers were submitted to *Physical Review Letters* almost at the same time. On November 18, a paper by Arecchi[185] arrived in which the experimental distribution $p(n,T)$ was given for an artificial Gaussian source and for a laser, together with the measurement of the first three moments of the photon counting distribution. Surprisingly good agreement, considering the operating conditions, was found with the theoretical predictions.

The following day (November 19), Freed and Haus[186] presented the photon statistics for a laser below threshold (that is a narrow-band Gaussian source).

The experiment described by Arecchi and performed by a group of researchers at CISE, Milan, was done by sending the light of an amplitude-stabilized single-mode He–Ne laser onto a moving, ground-glass disc. The effect of diffraction by the many tiny moving scattering centers was to convert the scattered beam into Gaussian noise whose bandwidth is determined by the velocity of the scattering plate.

A similar procedure was first introduced by Martienssen and Spiller.[187] Photoelectron distributions for this synthesized Gaussian source and for the laser source were given.

Freed and Haus in the paper[188] presented the first measurement of laser light below threshold. This work was stimulated by the determination of the probability distribution of photoelectron counts reported in Puerto Rico by Johnson et al.[189] Freed and Haus said:

A determination of the probability distributions of photoelectron counts has been reported by Johnson, McLean and Pike using an incoherent light source. Their work provided the initial impetus for the measurements reported here.

The laser, below threshold, was considered to produce a narrow-band Gaussian light. **Figure 10.12** shows the experimental results. If the laser output were to consist of a pure narrow-band Gaussian light centered at a single frequency, the probability of observing n photoelectrons within a time interval which was short compared with the inverse bandwidth would follow the Bose–Einstein distribution law[190] and, on the semilog plot of **Figure 10.12a** this should appear as a straight line down to counts of the order of four. By assuming that the deviation is due to admixture of other modes at other frequencies, Freed and Haus were able to calculate the full line also shown in Figure 10.12a.

The spectral width of the source was such that it was easy to satisfy the requirement $T \ll \tau_c$, as shown in Figure 10.12a. **Figure 10.12b** is, instead,

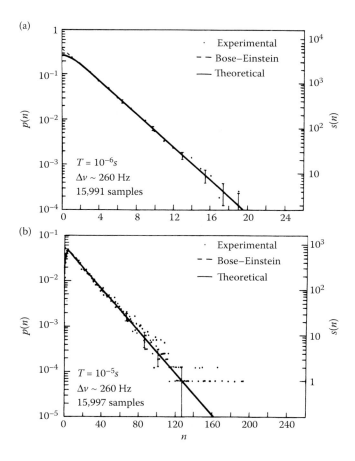

FIGURE 10.12 Probability distribution and number of observed samples versus photoelectron count for a laser below threshold: (a) counting interval $T < \tau_c$ (C Freed and H A Haus, *IEEE J. Quantum Electron.* **QE-2**, 190 (1966)), (b) $T > \tau_c$ (C Freed and H A Haus, *Phys. Rev. Lett.* **15**, 943 (1965); © 1966 IEEE).

undertaken with a counting interval longer than the coherence time. Under this condition, it was expected that the asymptotic expression 10.135, developed by Glauber,[191] would fit, as actually happened.

A more exhaustive account of the CISE measurements was published later,[192] together with a few other works,[193] until April 1966, when photon counting experiments were debated at the 4th International Quantum Electronics Conference, Phoenix, Arizona. Papers by Freed and Haus[194] and Arecchi[195] were submitted which, basically, gave a general review of the results previously obtained. Inexplicably, the Conference Committee did not accept the contribution from the Malvern group.

Meanwhile, the theoreticians were at work. Nearly all the theories of laser noise up to 1964 were based on a linear approach.[196] This approach predicts that the statistical properties of laser light are the same as those of light from thermal sources. The amplitude should have a Gaussian distribution. In 1964, a nonlinear approach was developed almost simultaneously by Haken[197] and Lamb.[168,198] A nonlinear laser equation which was able to predict the field fluctuations was thus derived, and it was demonstrated that laser light below threshold is narrow-band Gaussian and that above threshold the laser acts as a self-sustained oscillator with a high stabilized (classical) amplitude.

These results were confirmed by Armstrong and Smith[199] and Freed and Haus.[200]

The change in photon statistics between subthreshold and above threshold operation was quantitatively predicted by Risken in 1965,[201] and then confirmed by the experiments of Armstrong and Smith[202] and others.[203]

Later, in 1966, Lamb et al. eventually established the quantum theory of lasers.[204]

Light from a laser operating somewhat above threshold does exhibit intensity fluctuations and in this region the laser field may be approximated by a linear superposition of an amplitude-stabilized wave and a Gaussian noise wave. The case was first treated by Lachs[205] and Glauber[206] using the quantum theory, and by Mandel and Wolf[207] classically. An approach based on the calculation of the correlation functions was also developed by Morawitz[208] and generalized by Perina.[209]

The accurate experimental verification of the photon counting distribution was hindered by many practical difficulties. Most of them were solved in subsequent years and the work of the RRE group of Malvern was fundamental in this respect.[210] The group had, in fact, been working on the problem from the beginning, as we have seen; the importance of their work was quickly recognized by Kastler who, on seeing one of their experimental distributions displayed, commented with delight, *Ah, la loi des Bosons!* and he reproduced what is probably the first photon counting distribution to be published in a textbook in these very early days.[211]

Dead-time effects, which had strongly affected the first measurements of $F(n)$, Equation 10.143,[212] were corrected with a method discussed in Johnson et al.,[213] and were published in a paper received on May 31, 1966, by *Optica Acta*.[214] **Figure 10.13** shows the large, negative slopes of $F(n)$ versus n curves as obtained by Johnson et al.[215] for a stabilized single-mode gas laser (circles) and a stabilized tungsten lamp (triangles), and the dead-time modified Poisson distributions (broken lines) which fit the experimental data excellently.

A recollection of the early days of work at Malvern has been made by Pike.[216]

Other groups were also working on the problem of radiation fluctuations and photon statistics. In Rochester, Professor H Gamo and R E Grace were studying intensity fluctuations in single-mode gas lasers near the oscillation threshold with an ingenious two-photomultiplier technique. Their results were first presented in the *Second Rochester Conference on Coherence and Quantum Optics* June 24, 1966.[217]

Chang et al. in 1967[218] measured fluctuation statistics up to the fourth order and found good agreement with the nonlinear oscillator theory.

Using a three-detector coincidence-counting technique, Davidson and Mandel[219] observed time correlation functions of amplitude fluctuations up to the sixth order and confirmed earlier observation of fourth-order field statistics.[220] Later Meltzer, Davis, and Mandel[221] observed the counting distribution $p_T(n)$ for a range of values of the pump parameter encompassing the threshold region, and for a number of different ratios of counting time to correlation time.

Jakeman et al.[222] measured up to fourth order over a similar range of parameters.

In 1966, photon bunching in time of a thermal light beam was also demon-

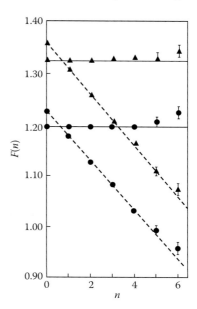

FIGURE 10.13 The experimentally measured and dead-time corrected value of $F(n)$ for a stabilized single-mode gas laser (circles) and a stabilized tungsten lamp (triangles). The data in each case were based on $\approx 10^8$ samples taken with 1 µs sampling I time. The broken lines are a fit of dead-time-modified Poisson distribution to the observed data. The horizontal sets of points are the experimental data corrected for dead-time effects and the associated full horizontal lines are for Poisson distributions of the same mean. The calculated dead-time in each case is 19.38 ns. (From F A Johnson, R Jones, T P McLean and E R Pike, *Opt. Acta* **14**, 35 (1967).)

strated. This bunching in time of a photon beam is completely analogous to the well-known tendency of a photon gas or indeed any boson gas to form clusters in space.[223] The bunching phenomenon in a photon beam was also analyzed by Dicke in 1954[224] in terms of the radiation process in an excited gas. He showed that there is a correlation between the emission direction of successive photons, such that the emission probability in the same direction is nearly twice that in an arbitrary direction.

Masers and Lasers

The first measurement of bunching of photons from a thermal source was done in 1966 by Morgan and Mandel[225] who used a low-pressure ^{198}Hg gas discharge lamp. Light was passed through a pinhole (diameter 0.54 mm), an optical filter that isolated the blue 5461 Å line, and a linear polarizer; it finally fell on a 56 AVP photomultiplier through a rectangular aperture (0.37×0.47 mm^2) whose dimensions were small enough to ensure a degree of coherence of at least 90% across the beam. The pulses from the photomultiplier were processed in such a way as to enable measurement of the conditional probability $p(t/\tau)$ as τ was varied between 1 and 6 ns.

Results are shown in **Figure 10.14a**. For comparison, the results of similar measurements carried out with a tungsten lamp as source are shown in

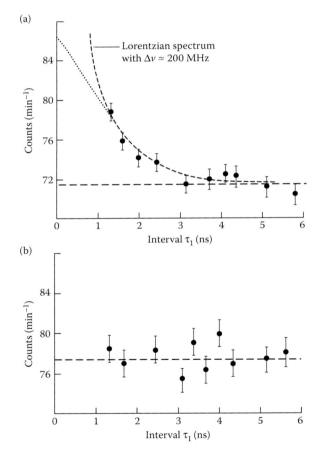

FIGURE 10.14 Counting rate illustrating the phenomenon of photon bunching. (a) Light from a ^{198}Hg source; (b) light from a tungsten lamp. The ordinates represent essentially a quantity which is proportional to the integral $\int_{\tau_1}^{\tau_2} p_c(\tau)$, where τ_2 is a constant. (From B L Morgan and L Mandel, *Phys. Rev. Lett.* **16**, 1012 (1966).)

446

Figure 10.14b. The results in this case showed no significant bunching of photon counts since the coherence time was far shorter than 1 ns.

Photo bunching is an aspect of the Bose–Einstein statistics, that is, of the statistics followed by photons emitted in equilibrium conditions. It is absent in light emitted by a coherent source (as in a laser) and it is also absent in chaotic sources when fluctuations are smeared out because of counting in time intervals larger than the coherence time.

In this respect, we observe that the Poisson distribution is to be expected when counting well-stabilized laser light and when counting the light from a chaotic source with $T \gg \tau_c$. This may at first sight appear somewhat confusing. In the case of a laser, a Poisson distribution corresponds to the genuine absence of intensity fluctuations; in the case of a chaotic source, the Poisson distribution occurs because in the measuring time T ($\gg \tau_c$) the detector averages out all the fluctuations which are in fact present in this case, and which appear as soon as $T \ll \tau_c$ and give rise to photon bunching, and so on.[226]

When photons are emitted in nonequilibrium conditions, other behavior may occur; an example is antibunching, which was first observed by Mandel and coworkers.[227]

The importance of the statistical properties of radiation is now widely appreciated in many different applications, for example, photon-correlation methods in scattering, propagation in the turbulent atmosphere, nonlinear optics, and so on.[228]

10.11 Nonclassical States of Light and Single-Photon Sources

When Glauber completed the model of radiation detection, discussing on a quantum point of view the interaction of radiation and matter, and constructed a quantum theory of coherence, a number of interesting states of radiation received a full reconnaissance as useful and possible states: among them the *coherent* states we have already discussed and other states such as *single photon*, *squeezed*, and *entangled* states, just to mention the principal ones. Full description of these states may be found in many excellent textbooks, for example, Mandel and Wolf.[229]

The generation of quantum states of the radiation field, other than the coherent states generated by lasers, started to receive great attention from the 1970s. *Single-photon* states, in particular, were considered as having—and nowadays have—possible applications in quantum communication, information processing, and quantum computing, such as quantum networks, secure quantum communications, and quantum cryptography.[230]

For example, the security in some schemes of quantum cryptography is based on the fact that each bit of information is coded on a single photon. The fundamental impossibility of duplicating the complete quantum state of a single particle (*no cloning theorem*)[231] prevents any potential eavesdropper from intercepting the message without the receiver's noticing.

Single-photon states are states with a prescribed number of photons and are called *number states* or *Fock states*.

They were first introduced and discussed by Fock.[232] A Fock state is strictly quantum-mechanical and contains a precisely definite number of quanta of field excitation; hence, its phase is completely undefined.

The eigenstate of the photon number operator $\tilde{n} = a^\dagger a$ which appear in Equation 10.83 has a perfectly fixed photon number n. Since $\tilde{n}$ is a Hermitian operator, the number n is real. If $|n\rangle$ is an eigenstate of $\tilde{n}$, than $\tilde{a}|n\rangle$ must be an eigenstate as well, with the eigenvalue $n - 1$.

In a similar way, $\tilde{a}^\dagger|n\rangle$ is an eigenstate of $\tilde{n}$ with eigenvalue $n + 1$. So the fundamental relations 10.84 and 10.85 are derived. The state $\tilde{a}|0\rangle = 0$ exists and it is the *vacuum state*.

States with a prescribed number of photons can be created by applying the creation operator to the vacuum state

$$|n\rangle = \tilde{a}^{\dagger n}/\sqrt{n!}|0\rangle. \tag{10.144}$$

The Fock states must be complete

$$\sum_{n=0}^{\infty} |n\rangle\langle n| = 1, \tag{10.145}$$

and orthonormal

$$\langle n|n'\rangle = \delta_{nn'}. \tag{10.146}$$

The expectation value of the electric field and its square are

$$\langle n|E|n\rangle = 0, \tag{10.147}$$

and

$$\langle n|E^2|n\rangle = (\hbar\omega/\varepsilon_o V)(n + \tfrac{1}{2}), \tag{10.148}$$

where ε_o is the vacuum dielectric constant and V is the volume.

The electromagnetic wave amplitude can be represented by the quantity

$$E_o = (2\hbar\omega/\varepsilon_o V)^{1/2}(n + \tfrac{1}{2})^{1/2}. \tag{10.149}$$

The phase of the wave is of course completely uncertain.

An ideal single-photon source would produce exactly one photon in a definite quantum state, in contrast with a "classical" source, such as attenuated laser pulses, for which the photon number follows a Poisson distribution.

Much progress has been made in the years towards such devices, especially in suppressing the probability of two photons in the same pulse.

The first experiment with probably nearly a photon at the time was performed by G I Taylor[233] at the Trinity College, Cambridge in the United Kingdom in 1909. Taylor performed an experiment in which single photons are supposed to play. From an historical point of view, it is interesting to note that although in 1909 the Einstein paper on the quantization of light was already published,

Taylor did not mention it but referred to a suggestion of J J Thomson[234] writing: "The phenomena of ionization by light and by Roentgen rays have led to a theory according to which energy is distributed unevenly over the wavefront. There are regions of maximum energy widely separated by large undisturbed areas. When the intensity of light is reduced these regions become more widely separated, but the amount of energy in any one of them does not change, that is, they are indivisible units. Sir J J Thomson suggested that if the intensity of light in a diffraction pattern were so greatly reduced that only a few of these indivisible units of energy should occur on a Huygens zone at once the ordinary phenomena of diffraction would be modified." Therefore, Taylor performed an experiment taking photographs of the shadow of a needle, illuminated by light coming from a narrow slit placed in front of a gas flame, and attenuating the intensity by means of smoked glass screens. Five diffraction photographs were taken, each one with a different intensity of incoming light. The last photograph was taken with an intensity so low that exposure was 2000 h or about three months. In no case there was any diminution in the sharpness of the pattern. The amount of energy falling on the plate during the longest exposure was derived to be the same as that due to a standard candle burning at a distance slightly exceeding a mile or about 5×10^{-6} erg per s on 1 cm^2. If we take this figure as for good, it corresponds to a few photons per microsecond per cm^2 so that we may assume they arrived mostly one at the time.

The result is then taken as a demonstration that each photon can produce its diffraction pattern by itself.

However, very attenuated thermal sources suffer of the circumstance that chaotic photons tend to bunch. Faint laser pulses and parametric sources of entangled pairs deliver Poisson distributions of photons (or of photon pairs), from which multiphoton events can never be entirely suppressed. Nevertheless, such sources are much easier to build and operate than single-photon sources and single photon number states may be approximated by coherent states with very low average photon number $\langle n \rangle$, using only standard pulsed lasers and calibrated attenuators. However, when n is small, most pulses are empty.

With an attenuated pulsed laser source, the probability of having 0, 1, 2, 3, or more photons present at a time is controlled by the Poisson statistics as given in Equation 10.126. One may introduce a probability p_m that a nonempty weak coherent pulse contains more than one photon

$$p_m = p_{n>2}/p_{n>1} = [1 - p^{(0)} - p^{(1)}]/[1 - p^{(0)}] \simeq \langle n \rangle/2, \qquad (10.150)$$

where $p_{n>1}$ and $p_{n>2}$ are the probabilities that a pulse contains at least one and at least two photons, respectively. The value of p_m can therefore be made arbitrary small by decreasing $\langle n \rangle$. However, when $\langle n \rangle$ is small, most pulses are empty:

$$p(n = 0) \simeq 1 - \langle n \rangle. \qquad (10.151)$$

Increasing the pulsed laser rate could help, but in this way also dark counts will increase and the ratio of detected photons to dark counts decreases with $\langle n \rangle$.[235]

After the Taylor experiment, the successive experiment in which single photons were involved was made using an atomic cascading process in which an excited atomic level decayed with the emission of two photons of different frequency.[236] The experiment was based on a cascade transition of mercury atoms. Each excited atom delivers a couple of two photons with different colors. A photon at one of the wavelengths is detected after spectral filtering and is used for the conditional detection of its companion at the other wavelength. Each single photon is thus herald for the presence of its companion.

One-photon state presents a peculiar anticorrelation effect, which does not exist for a classical wave. If we send a one-photon state on a beam splitter and place photon counting detectors on the reflected and transmitted beams in the disposition first used by Hanbury Brown and Twiss (Figure 10.4), we never observe any coincidence between counts measured by the two detectors, as this would violate energy conservation. The photon cannot split to be present at the two output ports contemporarily. As a consequence of the principles of quantum mechanics, the wavefunction of the photon has to collapse onto either one or the other of the two detectors. This absence of coincidences of detection events on the two detectors has been dubbed photon *antibunching* and is utterly irreconcilable with a classical description of light. This effect was predicted by Carmichael and Walls in 1976.[237] Antibunching can be assessed measuring the joint probability $p(t|t + \tau)$ of detecting two photoelectric pulses at time t within Δt and at time $t + \tau$ within $\Delta \tau$. The normalized quantity

$$p(t|t + \tau)/p(t) = \langle I(t)I(t + \tau)\rangle/\langle I(t)\rangle^2 = g^{(2)}(\tau) \qquad \textbf{(10.152)}$$

is the second-order correlation function $g^2(\tau)$. Antibunching means $g^{(2)}(0) = 0$.

Naively antibunching can be understood very simply. In the emission process from a single atom or molecule, the correlation function measures the joint probability for the arrival of a photon at time $t = 0$ and the arrival of a photon at $t > 0$. After the emission of a photon at $t = 0$, the quantum system is projected in its ground state since it just emitted a photon. Hence, the simultaneous emission of a second photon is impossible because the molecule cannot emit from the ground state. It needs some time to have a finite probability to be again in the excited state and emit a second photon. On an average, a time of half a Rabi period[238] has to elapse to have a finite probability for the molecule to be in the excited state and emit a second photon.

Antibunching was first observed in pioneering experiments by Mandel.[239]

To observe antibunching correlations, the second-order correlation function $g^{(2)}(\tau)$ is generally measured by determining the distribution of time delays $N(\tau)$ between the arrival of successive photons in a dual beam detector

$$g^{(2)}(\tau) = \langle n(t)n(t + \tau)\rangle/\langle n\rangle^2 = p(t/t + \tau)/p(t). \qquad \textbf{(10.153)}$$

In Equation 10.153, $n(t)$ is the number of photons counted at time t.

For a number state with n photons, it is

$$g^{(2)}(0) = 1 - 1/n. \qquad \textbf{(10.154)}$$

An example of the result obtained when antibunching is present is shown in **Figure 10.15**.[240]

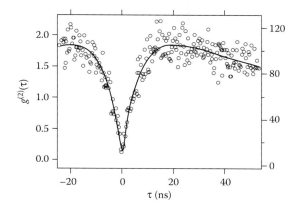

FIGURE 10.15 An example of antibunching. (From A Beveratos et al., *Phys. Rev. A 64*, 061802 (2001). With permission.)

Single-photon sources are based on the property of a single emitting dipole to emit only one photon at a time. When excited by a short and intense pulse, such an emitter delivers one and only one photon. Unfortunately, the time of emission cannot be predicted exactly because it stays in the lifetime of the transition and also the emission direction is at random. The single-photon emission is usually claimed when photon antibunching is observed.

In the experiment of Kimble, Degenais, and Mandel,[239] photon antibunching was obtained in resonant fluorescence on single atoms of sodium. The atoms in an atomic beam were excited by the light beam of a tunable dye laser. The emitted fluorescence light was collected at right angles to both the laser and the atomic beams by a microscope objective, divided by a beam splitter and sent to two photomultipliers. After amplification and pulse shaping, the pulses from the two detectors were fed to start and stop inputs of a time-to-digital converter, where the time intervals τ between start and stop pulses were digitized. The number of events $n(\tau)$ stored at address τ is therefore a measure of the joint photoelectric detection probability density $p(t, t + \tau)$. The value of $n(\tau)$ increased with τ from its smallest value at $\tau = 0$ (see **Figure 10.16**).

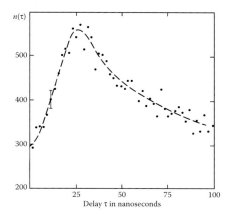

FIGURE 10.16 The number of recorded pulse pair $n(\tau)$ as a function of the time delay τ in nanoseconds. The growth of $n(\tau)$ from $\tau = 0$ shows antibunching. (From J Kimble, M Degenais and L Mandel, *Phys. Rev. Lett.* **39**, 691 (1977). With permission.)

Although both effects are nonclassical, are exhibited only by a quantum field, and are often associated, sub-Poissonian statistics and antibunching are distinct effects that need not necessarily occur together, one may occur without the other.[241]

However, for a stationary single-photon source, the nonclassical nature of the emitted radiation would lead to sub-Possonian photon statistics with $g^{(2)}(0) < 1$.

Observation of sub-Poissonian photon statistics was first made by Short and Mandel[242] using fluorescence from a single two-level atom which emitted single photons.

For some application, it is necessary that the single photons be indistinguishable. The produced packets should be fully coherent, characterized by a Fourier-transform relationship between their temporal and spectral profiles, that is

$$\Delta v \Delta \tau = 1/2\pi, \qquad (10.155)$$

where $\Delta v = 1/2\pi T_2$ and $\Delta \tau = 2T_1$ designate, respectively, the Lorentzian half-width and the spontaneous decay time of the field amplitude. This property is required for the implementation of any photon-based quantum information processing system.[243]

Single-photon sources are obtained, with the limitations above said, using a number of different systems like fluorescence from single atoms or molecules, semiconductor pn junctions, atoms inside a microcavity, semiconductor quantum dots, and color centers in different materials (mostly individual nitrogen vacancies in diamond), from entangled states and from various plasmonic sources.[244]

The *coherent states* we have already discussed are the eigenstates $|\alpha\rangle$ of the annihilation operator $\tilde{a}$ (Equation 10.88), with eigenvalue α, which is in general a complex number since $\tilde{a}$ is not a Hermitian operator. A coherent state is the closest analog to a classical light field and exhibits a Poisson photon number distribution with an average photon number $|\alpha^2|$, as discussed previously. Coherent states have relatively well-defined amplitude and phase, with minimal fluctuations permitted by the Heisenberg uncertainty principle. They are an overcomplete set.

Because the number states is a complete set, it is possible to express coherent states as a superposition of n states

$$|\alpha\rangle = \exp(-1/2|\alpha|^2 \sum \alpha^n/(n!)^{1/2} |n\rangle. \qquad (10.156)$$

And vice-versa n-states may be expressed as a superposition of coherent states

$$|n\rangle = (1/\pi)\int \exp(-|\alpha|^2/2)\alpha^{*n}/(n!)^{1/2}|\alpha\rangle \, d^2\alpha, \qquad (10.157)$$

where $d^2\alpha = d(\text{Re } \alpha)d(\text{Im } \alpha)$.

The coherent state is nearly a classical-like state because it not only yields the correct form for the field expectation values, but contains only the noise of the vacuum.

Coherent states were first discussed by Schroedinger in connection with the quantum harmonic oscillator[245] referring to them as states of minimum uncertain product. They were further investigated by Klauder.[246] Glauber gave them the name of *coherent states*.

The other states of interest here are *squeezed* and *entangled states*. The real and imaginary parts of the complex amplitude of the electromagnetic field fluctuate with equal dispersion in a coherent state. The phenomenon of vacuum fluctuations is a manifestation of this effect, because the vacuum state is an example of a particular coherent state. In a *squeezed state*, one part of the field fluctuates less and the other part fluctuates more than in the vacuum state.

Pauli asked which are the minimum uncertainty states[247] and found that apart from a displacement, the minimum-uncertainty states have Gaussian wave functions like coherent states. However, the variance should not necessary equal $1/2\hbar$, as is the case for coherent states. In other words, the statistical uncertainty of the position quadrature q may be squeezed below the vacuum level $1/2\hbar$ at the cost, however, of enhancing the uncertainty in the canonical conjugate quadrature p and vice versa.

To be more specific, a state is squeezed if either of the Hermitian operator q and p defined in Equations 10.76 and 10.77 has a variance

$$(\Delta q)^2 = \langle q^2 \rangle - \langle q \rangle^2, (\Delta p)^2 = \langle p^2 \rangle - \langle p \rangle^2$$

less than $(1/2)\hbar$ that is

$$(\Delta q)^2 < 1/2\hbar \quad \text{or} \quad (\Delta p)^2 < 1/2\hbar.$$

Such states are characterized by reduced quantum fluctuations in one quadrature component of the field at the expense of increased fluctuations in the other noncommuting component. The possibility of obtaining light whose amplitude fluctuations are less than those expected from shot noise considerations attracted a great deal of attention.

Squeezed states of light were first studied by theorists interested in their properties as generalized minimum-uncertainty states.[248] These properties were discovered independently by several researchers using different terminologies and have been described as pulsating wave packets (Takahasi), new coherent states (Lu), and wave packet states (Hollenhorst).

The squeezing effect was intensively studied in the years 1980s and the theoretical predictions indicated it should manifest itself in nonlinear optical processes as four-wave mixing,[249] parametric amplification,[250] harmonic generation,[251] and multiphoton absorption.[252] These theoretical predictions stimulated intensive experimental studies.[253]

Squeezing in a two-photon state was first discussed by Stoler and Yuen.[254]

Squeezed states could offer the possibility to go below the vacuum noise in a number of applications. As an example, we mention that systems based on light signals with phase-sensitive quantum noise were proposed by Yuen and Shapiro[255] in 1978.

They were suggested also as being able to play a role in increasing the sensitivity of a gravitational wave detector. Hollenhorst[256] discussed all possible measurements that could be applied on the measurements of resonant antenna as were proposed by Braginskii et al.[257] The proposal by these authors was discussed by several papers.[258] Considering a gravity-wave antenna consisting of a single-mode harmonic oscillator with no coupling to any other modes, Hollenhorst examined the most suitable states to be used. A set of initial states could be coherent states but another set of states was what he called *wave-packet states* that were generated in analogy to coherent states by a unitary displacement operator of the form

$$S(\zeta) = \exp(1/2\zeta^*a^2 - 1/2\zeta a^{\dagger 2}),$$

where ζ is the complex squeezed parameter.

If $|\varphi\rangle$ is the state of a system and r is a real number, then $S(r)|\varphi\rangle$ represents the same system compressed in position space by the factor $\alpha = e^{-r}$ and expanded in momentum space by the factor $1/\alpha = e^r$. For this reason, Hollenhorst called $S(r)$ the "squeezed" operator and the states were since called *squeezed*.

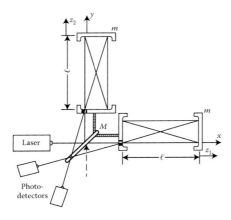

The following year (1980), Caves[259] discussed the quantum-mechanical limits on the performance of interferometers in the contest of gravitational wave detection. Basically detection occurs in a two-arm, multireflection Michelson system, powered by a laser (**Figure 10.17**). Each arm of the interferometer has a length l, and the displacements of the end mirrors from their positions are denoted z. The intensity in one of the output ports is measured by an ideal photodetector (quantum efficiency one) and this measurement provides information about the displacement between the positions of the end mirrors. A number of causes produce fluctuations in the output.

FIGURE 10.17 Schematic diagram of Michelson interferometer. (From C M Caves, *Phys. Rev.* **D23**, 1693 (1981). With permission.)

The quantum limit was assumed to be produced by fluctuating radiation pressure on the mirrors. Caves[259] pointed out that the relevant radiation-pressure force has nothing to do with input power fluctuations; instead, it can be attributed to vacuum (zero-point) fluctuations in the electromagnetic field which enters the interferometer from the unused input port. When superposed on the input laser light, these fluctuations produce a fluctuating force that perturbs z. In a following paper, Cave[260] suggested to modify the light entering the normally unused input port. Specifically, the unused port must see not the vacuum (ground) state of the electromagnetic field, but rather a squeezed state. Up to now, gravitational waves have not been detected using interferometers, but experiments to obtain squeezed states continued.

The first experimental realization of squeezed light was reported by Slusher et al.[261] using four-wave mixing in sodium atoms. They were able to reduce the optical noise below the vacuum fluctuation level by 7–10% using a combination of phase-stable laser excitation and cavity field enhancement.

Shelby et al.[262] obtained 12.5% reduction below vacuum noise using an optical fiber and Wu et al.[263] used a parametric down-conversion process.

Generation of squeezed vacuum (63% noise reduction) was obtained by Wu et al.[264] using a degenerate optical parametric amplifier (OPO). A nonlinear crystal of $LiNbO_3$ was pumped with a single-mode doubled YAG laser and placed in an optical cavity in order to enhance the nonlinearity. The observation of squeezing in the subharmonic field was done by homodyne detection: the light exiting the parametric amplifier was combined on a beam splitter with the light at the fundamental frequency from the YAG laser, which acted as a strong local oscillator. The noise in the observed signal was then related to the fluctuations of one quadrature component of the subharmonic field, depending on the phase of the local oscillator.

Another experiment was made by Debuisschert et al.[265] obtaining 69% noise reduction in a twin photon beams generation using a nondegenerate OPO. The parametric medium was inserted in a resonant cavity for the two signal fields. Above some pump threshold, the system oscillates and generated two intense beams having a high degree of quantum correlation. The photon noise reduction was observed in the intensity difference between the two beams.

We discussed *entangled states* in Chapter 8.

Notes

1. E Whittaker, *A History of the Theories of Aether and Electricity* (Harper and Brothers: New York, 1960).
2. M Jammer, *The Conceptual Development of Quantum Mechanics* (McGraw-Hill: New York, 1966).
3. M Planck gave his formula in a communication read on October 19, 1900, before the German Physical Society and published in *Verh. Deutsch. Phys. Ges.* **2**, 202 (1900).
4. Lord Rayleigh, *Phil. Mag.* **49**, 539 (1900); cf. *Nature* **72**, 54 (May 1905); *Nature* **72**, 243 (July 1905); J H Jeans, *Phil. Mag.* **10**, 91 (1905).
5. This result was read by Planck on December 14, 1900, before the German Physical Society and published in *Verh. Deutsch. Phys. Ges.* **2** 237 (1900). This and the paper in Note 3 were re-edited and printed in the new form in *Ann. Phys. Lpz.* **4**, 553 (1901). An excellent commentary to Planck's papers was given by M J Klein in vol 1 of the *Archive for History of Exact Sciences* (1962), p. 459.
6. A Einstein, *Ann. Phys. Lpz.* **17**, 132 (1905). English translations are by A B Arons and M B Peppard, *Am. J. Phys.* **33**, 367 (1965) and D ter Haar, *The Old Quantum Theory* (Pergamon Press: Oxford, New York, 1967), p. 91. It was nominally for this work that the Nobel Prize was awarded to Einstein.
7. The three papers are: (i) A Einstein, *Uber einen die Erzeugung und Verwandlung des Lichtes betreffenden heuristischen Gesichtspunkt* (see Note 6). (ii) A Einstein, *Die von der molekularkinetischen Theorie der Warme geforderte Bewegung von in ruhenden Flussigkeiten suspendierten Teilchen, Ann. Phys. Lpz.* **17**, 549 (1905).

This paper was reprinted in A Einstein, *Investigations on the Theory of the Brownian Movement*, translated by A D Cowper, with notes by R Furth (Dover: New York, 1956). (iii) A Einstein, Zur Elektrodynamik bewegter Koerper, *Ann. Phys. Lpz.* **17**, 891 (1905). This paper was reprinted in A Einstein, H A Lorentz, H Weyl and H Minkowski, *The Principle of Relativity: A Collection of Original Memories on the Special and General Theory of Relativity* translated by W Perrett and G B Jeffrey, with notes by A Sommerfeld (Dover: New York).

8. A deeper analysis of this paper is in Martin J Klein Einstein's First Paper on Quanta in the *Natural Philosopher* (Baisdell: New York, 1963), Vol. 2, p. 59 and in Einstein and the Wave-Particle Duality, ibid. Vol. 3 (1964), p. 1.

9. Planck proposed (*Verh. Deutsch. Phys. Ges.* **13**, 138 (1911)) a new hypothesis that, although emission of radiation always takes place discontinuously in quanta, absorption on the other hand is a continuous process which takes place according to the laws of the classical theory. Radiation while in transit might therefore be represented by Maxwell's theory, and the energy of an oscillator at any instant might have any value whatever. See Note 2 for a detailed discussion of these important steps.

10. G N Lewis, *Nature* **118**, 874 (1926).

11. P Lenard, *Ann. Phys. Lpz.* **8**, 149 (1902); see also J H Jeans, *Report on Radiation and the Quantum Theory* (The Electrician Printing and Publishing Co: London, 1914); R H Stuewer in *Historical and Philosophical Perspectives of Science,* edited by R H Stuewer (University of Minnesota Press: Minneapolis, 1970), p. 246.

12. O W Richardson and K T Compton, *Phil. Mag.* **24**, 575 (1912).

13. A L Hughes, *Phil. Trans.* **212**, 205 (1912).

14. R A Millikan, *Phys. Rev.* **2**, 109 (1913); **4**, 73 (1914); **7**, 355 (1916); *Philos. Mag.* **34**, 1 (1917).

15. C A Sadler and P Mesham, *Phil. Mag.* **24**, 138 (1912).

16. A H Compton, *Phil. Nat. Res. Council USA* **4** (October 20, 1922); see also *Phys. Rev.* **21**, 483 (1923).

17. Cf. also P Debye, *Phys. Z.* **24**, 161 (1923) who discovered the theory independently; A H Compton, *Phil. Mag.* **46**, 897 (1923); A H Compton and W A Simon, *Phys. Rev.* **25**, 306 (1925).

18. C Seelig, *Albert Einstein: A Documentary Biography* (London, 1956), pp. 143–145 reported by M J Klein in *Nat. Philo.* **2**, 59 (1963).

19. R A Millikan, *The Electron* (Chicago, IL, 1917), p. 238.

20. A Einstein, *Phys. Z.* **10**, 185 (1909).

21. H A Lorentz, *Les Theories Statistiques en Thermodynamique* (G B Teubner: Leipzig, 1916), p. 114.

22. A Einstein, *Phys. Z.* **18**, 121 (1917). For a brief general survey of these papers, see for example, W Pauli in *Albert Einstein: Philosopher-Scientist,* edited by P A Schilpp (Tudor: New York, 1957), p. 147.

23. P Ehrenfest, *Ann. Phys. Lpz.* **36**, 91 (1911).

24. A Joffé, *Ann. Phys. Lpz.* **36**, 534 (1911).

25. L Natanson, *Phys. Z.* **12**, 659 (1911).

26. G Krutkow, *Phys. Z.* **15**, 133 (1914).

27. L de Broglie, *Comptes Rendus* **175**, 811 (1922); *J. Phys.* **3**, 422 (1922).

28. W Bothe, *Z. Phys.* **20**, 145 (1923).

29. J H Webb, *Am. J. Phys.* **40**, 850 (1972) has observed that if stimulated emission is omitted in the Einstein derivation, the Wien formula results, and the fluctuations become equal to the single term n, which is characteristic of classical statistics. This shows that stimulated emission is the mechanism responsible for the appearance of the Bose–Einstein statistics in black-body radiation.

30. S N Bose, *Z. Phys.* **26**, 178 (1924); English translation in *Am. J. Phys.* **44**, 1056 (1976).

31. W A Blanpied, *Am. J. Phys.* **40**, 1212 (1972); M Jammer, *The Conceptual Development of Quantum Mechanics* (McGraw-Hill: New York, 1966), p. 248.

32. Bose was able to derive Planck's formula using only the light quanta hypothesis combined with statistical mechanics. He therefore succeeded in a task Einstein had failed in for 15 years, overcoming the logical defect existing in previous derivations where quantum concepts were used together with classical electrodynamics which was already demonstrated to be inconsistent with them (see Chapter 2).

33. A Einstein, *Berlin Sitz.* P. 261 (1924).

34. S N Bose, *Z. Phys.* **27**, 384 (1924). English translation by O Theimer and B Ram, *Am. J. Phys.* **45**, 242 (1977).

35. This new paper (see Note 34) by Bose treated the equilibrium of the radiation when it is in interaction with matter and is continuously emitted and absorbed. In Chapter 2, we have seen Einstein's derivation of Planck's law (see Note 22) which assumes that the rate of emission contains a term proportional to the photon number in one-phase cell for the rate of spontaneous transition (stimulated emission). This special model was extended to multilevel systems by Einstein and Ehrenfest (Chapter 2). On the other hand, Pauli (*Z. Phys.* **18**, 272 (1923)) had studied light scattering from electrons and had shown that if the Compton effect was taken into account, Planck's distribution was stationary; Bose in his paper (see Note 34) considered the general case and showed that in his formulation both Pauli processes and Einstein–Ehrenfest processes were included as special cases, and this alone was already a noteworthy result. Further, Bose observed that instead of Einstein's hypothesis of spontaneous and stimulated decays, for emission it is possible to consider only spontaneous emission if the absorption is taken to be proportional, not to the number of quanta per phase space cell, but to this number divided by the same number plus one. Until one considers radiative equilibrium, the hypothesis is as good as Einstein's because only these ratios appear.

36. A Einstein, Letters to P Ehrenfest dated September 29, December 2, 1924, Einstein Archive.

37. M J Klein, *Nat. Phil.* **3**, 26 (1963).

38. A Einstein, *Berlin Sitz.* pp. 3, 18 (1925).

39. For a discussion of the papers cited in Notes 33 and 38, see M J Klein in *The Natural Philosopher*, Vol. 3 (Blandell: New York, 1964), p. 1.

40. A Einstein, *Sitz. Preuss. Akad. Wiss., Phys. Math. Kl. Bericht* **1**, 3 (1925); ibidem 261 (1924).

41. V Bagnato, D E Pritchard, D Kleppner, *Phys. Rev.* **A35**, 4354 (1987).

42. W Pauli, *Z. Phys.* **31**, 765 (1925).

43. E Fermi, *Z. Phys.* **36**, 902 (1926).

44. P A M Dirac, *P. R. Soc.* **112**, 661 (1926).

45. See A Pais, *Subtle is the Lord* (Oxford Univ. Press: Oxford, 1982).

46. F London, *Nature* **141**, 643 (1938).

47. L Tisza, *Nature* **141**, 913 (1938).

48. H Cho and G A Williams, *Phys. Rev. Lett.* **75**, 1562 (1995).

49. J P Wolfe, J L Lin and D W Snoke, in *Bose–Einstein Condensation*, edited by A Griffin, D W Snoke and S Stringari (Cambridge Univ. Press: Cambridge, 1995), p. 281 and references therein.

50. W S Stwalley and L H Nosanov, *Phys. Rev. Lett.* **36**, 910 (1976).

51. *Bull. Am. Phys. Soc.* **23**, 85 (1978).
52. I F Silvera and J T M Walraven, *Phys. Rev. Lett.* **44**, 164 (1980); W N Hardy et al., *Physica* **109**, 1964 (1982); H J Hess et al., *Phys. Rev. Lett.* **51**, 483 (1983); B R Johnson et al., *Phys. Rev. Lett.* **52,** 1508 (1984); see also T J Greytak and D Kleppner in *New Trends in Atomic Physics*, edited by G Grynberg and R Stora (North-Holland: Amsterdam, 1984).
53. I F Silvera and J T M Walraven, *Phys. Rev. Lett.* **44**, 164 (1980).
54. H J Hess et al., *Phys. Rev. Lett.* **51**, 483 (1983).
55. H F Hess et al., *Phys. Rev. Lett.* **52**, 1520 (1984).
56. H F Hess, *Phys. Rev.* B34, 3476 (1986).
57. H F Hess et al., *Phys. Rev. Lett.* **59**, 672 (1987); N Masuhara et al., *Phys. Rev. Lett.* **61**, 935 (1988).
58. R V E Lovelace et al., *Nature* **318**, 30 (1985).
59. The idea that laser light could be used to cool atoms was suggested by D J Wineland and H Dehmelt, *Bull. Am. Phys. Soc.* **20**, 637 (1975); T W Hansch and A L Schawlow, *Opt. Comm.* **13**, 68 (1975) and V Letokhov *Pis'ma Zh. Eksp. Teor. Fiz.* **7**, 348 (1968). Early optical force experiments were performed by Askin (J F Bjorkholm et al., *Phys. Rev. Lett.* **41**, 1361 (1978). Trapped ions were laser-cooled by Neuhauser et al., *Phys. Rev. Lett.* **41**, 233 (1978)) and D J Wineland et al., *Phys. Rev. Lett.* **40**, 1639 (1978). Atomic beams were deflected and slowed by S V Andreev et al., *Eksp. Teor. Fiz.* **34**, 463 (1981), J Prodan et al., *Phys. Rev. Lett.* **54**, 463 (1985) and W Ertmer et al., *Phys. Rev. Lett.* **54**, 992 (1985). Optical molasses, where atoms are cooled to very low temperatures by six perpendicular intersecting laser beams were first studied by S Chu et al., *Phys. Rev. Lett.* **55**, 48 (1985). The magneto optical trap (MOT) was demonstrated by E L Raab et al., *Phys. Rev. Lett.* **59**, 2631 (1987). For more, see *Proceedings of the International School of Physics "Enrico Fernmi," Course CXVIII, Laser Manipulation of Atoms and Ions*; S Chu, *Rev. Mod. Phys.* **70**, 685 (1998); C N Cohen-Tannoudji, *Rev. Mod. Phys.* **70**, 707 (1998); W D Phillips, *Rev. Mod. Phys.* **70**, 721 (1998). Steven Chu, Claude Cohen-Tannoudji and William Phillips were awarded the Physics Nobel Prize in 1997 for the development of methods to cool and trap atoms with laser light.
60. C Monroe et al., *Phys. Rev. Lett.* **65**, 1571 (1990).
61. C Monroe et al., *Phys. Rev. Lett.* **65**, 1371 (1990); ibidem **70**, 414 (1993); E A Cornell, C Monroe and C E Wieman, *Phys. Rev. Lett.* **67**, 2439 (1991).
62. A more complete story was described by E A Cornell and C E Wieman in their Nobel Lecture, December 8, 2001, *Rev. Mod. Phys.* (2002) where also other references are given.
63. M H Anderson, J R Ensher, M R Matthews, C E Wieman and E A Cornell, *Science* **269**, 198 (1995).
64. W Ketterle and D E Pritchard, *Appl. Phys. B: Photophys. Laser Chem.* **B54**, 403 (1992).
65. D E Pritchard and W Ketterle, in *Proceedings of the International School of Physics "Enrico Fermi", Course CXVIII, Laser Manipulation of Atoms and Ions*, p. 473.
66. W Ketterle et al., *Phys. Rev. Lett.* **70**, 2253 (1993).
67. The story of Ketterle's achievement is recounted in the paper W Ketterle, *Rev. Mod. Phys.* **74**, 1131 (2002).
68. K B Davis, M D Mewes, M R Andrews, N J van Druten, D S Durfee, D M Kurn and W Ketterle, *Phys. Rev. Lett.* **75**, 3969 (1995).
69. M H Anderson, J R Ensher, M R Matthews, C E Wieman and E A Cornell *Science* **269**, 198 (1995).
70. C C Bradley, C A Sackett, J J Tollett and R G Hulet, *Phys. Rev. Lett.* **75**, 1687 (1995).

71. K B Davis, M D Mewes, M R Andrews, N J van Druten, D S Durfee, D M Kurn and W Ketterle, *Phys. Rev. Lett.* **75**, 3969 (1995).

72. D G Fried et al., *Phys. Rev. Lett.* **81**, 3811 (1998); T C Killian et al., *Phys. Rev. Lett.* **81**, 3807 (1998).

73. Kleppner et al., in Bose–Einstein Condensation in Atomic Gases, *Proceedings of the International School of Physics Enrico Fermi, Course CXL*, edited by M Inguscio, S Stringari and C E Wieman (IOS Press: Amsterdam, 1999) p. 177.

74. Andrews et al., *Science*, **275**, 637 (1997).

75. M O Mewes et al., *Phys. Rev. Lett.* **78**, 582 (1997).

76. A Smekal, *Z. Phys.* **37**, 319 (1926).

77. W Bothe, *Z. Phys.* **41**, 345 (1927).

78. R Furth, *Z. Phys.* **48**, 323 (1928); **50**, 310 (1928).

79. R Furth, *Z. Phys.* **50**, 310 (1928).

80. R E Burgess, *Proc. Phys. Soc.* **53**, 293 (1941); **58**, 313 (1946).

81. J M W Milatz and H A van der Velden, *Physica* **10**, 369 (1943); W B Lewis, *Proc. Phys. Soc.* **59**, 34 (1947); R C Jones, *J. Opt. Soc. Am.* **37**, 879 (1947); P Fellgett, *J. Opt. Soc. Am.* **39**, 970 (1949).

82. R E Burgess, *Proc. Phys. Soc.* **53**, 293 (1941).

83. H B Callen and T A Welton, *Phys. Rev.* **83**, 34 (1951); H Ekstein and N Rostoker, *Phys. Rev.* **100**, 1023 (1955); J Weber, *Phys. Rev.* **101**, 1620 (1956).

84. H Nyquist, *Phys. Rev.* **32**, 110 (1928); J B Johnson, *Phys. Rev.* **32**, 97 (1928).

85. W B Lewis, *Proc. Phys. Soc.* **59**, 34 (1947).

86. P Fellgett, *J. Opt. Soc. Am.* **39**, 970 (1949).

87. R C Jones, *Adv. Electron.* **5**, 1 (1953).

88. P Fellgett, *Nature* **179**, 956 (1957); R Q Twiss and R Hanbury Brown, *Nature* **179**, 1128 (1957); P Fellgett, R C Jones and R Q Twiss, *Nature* **184**, 967 (1959).

89. D K C MacDonald, *Rep. Prog. Phys.* **12**, 56 (1948).

90. E Verdet, *Ann. Sci. L'Ecole Normale Superieure* **2**, 291 (1865).

91. M Born and E Wolf, *Principles of Optics* (Pergamon: Oxford, 1965), ch 10.

92. L Mandel and E Wolf, *Rev. Mod. Phys.* **37**, 231 (1965).

93. The concept of an analytic signal was introduced by D Gabor, *J. Instrum. Electron. Engrs.* **93**, part 4, 429 (1946). See also V I Bunimovich, *J. Tech. Phys. USSR* **19**, 1231 (1949). It is well treated in the classical textbook by M Born and E Wolf, *Principles of Optics* (Pergamon: Oxford, 1965) ch 10.

94. F Zernike, *Physica* **5**, 785 (1938). Measure of correlation of light vibrations was introduced by M von Laue, *Ann. Phys. Lpz.* **23**, 1, 795 (1907). Joint probability distribution for the light disturbances was also determined by P H van Cittert, *Physica* **1**, 201 (1934); ibid. **6**, 1129 (1939). Actually Zernike introduced what is now called the mutual intensity function, that is, $\Gamma_{12}(\tau = 0)$. The function $\Gamma_{12}(\tau)$ was first introduced by E Wolf, *Proc. Roy. Soc.* **A230**, 246 (1955).

95. H H Hopkins, *Proc. R. Soc.* **A208**, 263 (1951); ibid. **A217**, 408 (1953).

96. E Wolf, *Proc. R. Soc.* **A230**, 246 (1955); ibid. **A225**, 96 (1954); *Nuovo Cim.* **12**, 884 (1954). Here the definitions in term of analytic signal are introduced.

97. A Blanc-Lapierre and P Dumontet, *Rev. D'Optique* **34**, 1 (1955).

98. The publication *Photoelectric Mixing of Incoherent Light* was in *Phys. Rev.* **99**, 1691 (1955) under the names A T Forrester, R A Gudmundsen and P O Johnson. Early suggestions were A T Forrester, W E Parkins and E Gerjnoy, *Phys. Rev.* **72**, 728 (1947); E Gerjnoy, A T Forrester and W E Parkins, *Phys. Rev.* **73**, 922 (1948).

99. C T J Alkemade, *Physica* **25**, 1145 (1959).

100. A T Forrester, *J. Opt. Soc. Am.* **51**, 253 (1961); see also *Advances in Quantum Electronics,* edited by J R Singer (Columbia University Press: New York, 1961), p. 233.

101. These two approaches are sometimes referred to in the literature under the names of *heterodyne* and *homodyne* detection. This terminology is, however, incorrect. The term *heterodyne* should be used when beats are produced between two different lasers. When, as it usually the case in a scattering experiment, beats are produced between light from the same laser which is split in a *scattered* beam and a *reference* beam, the term *homodyne* should be used, while when the same beam beats with itself one should speak of *self-beating* spectroscopy.

102. A Javan, W R Bennett and D R Herriott, *Phys. Rev. Lett.* **6**, 106 (1961) applied the technique to study He–Ne laser output; later A Javan, E W Ballik and W L Bond, *J. Opt. Soc. Am.* **52**, 96 (1962) used the beat note produced by the output of two independent lasers which simultaneously illuminated a single photomultiplier to determine the stability of the lasers. In the same year, S E Harris, B J McMurtry and A E Siegman, *Appl. Phys. Lett.* **1**. 37 (1962) used beats in solid-state devices and applied the method to study ruby laser beats, B J McMurtry and A E Siegan, *Appl. Opt.* **1**, 51 (1962).

103. See, for example, H Z Cummins and H L Swinney in *Prog. in Opt.*, edited by E Wolf **8**, 133 (1970); E Jakeman, C J Oliver and E R Pike, *Adv. Phys.* **24**, 349 (1975); H Z Cummins and E R Pike editors, *Photon Correlation Spectroscopy and Velocimetry* (Plenum Press: New York, 1977).

104. R Hanbury Brown, R C Jennison and M K Das Gupta, *Nature* **170**, 1061 (1952); R Hanbury Brown and R Q Twiss, *Phil. Mag. Ser.* 7 45, 663 (1954); R C Jennison and M K Das Gupta, *Phil. Mag. Ser.* 8 **1**, 55 (1956).

105. R Hanbury Brown and R Q Twiss, *Nature* **177**, 27 (1956); ibid. **178**, 1046 (1956); *Proc. R. Soc.* **A242**, 300 (1957); ibid. **A243**, 291 (1957); ibid. **A248**, 199 (1958); ibid. **A248,** 222 (1958).

106. R Hanbury Brown, *The Intensity Interferometer* (Taylor & Francis: London, 1974), pp. 3, 4, 6ff.

107. R Hanbury Brown and R Q Twiss, *Phil. Mag. Ser.* 7 **45**, 663 (1954).

108. R Hanbury Brown, R C Jennison and M K Das Gupta, *Nature* **170**, 1061 (1952).

109. A Adam, L Janossy and P Varga, *Acta Phys. Hung.* **4**, 301 (1955).

110. R Hanbury Brown and R Q Twiss, *Nature* **178**, 1447 (1956).

111. E Brannen and H I S Ferguson, *Nature* **178**, 481 (1956).

112. R Q Twiss, A G Little and R Hanbury Brown, *Nature* **180**, 324 (1957).

113. R Hanbury Brown and R Q Twiss, *Nature* **178**, 1046 (1956); *P. R. Soc.* **A248**, 222 (1958).

114. R Hanbury Brown and R Q Twiss, *Nature* **177**, 27 (1956).

115. R Hanbury Brown and R Q Twiss, *Proc. R. Soc.* **A243**, 291 (1957).

116. R Hanbury Brown and R Q Twiss, *Proc. R. Soc.* **A242**, 300 (1957); ibid. **A243**, 291 (1957).

117. This result comes from the Gaussian nature of thermal light. That white light of thermal origin has the properties of a Gaussian random process had been well established many years ago in several classical papers, see A Einstein, *Ann. Phys. Lpz.* **47**, 879 (1915). It was later shown by several authors; cf. I S Reed, *IRE Trans. Inform. Theory* **IT-8,** 194 (1962); C L Metha, *Lectures on Theoretical Physics,* edited by W E Brittin (University of Colorado Press: Boulder, Colorado, 1961) Vol. 7, p. 398; L Mandel and E Wolf, *Phys. Rev.* **124**, 1694 (1961).

118. The relation was later derived by L Mandel, *J. Opt. Soc. Am.* **51**, 1342 (1961).

119. D Gabor, *Phil. Mag. Ser.* 7, **41**, 1161 (1950); *Progress in Optics,* Vol. 1 (North-Holland: Amsterdam, 1961), p. 111; L Mandel, *J. Opt. Soc. Am.* **52**, 1407 (1962); ibid. **51**, 797 (1961).

120. G A Rebka and R V Pound, *Nature* **180**, 1035 (1957).

121. E Brannen, H I S Ferguson, W Wehlan, *Can. J. Phys.* **36**, 871 (1958); see also W Martienssen and E Spiller, *Am.J. Phys.* **32**, 919 (1964); G L Farkas, L Janossy, Z Naray and P Varga, *Acta Phys. Acad. Sc. Hung.* **18**, 199 (1965).

122. E M Purcell, *Nature* **178**, 1449 (1956); sentences reproduced by permission copyright © 1956 Macmillan Journals Limited.

123. L Mandel, *Proc. Phys. Soc.* **71**, 1037 (1958); **74**, 233 (1959); F D Kahn, *Opt. Acta* **5**, 93 (1958); U Fno, *Am. J. Phys.* **29**, 539 (1961).

124. L Mandel, *Proc. Phys. Soc.* **72**, 1037 (1958); **74**, 233 (1959).

125. L Mandel, *Proc. Phys. Soc.* **72**, 1037 (1958).

126. S O Rice, *Bell Syst. Tech. J.* **23**, 282 (1944); ibid. **24**, 46 (1945).

127. L Mandel, *J. Opt. Soc. Am.* **51**, 797 (1961).

128. A Einstein, *Congres Solvay* (1912).

129. L Mandel, E C G Sudarshan and E Wolf, *Proc. Phys Soc.* **84**, 435 (1964).

130. R J Glauber, *Phys. Rev. Lett.* **10**, 84 (1963).

131. L Mandel and E Wolf, *Phys. Rev.* **124,** 1696 (1961).

132. R J Glauber, *Phys. Rev.* **130**, 2529 (1963); **131**, 2766 (1963).

133. R J Glauber, in *Quantum Optics and Electronics*, Les Houches 1964 edited by C DeWitt, A Blandin and C Cohen-Tannoudji (Gordon & Breach: New York, 1964), p. 65.

134. The density operator was introduced by von Neumann, *Gesellschaft der Wissenschaften zu Gottinger Mth. Phy. Nachrichten* **245** (1927) and *Mathematical Foundations of Quantum Mechanics* (Princeton University Press: Princeton, NJ, 1955); see also P A M Dirac, *Proc. Camb. Phil. Soc.* **25**,62 (1929); **26**, 376 (1930); **27**, 240 (1930); D ter Haar, *Rep. Prog. Phys.* **24**, 304 (1961); see also W H Louisell, *Radiation and Noise in Quantum Electronics* (McGraw-Hill: New York, 1964).

135. R J Glauber, *Phys. Rev.* **131**, 2766 (1963).

136. The P representation had been already introduced almost contemporarily by E C G Sudarshan, *Phys. Rev. Lett.* **10,** 277 (1963); see also *Proc. Symposium on Optical Masers* (Wiley: New York, 1963), p. 45; J R Klauder, J McKenna and D G Currie, *J. Math. Phys.* **6**, 733 (1965); C L Metha and E C G Sudarshan, *Phys. Rev.* **138**, B274 (1965). See also J R Klauder and E C G Sudarshan, *Fundamentals of Quantum Optics* (Benjamin: New York, 1968).

137. Coherent states are just displace versions of the ground state of the harmonic oscillator (see R J Glauber, *Phys. Rev.* **131**, 2766 (1963)) and were first discussed by E Schroedinger, *Naturwiss* **14**, 644 (1926).

138. R J Glauber, *Phys. Rev.* **131**, 2766 (1963).

139. Functions of the type of the P functions were first discussed by E Wiger, *Phys. Rev.* **40**, 749 (1932). The name sometimes used of *quasi probability* was suggested by G A Backer, *Phys. Rev.* **109**, 2198 (1958).

140. R J Glauber, *Physics of Quantum Electronics* (Conf. Proc. San Juan, Puero Rico, June 28–30, 1965), edited by P L Kelly, B Lax and P E Tannenwald (McGrawHill: New York, 1966), p. 788.

141. R J Glauber, in *Quantum Optics and Electronics*, Les Houches 1964 edited by C DeWitt, A Blandin and C Cohen-Tannoudji (Gordon & Breach: New York, 1964), p. 65.

142. F Ghielmetti, *Phys. Lett.* **12**, 210 (1964).

143. L Mandel, *Proc. Phys. Soc.* **74**, 233 (1959).

144. J Perina, *Quantum Statistics of Linear and Nonlinear Optical Phenomena* (Reidel Pu. Co.: Dortrecht, 1984).

145. L Mandel, *Phys. Rev.* **136**, 647 (1964).

146. L Mandel, *Phys. Rev.* **152**, 438 (1966).

147. L Mandel and E Wolf, *Optical Coherence and Quantum Optics* (Cambridge Univ. Press: Cambridge, 1995).

148. LMandel and E Wolf, *Phys.Rev. Lett.* **10**, 276 (1963).

149. E M Purcell, *Nature* **178**, 1449 (1956).

150. C T J Alkemade, *Physica* **25**, 1145 (1925); L Mandel, *Proc. Phys. Soc.* **72**, 1037 (1958) ; ibidem 74, 233 (1959); E Wolf, *Proc. Phys. Soc.* **76**, 424 (1960).

151. E Wolf, in *Quantum Electronics,* edited by P Grivet and N Bloembergen (Dunod: Paris, 1964), p. 13.

152. L Mandel in *Quantum Electronics,* edited by P Grivet and N Bloembergen (Dunod: Paris, 1964), p. 101.

153. E C G Sudarshan, *Phys. Rev. Lett.* **10**, 277 (1963).

154. K E Cahill, *Phys. Rev.* **138**, B1566 (1965).

155. U M Titulaer and R J Glauber, *Phys. Rev.* **140**, B676 (1965).

156. M Bertolotti, B Crosignani and P Di Porto, *J. Phys. A: Gen. Phys.* **3**, L37 (1970).

157. M Bertolotti, B Crosignani, P Di Porto and D Sette, *Phys. Rev.* **157**, 146 (1967); see also Y R Shen, *Phys. Rev.* **155**, 921 (1967).

158. For a full discussion, see B Crosignani, P Di Porto and M Bertolotti, *Statistical Properties of Scattered Light* (Academic: New York, 1975).

159. P Di Porto, M Bertolotti and B Crosignani, *J. Appl. Phys.* **40**, 5083 (1969); for a deeper discussion, see L Mandel and E Wolf, *Optical Coherence and Quantum Optics* (Cambridge Univ. Press: Cambridge 1995) but see also J R Klauder and E C G Sudarsham, *Fundamentals of Quantum Optics* (Benjamin, Inc.: New York, 1968); J Perina, *Quantum Statistics of Linear and Nonlinear Optical Phenomena* (Reidel Pu. Co.: Dortrecht, 1984); J Perina, ed., *Coherence and Statistics of Photons and Atoms* (Wiley: New York, 2001); J Perina, Z Hradil and B Jurco, *Quantum Optics and Fundamentals of Physics* (Kluwer Ac. Pu.: Dortrecht, 1994); J Perina, *Coherence of Light* (Reidel Pu.Co.: Dortrecht,1985); P J Bourke, J Butterworth, L E Drain, P A Egelstaff, P Hutchinson, B Moss, P Schofield, A J Hughes, J J B O'Shaughnessy, E R Pike, E Jakeman and D A Jackson, *Phys. Lett.* **28**A, 692 (1969).

160. See the book by B Crosignani, P Di Porto and M Bertolotti, *Statistical Properties of Scattered Light* (Academic: New York, 1975); also E Jakeman and K D Ridley, *Modeling Fluctuations in Scattered Waves* (Taylor & Francis, 2006) can be seen.

161. Probably the most complete is L Mandel and E Wolf, *Optical Coherence and Quantum Optics* (Cambridge Univ. Press: Cambridge, 1995) but see also J R Klauder and E C G Sudarsham, *Fundamentals of Quantum Optics* (Benjamin, Inc.: New York, 1968); J Perina, *Quantum Statistics of Linear and Nonlinear Optical Phenomena* (Reidel Pu. Co.: Dortrecht, 1984); J Perina, ed., *Coherence and Statistics of Photons and Atoms* (Wiley: New York, 2001); J Perina, Z Hradil and B Jurco, *Quantum Optics and Fundamentals of Physics* (Kluwer Academic Publishers: Dortrecht, 1994); J Perina, *Coherence of Light* (Reidel Pu. Co.: Dortrecht, 1985).

162. R J Glauber, *Physics of Quantum Electronics* (Conf. Proc. San Juan, Puero Rico, June 28–30, 1965), edited by P L Kelly, B Lax and P E Tannenwald (McGraw-Hill: New York, 1966), pp. 788167.

163. R J Glauber, *Physics of Quantum Electronics* (Conf. Proc. San Juan, Puero Rico June 28–30, 1965) edited by P L Kelly, B Lax and P E Tannenwald (McGraw-Hill: New York, 1966), pp. 788158–167.

164. J A Bellisio, C Freed and H A Haus, *Appl. Phys. Lett.* **4**, 5 (1964); R L Bailey ad J H Sanders, *Phys. Lett.* **10**, 29 (1964); P T Bolwijn, C Th J Alkemade and G A Boschloo, *Phys. Lett.* **4**, 59 (1963).

165. L J Prescott and A Van der Ziel, *Phys. Lett.* **12**, 317 (1964).

166. C Freed and H A Haus, *Appl. Phys. Lett.* **6**, 85 (1965).

167. B van der Pol, *Phil. Mag.* **3**, 65 (1927).
168. W E Lamb Jr., *Phys. Rev.* **134**, A1429 (1964).
169. J A Armstrong and A W Smith, *Phys. Rev. Lett.* **14**, 68 (1965); in *Proc. Int. Conf. on the Physics of Quantum Electronics*, Puerto Rico, 1965, edited by P Kelly, B Lax and P Tannenwald (McGraw-Hill: New York, 1965), p. 701; A W Smith and J A Armstrong, *Phys. Lett.* **16**, 38 (1965).
170. A W Smith and J A Armstrong, *Phys. Lett.* **16**, 38 (1965).
171. J A Armstrong and A W Smith, *Phys. Rev.* **140**, A155 (1965).
172. J E Geusic, H M Marcos and L G van Uitert in *Physics of Quantum Electronics,* edited by P L Kelley, B Lax and P E Tannenwald (McGraw-Hill, New York, 1966), p. 725.
173. T P McLean and E R Pike, *Phys. Lett.* **15**, 318 (1965).
174. J Marguin, R Marcy, G Hepner and G Pircher, *Comptes Rendus* **260**, 1361 (1965).
175. R J Glauber, *Physics of Quantum Electronics* (Conf. Proc. San Juan, Puero Rico, June 28–30, 1965), edited by P L Kelly, B Lax and P E Tannenwald (McGraw-Hill: New York, 1966), p. 788.
176. F A Johnson, T P McLean and E R Pike, in *Physics of Quantum Electronics*, edited by P L Kelly et al. (McGraw-Hill: New York, 1966), p. 706.
177. C Freed and H A Haus, in *Physics of Quantum Electronics,* edited by P L Kelly et al., (McGraw-Hill: New York, 1966), p. 715.
178. J A Armstrong and A W Smith, in *Proc. Int. Conf. on the Physics of Quantum Electronics*, Puerto Rico, 1965, edited by P Kelly, B Lax and P Tannenwald (McGraw-Hill: New York, 1965), p. 701.
179. H A Haus, *Quarterly Progress Report, Research Laboratory of Electronics*, MIT 15 April 1964, p. 49.
180. E R Pike has written a recollection of the work at Malvern, E R Pike, *JEOS:RP* **5**, 100475 (2010).
181. *RRE Newsletter and Research Review* no March 4, 1965; there is a brief communication by T P McLean and E R Pike in which, by using the calculation later presented in *Phys. Lett.* **15**, 318 (1965), they report on an experiment under way.
182. C Freed and H A Haus, in *Physics of Quantum Electronics,* edited by P L Kelly et al. (McGraw-Hill: New York, 1966), p. 715.
183. F A Johnson, T P McLean and E R Pike, in *Physics of Quantum Electronics,* edited by P L Kelly et al (McGraw-Hill: New York, 1966), p. 706.
184. E Jakeman, C J Oliver and E Pike, *J. Phys. A: Gen. Phys.* **1**, 497 (1968).
185. F T Arecchi, *Phys. Rev. Lett.* **15**, 912 (1965).
186. C Freed and H A Haus, *Phys. Rev. Lett.* **15**, 943 (1965).
187. W Martienssen and E Spiller, *Am. J. Phys.* **32**, 919 (1964). Later, these authors (*Phys. Rev. Lett.* **16**, 531 (1966)) used a similar system to determine the probability densities for the light intensity of chaotic light if the detector averages over N cells of phase space.
188. C Freed and H A Haus, *Phys. Rev. Lett.* **15**, 943 (1965).
189. F A Johnson, T P McLean and E R Pike, in *Physics of Quantum Electronics*, edited by P L Kelly et al. (McGraw-Hill: New York, 1966), p. 706.
190. U M Titulaer and R J Glauber, *Phys. Rev.* **140**, B676 (1965).
191. R J Glauber, *Physics of Quantum Electronics* (Conf. Proc. San Juan, Puero Rico, June 28–30, 1965), edited by P L Kelly, B Lax and P E Tannenwald (McGraw-Hill: New York, 1966), p. 788.
192. F T Arecchi, E Gatti and A Sona, *Phys. Lett.* **20**, 27 (1966).
193. G Farkas, L Janossy, Z Naray and P Varga, *Acta Phys. Acad. Sci. Hung.* **18**, 199 (1965); Yu F Skachkov, *JETP* **21**, 1026 (1965); R Jones, F A Johnson, T P McLean and E R Pike, *Phys. Rev. Lett.* **16**, 589 (1966).

194. C Freed and H A Haus, *IEEE J. Quantum Electron.* **QE-2**, 190 (1966).

195. F T Arecchi, A Berne, A Sona and P Burlamacchi, *IEEE J. Quantum Electron.* **QE-2**, 341 (1966).

196. See J Weber, *Phys. Rev.* **101**, 1620 (1959); W G Wagner and G Birnbaum, *J. Appl. Phys.* **32**, 1185 (1961); J A Fleck Jr., *J. Appl. Phys.* **34**, 2997 (1963); R V Pound, *Ann. Phys. NY* **1**, 24 (1957); J Weber, *Rev. Mod. Phys.* **31**, 681 (1959); M P W Strandberg, *Phys. Rev.* **106**, 617 (1957); D E McCumber, *Phys. Rev.* **130**, 675 (1962); W H Wells, *Ann. Phys. NY* **12**, 1 (1961); G Kemeny, *Phys. Rev.* **133**, A69 (1964); F Schwabl and W Thirring, *Erg. Nat.* **36**, 219 (1964); however, some authors developed nonlinear cases, neglecting spontaneous emission, H Haken and H Sauermann, *Z. Phys.* **173**, 261 (1963); **176**, 47 (1963); R M Bevensee, *J. Math. Phys.* **5**, 308 (1964).

197. H Haken, *Z. Phys.* **181**, 96 (1964); **182**, 346 (1965); *Phys. Rev. Lett.* **13**, 329 (1964).

198. Lamb had previously circulated his ideas; see, for example, M Sargent III and M O Scully in *Laser Handbbook,* edited by F T Arecchi and E O Schultz-DuBois (North-Holland: Amsterdam, 1972), p. 45.

199. J A Armstrong and A W Smith, *Phys. Rev. Lett.* **14,** 68 (1965); *Phys. Lett.* **19**, 650 (1965).

200. C Freed and H A Haus, *Appl. Phys. Lett.* **6**, 85 (1965).

201. H Risken, *Z. Phys.* **186**, 85 (1965).

202. A W Smith and J A Armstrong, *Phys. Lett.* **19,** 650 (1966); *Phys. Rev. Lett.* **16**, 1169 (1966).

203. F T Arecchi, G S Rodari and A Sona, *Phys. Lett.* **25A**, 59 (1967); D Meltzer, W Davis and L Mandel, *Appl. Phys Lett.* **17**, 242 (1970); E Jakeman, C J Oliver, E R Pike, M Lax and M Zwanziger, *J. Phys. A: Gen Phys.* **3**, L52 (1970).

204. M Scully and W E Lamb Jr., *Phys. Rev. Lett.* **16,** 853 (1966); *Phys. Rev.* **159**, 208 (1967); *Phys. Rev.* **166**, 246 (1968).

205. G Lachs, *Phys. Rev.* **138**, B1012 (1965).

206. R J Glauber, *Physics of Quantum Electronics* (Conf. Proc. San Juan, Puero Rico, June 28–30, 1965), edited by P L Kelly, B Lax and P E Tannenwald (McGraw-Hill: New York, 1966), p. 788.

207. L Mandel and E Wolf, *Phys. Rev.* **149**, 1033 (1966).

208. H Morawitz, *Phys. Rev.* **139**, A1072 (1965); *Z. Phys.* **195**, 20 (1966).

209. J Perina, *Phys. Lett.* **24A**, 333 (1967); see also J Perina, *Coherence of Light* (Van Nostrand: London, 1972); E R Pike, *Riv. Nuovo Cim.* **1**, 277 (1969).

210. For example, dead-time effects were treated in R Jones et al., *Phys. Rev. Lett.* **16**, 589 (1966); but see also G Bedard, *Proc. Phys. Soc.* **90**, 131 (1967). A summary of all the problems encountered in photon counting measurements can be found in *Photo Correlation and Light Beating Specrocscopy,* edited by H Z Cummins and E R Pike, NATO-ASI Series B3 (Plenum Press: London, 1974) and *Photon Correlation Spectroscopy and Velocimetry,* edited by H Z Cummins and E R Pike, NATO-ASI Series B23 (Plenum Press: London, 1977).

211. Quasi-thermal and laser photon counting distributions compared directly on an oscillograph were published in Professor Kastler's revision of Bruhat's text *Thermodynamique*, Paris (1966); see also E R Pike, *Riv. Nuovo Cim.* **1**, 277 (1969).

212. F A Johnson, T P McLean and E R Pike, in *Physics of Quantum Electronics*, edited by P L Kelly et al., (McGraw-Hill: New York, 1966), p. 706.

213. R Jones, F A Johnson, T P McLean and E R Pike, *Phys. Rev. Lett.* **16**, 589 (1966).

214. F A Johnson, R Jones, T P McLean and E R Pike, *Opt. Acta* **14**, 35 (1967).

215. F A Johnson, T P McLean and E R Pike, in *Physics of Quantum Electronics*, edited by P L Kelly et al. (McGraw-Hill: New York, 1966), p. 706.

216. E R Pike, *JEOS:RP* **5**, 100475 (2010).

217. H Gamo, R E Grace and T J Walter, *2nd Rochester Conf. on Coherence and Quatum Optics,* Rochester, NY, June 1966, Conference Abstracts, p. 183.
218. R F Chang, R W Detenbeck, V Korenman, C O Alley Jr. and U Hochuli, *Phys. Lett.* **25A**, 272 (1967).
219. F Davidson and L Mandel, *Phys. Lett.* **27A**, 579 (1968).
220. F Davidson and L Mandel, *Phys. Lett.* **25A**, 700 (1967).
221. D Meltzer, W Davis and L Mandel, *Appl. Phys. Lett.* **17**, 242 (1970).
222. E Jakeman, C J Oliver and E R Pike, *J. Phys. A: Gen. Phys.* **3**, L45 (1970).
223. See L Brillouin, *Les Statistiques Quantiques* (Les Presses Universitaires France: Paris, 1930), ch 6. For the tendency of a boson gas to form clusters in space, see G E Uhlenbeck and L Gropper, *Phys. Rev.* **41**, 79 (1932); F Loudon, *Phys. Rev.* **54**, 947 (1938); *J. Chem. Phys.* **11**, 203 (1943).
224. R H Dicke, *Phys. Rev.* **93**, 99 (1954).
225. B L Morgan and L Mandel, *Phys Rev. Lett.* **16**, 1012 (1966).
226. Photon bunching is a typical case in which the P representation is missing. See, for example, J Perina, *Prog. Opt.* **18**, 129 (1980).
227. H J Kimble, M Degenais and L Mandel, *Phys. Rev. Lett.* **39**, 691 (1977); *Phys. Rev.* **A18**, 201 (1978); M Dagenais and L Mandel, *Phys. Rev.* **A18,** 2217 (1978); see also E Jakeman, E R Pike, P N Pusey and J M Vaughan, *J. Phys. A: Math. Gen.* **10**, L257 (1977); M M Tehrani and L Mandel, *Phys. Rev.* **A17,** 677, 694 (1978); L Mandel, *Opt. Lett.* **4**, 205 (1979). For a general review of these interesting effects, see J Perina, *Progr. Opt.* **18**, 129 (1980).
228. L Mandel and E Wolf, *Optical Coherence and Quantum Optics* (Cambridge Univ. Press: Cambridge, 1995); J Perina, *Quantum Statistics of Linear and Nonlinear Optical Phenomena* (Reidel Pu. Co.: Dortrecht, 1984); J Perina, ed., *Coherence and Statistics of Photons and Atoms* (Wiley: New York, 2001); J Perina, Z Hradil and B Jurco, *Quantum Optics and Fundamentals of Physics* (Kluwer Academic Publishers: Dortrecht, 1994); J Perina, *Coherence of Light* (Reidel Pu. Co.: Dortrecht, 1985); J Perina, *Coherence of Light* (Van Nostrand: London, 1972); E R Pike, *Riv. Nuovo Cim.* **1**, 277 (1969); B Saleh, *Photoelectron Statistics* (Springer: Berlin, 1978); B Crosignani, P Di Porto and M Bertolotti, *Statistical Properties of Scattered Light* (Academic: New York, 1975).
229. Probably the most complete is L Mandel and E Wolf, *Optical Coherence and Quantum Optics* (Cambridge Univ. Press: Cambridge, 1995) but see also J C Garrison and R Y Chiao, *Quantum Optics,* (Oxford Univ Press: Oxford 2008).
230. A Beveratos et al., *Phys. Rev. Lett.* **89**, 187901 (2002); S Ya Kilin, *Advanced Opt.* **42**, 1 (2001); N Gisin et al., *Rev. Mod. Phys.* **74**, 145 (2002); N Gisin and R Thew, *Nature Photon.* **1**, 165 (2007); M Dusek et al., in *Progress in Optics,* edited by E Wolf (North-Holland, Amsterdam, 2006), Vol. 49, 381 (2006); N J Cerf and J F Flurasek in *Progress in Optics,* edited by E Wolf (North-Holland: Amsterdam, 2006) **49**, 455.
231. W K Wootters and W H Zurek, *Nature* **299**, 802 (1982); D Diecks, *Phys. Lett.* **A92**, 271 (1982); G C Ghirardi and T Weber, *Nuovo Cimento* **B78**, 9 (1983); N J Cerf and J F Flurasek in *Progress in Optics,* edited by E Wolf (North-Holland, Amsterdam, 2006) **49**, 455 (2006).
232. V Fock, *Z. Phys.* **75**, 622 (1932); see also L D Faddeev et al., *Selected Works V A Fock* (Chapman & Hall/CRC, London 2004).
233. G I Taylor, *Proc. Camb. Philos. Soc.* **15**, 114 (1909).
234. J J Thomson, *Proc. Camb. Philos. Soc.* **14**, 417 (1907).
235. N Gisin et al., *Rev. Mod. Phys.* **74**, 145 (2002).
236. J F Clauser, *Phys. Rev.* **D9**, 853 (1974).
237. H J Carmichael and D F Walls, *J. Phys.* **B9**, L43 (1976).

238. The Rabi frequency $\Omega = |\mu_{12} \cdot E|/\hbar$, where μ_{12} is the transition dipole moment and E is the laser field with which the atom is excited.

239. H J Kimble, M Degenais and L Mandel, *Phys. Rev. Lett.* **39**, 691 (1977).

240. A Beveratos et al., *Phys. Rev.* **A64**, 061802 (2001).

241. S Singh, *Opt. Comm.* **44**, 254 (1983).

242. R Short and L Mandel, *Phys. Rev. Lett.* **51**, 384 (1983).

243. E. Knill et al., *Nature* **409**, 46 (2001).

244. For an historical overview of these sources, see M Bertolotti, F Bovino and C Sibilia, *Progress in Optics* (Elsevier 2015).

245. E Schroedinger, *Naturwiss* **14**, 664 (1926).

246. J R Klauder, *Ann. Phys.* **11**, 123 (1960).

247. W Pauli, in *Handbuch der Physik*, edited by H Geiger and K Scheel (Springer: Berlin, 1933).

248. H Takahasi, *Adv. Comm. Syst.* **1**, 227 (1965); D R Robinson, *Commun. Math. Phys.* **1**, 159 (1965); D Stoler, *Phys. Rev.* **D1**, 3217 (1970); **D4**, 1925 (1971); E Y C Lu, *Lett. Nuov. Cim.* **2**, 1241 (1971); ibidem **3**, 585 (1972); H P Yuen, *Phys. Rev.* **A13,** 2226 (1976); J N Hollenhorst, *Phys. Rev.* **D19**, 1669 (1979); C M Caves, *Phys. Rev.* **D13**, 1693 (1981).

249. H P Yuen and J H Shapiro, *Opt. Lett.* **4**, 334 (1979); M D Reid and D F Walls, *Phys. Rev.* **A31**, 1622 (1985); J Perina et al., *Opt. Comm.* **49**, 285 (1984).

250. G J Milburn and D F Walls, *Phys. Rev.* **A27**, 329 (1983); C M Caves and B L Shumaker, *Phys. Rev.* **A31**, 3068 (1985).

251. L Mandel, *Opt. Comm.* **42,** 437 (1982); M Kozierowski and S Kielich, *Phys. Lett.* **A94**, 213 (1983); J Perina et al., *Opt. Comm.* **49**, 210 (1984); S Kielich et al., *Optica Acta* **32**, 1023 (1985) and references therein; M J Collett and D F Walls, *Phys. Rev.* **A32**, 2887 (1987); S Kielich et al., *J. Mod. Opt.* **34**, 979 (1987).

252. M S Zubairy et al., *Phys. Lett.* **A98**, 168 (1983); R Loudon, *Opt. Comm.* **49**, 67 (1984).

253. M D Levenson et al., *Phys. Rev.* **A32**, 1550 (1985) ; M W Maeda et al., *Phys. Rev.* **A32**, 3803 (1985) ; R E Slusher et al., *Phys. Rev. Lett.* **55**, 2409 (1985).

254. H P Yuen, *Phys. Rev.* **A13**, 2226 (1976); D Stoler, *Phys. Rev.* **D1**, 3217 (1970); for a review of the initial state of research, see D F Walls, *Nature*, **306**, 141 (1983) and R Loudon and P L Knight, *J. Mod. Opt.* **34**, 709 (1987).

255. H P Yuen and J H Shapiro, *IEEE Trans. Inform. Theory* **IT-24**, 657 (1978); **IT-26**, 78 (1980); J H Shapiro, H P Yuen and J A Machado Maia, *IEEE Trans. Inform. Theory* **IT-25**, 179 (1979).

256. J N Hollenhorst, *Phys. Rev.* **D19**, 1669 (1979).

257. V B Braginskii et al., *Sov. Phys. JETP* **41**, 28 (1975); **46**, 705 (1977).

258. W G Unruh, *Phys. Rev.* **D18**, 1764 (1978) ; **D17**, 1180 (1978); V Moncrief, *Ann. Phys. New York*, **114**, 201 (1978); K S Thorne et al., *Phys. Rev. Lett.*, **40**, 667 (1978); V B Braginskii et al., *JETP Lett.* **27**, 277 (1978).

259. C M Caves, *Phys. Rev. Lett.* **45**, 75 (1980).

260. C M Caves, *Phys. Rev.* **D23**, 1693 (1981).

261. R E Slusher et al., *Phys. Rev. Lett.* **55**, 2409 (1985).

262. R M Shelby et al., *Phys. Rev. Lett.* **57**, 691 (1986).

263. L A Wu et al., *Phys. Rev. Lett.* **57**, 2520 (1986).

264. L A Wu et al., *J. Opt. Soc. Am.* **B4,** 1465 (1988).

265. T Debuisschert et al., *Quantum Opt.* **1**, 3 (1989).

Subject Index

Subject Index

Author Index

Author Index

Brignon A, 309
Broglie L de, 398
Brok E G, 211
Bromberg, Lisa, 207
Brossel, Jean, 72, 77, 78
Brovetto P, 60
Buehler E, 112
Bunkin F V, 319, 376
Burger H C, 10
Burgess R E, 409
Burnham R D, 274
Burns G, 224
Byer R L, 293

C

Cagnac B, 78
Cahill K E, 434
Camparo J, 70
Cao H, 351, 352
Capasso, Federico, 231, 233
Carver, Thomas R, 58
Casimir, H B G, 41, 55
Caves C M, 453, 454
Ceglio N M, 379
Chang R F, 445
Chang R P H, 352
Chang Z, 329
Chen C J, 206
Chesler R B, 256, 257
Chiao R Y, 301, 302, 305
Childs J J, 362
Cho A Y, 233
Christov II P, 325
Chu K R, 118
Chu, Steven, 70
Chumak G M, 216
Churchwell E, 129
Ciftan M, 174
Cini M, 60
Clairon, André, 70
Clarkson W A, 374
Claussen M J, 129
Codella C, 130
Cohen B G, 226
Collins R J, 147, 255, 261
Compton A H, 395
Compton A, 10
Compton K T, 395
Convert G, 206
Corkum P B, 317, 319
Corkum P, 329
Cornell, Eric, 406, 407
Crampton S B, 115
Crosby G A, 215
Crosignani B, 303

Cruz C H B, 324
Cummings F W, 119
Cummins, Herman Z, 147, 190
Czavinsky P, 211

D

D'Haenens I J, 114
Daido H, 379, 381
Damon E K, 319
Danilychev V A, 211
Dapkusand P D, 353
Das Gupta M K, 411
Dasari R R, 362
Davisson C J, 15
Dayhoff E S, 202
De Grasse R W, 111
Deacon D A G, 235
Debuisschert T, 455
Debye P J, 42
Debye P, 15, 202
Degenais M, 451
Dehmelt H, 68, 94
Delone N B, 318, 319
DeMaria A J, 256
Denk W, 285
Derr V E, 113
Derzhiev V I, 376
Desurvire E, 272
Detch N J L, 206
Detenbeck R W, 445
Devlin G E, 174
Devor D P, 114
Dey A, 130
Diamond P J, 130
Dianov E M, 314
Dicke R H, 78, 141, 154–156
Diddams S A, 71
Diels J C, 322
Dienes A, 261, 321
Digonnet M J F, 272
Dill Jr. F H, 224
Ding K, 362
Dingle R, 228
Dinkelacker O, 23
Dirac P A M, 10, 233
Domenico M Di, 260
Dresher M, 329
Drude P, 12, 15
DuBois, 111
Ducuing J, 295
Dumke, William P, 222, 224
Dumontet P, 410
Duncan Jr. R C, 174
Dunn J, 380
Dupre F K, 41, 55

474

Author Index

Author Index